Universitext

Universitext

Universitext is a series of textbooks that presents material from a wide variety of mathematical disciplines at master's level and beyond. The books, often well class-tested by their author, may have an informal, personal even experimental approach to their subject matter. Some of the most successful and established books in the series have evolved through several editions, always following the evolution of teaching curricula, to very polished texts.

Thus as research topics trickle down into graduate-level teaching, first textbooks written for new, cutting-edge courses may make their way into *Universitext*.

For further volumes:
http://www.springer.com/series/223

Paul A. Fuhrmann

A Polynomial Approach to Linear Algebra

Second Edition

 Springer

Paul A. Fuhrmann
Ben-Gurion University of the Negev
Beer Sheva
Israel

ISSN 0172-5939   e-ISSN 2191-6675
ISBN 978-1-4614-0337-1   e-ISBN 978-1-4614-0338-8
DOI 10.1007/978-1-4614-0338-8
Springer New York Dordrecht Heidelberg London

Library of Congress Control Number: 2011941877

Mathematics Subject Classification (2010): 15-02, 93-02

Printed on acid-free paper

Springer is part of Springer Science+Business Media (www.springer.com)

To Nilly

Preface

Linear algebra is a well-entrenched mathematical subject that is taught in virtually every undergraduate program, both in the sciences and in engineering. Over the years, many texts have been written on linear algebra, and therefore it is up to the author to justify the presentation of another book in this area to the public.

I feel that my jusification for the writing of this book is based on a different choice of material and a different approach to the classical core of linear algebra. The main innovation in it is the emphasis placed on functional models and polynomial algebra as the best vehicle for the analysis of linear transformations and quadratic forms. In pursuing this innovation, a long standing trend in mathematics is being reversed. Modern algebra went from the specific to the general, abstracting the underlying unifying concepts and structures. The epitome of this trend was represented by the Bourbaki school. No doubt, this was an important step in the development of modern mathematics, but it had its faults too. It led to several generations of students who could not compute, nor could they give interesting examples of theorems they proved. Even worse, it increased the gap between pure mathematics and the general user of mathematics. It is the last group, made up of engineers and applied mathematicians, which is concerned not only in understanding a problem but also in its computational aspects. A very similar development occurred in functional analysis and operator theory. Initially, the axiomatization of Banach and Hilbert spaces led to a search for general methods and results. While there were some significant successes in this direction, it soon became apparent, especially in trying to understand the structure of bounded operators, that one has to be much more specific. In particular, the introduction of functional models, through the work of Livsic, Beurling, Halmos, Lax, de Branges, Sz.-Nagy and Foias, provided a new approach to structure theory. It is these ideas that we have taken as our motivation in the writing of this book.

In the present book, at least where the structure theory is concerned, we look at a special class of shift operators. These are defined using polynomial modular arithmetic. The interesting fact about this class is its property of universality, in the sense that every cyclic operator is similar to a shift and every linear operator on a finite-dimensional vector space is similar to a direct sum of shifts. Thus, the shifts are the building blocks of an arbitrary linear operator.

Basically, the approach taken in this book is a variation on the study of a linear transformation via the study of the module structure induced by it over the ring of polynomials. While module theory provides great elegance, it is also difficult to grasp by students. Furthermore, it seems too far removed from computation. Matrix theory seems to be at the other extreme, that is, it is too much concerned with computation and not enough with structure. Functional models, especially the polynomial models, lie on an intermediate level of absraction between module theory and matrix theory.

The book includes specific chapters devoted to quadratic forms and the establishment of algebraic stability criteria. The emphasis is shared between the general theory and the specific examples, which are in this case the study of the Hankel and Bezout forms. This general area, via the work of Hermite, is one of the roots of the theory of Hilbert spaces. I feel that it is most illuminating to see the Euclidean algorithm and the associated Bezout identity not as isolated results, but as an extremely effective tool in the development of fast inversion algorithms for structured matrices.

Another innovation in this book is the inclusion of basic system-theoretic ideas. It is my conviction that it is no longer possible to separate, in a natural way, the study of linear algebra from the study of linear systems. The two topics have benefited greatly from cross-fertilization. In particular, the theory of finite-dimensional linear systems seems to provide an unending flow of problems, ideas, and concepts that are quickly assimilated in linear algebra. Realization theory is as much a part of linear algebra as is the long familiar companion matrix.

The inclusion of a whole chapter on Hankel norm approximation theory, or AAK theory as it is commonly known, is also a new addition as far as linear algebra books are concerned. This part requires very little mathematical knowledge not covered in the book, but a certain mathematical maturity is assumed. I believe it is very much within the grasp of a well-motivated undergraduate. In this part several results from early chapters are reconstructed in a context in which stability is central. Thus the rational Hardy spaces enter, and we have analytic models and shifts. Lagrange and Hermite interpolation are replaced by Nevanlinna-Pick interpolation. Finally, coprimeness and the Bezout identity reappear, but over a different ring. I believe the study of these analogies goes a long way toward demonsrating to the student the underlying unity of mathematics.

Let me explain the philosophy that underlies the writing of this book. In a way, I share the aim of Halmos (1958) in trying to treat linear transformations on finite-dimensional vector spaces by methods of more general theories. These theories were functional analysis and operator theory in Hilbert space. This is still the case in this book. However, in the intervening years, operator theory has changed remarkably. The emphasis has moved from the study of self-adjoint and normal operators to the study of non-self-adjoint operators. The hope that a general structure theory for linear operators might be developed seems to be too naive. The methods utilizing

Riesz-Dunford integrals proved to be too restrictive. On the other hand, a whole new area centering on the theory of invariant subspaces and the construction and study of functional models was developed. This new development had its roots not only in pure mathematics but also in many applied areas, notably scattering, network and control theories, and some areas of stochastic processes such as estimation and prediction theories.

I hope that this book will show how linear algebra is related to other, more advanced, areas of mathematics. Polynomial models have their root in operator theory, especially that part of operator theory that centered on invariant subspace theory and Hardy spaces. Thus the point of view adopted here provides a natural link with that area of mathematics, as well as those application areas I have already mentioned.

In writing this book, I chose to work almost exclusively with scalar polynomials, the one exception in this project being the invariant factor algorithm and its application to structure theory. My choice was influenced by the desire to have the book accessible to most undergraduates. Virtually all results about scalar polynomial models have polynomial matrix generalizations, and some of the appropriate references are pointed out in the notes and remarks.

The exercises at the end of chapters have been chosen partly to indicate directions not covered in the book. I have refrained from including routine computational problems. This does not indicate a negative attitude toward computation. Quite to the contrary, I am a great believer in the exercise of computation and I suggest that readers choose, and work out, their own problems. This is the best way to get a better grasp of the presented material.

I usually use the first seven chapters for a one-year course on linear algebra at Ben Gurion University. If the group is a bit more advanced, one can supplement this by more material on quadratic forms. The material on qudratic forms and stability can be used as a one-semester course of special topics in linear algebra. Also, the material on linear systems and Hankel norm approximations can be used as a basis for either a one term course or a seminar.

Paul A. Fuhrmann

Preface to the Second Edition

Linear algebra is one of the most active areas of mathematics, and its importance is ever increasing. The reason for this is, apart from its intrinsic beauty and elegance, its usefulness to a large array of applied areas. This is a two-way road, for applications provide a great stimulus for new research directions. However, the danger of a tower-of-Babel phenomenon is ever present. The broadening of the field has to confront the possibility that, due to differences in terminology, notation, and concepts, the communication between different parts of linear algebra may break down. I strongly believe, based on my long research in the theory of linear systems, that the polynomial techniques presented in this book provide a very good common ground. In a sense, the presentation here is just a commercial for subsequent publications stressing extensions of the scalar techniques to the context of polynomial and rational matrix functions.

Moreover, in the fifteen years since the original publication of this book, my perspective on some of the topics has changed. This, at least partially, is due to the mathematical research I was doing during that period. The most significant changes are the following. Much greater emphasis is put on interpolation theory, both polynomial and rational. In particular, we also approach the commutant lifting theorem via the use of interpolation. The connection between the Chinese remainder theorem and interpolation is explained, and an analytic version of the theorem is given. New material has been added on tensor products, both of vector spaces and of modules. Because of their importance, special attention is given to the tensor products of quotient polynomial modules. In turn, this leads to a conceptual clarification of the role of Bezoutians and the Bezout map in understanding the difference between the tensor products of functional models taken with respect to the underlying field and those taken with respect to the corresponding polynomial ring. This enabled the introduction of some new material on model reduction. In particular, some connections between the polynomial Sylvester equation and model reduction techniques, related to interpolation on the one hand and projection methods on the other, are clarified. In the process of adding material, I also tried to streamline theorem statements and proofs and generally enhance the readability of the book. It is my hope that this effort was at least partially successful.

I am greatly indebted to my friends and colleagues Uwe Helmke and Abie Feintuch for reading parts of the manuscript and making useful suggestions and to Harald Wimmer for providing many useful references to the history of linear algebra. Special thanks to my beloved children, Amir, Oded, and Galit, who not only encouraged and supported me in the effort to update and improve this book, but also enlisted the help of their friends to review the manuscript. To these friends, Shlomo Hoory, Alexander Ivri, Arie Matsliah, Yossi Richter, and Patrick Worfolk, go my sincere thanks.

Paul A. Fuhrmann

Contents

Chapter 1
Algebraic Preliminaries

1.1 Introduction

This book emphasizes the use of polynomials, and more generally, rational functions, as the vehicle for the development of linear algebra and linear system theory. This is a powerful and elegant idea, and the development of linear theory is leaning more toward the conceptual than toward the technical. However, this approach has its own weakness. The stumbling block is that before learning linear algebra, one has to know the basics of algebra. Thus groups, rings, fields, and modules have to be introduced. This we proceed to do, accompanied by some examples that are relevant to the content of the rest of the book.

1.2 Sets and Maps

Let S be a set. If between elements of the set a relation $a \simeq b$ is defined, so that either $a \simeq b$ holds or not, then we say we have a **binary relation**. If a binary relation in S satisfies the following conditions:

1. $a \simeq a$ holds for all $a \in S$,
2. $a \simeq b \Rightarrow b \simeq a$,
3. $a \simeq b$ and $b \simeq c \Rightarrow a \simeq c$,

then we say we have an **equivalence relation** in S. The three conditions are referred to as reflexivity, symmetry, and transitivity respectively.

For each $a \in S$ we define its equivalence class by $S_a = \{x \in S \mid x \simeq a\}$. Clearly $S_a \subset S$ and $S_a \neq \emptyset$. An equivalence relation leads to a partition of the set S. By a **partition** of S we mean a representation of S as the disjoint union of subsets. Since clearly, using transitivity, either $S_a \cap S_b = \emptyset$ or $S_a = S_b$, and $S = \cup_{a \in S} S_a$, the set of

P.A. Fuhrmann, *A Polynomial Approach to Linear Algebra*, Universitext,   1
DOI 10.1007/978-1-4614-0338-8_1, © Springer Science+Business Media, LLC 2012

equivalence classes is a partition of S. Similarly, any partition $S = \cup_\alpha S_\alpha$ defines an equivalence relation by letting $a \simeq b$ if for some α we have $a, b \in S_\alpha$.

A rule that assigns to each member $a \in A$ a unique member $b \in B$ is called a **map** or a **function** from A into B. We will denote this by $f : A \longrightarrow B$ or $A \overset{f}{\longrightarrow} B$. We denote by $f(A)$ the image of the set A defined by $f(A) = \{y \mid y \in B, \exists x \in A \text{ s.t. } y = f(x)\}$. The inverse image of a subset $M \subset B$ is defined by $f^{-1}(M) = \{x \mid x \in A, f(x) \in M\}$. A map $f : A \longrightarrow B$ is called **injective** or 1-to-1 if $f(x) = f(y)$ implies $x = y$. A map $f : A \longrightarrow B$ is called **surjective** or onto if $f(A) = B$, i.e., for each $y \in B$ there exists an $x \in A$ such that $y = f(x)$. A map f is called **bijective** if it is both injective and surjective.

Given maps $f : A \longrightarrow B$ and $g : B \longrightarrow C$, we can define a map $h : A \longrightarrow C$ by letting $h(x) = g(f(x))$. We call this map h the **composition** or **product** of the maps f and g. This will be denoted by $h = g \circ f$. Given three maps $A \overset{f}{\longrightarrow} B \overset{g}{\longrightarrow} C \overset{h}{\longrightarrow} D$, we compute $h \circ (g \circ f)(x) = h(g(f(x)))$ and $(h \circ g) \circ f(x) = h(g(f(x)))$. So the product of maps is associative, i.e.,

$$h \circ (g \circ f) = (h \circ g) \circ f.$$

Due to the associative law of composition, we can write $h \circ g \circ f$, and more generally $f_n \circ \cdots \circ f_1$, unambiguously.

Given a map $f : A \longrightarrow B$, we define an equivalence relation $\simeq$ in A by letting

$$x_1 \simeq x_2 \Leftrightarrow f(x_1) = f(x_2).$$

Thus the equivalence class of a is given by $A_a = \{x \mid x \in A, f(x) = f(a)\}$. We will denote by $A/\simeq$ the set of equivalence classes and refer to this as the quotient set by the equivalence relation.

Next we define three transformations

$$A \overset{f_1}{\longrightarrow} A/R \overset{f_2}{\longrightarrow} f(A) \overset{f_3}{\longrightarrow} B$$

with the f_i defined by

$$f_1(a) = A_a,$$
$$f_2(A_a) = f(a),$$
$$f_3(b) = b, \ b \in f(A).$$

Clearly the map f_1 is surjective, f_2 is bijective and f_3 injective. Moreover, we have $f = f_3 \circ f_2 \circ f_1$. This factorization of f is referred to as the **canonical factorization**. The canonical factorization can be described also via the following commutative diagram:

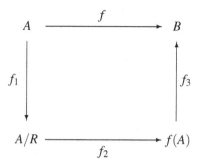

We note that $f_2 \circ f_1$ is surjective, whereas $f_3 \circ f_2$ is injective.

1.3 Groups

Given a set M, a **binary operation** in M is a map from $M \times M$ into M. Thus, an ordered pair (a,b) is mapped into an element of M denoted by ab. A set M with an associative binary operation is called a semigroup. Thus if $a,b \in M$ we have $ab \in M$, and the associative rule $a(bc) = (ab)c$ holds. Thus the product $a_1 \cdots a_n$ of elements of M is unambiguously defined.

We proceed to define the notion of a group, which is the cornerstone of most mathematical structures.

Definition 1.1. A **group** is a set G with a binary operation, called multiplication, that satisfies

1. $a(bc) = (ab)c$, i.e., the associative law.
2. There exists a left identity, or unit element, $e \in G$, i.e., $ea = a$ for all $a \in G$.
3. For each $a \in G$ there exists a left inverse, denoted by a^{-1}, which satisfies $a^{-1}a = e$.

A group G is called **abelian** if the group operation is commutative, i.e., if $ab = ba$ holds for all $a,b \in G$.

In many cases, an abelian group operation will be denoted using the additive notation, i.e., $a + b$ rather than ab, as in the case of the group of integers $\mathbb{Z}$ with addition as the group operation. Other useful examples are $\mathbb{R}$, the set of all real numbers under addition, and $\mathbb{R}_+$, the set of all positive real numbers with multiplication as the group operation.

Given a nonempty set S, the set of all bijective mappings of S onto itself forms a group with the group action being composition. The elements of G are called **permutations** of S. If $S = \{1,\ldots,n\}$, then the group of permutations of S is called the **symmetric group** of degree n and denoted by S_n.

Theorem 1.2. *1. Let G be a group and let a be an element of G. Then a left inverse*
   a^{-1} *of a is also a right inverse.*
2. *A left identity is also a right identity.*
3. *The identity element of a group is unique.*

Proof. 1. We compute

$$(a^{-1})^{-1}a^{-1}aa^{-1} = ((a^{-1})^{-1}a^{-1})(aa^{-1}) = e(aa^{-1}) = aa^{-1}$$
$$= (a^{-1})^{-1}(a^{-1}a)a^{-1} = (a^{-1})^{-1}(ea^{-1}) = (a^{-1})^{-1}a^{-1} = e.$$

So in particular, $aa^{-1} = e$.
2. Let $a \in G$ be arbitrary and let e be a left identity. Then

$$aa^{-1}a = a(a^{-1}a) = ae = (aa^{-1})a = ea = a.$$

Thus $ae = a$ for all a. So e is also a right identity.
3. Let e, e' be two identities in G. Then, using the fact that e is a left identity and e'
   a right identity, we get $e = ee' = e'$.   ∎

In a group G, equations of the form $axb = c$ are easily solvable, with the solution
given by $x = a^{-1}cb^{-1}$. Also, it is easily checked that we have the following rule for
inversion:
$$(a_1 \cdots a_n)^{-1} = a_n^{-1} \cdots a_1^{-1}.$$

Definition 1.3. A subset H of a group G is called a **subgroup** of G if it is a group
with the composition rule inherited from G. Thus H is a subgroup if with $a, b \in H$,
we have $ab \in H$ and $a^{-1} \in H$.

This can be made a bit more concise.

Lemma 1.4. *A subset H of a group G is a subgroup if and only if with $a, b \in H$,*
also $ab^{-1} \in H$.

Proof. If H is a subgroup of G, then with $a, b \in H$, it contains also b^{-1} and hence
also ab^{-1}.
   Conversely, if $a, b \in H$ implies $ab^{-1} \in H$, then $b^{-1} = eb^{-1} \in H$ and hence also
$ab = a(b^{-1})^{-1} \in H$, i.e., H is a subgroup of G.   ∎

Given a subgroup H of a group G, we say that two elements $a, b \in G$ are **H-
equivalent**, and write $a \simeq b$, if $b^{-1}a \in H$. It is easily checked that this is a bona fide
equivalence relation in G, i.e., it is a reflexive, symmetric, and transitive relation.
We denote by G_a the equivalence class of a, i.e.,

$$G_a = \{x \mid x \in G, x \simeq a\}$$

If we denote by aH the set $\{x \mid ah, h \in H\}$, then $G_a = aH$. We will refer to these as
right equivalence classes or as **right cosets**. In a completely analogous way, **left
equivalence classes**, or **left cosets**, Ha are defined.

Given a subgroup H of G, it is not usually the case that the sets of left and right cosets coincide. If this is the case, then we say that H is a **normal subgroup** of G. Assuming that a left coset aH is also a right coset Hb, we clearly have $a = ae \in H$ and hence $a \in Hb$ so necessarily $Hb = Ha$. Thus, for a normal subgroup, $aH = Ha$ for all $a \in G$. Equivalently, H is normal if and only if for all $a \in G$ we have $aHa^{-1} = H$.

Given a subgroup H of a group G, then any two cosets can be bijectively mapped onto each other. In fact, the map $\phi(ah) = bh$ is such a bijection between aH and bH. In particular, cosets aH all have the same cardinality as H.

We will define the **index** of a subgroup H in G, and denote it by $i_G(H)$, as the number of left cosets. This will be denoted by $[G : H]$. Given the trivial subgroup $E = \{e\}$ of G, the left and right cosets consist of single elements. Thus $[G : E]$, is just the number of elements of the group G. The number $[G : E]$ will be refered to also as the **order** $o(G)$ of the group G.

We can proceed now to study the connection between index and order.

Theorem 1.5 (Lagrange). *Given a subgroup H of a group G, we have*

$$[G : H][H : E] = [G : E],$$

or

$$o(G) = i_G(H)o(H).$$

∎

Homomorphisms are maps that preserve given structures. In the case of groups G_1 and G_2, a map $\phi : G_1 \longrightarrow G_2$ is called a **group homomorphism** if, for all $g_1, g_2 \in G_1$, we have

$$\phi(g_1 g_2) = \phi(g_1)\phi(g_2).$$

Lemma 1.6. *Let G_1 and G_2 be groups with unit elements e and e' respectively. Let $\phi : G_1 \longrightarrow G_2$ be a homomorphism. Then*

1. $\phi(e) = e'$.
2. $\phi(x^{-1}) = (\phi(x))^{-1}$.

Proof. 1. For any $x \in G_1$, we compute $\phi(x)e' = \phi(x) = \phi(xe) = \phi(x)\phi(e)$. Multiplying by $\phi(x)^{-1}$, we get $\phi(e) = e'$.

2. We compute

$$e' = \phi(e) = \phi(xx^{-1}) = \phi(x)\phi(x^{-1}).$$

This shows that $\phi(x^{-1}) = (\phi(x))^{-1}$.   ∎

A homomorphism $\phi : G_1 \longrightarrow G_2$ that is bijective is called an **isomorphism**. In this case G_1 and G_2 will be called isomorphic.

An example of a nontrivial isomorphism is the exponential function $x \mapsto e^x$, which maps $\mathbb{R}$ (with addition as the operation) isomorphically onto $\mathbb{R}_+$ (with multiplication as the group operation).

The general canonical factorization of maps discussed in Section 1.2 can now be applied to the special case of group homomorphisms. To this end we define, for a homomorphism $\phi : G_1 \longrightarrow G_2$, the kernel and image of ϕ by

$$\text{Ker}\,\phi = \phi^{-1}\{e'\} = \{g \in G_1 \mid \phi(g) = e'\},$$

and
$$\text{Im}\,\phi = \phi(G_1) = \{g' \in G_2 \mid \exists g \in G_1, \phi(g) = g'\}.$$

The kernel of a group homomorphism has a special property.

Lemma 1.7. *Let $\phi : G \longrightarrow G'$ be a group homomorphism and let $N = \text{Ker}\,\phi$. Then N is a normal subgroup of G.*

Proof. Let $x \in G$ and $n \in N$. Then

$$\begin{aligned}
\phi(xnx^{-1}) &= \phi(x)\phi(n)\phi(x^{-1}) = \phi(x)e'\phi(x^{-1}) \\
&= \phi(x)\phi(x^{-1}) = e'.
\end{aligned}$$

So $xnx^{-1} \in N$, i.e., $xNx^{-1} \subset N$ for every $x \in G$. This implies $N \subset x^{-1}Nx$. Since this inclusion holds for all $x \in G$, we get $N = xNx^{-1}$, or $Nx = xN$, i.e., the left and right cosets are equal. So N is a normal subgroup. ∎

Note now that given $x, y \in G$ and a normal subgroup N of G, we can define a product in the set of all cosets by

$$xN \cdot yN = xyN. \tag{1.1}$$

Theorem 1.8. *Let $N \subset G$ be a normal subgroup. Denote by G/N the set of all cosets and define the product of cosets by (1.1). Then G/N is a group called the* **factor group,** *or* **quotient group,** *of G by N.*

Proof. Clearly (1.1) shows that G/N is closed under multiplication. To check associativity, we note that

$$\begin{aligned}
(xN \cdot yN) \cdot zN &= (xyN) \cdot zN = (xy)zN \\
&= x(yz)N = xN \cdot (yzN) \\
&= xN \cdot (yN \cdot zN).
\end{aligned}$$

For the unit element e of G we have $eN = N$, and $eN \cdot xN = (ex)N = xN$. So $eN = N$ is the unit element in G/N. Finally, given $x \in G$, we check that the inverse element is $(xN)^{-1} = x^{-1}N$. ∎

Theorem 1.9. *N is a normal subgroup of the group G if and only if N is the kernel of a group homomorphism.*

Proof. By Lemma 1.7, it suffices to show that if N is a normal subgroup, then it is the kernel of a group homomorphism. We do this by constructing such a homomorphism. Let G/N be the factor group of G by N. Define $\pi : G \longrightarrow G/N$ by

$$\pi(g) = gN. \tag{1.2}$$

Clearly (1.1) shows that π is a group homomorphism. Moreover, $\pi(g) = N$ if and only if $gN = N$, and this holds if and only if $g \in N$. Thus $\operatorname{Ker} \pi = N$. ∎

The map $\pi : G \longrightarrow G/N$ defined by (1.2) is called the **canonical projection**.

A homomorphism ϕ whose kernel contains a normal subgroup can be factored through the factor group.

Proposition 1.10. *Let $\phi : G \longrightarrow G'$ be a homomorphism with $\operatorname{Ker} \phi \supset N$, where N is a normal subgroup of G. Let π be the canonical projection of G onto G/N. Then there exists a unique homomorphism $\overline{\phi} : G/N \longrightarrow G'$ for which $\phi = \overline{\phi} \circ \pi$, or equivalently, the following diagram is commutative:*

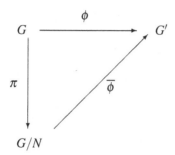

Proof. We define a map $\overline{\phi} : G/N \longrightarrow G'$ by $\overline{\phi}(xN) = \phi(x)$. This map is well defined, since $\operatorname{Ker} \phi \supset N$. It is a homomorphism because ϕ is, and of course,

$$\phi(x) = \overline{\phi}(xN) = \overline{\phi}(\pi(x)) = (\overline{\phi} \circ \pi)(x).$$

Finally, $\overline{\phi}$ is uniquely defined by $\overline{\phi}(xN) = \phi(x)$. ∎

We call $\overline{\phi}$ the **induced map** by ϕ on G/N. As a corollary we obtain the following important result, the prototype of many others, which classifies images of group homomorphisms.

Theorem 1.11. *Let $\phi : G \longrightarrow G'$ be a surjective group homomorphism with $\operatorname{Ker} \phi = N$. Then G' is isomorphic to the factor group G/N.*

Proof. The induced map is clearly injective. In fact, if $\overline{\phi}(xN) = e'$, we get $\phi(x) = e'$, or $x \in \operatorname{Ker} \phi = N$. It is also surjective by the assumption that ϕ is surjective. So we conclude that $\overline{\phi}$ is an isomorphism. ∎

A group G is **cyclic** if all its elements are powers of one element $a \in G$, i.e., of the form a^n. In this case, a is called a **generator** of G. Clearly, $\mathbb{Z}$ is a cyclic group with addition as group operation. Defining $n\mathbb{Z} = \{nk \mid k \in \mathbb{Z}\}$, it is obvious that $n\mathbb{Z}$ is a normal subgroup of $\mathbb{Z}$. We denote by Z_n the quotient group $\mathbb{Z}/n\mathbb{Z}$. We can identify Z_n with the set $\{0, \ldots, n-1\}$ with the group operation being addition modulo n.

1.4   Rings and Fields

Most of the mathematical structures that we will encounter in this book have, unlike groups, two operations, namely addition and multiplication. The simplest examples of these are rings and fields. These are introduced in this section.

Definition 1.12. A **ring** R is a set with two laws of composition called addition and multiplication that satisfy, for all elements of R, the following.

1. Laws of **addition**:

   a. Associative law: $a + (b+c) = (a+b) + c$.
   b. Commutative law: $a + b = b + a$.
   c. Solvability of the equation $a + x = b$.

2. Laws of **multiplication**:

   a. Associative law: $a(bc) = (ab)c$.
   b. Distributive laws:
   $$a(b+c) = ab + ac,$$
   $$(b+c)a = ba + ca.$$

3. R is called a **commutative ring** if the commutative law $ab = ba$ holds for all $a, b \in R$.

   Law 1 (c) implies the existence of a unique **zero element**, i.e., an element $0 \in R$ satisfying $0 + a = a + 0 = a$ for all $a \in R$. We call R a ring with **identity** if there exists an element $1 \in R$ such that for all $a \in R$, we have $1a = a1 = a$. An element a in a ring with identity has a **right inverse** b if $ab = 1$ and a **left inverse** if $ba = e$. If a has both left and right inverses they must be equal, and then we say that a is **invertible** and denote its inverse by a^{-1}.

   A **field** is a commutative ring with identity in which every nonzero element is invertible.

Definition 1.13. Let R_1 and R_2 be two rings. A **ring homomorphism** is a function $\phi : R_1 \longrightarrow R_2$ that satifies

$$\phi(x+y) = \phi(x) + \phi(y),$$
$$\phi(xy) = \phi(x)\phi(y).$$

If R_1 and R_2 are rings with identities, then we require also $\phi(e) = e'$.

As in the case of groups, the **kernel** of a ring homomorphism $\phi : R_1 \longrightarrow R_2$ is defined by $\operatorname{Ker} \phi = \phi^{-1}(0) = \{x \in R_1 \mid \phi(x) = 0\}$. The kernel of a ring homomorphism has special properties. In fact, if $x, y \in \operatorname{Ker} \phi$ and $r \in R_1$, then also $x + y, rx \in \operatorname{Ker} \phi$. This leads to the following definition.

Definition 1.14. A subset J of a ring R is called a **left ideal** if $x, y \in J$ and $r \in R$ implies $x + y, rx \in J$.

Thus, a subset J of R is a **left ideal** if it is an additive subgroup of R and $RJ \subset J$. If R contains an identity then $RJ = J$. For right ideals we replace the second condition by $JR = J$. A two-sided ideal, or just an **ideal**, is a subset of R that is both a left and right ideal. The sum of a finite number of ideals $J_1, \ldots, J_r$ in R is defined as the set $J_1 + \cdots + J_r = \{a_1 + \cdots + a_r \mid a_i \in J_i\}$.

Proposition 1.15. *Let R be a ring. Then*

1. *The sum of a finite number of left ideals is a left ideal.*
2. *The intersection of any set of left ideals in R is a left ideal.*

Proof. 1. Let $J_1, \ldots, J_k$ be left ideals in R. Then $J = J_1 + \cdots + J_k = \{a_1 + \cdots + a_k \mid a_i \in J_i\}$. Clearly if $a_i, b_i \in J_i$ and $r \in R$, then

$$\sum_{i=1}^{k} a_i + \sum_{i=1}^{k} b_i = \sum_{i=1}^{k} (a_i + b_i) \in J,$$

and

$$r \sum_{i=1}^{k} a_i = \sum_{i=1}^{k} r a_i \in J.$$

2. Let $J = \cap_\alpha J_\alpha$ with J_α left ideals. If $a, b \in J$ then $a, b \in J_\alpha$ for all α. Hence $a + b \in J_\alpha$ for all α, which implies $a + b \in J$. A similar argument holds to show that $ra \in J$. ∎

Given a two-sided ideal J in a ring R, we can construct the **quotient ring**, denoted by R/J, whose elements are the cosets $a + J$ by defining the operations of addition and multiplication by

$$\begin{cases} (a+J) + (b+J) = (a+b) + J, \\ (a+J)(b+J) \quad = ab + J. \end{cases}$$

It is easy to check that with the arithmetic operations so defined, R/J is indeed a ring.

The following theorem gives a complete characterization of ideals. It is the counterpart, in the setting of rings, of Theorem 1.9.

Theorem 1.16. *Let R be a ring. A subset $J \subset R$ is a two-sided ideal if and only if it is the kernel of a ring homomorphism.*

Proof. We saw already that the kernel of a ring homomorphism is a two-sided ideal.

Conversely, let J be a two-sided ideal. We define the canonical projection $\pi : R \longrightarrow R/J$ by $\pi(a) = a + J$. It is easy to check that π is a surjective ring homomorphism, with $\mathrm{Ker}\, \pi = J$.   ∎

An ideal in a ring R that is generated by a single element, i.e., of the form $J = (d) = \{rd \mid r \in R\}$, is called a **principal ideal**. The element d is called a **generator** of the ideal. More generally, given $a_1, \ldots, a_k \in R$, the set $J = (a_i, \ldots, a_k) = \{\sum_{i=1}^{k} r_i a_i \mid r_i \in R\}$ is obviously an ideal. We say that $a_1, \ldots, a_k$ are generators of this ideal.

In a ring R a nonzero element a is called a **zero divisor** if there exists another nonzero element $b \in R$ such that $ab = 0$. A commutative ring without a zero divisors is called an **entire ring** or an **integral domain**. A commutative ring R with identity and no zero divisors is called a **principal ideal domain**, or **PID** for short, if every ideal in R is principal.

In a ring R we have a division relation. If $c = ab$ we say that a **divides** c or a is a **left divisor** or **left factor** of c. Given $a_1, \ldots, a_n \in R$, we say that a is a common left divisor of the a_i if it is a left divisor of all a_i. We say that a is a **greatest common left divisor**, or g.c.l.d., if it is a common left divisor and is divisible by any other common left divisor. Two greatest common left divisors differ by a right factor that is invertible in R. We say that $a_1, \ldots, a_n \in R$ are **left coprime** if any greatest common left divisor is invertible in R. Right divisors are similarly defined. If R is a commutative ring, a greatest common left divisor is also a greatest common right divisor and will be refered to as a **greatest common divisor**, or a **g.c.d.** for short.

Proposition 1.17. *Let R be a principal ideal domain with identity. Then $a_1, \ldots, a_n \in R$ are coprime if and only if there exist $b_i \in R$ for which the* **Bezout identity**

$$a_1 b_1 + \cdots + a_n b_n = 1 \tag{1.3}$$

holds.

Proof. If there exist $b_i \in R$ for which the Bezout identity holds, then any common divisor of the a_i is a divisor of 1, hence necessarily invertible.

Conversely, we consider the ideal J generated by the a_i. Since R is a principal ideal domain, J is generated by a single element d, which necessarily is invertible. So $1 \in J$ and hence there exist b_i such that the Bezout identity (1.3) holds.   ∎

We present now a few examples of rings. We pay special attention to the ring of polynomials, due to the central role it plays in this book. Many of the results concerning ideals, factorizations, the Chinese remainder theorem, etc. hold also in the ring of integers. We do not give for those separate proofs nor, for the sake of concreteness, do we give proofs in the general context of Euclidean domains.

1.4.1   The Integers

The set of integers $\mathbb{Z}$ is a commutative ring under the usual operations of addition and multiplication.

1.4.2   The Polynomial Ring

A **polynomial** is an expression of the form

$$p(z) = \sum_{i=0}^{n} a_i z^i, \qquad a_i \in \mathbb{F},\ 0 \le n \in \mathbb{Z}.$$

The numbers a_i are called the **polynomial coefficients**. We shall denote by $\mathbb{F}[z]$ the set of all polynomials with coefficients in the field $\mathbb{F}$, i.e., $\mathbb{F}[z] = \{\sum_{i=0}^{n} a_i z^i \mid a_i \in \mathbb{F}, 0 \le n \in \mathbb{Z}\}$. Two polynomials are called equal if all their coefficients coincide. If $p(z) = \sum_{i=0}^{n} a_i z^i$ and $a_n \ne 0$, then we say that n is the **degree** of the polynomial $p(z)$, which we denote by $\deg p$. The coefficient a_n is called the **leading coefficient** of the polynomial $p(z)$. We define the degree of the zero polynomial to be $-\infty$. For a polynomial $p(z) = \sum_{i=0}^{n} a_i z^i$, we assume $a_k = 0$ for $k > n = \deg p$. Given two polynomials $p(z) = \sum_{i=0}^{n} a_i z^i$ and $q(z) = \sum_{i=0}^{m} b_i z^i$, we define their sum $p(z) + q(z)$ by

$$(p+q)(z) = \sum_{i=0}^{\max(m,n)} (a_i + b_i) z^i. \tag{1.4}$$

The product, $p(z)q(z)$, is defined by

$$(pq)(z) = \sum_{i=0}^{m+n} c_i z^i, \tag{1.5}$$

where

$$c_i = \sum_{j=0}^{i} a_j b_{i-j}. \tag{1.6}$$

It is easily checked that with these operations of addition and multiplication, $\mathbb{F}[z]$ is a commutative ring with an identity and has no zero divisors.

The next theorem sums up the most elementary properties of polynomials.

Theorem 1.18. *Let $p(z), q(z)$ be polynomials in $\mathbb{F}[z]$. Then*

1. $\deg(pq) = \deg p + \deg q$.
2. $\deg(p+q) \le \max\{\deg p, \deg q\}$.

Proof. 1. If $p(z)$ or $q(z)$ is the zero polynomial, then both sides of the equality are equal to $-\infty$. So we assume that both $p(z)$ and $q(z)$ are nonzero. Let

$p(z) = \sum_{i=0}^{n} a_i z^i$ and $q(z) = \sum_{i=0}^{m} b_i z^i$, with $a_n, b_m \neq 0$. Then $c_{n+m} = a_n b_m \neq 0$ but $c_k = 0$ for $k > m+n$.

2. This is immediate.   ∎

Note that the inequality $\deg(p+q) < \max\{\deg p, \deg q\}$ may occur due to cancellations.

Corollary 1.19. *If $p(z), q(z) \in \mathbb{F}[z]$ and $p(z)q(z) = 0$, then $p(z) = 0$ or $q(z) = 0$.*

Proof. If $p(z)q(z) = 0$, then

$$-\infty = \deg pq = \deg p + \deg q.$$

So either $\deg p = -\infty$ or $\deg q = -\infty$.   ∎

In $\mathbb{F}[z]$, as in the ring of integers $\mathbb{Z}$, we have a process of division with remainder.

Proposition 1.20. *Given a nonzero polynomial $p(z) = \sum_{i=0}^{m} p_i z^i$ with $p_m \neq 0$, then an arbitrary polynomial $q(z) \in \mathbb{F}[z]$ can be written uniquely in the form*

$$q(z) = a(z)p(z) + r(z) \tag{1.7}$$

with $\deg r < \deg p$.

Proof. If the degree of $q(z)$ is less than the degree of $p(z)$, then we write $q(z) = 0 \cdot p(z) + q(z)$ and this is the required representation. So we may assume $\deg q = n \geq m = \deg p$. The proof will proceed by induction on the degree of $q(z)$. We assume that for all polynomials of degree less than n such a representation exists. Clearly,

$$q_1(z) = q(z) - q_n p_m^{-1} z^{n-m} p(z)$$

is a polynomial of degree $\leq n-1$. Hence by the induction hypothesis,

$$q_1(z) = a_1(z)p(z) + r_1(z),$$

with $\deg(r_1) < \deg(p)$. But this implies

$$q(z) = (q_n p_m^{-1} z^{n-m} + a_1(z))p(z) + r(z) = a(z)p(z) + r(z),$$

with

$$a(z) = q_n p_m^{-1} z^{n-m} + a_1(z).$$

To show uniqueness, let

$$q(z) = a_1(z)p(z) + r_1(z) = a_2(z)p(z) + r_2(z).$$

This implies

$$(a_1(z) - a_2(z))p(z) = r_2(z) - r_1(z).$$

A consideration of the degrees of both sides shows that, necessarily, they are both equal to zero. Hence, the uniqueness of the representation (1.7) follows. ∎

The properties of the degree function in the ring $\mathbb{F}[z]$ can be abstracted to general rings.

Definition 1.21. A ring R is called a **Euclidean ring** if there exists a function δ from the set of nonzero elements in R into the set of nonnegative integers that satisfies

1. For all $a, b \neq 0$, we have $ab \neq 0$ and $\delta(ab) \geq \delta(a)$.
2. For all $f, g \in R$, with $g \neq 0$, there exist $a, r \in R$ such that $f = ag + r$ and $\delta(r) < \delta(g)$.

We can define $\delta(0) = -\infty$.

Obviously, with this definition, Proposition 1.20 implies that the ring of polynomials $\mathbb{F}[z]$ is a Euclidean ring. We note that in a Euclidean ring there are no zero divisors.

It is convenient to have a notation for the **remainder** of a polynomial $f(z)$ divided by $q(z)$. If $f(z) = a(z)q(z) + r(z)$ with $\deg r < \deg q$, we shall write $r = \pi_q f$.

We give several properties of the operation of taking a remainder.

Lemma 1.22. *Let* $q(z), a(z), b(z) \in \mathbb{F}[z]$ *with* $q(z)$ *nonzero. Then*

$$\pi_q(a\pi_q b) = \pi_q(ab). \tag{1.8}$$

Proof. Let $b(z) = b_1(z)q(z) + \pi_q(z)b(z)$. Then $a(z)b(z) = a(z)b_1(z)q(z) + a(z)\pi_q b$. Obviously, $\pi_q(ab_1 q) = 0$, and hence (1.8) follows. ∎

Corollary 1.23. *Given polynomials* $a_i(z) \in \mathbb{F}[z]$, $i = 1, \dots, k$, *then*

$$\pi_q(a_1 \cdots a_k) = \pi_q(a_1 \pi_q(a_2 \cdots \pi_q(a_k) \cdots)).$$

Proof. By induction. ∎

The following result simplifies in some important cases the computation of the remainder.

Lemma 1.24. *Let* $f(z), p(z), q(z) \in \mathbb{F}[z]$, *with* $p(z), q(z)$ *nonzero. Then*

$$\pi_{pq}(pf) = p\pi_q(f). \tag{1.9}$$

Proof. Let $r = \pi_q f$, i.e., for some polynomial $a(z)$, we have $f(z) = a(z)q(z) + r(z)$ and $\deg r < \deg q$. The representation of $f(z)$ implies $p(z)f(z) = a(z)(p(z)q(z)) + p(z)r(z)$. Since

$$\deg(pr) = \deg p + \deg r < \deg p + \deg q = \deg(pq),$$

it follows that $\pi_{pq}(pf) = pr = p\pi_q f$, and hence (1.9) holds. ∎

Definition 1.25. Let $p(z), q(z) \in \mathbb{F}[z]$. We say that $p(z)$ **divides** $q(z)$, or that $p(z)$ is a **factor** of $q(z)$, and write $p(z) \mid q(z)$, if there exists a polynomial $a(z)$ such that $q(z) = p(z)a(z)$.

If $p(z) \in \mathbb{F}[z]$ and $p(z) = \sum_{i=0}^{n} a_i z^i$, then $p(z)$ defines a function on $\mathbb{F}$ given by

$$p(\alpha) = \sum_{i=0}^{n} a_i \alpha^i, \quad \alpha \in \mathbb{F}.$$

Then $p(\alpha)$ is called the value of $p(z)$ at α. An element $\alpha \in \mathbb{F}$ is called a **zero** of $p(z)$ if $p(\alpha) = 0$. We never identify the polynomial with the function defined by it.

Theorem 1.26. *Let $p(z) \in \mathbb{F}[z]$. Then α is a zero of $p(z)$ if and only if $(z - \alpha) \mid p(z)$.*

Proof. If $(z - \alpha) \mid p(z)$, then $p(z) = (z - \alpha)a(z)$, and hence $p(\alpha) = 0$.

Conversely, by the division rule, we have

$$p(z) = a(z)(z - \alpha) + r(z)$$

with $r(z)$ necessarily a constant. Substituting α in this equality implies $r = 0$. ∎

Theorem 1.27. *Let $p(z) \in \mathbb{F}[z]$ be a polynomial of degree n. Then $p(z)$ has at most n zeros in $\mathbb{F}$.*

Proof. The proof is by induction. The statement is certainly true for zero-degree polynomials. Assume that we have proved it for all polynomials of degree less than n. Suppose that $p(z)$ is a polynomial of degree n. Either it has no zeros and the statement holds, or there exists a zero α. But then $p(z) = (z - \alpha)a(z)$, and $a(z)$ has, by the induction hypothesis, at most $n - 1$ zeros. ∎

Theorem 1.28. *Let $\mathbb{F}$ be a field and $\mathbb{F}[z]$ the ring of polynomials over $\mathbb{F}$. Then $\mathbb{F}[z]$ is a principal ideal domain.*

Proof. We have already shown that $\mathbb{F}[z]$ is a commutative ring with identity that contains no zero divisors. Let J be any ideal in $\mathbb{F}[z]$. If $J = \{0\}$, then J is generated by 0. So let us assume that $J \neq \{0\}$. Let $d(z)$ be any nonzero polynomial in J of minimal degree. We will show that $J = d\mathbb{F}[z]$.

To this end, let $f(z) \in J$ be arbitrary. By the division rule of polynomials we have $f(z) = a(z)d(z) + r(z)$ with $\deg r < \deg d$. Now $r(z) = f(z) - a(z)d(z) \in J$, since both $f(z)$ and $d(z)$ are in J. Since $d(z)$ was a nonzero element of smallest degree, we must have $r(z) = 0$. So $f(z) \in d\mathbb{F}[z]$ and hence $J \subset d\mathbb{F}[z]$. Conversely, since $d(z) \in J$, we have $d\mathbb{F}[z] \subset J$, and so equality follows. ∎

Definition 1.29. 1. Let $p_1(z), \ldots, p_n(z) \in \mathbb{F}[z]$. A polynomial $d(z) \in \mathbb{F}[z]$ will be called a **greatest common divisor** of $p_1(z), \ldots, p_n(z) \in \mathbb{F}[z]$ if

   a. We have the division relation $d(z) \mid p_i(z)$, for all $i = 1, \ldots, n$.
   b. If $d_1(z) \mid p_i(z)$, for all $i = 1, \ldots, n$, then $d_1(z) \mid d(z)$.

2. Let $p_1(z), \ldots, p_n(z) \in \mathbb{F}[z]$. A polynomial $d(z) \in \mathbb{F}[z]$ will be called a **least common multiple** of $p_1(z), \ldots, p_n(z) \in \mathbb{F}[z]$ if

   a. We have the division relation $p_i(z) \mid d(z)$, for all $i = 1, \ldots, n$.
   b. If $p_i(z) \mid d'(z)$, for all $i = 1, \ldots, n$, then $d \mid d'$.

Given polynomials $p_1(z), p_2(z)$, we denote by $p_1(z) \wedge p_2(z)$ their greatest common divisor and by $p_1(z) \vee p_2(z)$ their least common multiple. It is easily shown that a greatest common divisor is unique up to a constant multiple.

We will say that polynomials $p_1(z), \ldots, p_n(z) \in \mathbb{F}[z]$ are **coprime** if 1 is a greatest common divisor.

The availability of the division process in $\mathbb{F}[z]$ leads directly to the **Euclidean algorithm**. This gives an algorithm for the computation of a g.c.d. of two polynomials.

Theorem 1.30. *Let $p(z), q(z) \in \mathbb{F}[z]$. Set $q_{-1}(z) = q(z)$ and $q_0(z) = p(z)$. Define inductively, using the division rule for polynomials, a sequence of polynomials q_i by*

$$q_{i+1}(z) = a_{i+1}(z)q_i(z) - q_{i-1}(z) \tag{1.10}$$

with $\deg q_{i+1} < \deg q_i$. Let $q_s(z)$ be the last nonzero remainder, i.e., $q_{s+1}(z) = 0$. Then $q_s(z)$ is a g.c.d. of $q(z)$ and $p(z)$.

Proof. First we show that $q_s(z)$ is a common divisor of $q(z)$ and $p(z)$. Indeed, since $q_{s+1}(z) = 0$, we have $q_{s-1}(z) = a_{s+1}(z)q_s(z)$, that is, $q_s(z) \mid q_{s-1}(z)$. Assume that we proved $q_s(z) \mid q_{s-1}(z), \ldots, q_i(z)$. Since $q_{i-1}(z) = a_{i+1}(z)q_i(z) - q_{i+1}(z)$ it follows that $q_s(z) \mid q_{i-1}(z)$ and hence, by induction $q_s(z)$ divides all $q_i(z)$, and in particular $q_0(z), q_{-1}(z)$. So $q_s(z)$ is a common divisor of $q(z)$ and $p(z)$.

Let $J(q_i, q_{i-1}) = \{aq_i + bq_{i-1} \mid a(z), b(z) \in \mathbb{F}[z]\}$ be the ideal generated by $q_i(z), q_{i-1}(z)$. Clearly, equation (1.10) shows, again using an induction argument, that

$$q_{i+1}(z) \in J(q_i, q_{i-1}) \subset J(q_{i-1}, q_{i-2}) \subset \cdots \subset J(q_0, q_{-1}).$$

In particular, $q_s(z) \in J(q_0, q_{-1})$. So there exist polynomials $a(z), b(z)$ such that $q_s(z) = a(z)p(z) + b(z)q(z)$. This shows that any common divisor $r(z)$ of $p(z)$ and $q(z)$ divides also $q_s(z)$. Thus $q_s(z)$ is a g.c.d. of $p(z)$ and $q(z)$. ∎

We remark that the polynomials $a(z), b(z)$ in the representation $q_s(z) = a(z)p(z) + b(z)q(z)$ can be easily calculated from the polynomials $a_i(z)$. We will return to this in Chapter 8.

Corollary 1.31. *Let $p(z), q(z) \in \mathbb{F}[z]$. Then*

*1. $p(z)$ and $q(z)$ are coprime if and only if the **Bezout equation***

$$a(z)p(z) + b(z)q(z) = 1 \tag{1.11}$$

is solvable in $\mathbb{F}[z]$.

2. *A solution pair $a(z), b(z)$ is uniquely determined if we require additionally that*
   $\deg a < \deg q$. *In that case, we have also* $\deg b < \deg p$.

Proof. 1. Solvability of (1.11) is clearly a sufficient condition for the coprimeness
   of $p(z), q(z)$. Necessity is a consequence of Theorem 1.30.
2. The solution polynomials $a(z), b(z)$ of the Bezout equation (1.11) are not unique.
   In fact, if $a(z), b(z)$ is a solution pair and $r(z) \in \mathbb{F}[z]$ is arbitrary, then also $(a(z) -
   r(z)q(z)), (b(z) + r(z)p(z))$ is a solution. By the division rule, we can choose
   $r(z)$ such that $\deg(a - rq) < \deg q$, which also implies $\deg(b + rp) < \deg p$. This
   proves uniqueness.   ∎

In view of Theorem 1.16, the easiest way to construct ideals is by taking sums
and intersections of kernels of ring homomorphisms. The case of interest for us is
for the ring of polynomials.

Definition 1.32. We define, for each $\alpha \in \mathbb{F}$, a map $\phi_\alpha : \mathbb{F}[z] \longrightarrow \mathbb{F}$ by

$$\phi_\alpha(p) = p(\alpha). \tag{1.12}$$

Theorem 1.33. *A map $\phi : \mathbb{F}[z] \longrightarrow \mathbb{F}$ is a ring homomorphism if and only if $\phi(p) =
p(\alpha)$ for some $\alpha \in \mathbb{F}$.*

Proof. If ϕ_α is defined by (1.12), then clearly it is a ring homomorphism.
    Conversely, let $\phi : \mathbb{F}[z] \longrightarrow \mathbb{F}$ be ring homomorphism. Set $\alpha = \phi(z)$. Then, given
$p(z) = \sum_{i=0}^k p_i z^i$, we have

$$\phi(p) = \phi \sum_{i=0}^k p_i z^i = \sum_{i=0}^k p_i \phi(z^i) = \sum_{i=0}^k p_i \phi(z)^i = \sum_{i=0}^k p_i \alpha^i = p(\alpha).$$

∎

Corollary 1.34. *Given $\alpha_1, \dots, \alpha_n \in \mathbb{F}$, the set*

$$J_{\alpha_1,\dots,\alpha_n} = \{p(z) \in \mathbb{F}[z] \mid p(\alpha_1) = \dots = p(\alpha_n) = 0\}$$

is an ideal in $\mathbb{F}[z]$. Moreover, $J_{\alpha_1,\dots,\alpha_n} = d\mathbb{F}[z]$, where $d(z) = \Pi_{i=1}^n (z - \alpha_i)$.

Proof. For ϕ_α defined by (1.12), we have $\text{Ker}\,\phi_\alpha = \{p \in \mathbb{F}[z] \mid p(\alpha) = 0\} = J_\alpha$,
which is an ideal. Clearly $J_{\alpha_1,\dots,\alpha_n} = \cap_{i=1}^n J_{\alpha_i}$, and the intersection of ideals is an
ideal.
    Obviously, for $d(z)$ defined as above and an arbitrary polynomial $f(z)$ we have
$(df)(\alpha_i) = d(\alpha_i)f(\alpha_i) = 0$, so $d(z)f(z) \in J_{\alpha_1,\dots,\alpha_n}$. Conversely, if $g(z) \in J_{\alpha_1,\dots,\alpha_n}$,
we have $g(\alpha_i) = 0$, and hence $g(z)$ is divisible by $z - \alpha_i$. Since the α_i are distinct,
$g(z)$ is divisible by $d(z)$, or $g(z) = d(z)f(z)$.   ∎

Proposition 1.35. *Let $d(z) \in \mathbb{F}[z]$. Then*

$$d\mathbb{F}[z] = \{d(z)p(z) \mid p(z) \in \mathbb{F}[z]\}$$

is an ideal.

The next, important, result relates the generator of an ideal to division properties.

Theorem 1.36. *The ideal J generated by $p_1(z), \dots, p_n(z) \in \mathbb{F}[z]$, namely*

$$J = \left\{ \sum_{i=1}^{n} r_i(z)p_i(z) \mid r_i(z) \in \mathbb{F}[z] \right\}, \tag{1.13}$$

has the representation $J = d\mathbb{F}[z]$ if and only if $d(z)$ is a greatest common divisor of $p_1(z), \dots, p_n(z) \in \mathbb{F}[z]$.

Proof. By Theorem 1.28, there exists a $d(z) \in J$ such that $J = (d)$. Since $p_i(z) \in J$, there exist polynomials $q_i(z)$ such that $p_i(z) = d(z)q_i(z)$ for $i = 1, \dots, n$. So $d(z)$ is a common divisor of the $p_i(z)$. We will show that it is maximal. Assume that $d'(z)$ is another common divisor of the $p_i(z)$, i.e., $p_i(z) = d'(z)s_i(z)$. Since $d(z) \in J$, there exist polynomials $a_i(z)$ such that $d(z) = \sum_{i=1}^{n} a_i(z)p_i(z)$. Therefore

$$d(z) = \sum_{i=1}^{n} a_i(z)p_i(z) = \sum_{i=1}^{n} a_i(z)d'(z)q_i(z) = d'(z)\sum_{i=1}^{n} a_i(z)q_i(z).$$

But this means that $d'(z) \mid d(z)$, and so $d(z)$ is a g.c.d. ∎

Corollary 1.37. *Let $p_1(z), \dots, p_n(z) \in \mathbb{F}[z]$, and let $d(z)$ be their greatest common divisor. Then there exist polynomials $a_1(z), \dots, a_n(z) \in \mathbb{F}[z]$ such that*

$$d(z) = \sum_{i-1}^{n} a_i(z)p_i(z). \tag{1.14}$$

Proof. Let $d(z)$ be the g.c.d. of the $p_i(z)$. Obviously, $d(z) \in J = \{\sum_{i=1}^{n} r_i(z)p_i(z) \mid r_i(z) \in \mathbb{F}[z]\}$. Therefore there exist $a_i(z) \in \mathbb{F}[z]$ for which (1.14) holds. ∎

Corollary 1.38. *Polynomials $p_1(z), \dots, p_n(z) \in \mathbb{F}[z]$ are coprime if and only if there exist polynomials $a_1(z), \dots, a_n(z) \in \mathbb{F}[z]$ such that*

$$\sum_{i=1}^{n} a_i(z)p_i(z) = 1. \tag{1.15}$$

Equation (1.15), one of the most important equations in mathematics, will be refered to as the **Bezout equation**.

The importance of polynomials in linear algebra stems from the strong connection between factorization of polynomials and the structure of linear transformations. The primary decomposition theorem is of particular applicability.

Definition 1.39. A polynomial $p(z) \in \mathbb{F}[z]$ is **factorizable**, or **reducible**, if there exist polynomials $f(z), g(z) \in \mathbb{F}[z]$ of degree ≥ 1 such that $p(z) = f(z)g(z)$. If $p(z)$ is not factorizable, it is called a **prime** or an **irreducible** polynomial.

Note that the reducibility of a polynomial is dependent on the field $\mathbb{F}$.

Theorem 1.40. *Let* $p(z), f(z), g(z) \in \mathbb{F}[z]$, *and assume that* $p(z)$ *is irreducible and* $p(z) \mid (f(z)g(z))$. *Then either* $p(z) \mid f(z)$ *or* $p(z) \mid g(z)$.

Proof. Assume that $p(z) \mid (f(z)g(z))$ but $p(z)$ does not divide $f(z)$. Then a g.c.d. of $p(z)$ and $f(z)$ is 1. There exist therefore polynomials $a(z), b(z)$ such that the Bezout equation $1 = a(z)f(z) + b(z)p(z)$ holds. From this it follows that

$$g(z) = a(z)(f(z)g(z)) + (b(z)g(z))p(z).$$

This implies $p(z) \mid g(z)$.   ∎

Corollary 1.41. *Let* $p(z)$ *be an irreducible polynomial and assume* $p(z) \mid (f_1(z) \cdots f_n(z))$. *Then there exists an index i for which* $p(z) \mid f_i(z)$.

Proof. By induction.   ∎

Lemma 1.42. *Let* $p(z)$ *and* $q(z)$ *be coprime. Then if* $p(z) \mid q(z)s(z)$, *it follows that* $p(z) \mid s(z)$.

Proof. Follows immediately from Theorem 1.40.   ∎

Lemma 1.43. *Let* $p(z), q(z)$ *be coprime. Then* $p(z)q(z)$ *is their l.c.m.*

Proof. Clearly, $p(z)q(z)$ is a common multiple. Let $s(z)$ be an arbitrary common multiple. In particular, we can write $s(z) = q(z)t(z)$. Since $p(z)$ and $q(z)$ are coprime and $p(z) \mid s(z)$, it follows that $p(z) \mid t(z)$ or that $t(z) = p(z)t'(z)$. Thus $s(z) = (p(z)q(z))t'(z)$, and therefore $p(z)q(z)$ is a least common multiple.   ∎

A polynomial $p(z) \in \mathbb{F}[z]$ is called **monic** if its highest nonzero coefficient is 1.

Theorem 1.44. *Let* $p(z)$ *be a monic polynomial in* $\mathbb{F}[z]$. *Then* $p(z)$ *has a unique, up to ordering, factorization into a product of prime monic polynomials.*

Proof. We prove the theorem by induction on the degree of $p(z)$. If $\deg p = 1$, the statement is trivially true.

Assume that we proved the statement for all polynomials of degree $< n$. Let $p(z)$ be of degree n. Then either $p(z)$ is prime or $p(z) = f(z)g(z)$ with $1 < \deg f, \deg g < n$. By the induction hypothesis, both $f(z)$ and $g(z)$ are decomposable into a product of prime monic polynomials. This implies the existence of a decomposition of $p(z)$ into the product of monic primes. It remains to prove uniqueness. Suppose $p_1(z), \ldots, p_m(z)$ and $q_1(z), \ldots, q_n(z)$ are all prime and monic and

$$p_1(z) \cdots p_m(z) = q_1(z) \cdots q_n(z).$$

Clearly, $p_m(z) \mid q_1(z) \cdots q_n(z)$, so, by Corollary 1.41, there exists an i such that $p_m(z) \mid q_i(z)$. Since both are monic and irreducible, it follows that $p_m(z) = q_i(z)$. Without loss of generality we may assume $p_m(z) = q_n(z)$ Since there are no zero divisors in $\mathbb{F}[z]$, we get $p_1(z) \cdots p_{m-1}(z) = q_1(z) \cdots q_{n-1}(z)$. We complete the proof by using the induction hypothesis. ∎

Corollary 1.45. *Given a monic polynomial $p(z) \in \mathbb{F}[z]$. There exist monic primes $p_i(z)$ and positive integers n_i, $i = 1, \ldots, s$, such that*

$$p = p_1(z)^{n_1} \cdots p_s(z)^{n_s}. \tag{1.16}$$

The primes $p_i(z)$ and the integers n_i are uniquely determined.

Proof. Follows from the previous theorem. ∎

The factorization in (1.16) is called the **primary decomposition** of $p(z)$. The monicity assumption is necessary only to get uniqueness. Without it, the theorem still holds, but the primes are determined only up to constant factors.

The next result relates division in the ring of polynomials to the geometry of ideals. We saw already, in Proposition 1.15, that in a ring the sum and intersection of ideals are also ideals. the next proposition makes this specific for the ring of polynomials.

Proposition 1.46. *1. Let $p(z), q(z) \in \mathbb{F}[z]$. Then $q\mathbb{F}[z] \subset p\mathbb{F}[z]$ if and only if $p(z) \mid q(z)$.*
2. Let $p_i(z) \in \mathbb{F}[z]$ for $i = 1, \ldots, n$. Then

$$\cap_{i=1}^{n} p_i \mathbb{F}[z] = p\mathbb{F}[z],$$

where $p(z)$ is the l.c.m. of the $p_i(z)$.
3. Let $p_i(z) \in \mathbb{F}[z]$ for $i = 1, \ldots, n$. Then

$$\sum_{i=1}^{n} p_i \mathbb{F}[z] = p\mathbb{F}[z],$$

where $p(z)$ is the g.c.d. of the $p_i(z)$.

Proof. 1. Assume $q\mathbb{F}[z] \subset p\mathbb{F}[z]$. Thus there exists a polynomial $f(z)$ for which $q(z) = p(z)f(z)$, i.e., $p(z) \mid q(z)$.

Conversely, assume $p(z) \mid q(z)$, i.e., $q(z) = p(z)f(z)$ for some polynomial $f(z)$. Then

$$q\mathbb{F}[z] = \{q \cdot g \mid g \in \mathbb{F}[z]\} = \{pfq \mid g \in \mathbb{F}[z]\} \subset \{ph \mid h \in \mathbb{F}[z]\} = p\mathbb{F}[z].$$

2. By Theorem 1.28, the intersection $\cap_{i=1}^{m} p_i \mathbb{F}[z] = p\mathbb{F}[z]$ is a principal ideal, i.e., for some $p(z) \in \mathbb{F}[z]$ we have $\cap_{i=1}^{m} p_i \mathbb{F}[z] = p\mathbb{F}[z]$. Clearly, $p\mathbb{F}[z] \subset p_i \mathbb{F}[z]$ for all i. By part (i), this implies $p_i(z) \mid p(z)$. So $p(z)$ is a common multiple of the $p_i(z)$.

Suppose $q(z)$ is any common multiple, i.e., $q(z) = p_i(z)q_i(z)$ and so $q\mathbb{F}[z] \subset p_i\mathbb{F}[z]$ for all i, which implies that $q\mathbb{F}[z] \subset \cap_{i=1}^m p_i\mathbb{F}[z] = p\mathbb{F}[z]$. But this inclusion shows that $p(z) \mid q(z)$, and hence $p(z)$ is the l.c.m of the $p_i(z)$.

3. Again, by Theorem 1.28, $\sum_{i=1}^m p_i\mathbb{F}[z]$ is a principal ideal, i.e., for some $p(z) \in \mathbb{F}[z]$ we have $\sum_{i=1}^m p_i\mathbb{F}[z] = p\mathbb{F}[z]$. Obviously, $p_i\mathbb{F}[z] \subset p\mathbb{F}[z]$ for all i, and hence $p(z) \mid p_i(z)$. Thus $p(z)$ is a common divisor for the $p_i(z)$. Let $q(z)$ be any other common divisor. Then $p_i\mathbb{F}[z] \subset q\mathbb{F}[z]$, and hence $p\mathbb{F}[z] = \sum_{i=1}^m p_i\mathbb{F}[z] \subset q\mathbb{F}[z]$, which shows that $q(z) \mid p(z)$, that is, $p(z)$ is the g.c.d. ∎

For the case of two polynomials we have the following.

Proposition 1.47. *Let $p(z), q(z)$ be nonzero monic polynomials and let $r(z)$ and $s(z)$ be their g.c.d. and l.c.m. respectively, both taken to be monic. Then we have $p(z)q(z) = r(z)s(z)$.*

Proof. Write $p(z) = r(z)p_1(z), q(z) = r(z)q_1(z)$, with $p_1(z), q_1(z)$ coprime and monic. Clearly $s(z) = r(z)p_1(z)q_1(z)$ is a common multiple of $p(z), q(z)$. Let $s'(z)$ be any other multiple. Since $p(z) \mid s'(z)$, we have $s'(z) = r(z)p_1(z)t(z)$ for some polynomial $t(z)$. Since $q(z) \mid s'(z)$, we have $q_1(z) \mid p_1(z)t(z)$. Since $p_1(z), q_1(z)$ are coprime, it follows from Lemma 1.42 that $q_1(z) \mid t(z)$. This shows that $s(z) = r(z)p_1(z)q_1(z)$ is the unique monic l.c.m. of $p(z)$ and $q(z)$. The equality $p(z)q(z) = r(z)s(z)$ is now obvious. ∎

1.4.3   Formal Power Series

For a given field $\mathbb{F}$, we denote by $\mathbb{F}[[z]]$ the set of all formal power series, i.e., of formal sums of the form $f(z) = \sum_{j=0}^\infty f_j z^j$. Addition and multiplication are defined by

$$\sum_{j=0}^\infty f_j z^j + \sum_{j=0}^\infty g_j z^j = \sum_{j=0}^\infty (f_j + g_j) z^j \tag{1.17}$$

and

$$\sum_{j=0}^\infty f_j z^j \cdot \sum_{j=0}^\infty g_j z^j = \sum_{k=0}^\infty h_k z^k, \tag{1.18}$$

with

$$h_k = \sum_{j=0}^k f_j g_{k-j}. \tag{1.19}$$

With these operations, $\mathbb{F}[[z]]$ is a ring with identity 1.

An element $f(z) = \sum_{j=0}^\infty f_j z^j \in \mathbb{F}[[z]]$ is invertible if and only if $f_0 \neq 0$. To see this let $g(z) = \sum_{j=0}^\infty g_j z^j$. Then $g(z)$ is an inverse of $f(z)$ if and only if

$$1 = (fg)(z) = \sum_{k=0}^{\infty} \left\{ \sum_{j=0}^{k} f_j g_{k-j} \right\} z^k.$$

This is equivalent to the solvability of the infinite system of equations

$$\sum_{j=0}^{k} f_j g_{k-j} = \begin{cases} 1 & k = 0, \\ 0 & k > 0. \end{cases}$$

The first equation is $f_0 g_0 = 1$, which shows the necessity of the condition $f_0 \neq 0$. This is also sufficient, since the system of equations can be solved recursively.

The following result analyzes the ideal structure in $\mathbb{F}[[z]]$.

Proposition 1.48. $J \subset \mathbb{F}[[z]]$ *is a nonzero ideal if and only if for some nonnegative integer n, we have* $J = z^n \mathbb{F}[[z]]$. *Thus* $\mathbb{F}[[z]]$ *is a principal ideal domain.*

Proof. Clearly, any set of the form $z^n \mathbb{F}[[z]]$ is an ideal. To prove the converse, we set, for $f(z) = \sum_{j=0}^{\infty} f_j z^j \in \mathbb{F}[[z]]$,

$$\delta(f) = \begin{cases} \min\{n \mid f_n \neq 0\} & f \neq 0 \\ -\infty & f = 0 \end{cases}$$

Let now $f \in J$ be any nonzero element that minimizes $\delta(f)$. Then $f(z) = z^n h(z)$ with h invertible. Thus z^n belongs to J and generates it. ∎

In the sequel we find it convenient to work with the ring $\mathbb{F}[[z^{-1}]]$ of formal power series in z^{-1}.

We study now an important construction that allows us to construct from some ring larger rings. The prototype of this situation is the construction of the field of rational numbers out of the ring of integers.

Given rings R and $\overline{R}$, we say that R is **embedded** in $\overline{R}$ if there exists an injective homomorphism of R into $\overline{R}$.

A set S in a ring R with identity is called a **multiplicative set** if $0 \notin S$, $1 \in S$, and $a, b \in S$ implies also $ab \in S$. Given a commutative ring with identity R and a multiplicative set S in R, we proceed to construct a new ring. Let $\mathscr{M}$ be the set of ordered paits (r, s) with $r \in R$ and $s \in S$. We introduce a relation in $\mathscr{M}$ by saying that $(r, s) \simeq (r', s')$ if there exists $\sigma \in S$ for which $\sigma(s'r - sr') = 0$. We claim that this is indeed an equivalence relation. Reflexivity and symmetry are trivial to check. To check transitivity, assume $\sigma(s'r - sr') = 0$ and $\tau(s''r' - s') = 0$ with $\sigma, \tau \in S$. We compute

$$\tau \sigma s'(s''r) = \tau s'' \sigma(s'r) = \tau s'' \sigma(sr') = \sigma s \tau(s''r') = \sigma s \tau(s'r'') = \sigma s' \tau(sr''),$$

or $\tau \sigma s'(s''r - sr'') = 0$.

We denote by r/s the equivalence class of the pair (r, s). We denote by R/S the set of all equivalence classes in $\mathscr{M}$. In R/S we define operations of addition and multiplication by

$$\frac{r}{s} + \frac{r'}{s'} = \frac{rs' + sr'}{ss'},$$

$$\frac{r}{s} \cdot \frac{r'}{s'} = \frac{rr'}{ss'}.$$

(1.20)

It can be checked that these operations are well defined, i.e., they are independent of the equivalence class representatives. Also, it is straightforward to verify that with these operations R/S is a commutative ring. This is called a **ring of quotients** of R. Of course, there may be many such rings, depending on the multiplicative sets we are taking.

In case R is an entire ring, then the map $\phi : R \longrightarrow R/S$ defined by $\phi(r) = r/1$ is an injective ring homomorphism. In this case more can be said.

Theorem 1.49. *Let R be an entire ring. Then R can be embedded in a field.*

Proof. Let $S = R - \{0\}$, that is, S is the set of all nonzero elements in R. Obviously S is a multiplicative set. We let $\mathbb{F} = R/S$. Then $\mathbb{F}$ is a commutative ring with identity. To show that $\mathbb{F}$ is a field, it suffices to show that every nonzero element is invertible. If $a/b \neq 0$, this implies $a \neq 0$ and hence $(a/b)^{-1} = b/a$. Moreover, the map ϕ given before provides an embedding of R in $\mathbb{F}$. $\blacksquare$

The field $\mathbb{F}$ constructed by the previous theorem is called the **field of quotients** of R.

For our purposes the most important example of a field of quotients is that of the field of rational functions, denoted by $\mathbb{F}(z)$, which is obtained as the field of quotients of the ring of polynomials $\mathbb{F}[z]$.

Let us consider now an entire ring that is also a principal ideal domain. Let $\mathbb{F}$ be the field of fractions of R. Given any $f \in \mathbb{F}$, we can consider $J = \{r \in R \mid rf \in R\}$. Obviously, J is an ideal, hence generated by an element $q \in R$ that is uniquely defined up to an invertible factor. Thus $qf = p$ for $p \in R$ and $f = p/q$. Obviously, p, q are coprime, and $f = p/q$ is called a **coprime factorization** of f.

We give now a few examples of rings and fields that are a product of the process of taking a ring of quotients.

1.4.4   Rational Functions

For a field $\mathbb{F}$ we saw that the ring of polynomials $\mathbb{F}[z]$ is a principal ideal domain. Its field of quotients is called the field of rational functions and denoted by $\mathbb{F}(z)$. Its elements are called **rational functions**.

Every rational function has a representation of the form $p(z)/q(z)$ with $p(z), q(z)$ coprime. We can make the coprime factorization unique if we require the polynomial $q(z)$ to be monic. By Proposition 1.20, every polynomial $p(z)$ has a unique representation in the form $p(z) = a(z)q(z) + r(z)$ with $\deg r < \deg q$. This implies

$$\frac{p(z)}{q(z)} = a(z) + \frac{r(z)}{q(z)}. \tag{1.21}$$

A rational function $r(z)/q(z)$ is called **proper** if $\deg r \le \deg q$ and **strictly proper** if $\deg r < \deg q$. The set of all strictly proper rational functions is denoted by $\mathbb{F}_-(z)$. Thus we have

$$\mathbb{F}(z) = \mathbb{F}[z] \oplus \mathbb{F}_-(z).$$

Here the direct sum representation refers to the uniqueness of the representation (1.21).

1.4.5 Proper Rational Functions

We denote by $\mathbb{F}_{pr}(z)$ the subring of $\mathbb{F}(z)$ defined by

$$\mathbb{F}_{pr}(z) = \left\{ f(z) \in \mathbb{F}(z) \mid f(z) = \frac{p(z)}{q(z)}, \deg p \le \deg q \right\}. \tag{1.22}$$

It is easily checked that $\mathbb{F}_{pr}(z)$ is a commutative ring with identity. An element $f(z) \in \mathbb{F}_{pr}(z)$ is a unit if and only if in any representation $f(z) = p(z)/q(z)$, we have $\deg p = \deg q$. We define the **relative degree** ρ by $\rho(p/q) = \deg q - \deg p$, and $\rho(0) = -\infty$.

Theorem 1.50. $\mathbb{F}_{pr}(z)$ *is a Euclidean domain with respect to the relative degree.*

Proof. We verify the conditions of a Euclidean ring as given in Definition 1.21. Let ρ be the relative degree. Let $f(z) = p_1(z)/q_1(z)$ and $g(z) = p_2(z)/q_2(z)$ be nonzero. Then

$$\begin{aligned}
\rho(fg) = \rho\left(\frac{p_1 p_2}{q_1 q_2}\right) &= \deg(q_1 q_2) - \deg(p_1 p_2) \\
&= \deg q_1 + \deg q_2 - \deg p_1 - \deg p_2 \\
&= (\deg q_1 - \deg p_1) + (\deg q_2 - \deg p_2) \\
&= \rho(f) + \rho(g) \ge \rho(f).
\end{aligned}$$

Next, we turn to division. If $\rho(f) < \rho(g)$, we write $f(z) = 0 \cdot g(z) + f(z)$. If $g(z) \ne 0$ and $\rho(f) \ge \rho(g)$, we show that $g(z)$ divides $f(z)$. Let σ and τ be the relative degrees of $f(z)$ and $g(z)$; thus $\sigma \ge \tau$. So we can write

$$f(z) = \frac{1}{z^\sigma} f_1(z), \quad g(z) = \frac{1}{z^\tau} g_1(z),$$

with $f_1(z), g_1(z)$ units in $\mathbb{F}_{pr}(z)$, i.e., satisfying $\rho(f_1) = \rho(g_1) = 0$. Then we can write

$$f(z) = \left(\frac{1}{z^\tau}g_1(z)\right)\left(\frac{1}{z^{\sigma-\tau}}g_1(z)^{-1}f_1(z)\right).$$

Of course $\frac{1}{z^{\sigma-\tau}}g_1(z)^{-1}f_1(z)$ belongs to $\mathbb{F}_{pr}(z)$ and has relative degree $\sigma - \tau$.   ∎

Since $\mathbb{F}_{pr}(z)$ is Euclidean, it is a principal ideal domain. We proceed to characterize all ideals in $\mathbb{F}_{pr}(z)$.

Theorem 1.51. *A subset $J \subset \mathbb{F}_{pr}(z)$ is a nonzero ideal if and only if it is of the form $J = \frac{1}{z^\sigma}\mathbb{F}_{pr}(z)$ for some nonnegative integer σ.*

Proof. Clearly, $J = \frac{1}{z^\sigma}\mathbb{F}_{pr}(z)$ is an ideal.

Conversely, let J be a nonzero ideal. Let $f(z)$ be a nonzero element in J of least relative degree, say σ. We may assume without loss of generality that $f(z) = \frac{1}{z^\sigma}$. By the proof of Theorem 1.50, $f(z)$ divides every element with relative degree greater than or equal to σ. So $f(z)$ is a generator of J.   ∎

Heuristically speaking, and adopting the language of complex analysis, we see that ideals are determined by the zeros at infinity, multiplicities counted. In terms of singularities, $\mathbb{F}_{pr}(z)$ is the set of all rational functions that have no singularity (pole) at infinity. In comparison, $\mathbb{F}[z]$ is the set of all rational functions whose only singularity is at infinity. Of course, there are many intermediate situations and we turn to these.

1.4.6   Stable Rational Functions

Fix the field in what follows to be the real field $\mathbb{R}$. We can identify many subrings of $\mathbb{R}(z)$ simply by taking the ring of quotients of $\mathbb{R}[z]$ with respect to multiplicative sets that are smaller than the set of all nonzero polynomials. We shall say that a polynomial is **stable** if all its (complex) zeros are in a given subset Σ of the complex plane. It is antistable if all its zeros lie in the complement of Σ. We will assume now that the domain of stability is the open left half-plane. Let S be the set of all stable polynomials. Then S is obviously a multiplicative subset of $\mathbb{R}[z]$. The ring of quotients $\mathscr{S} = \mathbb{R}[z]/S \subset \mathbb{R}(z)$ is called the ring of stable rational functions. As a ring of fractions it is a commutative ring with identity. An element $f(z) \in \mathscr{S}$ is a unit if in an irreducible representation $f(z) = p(z)/q(z)$ the numerator $p(z)$ is stable. Thus we may say that $f(z) \in \mathscr{S}$ is a unit if it has no zeros in the closed left half-plane. Such functions are sometimes called **minimum phase** functions.

From the point of view of complex analysis, the ring of stable rational functions is the set of rational functions that have all their singularities in the open left half-plane or at the point at infinity.

In $\mathscr{S}$ we define a degree function δ as follows. Let $f(z) = p(z)/q(z)$ with $p(z), q(z)$ coprime. We let $\delta(f)$ be the number of antistable zeros of $p(z)$, and hence of $f(z)$. Note that δ does not take into account zeros at infinity.

Theorem 1.52. $\mathscr{S}$ *is a Euclidean ring with respect to the degree function* δ.

Proof. That $\delta(fg) = \delta(f) + \delta(g) \geq \delta(f)$ is obvious. Let now $f(z), g(z) \in \mathscr{S}$ with $g(z) \neq 0$. Assume without loss of generality that $\delta(f) \geq \delta(g)$. Let $g(z) = \alpha_g(z)/\beta_g(z)$ with $\alpha_g(z), \beta_g(z)$ coprime polynomials. Similarly, let $f(z) = \alpha_f(z)/\beta_f(z)$ with $\alpha_f(z), \beta_f(z)$ coprime. Factor $\alpha_g(z) = \alpha_+(z)\alpha_-(z)$, with $\alpha_-(z)$ stable and $\alpha_+(z)$ antistable. Then

$$g(z) = \frac{\alpha_+(z)\alpha_-(z)}{\beta_g(z)} = \frac{\alpha_-(z)(z+1)^v}{\beta_g(z)} \cdot \frac{\alpha_+(z)}{(z+1)^v} = e(z) \cdot \frac{\alpha_+(z)}{(z+1)^v},$$

with $e(z)$ a unit and $v = \delta(g)$. Since $\beta_f(z)$ is stable, $\beta_f(z), \alpha_+(z)$ are coprime in $\mathbb{R}[z]$ and there exist polynomials $\phi(z), \psi(z)$ for which $\phi(z)\alpha_+(z) + \psi(z)\beta_f(z) = \alpha_f(z)(z+1)^{v-1}$, and we may assume without loss of generality that $\deg \psi < \deg \alpha_+$. Dividing both sides by $\beta_f(z)(z+1)^{v-1}$, we get

$$\frac{\alpha_f(z)}{\beta_f(z)} = \left(\frac{\phi(z+1)}{\beta_f(z)} \right) \left(\frac{\alpha_+(z)}{(z+1)^{v-1}} \right) + \left(\frac{\psi(z)}{(z+1)^{v-1}} \right),$$

and we check that

$$\delta\left(\frac{\alpha_+}{(z+1)^{v-1}} \right) \leq \deg \psi < \deg \alpha_+ = \delta(g).$$

∎

Ideals in $\mathscr{S}$ can be represented by generators that are, up to unit factors, antistable polynomials. For two antistable polynomials $p_1(z), p_2(z)$ we have $p_2\mathscr{S} \subset p_1\mathscr{S}$ if and only if $p_1(z) \mid p_2(z)$. More generally, for $f_1(z), f_2(z) \in \mathscr{S}$, we have $f_2\mathscr{S} \subset f_1\mathscr{S}$ if and only if $f_1(z) \mid f_2(z)$. However, the division relation $f_1(z) \mid f_2(z)$ means that every zero of $f_1(z)$ in the closed right half-plane is also a zero of $f_2(z)$ with at least the same multiplicity. Here zeros should be interpreted as complex zeros.

Since the intersection of subrings is itself a ring, we can consider the ring $\mathbb{R}_{pr}(z) \cap \mathscr{S}$. This is equivalent to excluding from $\mathscr{S}$ elements with a singularity at infinity. We denote this ring by $\mathbf{RH}_+^\infty$. Because of its great importance to system-theoretic problems, we shall defer the study of this ring to Chapters 11 and 12.

1.4.7 Truncated Laurent Series

Let $\mathbb{F}$ be a field. We denote by $\mathbb{F}((z^{-1}))$ the set of all formal sums of the form $f(z) = \sum_{j=-\infty}^{n_f} f_j z^j$ with $n_f \in \mathbb{Z}$. The operations of addition and multiplication are defined by

$$(f+g)(z) = \sum_{j=-\infty}^{\max\{n_f,n_g\}} (f_j+g_j)z^j,$$

and

$$(fg)(z) = \sum_{k=-\infty}^{\{n_f+n_g\}} h_k z^k,$$

with

$$h_k = \sum_{j=-\infty}^{\infty} f_j g_{k-j}.$$

Notice that the last sum is well defined, since it contains only a finite number of nonzero terms. We can check that all the field axioms are satisfied; in particular, all nonzero elements are invertible. We call this the **field of truncated Laurent series**. Note that $\mathbb{F}((z^{-1}))$ can be considered the field of fractions of $\mathbb{F}[[z^{-1}]]$. Clearly, $\mathbb{F}(z)$ can be considered a subfield of $\mathbb{F}((z^{-1}))$.

We introduce for later use the projection maps

$$\pi_+ \sum_{j=-\infty}^{n_f} f_j z^j = \sum_{j=0}^{n_f} f_j z^j,$$
$$\pi_- \sum_{j=-\infty}^{n_f} f_j z^j = \sum_{j=-\infty}^{-1} f_j z^j. \tag{1.23}$$

1.5   Modules

The module structure is one of the most fundamental algebraic concepts. Most of the rest of this book is, to a certain extent, an elaboration on the module theme. This is particularly true for the case of linear transformations and linear systems.

Let R be a ring with identity. A **left module** M over the ring R is a commutative group together with an operation of R on M that for all $r, s \in R$ and $x, y \in M$, satisfies

$$r(x+y) = rx + ry,$$
$$(r+s)x = rx + sx,$$
$$r(sx) = (rs)x,$$
$$1x = x.$$

Right modules are defined similarly. Let M be a left R-module. A subset N of M is a **submodule** of M if it is an additive subgroup of M that further satisfies $RN \subset M$.

Given two left R-modules M_1 and M_2, a map $\phi : M_1 \longrightarrow M_2$ is an R-module homomorphism if for all $x, y \in M_1$ and $r \in R$,

$$\phi(x+y) = \phi x + \phi y,$$
$$\phi(rx) = r\phi(x).$$

Given an R-module homomorphism $\phi : M_1 \longrightarrow M_2$, then $\text{Ker}\,\phi$ and $\text{Im}\,\phi$ are submodules of M_1 and M_2 respectively. Given R-modules $M_0, \ldots, M_n$, a sequence of R-module homomorphisms

$$\longrightarrow M_{i-1} \xrightarrow{\phi_{i-1}} M_i \xrightarrow{\phi_i} M_{i+1} \longrightarrow$$

is called an **exact sequence** if $\text{Im}\,\phi_{i-1} = \text{Ker}\,\phi_i$. An exact sequence of the form

$$0 \longrightarrow M_1 \xrightarrow{\phi} M_2 \xrightarrow{\psi} M_3 \longrightarrow 0$$

is called a **short exact sequence**. This means that ϕ is injective and ψ is surjective.

Given modules M_i over the ring R, we can make the Cartesian product $M_1 \times \cdots \times M_k$ into an R-module by defining, for $m_i, n_i \in M_i$ and $r \in R$,

$$(m_1, \ldots, m_k) + (n_1, \ldots, n_k) = (m_1 + n_1, \ldots, m_k + n_k),$$
$$r(m_1, \ldots, m_k) = (rm_1, \ldots, rm_k). \tag{1.24}$$

Given a module M and submodules M_i, we say that M is the direct sum of the M_i, and write $M = M_1 \oplus \cdots \oplus M_k$, if every $m \in M$ has a unique representation of the form $m = m_1 + \cdots + m_k$ with $m_i \in M_i$.

Given a submodule N of a left R-module M, we can construct a module structure in the same manner in which we constructed quotient groups. We say that two elements $x, y \in M$ are equivalent if $x - y \in N$. The equivalence class of x is denoted by $[x]_N = x + N$. The set of equivalence classes is denoted by M/N. We make M/N into an R-module by defining, for all $r \in R$ and $x, y \in M$,

$$[x]_N + [y]_N = [x+y]_N,$$
$$r[x]_N = [rx]_N. \tag{1.25}$$

It is easy to check that these operations are well defined, that is, independent of the equivalence class representative. We state without proof the following.

Proposition 1.53. *With the operations defined in (1.25), the set of equivalence classes M/N is a module over R. This is called the* **quotient module** *of M by N.*

We shall make use of the following result. This is the counterpart of Theorem 1.9.

Proposition 1.54. *Let M be a module over the ring R. Then $N \subset M$ is a submodule if and only if it is the kernel of a R-module homomorphism.*

Proof. Clearly the kernel of a module homomorphism is a submodule.

Conversely, let $\pi : M \longrightarrow M/N$ be the canonical projection. Then π is a module homomorphism and $\text{Ker}\,\pi = N$. ∎

Note that with $i : N \longrightarrow M$ the natural embedding, then

$$\{0\} \longrightarrow N \overset{i}{\longrightarrow} M \overset{\pi}{\longrightarrow} M/N \longrightarrow \{0\}$$

is a short exact sequence.

Proposition 1.55. *Let M, N be R-modules and let $\phi : M \longrightarrow N$ be a surjective R-module homomorphism. Then we have the module isomorphism*

$$N \simeq M/\mathrm{Ker}\,\phi.$$

Proof. Let π be the canonical projection of N onto $N/\mathrm{Ker}\,\phi$, that is, $\pi(m) = m + \mathrm{Ker}\,\phi$. We define now a map $\tau : N/\mathrm{Ker}\,\phi \longrightarrow M$ by $\tau(m + \mathrm{Ker}\,\phi) = \phi(m)$. The map τ is well defined. For if m_1, m_2 are representatives of the same coset, then $m_1 - m_2 \in \mathrm{Ker}\,\phi$. This implies $\phi(m_1) = \phi(m_2)$. That $\tau\phi$ is a homomorphism follows from the fact that ϕ is one. Clearly, the surjectivity of ϕ implies that of τ. Finally, $\tau(m + \mathrm{Ker}\,\phi) = 0$ if and only if $\phi(m) = 0$, i.e., $m \in \mathrm{Ker}\,\phi$. Thus τ is also injective, hence an isomorphism. ∎

Let M be a module over the ring R and let $S \subset M$. By a **linear combination**, with coefficients in R, we mean a sum $\sum_{\alpha \in S'} a_\alpha m_\alpha$, where $a_\alpha \in R$, $m_\alpha \in M$ and S' a finite subset of S. The elements a_α are called the coefficients of the linear combination. A subset S of M is called **linearly independent** if whenever $\sum_{\alpha \in S'} a_\alpha m_\alpha = 0$, we have $a_\alpha = 0$ for all $\alpha \in S'$. A subset S of M **generates** M if the set of all finite linear combinations of elements in S is equal to M. Equivalently, if the smallest submodule of M that includes S is M itself. S is a **basis** for M if it is non-empty subset, linearly independent and generates M.

We end by giving a few examples of important module structures. Every abelian group G is a module over the ring of integers $\mathbb{Z}$. Every ring is a module over itself, as well as over any subring. A vector space, as we shall see in Chapter 2, is a module over a field. Thus $\mathbb{F}(z)$ is a module over $\mathbb{F}[z]$; $\mathbb{F}((z^{-1}))$ is a module over each of the subrings $\mathbb{F}[z]$ and $\mathbb{F}[[z^{-1}]]$; $z^{-1}\mathbb{F}[[z^{-1}]]$ has an induced $\mathbb{F}[z]$-module structure, being isomorphic to $\mathbb{F}((z^{-1}))/\mathbb{F}[z]$. This module structure is defined by

$$z \cdot \sum_{j=1}^\infty \frac{h_j}{z^j} = \sum_{j=1}^\infty \frac{h_{j+1}}{z^j}.$$

As $\mathbb{F}$-vector spaces, we have the direct sum representation

$$\mathbb{F}((z^{-1})) = \mathbb{F}[z] \oplus z^{-1}\mathbb{F}[[z^{-1}]]. \qquad (1.26)$$

We denote by π_+ and π_- the projections of $\mathbb{F}((z^{-1}))$ on $\mathbb{F}[z]$ and $z^{-1}\mathbb{F}[[z^{-1}]]$ respectively, i.e., given by

$$\pi_- \sum_{j=-\infty}^{N} h_j z^j = \sum_{j=-\infty}^{-1} h_j z^j$$

$$\pi_+ \sum_{j=-\infty}^{N} h_j z^j = \sum_{j=0}^{N} h_j z^j \tag{1.27}$$

Clearly, π_+ and π_- are complementary projections, i.e., satisfy $\pi_\pm^2 = \pi_\pm$ and $\pi_+ + \pi_- = I$.

Since $\mathbb{F}((z^{-1}))$ is a 1-dimensional vector space over the field $\mathbb{F}((z^{-1}))$, every $\mathbb{F}((z^{-1}))$-linear map from $\mathbb{F}((z^{-1}))$ to $\mathbb{F}((z^{-1}))$ has a representation

$$(L_A f)(z) = A(z) f(z), \tag{1.28}$$

for some $A(z) \in \mathbb{F}((z^{-1}))$. The map L_A defined by (1.28) is called the **Laurent operator**, and $A(z)$ is called the **symbol** of the Laurent operator L_A. Of course in terms of the expansions of $A(z)$ and $f(z)$, we have $g = L_A f = \sum_k g_k z^k$ with $g_k = \sum_{j=-\infty}^{\infty} A_j f_{k-j}$. The sum is well defined, since there are only a finitely many nonzero terms in it.

Of special importance is the Laurent operator acting in $\mathbb{F}((z^{-1}))$ with symbol z. Because of the natural interpretation in terms of the expansion coefficients, we call it the **bilateral shift**, or simply the **shift**, and denote it by S. The shift S is clearly an invertible map in $\mathbb{F}((z^{-1}))$, and we have

$$(Sf)(z) = zf(z),$$
$$(S^{-1}f)(z) = z^{-1}f(z). \tag{1.29}$$

The subsets $\mathbb{F}[z]$ and $\mathbb{F}[[z^{-1}]]$ of $\mathbb{F}((z^{-1}))$ are closed under addition and multiplication; hence they inherit natural ring structures. Units in $\mathbb{F}[[z^{-1}]]$ are biproper, i.e., are proper and have a proper inverse.

Similarly $\mathbb{F}((z^{-1}))$ is a module over both $\mathbb{F}[z]$ and $\mathbb{F}[[z^{-1}]]$; $\mathbb{F}[z] \subset \mathbb{F}((z^{-1}))$ is an $\mathbb{F}[z]$ submodule and similarly $z^{-1}\mathbb{F}[[z^{-1}]] \subset \mathbb{F}((z^{-1}))$ is an $\mathbb{F}[[z^{-1}]]$ submodule. Furthermore, $\mathbb{F}((z^{-1}))$ becomes a module over various rings including $\mathbb{F}$, $\mathbb{F}[z]$, $\mathbb{F}((z^{-1}))$ and $\mathbb{F}[[z^{-1}]]$. Various module structures will be of interest in the sequel. In particular, we note the following.

Proposition 1.56. *1. $\mathbb{F}[z]$ is an $\mathbb{F}[z]$-submodule of $\mathbb{F}((z^{-1}))$.*
2. $\mathbb{F}[[z^{-1}]]$ and $z^{-1}\mathbb{F}[[z^{-1}]]$ are $\mathbb{F}[[z^{-1}]]$ submodules of $\mathbb{F}((z^{-1}))$.
3. As $\mathbb{F}[z]$ modules we have the following short exact sequence of module homomorphisms:

$$0 \longrightarrow \mathbb{F}[z] \xrightarrow{j} \mathbb{F}((z^{-1})) \xrightarrow{\pi} \mathbb{F}((z^{-1}))/\mathbb{F}[z] \longrightarrow 0,$$

with j the embedding of $\mathbb{F}[z]$ into $\mathbb{F}((z^{-1}))$ and π the canonical projection onto the quotient module. ∎

Elements of $\mathbb{F}((z^{-1}))/\mathbb{F}[z]$ are equivalence classes, and two elements are in the same equivalence class if and only if they differ in their polynomial terms only. For $h \in \mathbb{F}((z^{-1}))$, we denote by $[h]_{\mathbb{F}[z]}$ its equivalence class. A natural choice of representative in each equivalence class is the one element whose polynomial terms are all zero. This leads to the isomorphism

$$\mathbb{F}((z^{-1}))/\mathbb{F}[z] \simeq z^{-1}\mathbb{F}[[z^{-1}]], \tag{1.30}$$

given by $[h]_{\mathbb{F}[z]} \mapsto \pi_- h$.

There are important linear transformations induced by S in the spaces $\mathbb{F}[z]$ and $z^{-1}\mathbb{F}[[z^{-1}]]$. Noting that $\mathbb{F}[z]$ is an $\mathbb{F}[z]$-submodule of $\mathbb{F}((z^{-1}))$, then $\mathbb{F}[z]$ is S-invariant. Thus, we can define a map $S_+ : \mathbb{F}[z] \longrightarrow \mathbb{F}[z]$ by restricting S to $\mathbb{F}[z]$, i.e.,

$$S_+ = S \mid \mathbb{F}[z]. \tag{1.31}$$

Similarly, S induces a map in the quotient module $\mathbb{F}((z^{-1}))/\mathbb{F}[z]$ given by $[h]_{\mathbb{F}[z]} \mapsto [Sh]_{\mathbb{F}[z]}$. Using the isomorphism (1.33), we also define $S_- : z^{-1}\mathbb{F}[[z^{-1}]] \longrightarrow z^{-1}\mathbb{F}[[z^{-1}]]$ by

$$S_- h = \pi_- z h. \tag{1.32}$$

We note that S_+ is injective but not surjective, whereas S_- is surjective but not injective. Also $\operatorname{codim} \operatorname{Im} S_+ = \dim \operatorname{Ker} S_- = \dim \mathbb{F} = 1$. We will refer to S_+ as the **forward shift operator** and to S_- as the **backward shift operator**.

Clearly, $\mathbb{F}((z^{-1}))$, being a module over $\mathbb{F}((z^{-1}))$, is also a module over any subring of $\mathbb{F}((z^{-1}))$. In particular, it has a module structure over $\mathbb{F}[z]^{m \times m}$ and $\mathbb{F}[[z^{-1}]]$, as well as over the rings $\mathbb{F}[z]$ and $\mathbb{F}[[z^{-1}]]$. We can state the following.

Proposition 1.57. *1. $\mathbb{F}[z]$ is a left $\mathbb{F}[z]$ submodule, and hence also a left $\mathbb{F}[z]$ submodule of $\mathbb{F}((z^{-1}))$.*
2. $z^{-1}\mathbb{F}[[z^{-1}]]$ is an $\mathbb{F}[[z^{-1}]]$ submodule and hence also an $\mathbb{F}[[z^{-1}]]$ submodule of $\mathbb{F}((z^{-1}))$.
3. We have the isomorphism

$$z^{-1}\mathbb{F}[[z^{-1}]] \simeq \mathbb{F}((z^{-1}))/\mathbb{F}[z], \tag{1.33}$$

and $z^{-1}\mathbb{F}[[z^{-1}]]$ has therefore the naturally induced $\mathbb{F}[z]$-module structure. Also, it has an $\mathbb{F}[z]$-module structure given, for $A \in \mathbb{F}[z]$, by

$$A \cdot h = \pi_- A h, \qquad h \in z^{-1}\mathbb{F}[[z^{-1}]]. \tag{1.34}$$

4. Given $A \in \mathbb{F}[z]^{p \times m}$, with the $\mathbb{F}[z]$-module structures, the map $X : z^{-1}\mathbb{F}[[z^{-1}]] \longrightarrow z^{-1}\mathbb{F}[[z^{-1}]]^p$ defined by $h \mapsto \pi_- A h$ is a module homomorphism.
5. $\mathbb{F}[z]$ has an $\mathbb{F}[[z^{-1}]]$-module structure given, for $A \in \mathbb{F}[[z^{-1}]]$, by

$$A \cdot f = \pi_+ A f, \qquad f \in \mathbb{F}[z]. \tag{1.35}$$

1.6 Exercises

1. Show that the order $o(S_n)$ of the symmetric group S_n is $n!$.
2. Show that any subgroup of a cyclic group is cyclic, and so is any homomorphic image of G.
3. Show that every group of order ≤ 5 is abelian.
4. Prove the isomorphism $\mathbb{C} \simeq \mathbb{R}[z]/(z^2+1)\mathbb{R}[z]$.
5. Given a polynomial $p(z) = \sum_{k=0}^{n} p_k z^k$, we define its **formal derivative** by $p'(z) = \sum_{k=0}^{n} k p_k z^{k-1}$. Show that

$$
\begin{aligned}
(p(z)+q(z))' &= p'(z)+q'(z),\\
(p(z)q(z))' &= p(z)q'(z)+q(z)p'(z),\\
(p(z)^m)' &= mp'(z)p(z)^{m-1}.
\end{aligned}
$$

   Show that over a field of characteristic 0, a polynomial $p(z)$ factors into the product of distinct irreducible factors if and only if the greatest common divisor of $p(z)$ and $p'(z)$ is 1.
6. Let M, M_1 be R-modules and $N \subset M$ a submodule. Let $\pi : M \longrightarrow M/N$ be the canonical projection and $\phi : M \longrightarrow M_1$ an R-homomorphism. Show that if $\operatorname{Ker}\phi \supset N$, there exists a unique R-homomorphism, called the induced homomorphism, $\phi|_{M/N} : M/N \longrightarrow M_1$ for which $\phi|_{M/N} = \phi \circ \pi$. Show that

$$
\begin{aligned}
\operatorname{Ker}\phi|_{M/N} &= (\operatorname{Ker}\phi)/N,\\
\operatorname{Im}\phi|_{M/N} &= \operatorname{Im}\phi.
\end{aligned}
$$

7. Let M be a module and M_i submodules. Show that if

$$
\begin{cases}
M = M_1 + \cdots + M_k\\
M_i \cap \sum_{j \neq i} M_j = \{0\},
\end{cases}
$$

   then the map $\phi : M_1 \times \cdots \times M_k \longrightarrow M_1 + \cdots + M_k$ defined by

$$
\phi(m_1,\ldots,m_k) = m_1 + \cdots + m_k
$$

   is a module isomorphism.
8. Let M be a module and M_i submodules. Show that if $M = M_1 + \cdots + M_k$ and

$$
\begin{aligned}
M_1 \cap M_2 &= 0,\\
(M_1 + M_2) \cap M_3 &= 0,\\
&\vdots\\
(M_1 + \cdots + M_{k-1}) \cap M_k &= 0,
\end{aligned}
$$

   then $M = M_1 \oplus \cdots \oplus M_k$.

9. Let M be a module and K, L submodules.

   a. Show that
   $$(K+L)/K \simeq L/(K \cap L).$$

   b. If $K \subset L \subset M$, then
   $$M/L \simeq (M/K)/(L/K).$$

10. A module M over a ring R is called **free** if it is the zero module or has a basis. Show that if M is a free module over a principal ideal domain R having n basis elements and N is a submodule, then N is free and has at most n basis elements.

11. Show that $\mathbb{F}[z]^n, z^{-1}\mathbb{F}[[z^{-1}]]^n, \mathbb{F}((z^{-1}))^n$ are $\mathbb{F}[z]$-modules.

1.7 Notes and Remarks

The development of algebra can be traced to antiquity, in particular to the Babylonians and the Chinese. For an interesting survey of early non-European mathematics, see Joseph (2000).

Galois, one of the more brilliant and colorful mathematicians, introduced the term *group* in the context of solvability of polynomial equations by radicals. This was done in 1830, but Galois' writings were first published, by Liouville, only in 1846, fifteen years after Galois died in a duel. The abstract definition of a group is apparently due to Cayley.

The development of ring theory as a generalization of integer arithmetic owes much to the efforts to prove Fermat's last theorem. Although, in the context of algebraic number theory, ideals appear already in the work of Kummer, the concepts of rings and fields are probably due to Dedekind, Kronecker, and Hilbert. An axiomatic approach to the theory of rings and modules was first developed by E. Noether. Modern expositions of algebra all stem from the classical book by van der Waerden (1931), which in turn was based on lectures by E. Noether and E. Artin. Incidentally, this seems to be the first book having a chapter devoted to linear algebra. For a modern, general book on algebra, Lang (1965) is recommended.

Our emphasis on the ring of polynomials is not surprising, considering their well-entrenched role in the study of linear transformations in finite-dimensional vector spaces. The similar exposure given to the field of rational functions, and in particular to the subrings of stable rational functions and bounded stable rational functions, is motivated by the role they play in system theory. This will be treated in Chapters 10 to 12. For more information in this direction, the reader is advised to consult Vidyasagar (1985).

Chapter 2
Vector Spaces

2.1 Introduction

Vector spaces provide the setting in which the rest of the topics that are to be presented in this book are developed. The content is geometrically oriented and we focus on linear combinations, linear independence, bases, dimension, coordinates, subspaces, and quotient spaces. Change of basis transformations provide us with a first instance of linear transformations.

2.2 Vector Spaces

Vector spaces are modules over a field $\mathbb{F}$. For concreteness, we give an ab initio definition.

Definition 2.1. Let $\mathbb{F}$ be a field. A **vector space** $\mathscr{V}$ over $\mathbb{F}$ is a set, whose elements are called **vectors**, that satisfies the following set of axioms:

1. For each pair of vectors $x, y \in \mathscr{V}$ there exists a vector $x + y \in \mathscr{V}$, called the sum of x and y, and the following holds:

   a. The **commutative law:**
   $$x + y = y + x.$$

   b. The **associative law:**
   $$x + (y + z) = (x + y) + z.$$

   c. There exists a unique **zero vector** 0 that satisfies
   $$0 + x = x + 0,$$

   for all $x \in \mathscr{V}$.

   d. For each $x \in \mathscr{V}$ there exists a unique element $-x$ such that

P.A. Fuhrmann, *A Polynomial Approach to Linear Algebra*, Universitext,
DOI 10.1007/978-1-4614-0338-8_2, © Springer Science+Business Media, LLC 2012

$$x + (-x) = 0,$$

that is, $\mathcal{V}$ is a commutative group under addition.

2. For all $x \in \mathcal{V}$ and $\alpha \in \mathbb{F}$ there exists a vector $\alpha x \in \mathcal{V}$, called the **product** of α and x, and the following are satisfied:

   a. The **associative law**:

   $$\alpha(\beta x) = (\alpha\beta)x.$$

   b. For the unit $1 \in \mathbb{F}$ and all $x \in \mathcal{V}$ we have

   $$1 \cdot x = x.$$

3. The **distributive laws**:

   a. $(\alpha + \beta)x = \alpha x + \beta x,$
   b. $\alpha(x + y) = \alpha x + \alpha y.$

Examples:

1. Let

$$\mathbb{F}^n = \left\{ \begin{pmatrix} a_1 \\ \cdot \\ \cdot \\ \cdot \\ a_n \end{pmatrix} \middle| a_i \in \mathbb{F} \right\}.$$

We define the operations of addition and multiplication by a scalar $\alpha \in \mathbb{F}$ by

$$\begin{pmatrix} a_1 \\ \cdot \\ \cdot \\ a_n \end{pmatrix} + \begin{pmatrix} b_1 \\ \cdot \\ \cdot \\ b_n \end{pmatrix} = \begin{pmatrix} a_1 + b_1 \\ \cdot \\ \cdot \\ a_n + b_n \end{pmatrix}, \quad \alpha \begin{pmatrix} a_1 \\ \cdot \\ \cdot \\ a_n \end{pmatrix} = \begin{pmatrix} \alpha a_1 \\ \cdot \\ \cdot \\ \alpha a_n \end{pmatrix}.$$

With these definitions, $\mathbb{F}^n$ is a vector space.

2. An $m \times n$ **matrix** over the field $\mathbb{F}$ is a set of mn elements a_{ij} arranged in rows and columns, i.e.,

$$A = \begin{pmatrix} a_{11} & \cdots & a_{1n} \\ \cdot & \cdots & \cdot \\ \cdot & \cdots & \cdot \\ \cdot & \cdots & \cdot \\ a_{m1} & \cdots & a_{mn} \end{pmatrix}.$$

We denote by $\mathbb{F}^{m \times n}$ the set of all such matrices. We define in $\mathbb{F}^{m \times n}$ the operations of addition and multiplication by scalars by

$$(a_{ij}) + (b_{ij}) = (a_{ij} + b_{ij}),$$
$$\alpha(a_{ij}) = (\alpha a_{ij}).$$

These definitions make $\mathbb{F}^{m \times n}$ into a vector space. Given the matrix $A = (a_{ij})$, we define its **transpose**, which we denote by $\tilde{A}$, as the $n \times m$ matrix given by $(\tilde{a}_{ij}) = (a_{ji})$.

Given matrices $A \in \mathbb{F}^{p \times m}, B \in \mathbb{F}^{m \times n}$, we define the product $AB \in \mathbb{F}^{p \times n}$ of the matrices by

$$(AB)_{ij} = \sum_{k=1}^{m} a_{ik} b_{kj}. \tag{2.1}$$

It is easily checked that matrix multiplication is associative and distributive, i.e., we have

$$A(BC) = (AB)C,$$

$$A(B_1 + B_2) = AB_1 + AB_2,$$

$$(A_1 + A_2)B = A_1 B + A_2 B. \tag{2.2}$$

In $\mathbb{F}^{n \times n}$ we define the **identity matrix** I_n by

$$I_n = (\delta_{ij}). \tag{2.3}$$

Here δ_{ij} denotes the **Kronecker delta function**, defined by

$$\delta_{ij} = \begin{cases} 0 & \text{if } i \neq j, \\ 1 & \text{if } i = j. \end{cases} \tag{2.4}$$

3. The rings $\mathbb{F}[z], \mathbb{F}((z^{-1})), \mathbb{F}[[z]], \mathbb{F}[[z^{-1}]]$ are all vector spaces over $\mathbb{F}$. So is $\mathbb{F}(z)$ the field of rational functions.
4. Many interesting examples are obtained by considering function spaces. A typical example is $C_{\mathbb{R}}(X)$, the space of all real-valued continuous functions on a topological space X. With addition and multiplication by scalars are defined by

$$(f + g)(x) = f(x) + g(x),$$
$$(\alpha f)(x) = \alpha f(x),$$

$C_{\mathbb{R}}(X)$ is a vector space.

We note the following simple rules.

Proposition 2.2. *Let $\mathcal{V}$ be a vector space over the field $\mathbb{F}$. Let $\alpha \in \mathbb{F}$ and $x \in \mathcal{V}$. Then*

1. $0x = 0$.
2. $\alpha x = 0$ *implies* $\alpha = 0$ *or* $x = 0$.

2.3 Linear Combinations

Definition 2.3. Let $\mathcal{V}$ be a vector space over the field $\mathbb{F}$. Let $x_1, \ldots, x_n \in \mathcal{V}$ and $\alpha_1, \ldots, \alpha_n \in \mathbb{F}$. The vector $\alpha_1 x_1 + \cdots + \alpha_n x_n$ is an element of $\mathcal{V}$ and is called a **linear combination** of the vectors $x_1, \ldots, x_n$. The scalars $\alpha_1, \ldots, \alpha_n$ are called the **coefficients** of the linear combination. A vector $x \in \mathcal{V}$ is called a linear combination of the vectors $x_1, \ldots, x_n$ if there exist scalars $\alpha_1, \ldots, \alpha_n$ for which $x = \sum_{i=1}^{n} \alpha_i x_i$. A linear combination in which all coefficients are zero is called a **trivial linear combination**.

2.4 Subspaces

Definition 2.4. Let $\mathcal{V}$ be a vector space over the field $\mathbb{F}$. A nonempty subset $\mathcal{M}$ of $\mathcal{V}$ is called a **subspace** of $\mathcal{V}$ if for any pair of vectors $x, y \in \mathcal{M}$ and any pair of scalars $\alpha, \beta \in \mathbb{F}$, we have $\alpha x + \beta y \in \mathcal{M}$.

Thus a subset $\mathcal{M}$ of $\mathcal{V}$ is a subspace if and only if it is closed under linear combinations. An equivalent description of subspaces is subsets closed under addition and multiplication by scalars.

Examples:

1. For an arbitrary vector space $\mathcal{V}$, $\mathcal{V}$ itself and $\{0\}$ are subspaces. These are called the **trivial subspaces**.
2. Let

$$
\mathcal{M} = \left\{ \begin{pmatrix} a_1 \\ \cdot \\ \cdot \\ \cdot \\ a_n \end{pmatrix} \middle| a_n = 0 \right\}.
$$

   Then $\mathcal{M}$ is a subspace of $\mathbb{F}^n$.
3. Let A be an $m \times n$ matrix. Then

$$
\mathcal{M} = \{ x \in \mathbb{F}^n | Ax = 0 \}
$$

   is a subspace of $\mathbb{F}^n$. This is the space of solutions of a system of linear homogeneous equations.

Theorem 2.5. *Let $\{\mathcal{M}_\alpha\}_{\alpha \in A}$ be a collection of subspaces of $\mathcal{V}$. Then $\mathcal{M} = \cap_{\alpha \in A} \mathcal{M}_\alpha$ is a subspace of $\mathcal{V}$.*

Proof. Let $x, y \in \mathcal{M}$ and $\alpha, \beta \in \mathbb{F}$. Clearly, $\mathcal{M} \subset \mathcal{M}_\alpha$ for all α, and therefore $x, y \in \mathcal{M}_\alpha$, and since $\mathcal{M}_\alpha$ is a subspace, we have that $\alpha x + \beta y$ belongs to $\mathcal{M}_\alpha$, for all α, and hence to the intersection. So $\alpha x + \beta y \in \mathcal{M}$. $\blacksquare$

Definition 2.6. Let S be a subset of a vector space $\mathscr{V}$. The set $L(S)$, or span (S), the **subspace spanned by** S, is defined as the intersection of the (nonempty) set of all subspaces containing S. This is therefore the smallest subspace of $\mathscr{V}$ containing S.

Theorem 2.7. *Let S be a subset of a vector space $\mathscr{V}$. Then* span (S), *the subspace spanned by S, is the set of all finite linear combinations of elements of S.*

Proof. Let $\mathscr{M} = \{\sum_{i=1}^{n} \alpha_i x_i | \alpha_i \in \mathbb{F}, x_i \in S, n \in \mathbb{N}\}$. Clearly $S \subset \mathscr{M}$, and $\mathscr{M}$ is a subspace of $\mathscr{V}$, since linear combinations of linear combinations of elements of S are also linear combinations of elements of S. Thus span $(S) \subset \mathscr{M}$.

Conversely, we have $S \subset$ span (S). So necessarily span (S) contains all finite linear combinations of elements of S. Hence $\mathscr{M} \subset$ span (S), and equality follows. ∎

We saw that if $\mathscr{M}_1, \ldots, \mathscr{M}_k$ are subspaces of $\mathscr{V}$, then so is $\mathscr{M} = \cap_{i=1}^{k} \mathscr{M}_i$. This, in general, is not true for unions of subspaces. The natural concept is that of sum of subspaces.

Definition 2.8. Let $\mathscr{M}_1, \ldots, \mathscr{M}_k$ be subspaces of a vector space $\mathscr{V}$. The sum of these subspaces, $\sum_{i=1}^{p} \mathscr{M}_i$, is defined by

$$\sum_{i=1}^{p} \mathscr{M}_i = \left\{ \sum_{i=1}^{p} \alpha_i x_i | \alpha_i \in \mathbb{F}, x_i \in \mathscr{M}_i \right\}.$$

2.5 Linear Dependence and Independence

Definition 2.9. Vectors $x_1, \ldots, x_k$ in a vector space $\mathscr{V}$ are called **linearly dependent** if there exist $\alpha_1, \ldots, \alpha_k \in \mathbb{F}$, not all zero, such that

$$\alpha_1 x_1 + \cdots + \alpha_k x_k = 0.$$

Vectors $x_1, \ldots, x_k$ in a vector space $\mathscr{V}$ are called **linearly independent** if they are not linearly dependent.

Thus $x_1, \ldots, x_k$ are linearly dependent if there exists a nontrivial, vanishing linear combination. On the other hand, $x_1, \ldots, x_k$ are linearly independent if and only if $\alpha_1 x_1 + \cdots + \alpha_k x_k = 0$ implies $\alpha_1 = \cdots = \alpha_k = 0$. That is, the only vanshing linear combination of linearly independent vectors is the trivial one. We note that, as a consequence of Proposition 2.2, a set containing a single vector $x \in \mathscr{V}$ is linearly independent if and only if $x \neq 0$.

Definition 2.10. Vectors $x_1, \ldots, x_k$ in a vector space $\mathscr{V}$ are called a **spanning set** if $\mathscr{V} =$ span $(x_1, \ldots, x_k)$.

Theorem 2.11. *Let $S \subset \mathscr{V}$.*

1. If S is a spanning set and $S \subset S_1 \subset \mathscr{V}$, then S_1 is also a spanning set.

2. *If S is a linearly independent set and $S_0 \subset S$, then S_0 is also a linearly independent set.*
3. *If S is linearly dependent and $S \subset S_1 \subset \mathscr{V}$, then S_1 is also linearly dependent.*
4. *Every subset of $\mathscr{V}$ that includes the zero vector is linearly dependent.*

As a consequence, a spanning set must be sufficiently large, whereas for linear independence the set must be sufficiently small. The case in which these two properties are in balance is of special importance and this leads to the following.

Definition 2.12. A subset B of vectors in $\mathscr{V}$ is called a **basis** if

1. B is a spanning set.
2. B is linearly independent.

$\mathscr{V}$ is called a **finite-dimensional space** if there exists a basis in $\mathscr{V}$ having a finite number of elements.

Note that we defined the concept of finite dimensionality before having defined dimension.

Example: Let $\mathscr{V} = \mathbb{F}^n$. Let

$$e_1 = \begin{pmatrix} 1 \\ 0 \\ \cdot \\ \cdot \\ \cdot \\ 0 \end{pmatrix}, \; e_2 = \begin{pmatrix} 0 \\ 1 \\ 0 \\ \cdot \\ \cdot \\ 0 \end{pmatrix}, \; \ldots, \; e_n = \begin{pmatrix} 0 \\ \cdot \\ \cdot \\ \cdot \\ 0 \\ 1 \end{pmatrix}.$$

Then $\mathscr{B} = \{e_1, \ldots, e_n\}$ is a basis for $\mathbb{F}^n$.

The following result is the main technical instrument in the study of bases.

Theorem 2.13 (Steinitz). *Let $x_1, \ldots, x_m \in \mathscr{V}$ and let $e_1, \ldots, e_p$ be linearly independent vectors that satisfy $e_i \in \mathrm{span}\,(x_1, \ldots, x_m)$ for all i. Then there exist p vectors in $\{x_i\}$, without loss of generality we may assume they are the first p vectors, such that*

$$\mathrm{span}\,(e_1, \ldots, e_p, x_{p+1}, \ldots, x_m) = \mathrm{span}\,(x_1, \ldots, x_m).$$

Proof. We prove this by induction on p. For $p = 1$, we must have $e_1 \neq 0$ by linear independence. Therefore there exist α_i such that $e_1 = \sum_{i=1}^{m} \alpha_i x_i$. Necessarily $\alpha_i \neq 0$ for some i. Without loss of generality we assume $\alpha_1 \neq 0$. Therefore we can write

$$x_1 = \alpha_1^{-1} e_1 - \alpha_1^{-1} \sum_{i=2}^{m} \alpha_i x_i.$$

This means that $x_1 \in \mathrm{span}\,(e_1, x_2, \ldots, x_m)$. Of course we also have $x_i \in \mathrm{span}\,(e_1, x_2, \ldots, x_m)$ for $i = 2, \ldots, m$. Therefore

$$\text{span}(x_1, x_2, \ldots, x_m) \subset \text{span}(e_1, x_2, \ldots, x_m).$$

On the other hand, by our assumption, $e_1 \in \text{span}(x_1, x_2, \ldots, x_m)$ and hence

$$\text{span}(e_1, x_2, \ldots, x_m) \subset \text{span}(x_1, x_2, \ldots, x_m).$$

From these two inclusion relations, the following equality follows:

$$\text{span}(e_1, x_2, \ldots, x_m) = \text{span}(x_1, x_2, \ldots, x_m).$$

Assume that we have proved the assertion for up to $p - 1$ elements and assume that $e_1, \ldots, e_p$ are linearly independent vectors that satisfy $e_i \in \text{span}(x_1, \ldots, x_m)$ for all i. By the induction hypothesis, we have

$$\text{span}(e_1, \ldots, e_{p-1}, x_p, \ldots, x_m) = \text{span}(x_1, x_2, \ldots, x_m).$$

Therefore, $e_p \in \text{span}(e_1, \ldots, e_{p-1}, x_p, \ldots, x_m)$, and hence there exist α_i such that

$$e_p = \alpha_1 x_1 + \cdots + \alpha_{p-1} + \alpha_p x_p + \cdots + \alpha_m x_m.$$

It is impossible that $\alpha_p, \ldots, \alpha_m$ are all 0, for that implies that e_p is a linear combination of $e_1, \ldots, e_{p-1}$, contradicting the assumption of linear independence. So at least one of these numbers is nonzero, and without loss of generality, reordering the elements if necessary, we assume $\alpha_p \neq 0$. Now

$$x_p = \alpha_p^{-1}(\alpha_1 e_1 + \cdots + \alpha_{p-1} e_{p-1} - e_p + \alpha_{p+1} x_{p+1} + \cdots + \alpha_n x_n),$$

that is, $x_p \in \text{span}(e_1, \ldots, e_p x_{p+1}, \ldots, x_m)$. Therefore,

$$\text{span}(x_1, x_2, \ldots, x_m) \subset \text{span}(e_1, \ldots, e_p x_{p+1}, \ldots, x_m) \subset \text{span}(x_1, x_2, \ldots, x_m),$$

and hence the equality

$$\text{span}(e_1, \ldots, e_p x_{p+1}, \ldots, x_m) = \text{span}(x_1, x_2, \ldots, x_m).$$

■

Corollary 2.14. *The following assertions hold.*

1. Let $\{e_1, \ldots, e_n\}$ be a basis for the vector space $\mathscr{V}$, and let $\{f_1, \ldots, f_m\}$ be linearly independent vectors in $\mathscr{V}$. Then $m \leq n$.
2. Let $\{e_1, \ldots, e_n\}$ and $\{f_1, \ldots, f_m\}$ be two bases for $\mathscr{V}$. Then $n = m$.

Proof. 1. Apply Theorem 2.13.

2. By the first part we have both $m \leq n$ and $n \leq m$, so equality follows.   ∎

Thus two different bases in a finite-dimensional vector space have the same number of elements. This leads us to the following definition.

Definition 2.15. Let $\mathcal{V}$ be a vector space over the field $\mathbb{F}$. The **dimension** of $\mathcal{V}$ is defined as the number of elements in an arbitrary basis. We denote the dimension of $\mathcal{V}$ by $\dim \mathcal{V}$.

Theorem 2.16. *Let $\mathcal{V}$ be a vector space of dimension n. Then*

1. Every subset of $\mathcal{V}$ containing more than n vectors is linearly dependent.
2. A set of $p < n$ vectors in $\mathcal{V}$ cannot be a spanning set.

2.6 Subspaces and Bases

Theorem 2.17. *1. Let $\mathcal{V}$ be a vector space of dimension n and let $\mathcal{M}$ be a subspace. Then $\dim \mathcal{M} \leq \dim \mathcal{V}$.*
2. Let $\{e_1, \ldots, e_p\}$ be a basis for $\mathcal{M}$. Then there exist vectors $\{e_{p+1}, \ldots, e_n\}$ in $\mathcal{V}$ such that $\{e_1, \ldots, e_n\}$ is a basis for $\mathcal{V}$.

Proof. It suffices to prove the second assertion. Let $\{e_1, \ldots, e_p\}$ be a basis for $\mathcal{M}$ anf $\{f_1, \ldots, f_n\}$ a basis for $\mathcal{V}$. By Theorem 2.13 we can replace p of the f_i by the $e_j, j = 1, \ldots, p$, and get a spanning set for $\mathcal{V}$. But a spanning set with n elements is necessarily a basis for $\mathcal{V}$.   ∎

From two subspaces $\mathcal{M}_1, \mathcal{M}_2$ of a vector space $\mathcal{V}$ we can construct the subspaces $\mathcal{M}_1 \cap \mathcal{M}_2$ and $\mathcal{M}_1 + \mathcal{M}_2$. The next theorem studies the dimensions of these subspaces.

For the sum of two subspaces of a vector space $\mathcal{V}$ we have the following.

Theorem 2.18. *Let $\mathcal{M}_1, \mathcal{M}_2$ be subspaces of a vector space $\mathcal{V}$. Then*

$$\dim(\mathcal{M}_1 + \mathcal{M}_2) + \dim(\mathcal{M}_1 \cap \mathcal{M}_2) = \dim \mathcal{M}_1 + \dim \mathcal{M}_2. \qquad (2.5)$$

Proof. Let $\{e_1, \ldots, e_r\}$ be a basis for $\mathcal{M}_1 \cap \mathcal{M}_2$. Then there exist vectors $\{f_{r+1}, \ldots, f_p\}$ and $\{g_{r+1}, \ldots, g_q\}$ such that the set $\{e_1, \ldots, e_r, f_{r+1}, \ldots, f_p\}$ is a basis for $\mathcal{M}_1$ and the set $\{e_1, \ldots, e_r, g_{r+1}, \ldots, g_q\}$ is a basis for $\mathcal{M}_2$. We will proceed to show that the set $\{e_1, \ldots, e_r, f_{r+1}, \ldots, f_p, g_{r+1}, \ldots, g_q\}$ is a basis for $\mathcal{M}_1 + \mathcal{M}_2$.

Clearly $\{e_1, \ldots, e_r, f_{r+1}, \ldots, f_p, g_{r+1}, \ldots, g_q\}$ is a spanning set for $\mathcal{M}_1 + \mathcal{M}_2$. So it remains to show that it is linearly independent. Assume that they are linearly dependent, i.e., there exist $\alpha_i, \beta_i, \gamma_i$ such that

$$\sum_{i=1}^{r} \alpha_i e_i + \sum_{i=r+1}^{p} \beta_i f_i + \sum_{i=r+1}^{q} \gamma_i g_i = 0, \qquad (2.6)$$

or

$$\sum_{i=r+1}^{q} \gamma_i g_i = - \left[\sum_{i=1}^{r} \alpha_i e_i + \sum_{i=r+1}^{p} \beta_i f_i \right].$$

So it follows that $\sum_{i=r+1}^{q} \gamma_i g_i \in \mathcal{M}_1$. On the other hand, $\sum_{i=r+1}^{q} \gamma_i g_i \in \mathcal{M}_2$ as a linear combination of some of the basis elements of $\mathcal{M}_2$. Therefore $\sum_{i=r+1}^{q} \gamma_i g_i$ belongs to $\mathcal{M}_1 \cap \mathcal{M}_2$ and hence can be expressed as a linear combination of the e_i. So there exist numbers $\varepsilon_1, \ldots, \varepsilon_r$ such that

$$\sum_{i=1}^{r} \varepsilon_i e_i + \sum_{i=r+1}^{q} \gamma_i g_i = 0.$$

However, this is a linear combination of the basis elements of $\mathcal{M}_2$; hence $\varepsilon_i = \gamma_j = 0$. Now (2.6) reduces to

$$\sum_{i=1}^{r} \alpha_i e_i + \sum_{i=r+1}^{p} \beta_i f_i = 0.$$

From this, we conclude by the same reasoning that $\alpha_i = \beta_j = 0$. This proves the linear independence of the vectors $\{e_1, \ldots, e_r, f_{r+1}, \ldots, f_p, g_{r+1}, \ldots, g_q\}$, and so they are a basis for $\mathcal{M}_1 + \mathcal{M}_2$. Now

$$\dim(\mathcal{M}_1 + \mathcal{M}_2) = p + q - r = \dim \mathcal{M}_1 + \dim \mathcal{M}_2 - \dim(\mathcal{M}_1 \cap \mathcal{M}_2).$$

■

2.7 Direct Sums

Definition 2.19. Let $\mathcal{M}_i, i = 1, \ldots, p$, be subspaces of a vector space $\mathcal{V}$. We say that $\sum_{i=1}^{p} \mathcal{M}_i$ is a **direct sum** of the subspaces $\mathcal{M}_i$, and write $\mathcal{M} = \mathcal{M}_1 \oplus \cdots \oplus \mathcal{M}_p$, if for every $x \in \sum_{i=1}^{p} \mathcal{M}_i$ there exists a unique representation $x = \sum_{i=1}^{p} x_i$ with $x_i \in \mathcal{M}_i$.

Proposition 2.20. *Let $\mathcal{M}_1, \mathcal{M}_2$ be subspaces of a vector space $\mathcal{V}$. Then $\mathcal{M} = \mathcal{M}_1 \oplus \mathcal{M}_2$ if and only if $\mathcal{M} = \mathcal{M}_1 + \mathcal{M}_2$ and $\mathcal{M}_1 \cap \mathcal{M}_2 = \{0\}$.*

Proof. Assume $\mathcal{M} = \mathcal{M}_1 \oplus \mathcal{M}_2$. Then for $x \in \mathcal{M}$ we have $x = x_1 + x_2$, with $x_i \in \mathcal{M}_i$. Suppose there exists another representation of x in the form $x = y_1 + y_2$, with $y_i \in \mathcal{M}_i$. From $x_1 + x_2 = y_1 + y_2$ we get $z = x_1 - y_1 = y_2 - x_2$. Now $x_1 - y_1 \in \mathcal{M}_1$ and $y_2 - x_2 \in \mathcal{M}_2$. So, since $\mathcal{M}_1 \cap \mathcal{M}_2 = \{0\}$, we have $z = 0$, that is, $x_1 = y_1$ and $x_2 = y_2$.

Conversely, suppose every $x \in \mathcal{M}$ has a unique representation $x = x_1 + x_2$, with $x_i \in \mathcal{M}_i$. Thus $\mathcal{M} = \mathcal{M}_1 + \mathcal{M}_2$. Let $x \in \mathcal{M}_1 \cap \mathcal{M}_2$. Then $x = x + 0 = 0 + x$, which implies, by the uniqueness of the representation, that $x = 0$. ■

We consider next some examples.

1. The space $\mathbb{F}((z^{-1}))$ of truncated Laurent series has the spaces $\mathbb{F}[z]$ and $\mathbb{F}[[z^{-1}]]$ as subspaces. We clearly have the direct sum decomposition

$$\mathbb{F}((z^{-1})) = \mathbb{F}[z] \oplus z^{-1}\mathbb{F}[[z^{-1}]]. \tag{2.7}$$

The factor z^{-1} guarantees that the constant elements appear in one of the subspaces only.

2. We denote by $\mathbb{F}_n[z]$ the space of all polynomials of degree $< n$, i.e., $\mathbb{F}_n[z] = \{p(z) \in \mathbb{F}[z] \mid \deg p < n\}$. The following result is based on Proposition 1.46.

Proposition 2.21. *Let $p(z), q(z) \in \mathbb{F}[z]$ with $\deg p = m$ and $\deg q = n$. Let $r(z)$ be the g.c.d. of $p(z)$ and $q(z)$ and let $s(z)$ be their l.c.m. Let $\deg r = \rho$. Then*

1. $p\mathbb{F}_n[z] + q\mathbb{F}_m[z] = r\mathbb{F}_{m+n-\rho}[z]$.

2. $p\mathbb{F}_n[z] \cap q\mathbb{F}_m[z] = s\mathbb{F}_\rho[z]$.

Proof. 1. We know that with $r(z)$ the g.c.d. of $p(z)$ and $q(z)$, $p\mathbb{F}[z] + q\mathbb{F}[z] = r\mathbb{F}[z]$. So, given $f(z), g(z), \in \mathbb{F}[z]$, there exists $h(z) \in \mathbb{F}[z]$ such that $p(z)f(z) + q(z)g(z) = r(z)h(z)$. Now we take remainders after division by $p(z)q(z)$, i.e., we apply the map π_{pq}. Now $\pi_{pq}pf = p\pi_q f$ and $\pi_{pq}qg = q\pi_p g$. Finally, since by Proposition 1.47 we have $p(z)q(z) = r(z)s(z)$, it follows that $\pi_{pq}rh = \pi_{rs}rh = r\pi_s h$. Now $\deg s = \deg p + \deg q - \deg r = n + m - \rho$. So we get the equality $p\mathbb{F}_m[z] + q\mathbb{F}_m[z] = r\mathbb{F}_{m+n-\rho}[z]$.
2. We have $p\mathbb{F}[z] \cap q\mathbb{F}[z] = s\mathbb{F}[z]$. Again we apply π_{pq}. If $f \in p\mathbb{F}[z]$ then $f = pf'$, and hence $\pi_{pq}pf' = p\pi_q f' \in p\mathbb{F}_n[z]$. Similarly, if $f(z) \in q\mathbb{F}[z]$ then $f(z) = q(z)f''(z)$ and $\pi_{pq}f = \pi_{pq} \cdot q(z)f''(z) = q(z)\pi_p f'' \in q\mathbb{F}_m[z]$. On the other hand, $f(z) = s(z)h(z)$ implies $\pi_{pq}sh = \pi_{rs}sh = s\pi_r h \in s\mathbb{F}_\rho[z]$. ∎

Corollary 2.22. *Let $p(z), q(z) \in \mathbb{F}[z]$ with $\deg p = m$ and $\deg q = n$. Then $p(z) \wedge q(z) = 1$ if and only if*

$$\mathbb{F}_{m+n}[z] = p\mathbb{F}_n[z] \oplus q\mathbb{F}_m[z].$$

We say that the polynomials $p_1(z), \ldots, p_k(z)$ are **mutually coprime** if for $i \neq j$, $p_i(z)$ and $p_j(z)$ are coprime.

Theorem 2.23. *Let $p_1(z), \ldots, p_k(z) \in \mathbb{F}[z]$ with $\deg p_i = n_i$ with $\sum_{i=1}^{k} n_i = n$. Then the $p_i(z)$ are mutually coprime if and only if*

$$\mathbb{F}_n[z] = \pi_1(z)\mathbb{F}_{n_1}[z] \oplus \cdots \oplus \pi_k(z)\mathbb{F}_{n_k}[z],$$

where the $\pi_j(z)$ are defined by

$$\pi_i(z) = \Pi_{j \neq i} p_j(z).$$

Proof. The proof is by induction. Assume the $p_i(z)$ are mutually coprime. For $k = 2$ the result was proved in Proposition 2.21. Assume it has been proved for positive integers $\leq k - 1$. Let $\tau_i(z) = \Pi_{\substack{j=1 \\ j \neq i}}^{k-1} p_j(z)$. Then

$$\mathbb{F}_{n_1 + \cdots + n_{k-1}}[z] = \tau_1(z)\mathbb{F}_{n_1}[z] \oplus \cdots \oplus \tau_{k-1}(z)\mathbb{F}_{n_{k-1}}[z].$$

Now $\pi_k(z) \wedge p_k(z) = 1$, so

$$\begin{aligned}
\mathbb{F}_n[z] &= \pi_k(z)\mathbb{F}_{n_k}[z] \oplus p_k(z)\mathbb{F}_{n_1 + \cdots + n_{k-1}}[z] \\
&= p_k(z)\left\{ \tau_1(z)\mathbb{F}_{n_1}[z] \oplus \cdots \oplus \tau_{k-1}(z)\mathbb{F}_{n_{k-1}}[z] \right\} \oplus \pi_k(z)\mathbb{F}_{n_k}(z) \\
&= \pi_1(z)\mathbb{F}_{n_1}[z] \oplus \cdots \oplus \pi_k(z)\mathbb{F}_{n_k}[z].
\end{aligned}$$

Conversely, if $\mathbb{F}_n[z] = \pi_1(z)\mathbb{F}_{n_1}[z] \oplus \cdots \oplus \pi_k(z)\mathbb{F}_{n_k}[z]$, then there exist polynomials $f_i(z)$ such that $1 = \sum \pi_i f_i$. The coprimeness of the $\pi_i(z)$ implies the mutual coprimeness of the $p_i(z)$. ∎

Corollary 2.24. *Let $p(z) = p_1(z)^{n_1} \cdots p_k(z)^{n_k}$ be the primary decomposition of $p(z)$, with $\deg p_i = r_i$ and $n = \sum_{i=1}^{k} n_i$. Then*

$$\mathbb{F}_n[z] = p_2(z)^{n_2} \cdots p_k(z)^{n_k} \mathbb{F}_{r_1 n_1}[z] \oplus \cdots \oplus p_1(z)^{n_1} \cdots p_{k-1}(z)^{n_{k-1}} \mathbb{F}_{r_k n_k}[z].$$

Proof. Follows from Theorem 2.23, replacing $p_i(z)$ by $p_i(z)^{n_i}$. ∎

2.8 Quotient Spaces

We begin by introducing the concept of codimension.

Definition 2.25. We say that a subspace $\mathcal{M} \subset \mathcal{X}$ has **codimension** k, denoted by $\operatorname{codim} \mathcal{M} = k$, if

1. There exist k vectors $\{x_1, \ldots, x_k\}$, linearly independent over $\mathcal{M}$, i.e., for which $\sum_{i=1}^{k} \alpha_i x_i \in \mathcal{M}$ if and only if $\alpha_i = 0$ for all $i = 1, \ldots, k$.
2. $\mathcal{X} = L(\mathcal{M}, x_1, \ldots, x_k)$.

Now let $\mathcal{X}$ be a vector space over the field $\mathbb{F}$ and let $\mathcal{M}$ be a subspace. In $\mathcal{X}$ we define a relation

$$x \simeq y \text{ if } x - y \in \mathcal{M}. \tag{2.8}$$

It is easy to check that this is indeed an equivalence relation, i.e., it is reflexive, symmetric, and transitive. We denote by $[x]_{\mathcal{M}} = x + \mathcal{M} = \{x + m \mid m \in \mathcal{M}\}$ the equivalence class of $x \in \mathcal{X}$. We denote by $\mathcal{X}/\mathcal{M}$ the set of equivalence classes with respect to the equivalence relation induced by $\mathcal{M}$ as in (2.8).

So far, $\mathscr{X}/\mathscr{M}$ is just a set. We introduce in $\mathscr{X}/\mathscr{M}$ two operations, addition and multiplication by a scalar, as follows:

$$[x]_\mathscr{M} + [y]_\mathscr{M} = [x+y]_\mathscr{M}, \quad x,y \in \mathscr{X},$$
$$\alpha[x]_\mathscr{M} = [\alpha x]_\mathscr{M}.$$

Proposition 2.26. *1. The operations of addition and multiplication by a scalar are well defined, i.e., independent of the representatives x, y.*
2. With these operations, $\mathscr{X}/\mathscr{M}$ is a vector space over $\mathbb{F}$.
3. If $\mathscr{M}$ has codimension k in $\mathscr{X}$, then $\dim \mathscr{X}/\mathscr{M} = k$.

Proof. 1. Let $x' \simeq x$ and $y' \simeq y$. Thus $[x]_\mathscr{M} = [x']_\mathscr{M}$. This means that $x' = x + m_1, y' = y + m_2$ with $m_i \in \mathscr{M}$. Hence $x' + y' = x + y + (m_1 + m_2)$, which shows that $[x' + y']_\mathscr{M} = [x+y]_\mathscr{M}$.

Similarly, $\alpha x' = \alpha x + \alpha m$. So $\alpha x' - \alpha x = \alpha m \in \mathscr{M}$; hence $[\alpha x']_\mathscr{M} = [\alpha x]_\mathscr{M}$.
2. The axioms of a vector space for $\mathscr{X}/\mathscr{M}$ are easily shown to result from those in $\mathscr{X}$.
3. Let $x_1, \ldots, x_k$ be linearly independent over $\mathscr{M}$ and such that $L(x_1, \ldots, x_k, \mathscr{M}) = \mathscr{X}$. We claim that $\{[x_1], \ldots, [x_k]\}$ are linearly independent, for if $\sum_{i=1}^n \alpha_i [x_i] = 0$, it follows that $[\sum \alpha_i x_i]_\mathscr{M} = 0$, i.e., $\sum_{i=1}^n \alpha_i x_i \in \mathscr{M}$. Since $x_1, \ldots, x_n$ are linearly independent over $\mathscr{M}$, necessarily $\alpha_i = 0, i = 1, \ldots, k$. Let now $[x]_\mathscr{M}$ be an arbitrary equivalence class in $\mathscr{X}/\mathscr{M}$. Assume $x \in [x]_\mathscr{M}$. Then there exist $\alpha_i \in \mathbb{F}$ such that $x = \alpha_1 x_1 + \cdots + \alpha_n x_n + m$ for some $m \in \mathscr{M}$. This implies

$$[x]_\mathscr{M} = \alpha_1 [x_1]_\mathscr{M} + \cdots + \alpha_k [x_k]_\mathscr{M}.$$

So indeed $\{[x_1]_\mathscr{M}, \ldots, [x_k]_\mathscr{M}\}$ is a basis for $\mathscr{X}/\mathscr{M}$, and hence

$$\dim \mathscr{X}/\mathscr{M} = k.$$

Corollary 2.27. *Let $q \in \mathbb{F}[z]$ be a polynomial of degree n. Then*

1. $q\mathbb{F}[z]$ is a subspace of $\mathbb{F}[z]$ of codimension n.
2. $\dim \mathbb{F}[z]/q\mathbb{F}[z] = n$.

Proof. The polynomials $1, z, \ldots, z^{n-1}$ are obviously linearly independent over $q\mathbb{F}[z]$. Moreover, applying the division rule of polynomials, it is clear that $1, z, \ldots, z^{n-1}$ together with $q\mathbb{F}[z]$ span all of $\mathbb{F}[z]$. ■

Proposition 2.28. *Let $\mathbb{F}((z^{-1}))$ be the space of truncated Laurent series and $\mathbb{F}[z]$ and $z^{-1}\mathbb{F}[[z^{-1}]]$ the corresponding subspaces. Then we have the isomorphisms*

$$\mathbb{F}[z] \simeq \mathbb{F}((z^{-1}))/z^{-1}\mathbb{F}[[z^{-1}]]$$

and

$$z^{-1}\mathbb{F}[[z^{-1}]] \simeq \mathbb{F}((z^{-1}))/\mathbb{F}[z].$$

Proof. Follows from the direct sum representation (2.7). ■

2.9   Coordinates

Lemma 2.29. *Let $\mathcal{V}$ be a finite-dimensional vector space of dimension n and let $\mathcal{B} = \{e_1, \ldots, e_n\}$ be a basis for $\mathcal{V}$. Then every vector $x \in \mathcal{V}$ has a unique representation as a linear combination of the e_i. That is,*

$$x = \sum_{i=1}^{n} \alpha_i e_i. \tag{2.9}$$

Proof. Since $\mathcal{B}$ is a spanning set, such a representation exists. Since $\mathcal{B}$ is linearly independent, the representation (2.9) is unique.   ∎

Definition 2.30. The scalars $\alpha_1, \ldots, \alpha_n$ will be called the **coordinates** of x with respect to the basis $\mathcal{B}$, and we will use the notation

$$[x]^{\mathcal{B}} = \begin{pmatrix} \alpha_1 \\ \cdot \\ \cdot \\ \cdot \\ \alpha_n \end{pmatrix}.$$

The vector $[x]^{\mathcal{B}}$ will be called the **coordinate vector** of x with respect to $\mathcal{B}$. We will always write it in column form.

The map $x \mapsto [x]^{\mathcal{B}}$ is a map from $\mathcal{V}$ to $\mathbb{F}^n$.

Proposition 2.31. *Let $\mathcal{V}$ be a finite-dimensional vector space of dimension n and let $\mathcal{B} = \{e_1, \ldots, e_n\}$ be a basis for $\mathcal{V}$. The map $x \mapsto [x]^{\mathcal{B}}$ has the following properties:*

1.

$$[x+y]^{\mathcal{B}} = [x]^{\mathcal{B}} + [y]^{\mathcal{B}}.$$

2.

$$[\alpha x]^{\mathcal{B}} = \alpha [x]^{\mathcal{B}}.$$

3. $[x]^{\mathcal{B}} = 0$ if and only if $x = 0$.
4. For every $\alpha_1, \ldots, \alpha_n \in \mathbb{F}$ there exists a vector $x \in \mathcal{V}$ for which

$$[x]^{\mathcal{B}} = \begin{pmatrix} \alpha_1 \\ \cdot \\ \cdot \\ \cdot \\ \alpha_n \end{pmatrix}.$$

Proof. 1. Let $x = \sum_{i=1}^{n} \alpha_i e_i$, $y = \sum_{i=1}^{n} \beta_i e_i$. Then

$$x + y = \sum_{i=1}^{n} \alpha_i e_i + \sum_{i=1}^{n} \beta_i e_i = \sum_{i=1}^{n} (\alpha_i + \beta_i) e_i.$$

So

$$[x + y]^{\mathscr{B}} = \begin{pmatrix} \alpha_1 + \beta_1 \\ \cdot \\ \cdot \\ \cdot \\ \alpha_n + \beta_n \end{pmatrix} = \begin{pmatrix} \alpha_1 \\ \cdot \\ \cdot \\ \cdot \\ \alpha_n \end{pmatrix} + \begin{pmatrix} \beta_1 \\ \cdot \\ \cdot \\ \cdot \\ \beta_n \end{pmatrix} = [x]^{\mathscr{B}} + [y]^{\mathscr{B}}.$$

2. Let $\alpha \in \mathbb{F}$. Then $\alpha x = \sum_{i=1}^{n} \alpha \alpha_i e_i$, and therefore

$$[\alpha x]^{\mathscr{B}} = \begin{pmatrix} \alpha \alpha_1 \\ \cdot \\ \cdot \\ \cdot \\ \alpha \alpha_n \end{pmatrix} = \alpha \begin{pmatrix} \alpha_1 \\ \cdot \\ \cdot \\ \cdot \\ \alpha_n \end{pmatrix} = \alpha [x]^{\mathscr{B}}.$$

3. If $x = 0$ then $x = \sum_{i=1}^{n} 0 e_i$, and

$$[x]^{\mathscr{B}} = \begin{pmatrix} 0 \\ \cdot \\ \cdot \\ \cdot \\ 0 \end{pmatrix}.$$

Conversely, if $[x]^{\mathscr{B}} = 0$ then $x = \sum_{i=1}^{n} 0 e_i = 0$.

4. Let $\alpha_1, \ldots, \alpha_n \in \mathbb{F}$. Then we define a vector $x \in \mathscr{V}$ by $x = \sum_{i=1}^{n} \alpha_i e_i$. Then clearly

$$[x]^{\mathscr{B}} = \begin{pmatrix} \alpha_1 \\ \cdot \\ \cdot \\ \cdot \\ \alpha_n \end{pmatrix}.$$

∎

2.10   Change of Basis Transformations

Let $\mathscr{V}$ be a finite-dimensional vector space of dimension n and let $\mathscr{B} = \{e_1, \ldots, e_n\}$ and $\mathscr{B}_1 = \{f_1, \ldots, f_n\}$ be two different bases in $\mathscr{V}$. We will explore the connection between the coordinate vectors with respect to the two bases.

For each $x \in \mathcal{V}$ there exist unique $\alpha_j, \beta_j \in \mathbb{F}$ for which

$$x = \sum_{j=1}^{n} \alpha_j e_j = \sum_{j=1}^{n} \beta_j f_j. \tag{2.10}$$

Since for each $j = 1, \ldots, n$, the basis vector e_j has a unique expansion as a linear combination of the f_i, we set

$$e_j = \sum_{i=1}^{n} t_{ij} f_i.$$

These equations define n^2 numbers, which we arrange in an $n \times n$ matrix. We denote this matrix by $[I]_{\mathscr{B}}^{\mathscr{B}_1}$. We refer to this matrix as the **basis transformation matrix** from the basis $\mathscr{B}$ to the basis $\mathscr{B}_1$. Substituting back into (2.10), we get

$$x = \sum_{i=1}^{n} \beta_i f_i = \sum_{j=1}^{n} \alpha_j e_j = \sum_{j=1}^{n} \alpha_j \sum_{i=1}^{n} t_{ij} f_i$$

$$= \sum_{j=1}^{n} \sum_{i=1}^{n} t_{ij} \alpha_j f_i = \sum_{i=1}^{n} \left(\sum_{j=1}^{n} t_{ij} \alpha_j \right) f_i.$$

Equating coefficients of the f_i, we obtain

$$\beta_i = \sum_{j=1}^{n} t_{ij} \alpha_j.$$

Thus we conclude that

$$[x]^{\mathscr{B}_1} = [I]_{\mathscr{B}}^{\mathscr{B}_1} [x]^{\mathscr{B}}.$$

So the basis transformation matrix transforms the coordinate vector with respect to the basis $\mathscr{B}$ to the coordinate vector with respect to the basis $\mathscr{B}_1$.

Theorem 2.32. *Let $\mathcal{V}$ be a finite-dimensional vector space of dimension n and let $\mathscr{B} = \{e_1, \ldots, e_n\}$, $\mathscr{B}_1 = \{f_1, \ldots, f_n\}$, and $\mathscr{B}_2 = \{g_1, \ldots, g_n\}$ be three bases for $\mathcal{V}$. Then we have*

$$[I]_{\mathscr{B}}^{\mathscr{B}_2} = [I]_{\mathscr{B}_1}^{\mathscr{B}_2} [I]_{\mathscr{B}}^{\mathscr{B}_1}.$$

Proof. Let $[I]_{\mathscr{B}}^{\mathscr{B}_2} = (r_{ij})$, $[I]_{\mathscr{B}_1}^{\mathscr{B}_2} = (s_{ij})$, $[I]_{\mathscr{B}}^{\mathscr{B}_1} = (t_{ij})$. This means that

$$\begin{cases} e_j = \sum_{i=1}^{n} r_{ij} g_i, \\[2mm] f_k = \sum_{i=1}^{n} s_{ik} g_i, \\[2mm] e_j = \sum_{i=1}^{n} t_{kj} f_k. \end{cases}$$

Therefore $r_{ij} = \sum_{i=1}^{n} s_{ik} t_{kj}$. ∎

Corollary 2.33. *Let $\mathcal{V}$ be a finite-dimensional vector space with two bases $\mathcal{B}$ and $\mathcal{B}_1$. Then*

$$[I]_{\mathcal{B}}^{\mathcal{B}_1} = ([I]_{\mathcal{B}_1}^{\mathcal{B}})^{-1}.$$

Proof. Clearly $[I]_{\mathcal{B}}^{\mathcal{B}} = I$, where I denotes the identity matrix. By Theorem 2.32, we get

$$I = [I]_{\mathcal{B}}^{\mathcal{B}} = [I]_{\mathcal{B}_1}^{\mathcal{B}}[I]_{\mathcal{B}}^{\mathcal{B}_1}.$$
∎

Theorem 2.34. *The mapping in $\mathbb{F}^n$ defined by the matrix $[I]_{\mathcal{B}}^{\mathcal{B}_1}$ as*

$$[x]^{\mathcal{B}} \mapsto [I]_{\mathcal{B}}^{\mathcal{B}_1}[x]^{\mathcal{B}} = [x]^{\mathcal{B}_1}$$

has the following properties:

$$[I]_{\mathcal{B}}^{\mathcal{B}_1}([x]^{\mathcal{B}} + [y]^{\mathcal{B}}) = [I]_{\mathcal{B}}^{\mathcal{B}_1}[x]^{\mathcal{B}} + [I]_{\mathcal{B}}^{\mathcal{B}_1}[y]^{\mathcal{B}},$$

$$[I]_{\mathcal{B}}^{\mathcal{B}_1}(\alpha[x]^{\mathcal{B}}) = \alpha[I]_{\mathcal{B}}^{\mathcal{B}_1}[x]^{\mathcal{B}}.$$

Proof. We compute

$$[I]_{\mathcal{B}}^{\mathcal{B}_1}([x]^{\mathcal{B}} + [y]^{\mathcal{B}}) = [I]_{\mathcal{B}}^{\mathcal{B}_1}[x+y]^{\mathcal{B}} = [x+y]^{\mathcal{B}_1}$$

$$= [x]^{\mathcal{B}_1} + [y]^{\mathcal{B}_1} = [I]_{\mathcal{B}}^{\mathcal{B}_1}[x]^{\mathcal{B}} + [I]_{\mathcal{B}}^{\mathcal{B}_1}[y]^{\mathcal{B}}.$$

Similarly,

$$[I]_{\mathcal{B}}^{\mathcal{B}_1}(\alpha[x]^{\mathcal{B}}) = [I]_{\mathcal{B}}^{\mathcal{B}_1}[\alpha x]^{\mathcal{B}} = [\alpha x]^{\mathcal{B}_1}$$

$$= \alpha[x]^{\mathcal{B}_1} = \alpha[I]_{\mathcal{B}}^{\mathcal{B}_1}[x]^{\mathcal{B}}.$$
∎

These properties characterize linear maps. We will discuss those in detail in Chapter 4.

2.11 Lagrange Interpolation

Let $\mathbb{F}_n[z] = \{p(z) \in \mathbb{F}[z] \mid \deg p < n\}$. Clearly, $\mathbb{F}_n[z]$ is an n-dimensional subspace of $\mathbb{F}[z]$. In fact, $\mathcal{B}_{st} = \{1, z, \dots, z^{n-1}\}$ is a basis for $\mathbb{F}_n[z]$, and we will refer to this as the **standard basis** of $\mathbb{F}_n[z]$. Obviously, if $p(z) = \sum_{i=0}^{n-1} p_i z^i$, then

$$[p]^{st} = \begin{pmatrix} p_0 \\ \cdot \\ \cdot \\ \cdot \\ p_{n-1} \end{pmatrix}.$$

We exhibit next another important basis that is intimately related to polynomial interpolation. To this end we introduce

The Lagrange interpolation problem: Given distinct numbers $\alpha_i \in \mathbb{F}$, $i = 1,\ldots,n$ and another set of arbitrary numbers $c_i \in \mathbb{F}$, $i = 1,\ldots,n$, find a polynomial $p(z) \in \mathbb{F}_n[z]$ such that

$$p(\alpha_i) = c_i, \quad i = 1,\ldots,n.$$

We can replace this problem by a set of simpler

Special Interpolation Problems:

Find polynomials $l_i(z) \in \mathbb{F}_n[z]$ such that for all $i = 1,\ldots,n$,

$$l_i(\alpha_j) = \delta_{ij}, \quad j = 1,\ldots,n.$$

Here δ_{ij} is the Kronecker delta function.

If $l_i(z)$ are such polynomials, then a solution to the Lagrange interpolation problem is given by

$$p(z) = \sum_{i=1}^{n} c_i l_i(z).$$

This is seen by a direct evaluation at all α_i.

The existence and properties of the solution to the special interpolation problem are summarized in the following.

Proposition 2.35. *Given distinct numbers $\alpha_i \in \mathbb{F}$, $i = 1,\ldots,n$, let the* **Lagrange interpolation polynomials** *$l_i(z)$ be defined by*

$$l_i(z) = \frac{\Pi_{j \neq i}(z - \alpha_j)}{\Pi_{j \neq i}(\alpha_i - \alpha_j)}.$$

Then

1. The polynomials $l_i(z)$ are in $\mathbb{F}_n[z]$ and

$$l_i(\alpha_j) = \delta_{ij}, \quad j = 1,\ldots,n.$$

2. The set $\{l_1(z),\ldots,l_n(z)\}$ forms a basis for $\mathbb{F}_n[z]$.
3. For all $p(z) \in \mathbb{F}_n[z]$ we have

$$p(z) = \sum_{i=1}^{n} p(\alpha_i) l_i(z). \tag{2.11}$$

Proof. 1. Clearly, $l_i(z)$ has degree $n-1$; hence $l_i(z) \in \mathbb{F}_n[z]$. They are well defined, since by the distinctness of the α_i, all the denominators are different from zero. Moreover, by construction, we have

$$l_i(\alpha_j) = \delta_{ij}, \qquad j = 1, \ldots, n.$$

2. We show first that the polynomials $l_1(z), \ldots, l_n(z)$ are linearly independent. For this, assume that $\sum_{i=1}^{n} c_i l_i(z) = 0$. Evaluating at α_j, we get

$$0 = \sum_{i=1}^{n} c_i l_i(\alpha_j) = \sum_{i=1}^{n} c_i \delta_{ij} = c_j.$$

Since $\dim \mathbb{F}_n[z] = n$, they actually form a basis.

3. Since $\{l_1(z), \ldots, l_n(z)\}$ forms a basis for $\mathbb{F}_n[z]$, then for every $p(z) \in \mathbb{F}_n[z]$ there exist c_j such that

$$p(z) = \sum_{i=1}^{n} c_i l_i(z).$$

Evaluating at α_j, we get

$$p(\alpha_j) = \sum_{i=1}^{n} c_i l_i(\alpha_j) = \sum_{i=1}^{n} c_i \delta_{ij} = c_j.$$

■

The solution of the Lagrange interpolation problem presented above was achieved by an ad hoc construction of the Lagrange polynomials. In Chapter 5, we will see that more general interpolation problems can be solved by applying the Chinese remainder theorem.

Corollary 2.36. *Let $\alpha_1, \ldots, \alpha_n \in \mathbb{F}$ be distinct. Let $\mathcal{B}_{in} = \{l_1(z), \ldots, l_n(z)\}$ be the Lagrange interpolation basis and let $\mathcal{B}_{st}$ be the standard basis in $\mathbb{F}_n[z]$. Then the change of basis transformation from the standard basis to the interpolation basis is given by*

$$[I]_{st}^{in} = \begin{pmatrix} 1 & \alpha_1 & \cdot \cdot & \alpha_1^{n-1} \\ \cdot & \cdot & \cdot \cdot & \cdot \\ \cdot & \cdot & \cdot \cdot & \cdot \\ \cdot & \cdot & \cdot \cdot & \cdot \\ 1 & \alpha_n & \cdot \cdot & \alpha_n^{n-1} \end{pmatrix}. \tag{2.12}$$

Proof. From the equality (2.11) we get as special cases, for $i = 0, \ldots, n-1$, $z^j = \sum_{i=1}^{n} \alpha_i^j l_i(z)$, and (2.12) follows. ■

The matrix in (2.12) is called the **Vandermonde matrix**.

2.12 Taylor Expansion

We proceed to study, algebraically, the Taylor expansion of polynomials, and we do it within the framework of change of basis transformations.

Let $p(z) \in \mathbb{F}_n[z]$, i.e., $\deg p < n$, and let $\alpha \in \mathbb{F}$. We would like to have a representation of $p(z)$ in the form

$$p(z) = \sum_{j=0}^{n-1} p_j(z - \alpha)^j.$$

We refer to this as the **Taylor expansion** of $p(z)$ at the point α. Naturally, the standard representation of a polynomial, namely $p(z) = \sum_{j=0}^{n-1} p_j z^j$, is the Taylor expansion at 0.

Proposition 2.37. *Given a positive integer n and $\alpha \in \mathbb{F}$. Then*

1. *The set of polynomials $\mathscr{B}_\alpha = \{1, z - \alpha, \dots, (z - \alpha)^{n-1}\}$ forms a basis for $\mathbb{F}_n[z]$.*
2. *For every $p(z) \in \mathbb{F}_n[z]$ there exist unique numbers $p_{i,\alpha}$ such that*

$$p(z) = \sum_{j=0}^{n} p_{j,\alpha}(z - \alpha)^j. \tag{2.13}$$

3. *The change of basis transformation is given by*

$$[I]_\alpha^{st} = \begin{pmatrix} 1 & -\alpha & \alpha^2 & & (-\alpha)^{n-1} \\ 0 & 1 & -2\alpha & & \cdot \\ \cdot & & 1 & & \cdot \\ \cdot & & & \cdot & \cdot \\ \cdot & & & \cdot & -(n-1)\alpha \\ 0 & \cdot & \cdot & \cdot & 0 & 1 \end{pmatrix},$$

i.e., the ith column entries come from the binomial expansion of $(z - \alpha)^{i-1}$.

Proof. 1. In $\mathscr{B}_\alpha$ we have one polynomial of each degree from 0 to $n - 1$; hence these polynomials are linearly independent. Since $\dim \mathbb{F}_n[z] = n$, they form a basis.
2. Follows from the fact that $\mathscr{B}_\alpha$ is a basis.
3. We use the binomial expansion of $(z - \alpha)^{i-1}$.
∎

The Taylor expansion can be generalized by replacing the basis $\{(z - \lambda)^i\}_{i=0}^{n-1}$ with $\{\Pi_{j=0}^{i-1}(z - \lambda_j)\}_{i=0}^{n-1}$, where the $\lambda_i \in \mathbb{F}$ are assumed to be distinct. The proof of the following proposition is analogous to the preceeding one and we omit it.

Proposition 2.38. *Let $\lambda_i \in \mathbb{F}$, $i = 0, \dots, n - 1$, be distinct. Then*

1. *The set of polynomials $\{\Pi_{j=0}^{i-1}(z-\lambda_j)\}_{i=0}^{n-1}$ forms a basis for $\mathbb{F}_n[z]$. Here we use the convention that $\Pi_{j=0}^{-1}(z-\lambda_j) = 1$.*
2. *For every $p(z) \in \mathbb{F}_n[z]$ there exist unique numbers c_i such that*

$$p(z) = \sum_{i=0}^{n-1} c_i \Pi_{j=0}^{i-1}(z-\lambda_j). \qquad (2.14)$$

We will refer to (2.14) as the **Newton expansion** of $p(z)$. This expansion is important for solving polynomial interpolation problems, and we will return to it in Section 5.5.3.

2.13   Exercises

1. Let $\mathcal{K}, \mathcal{L}, \mathcal{M}$ be subspaces of a vector space. Show that

$$\mathcal{K} \cap (\mathcal{K} \cap \mathcal{L} + \mathcal{M}) = \mathcal{K} \cap \mathcal{L} + \mathcal{K} \cap \mathcal{M},$$
$$(\mathcal{K} + \mathcal{L}) \cap (\mathcal{K} + \mathcal{M}) = \mathcal{K} + (\mathcal{K} + \mathcal{L}) \cap \mathcal{M}.$$

2. Let $\mathcal{M}_i$ be subspaces of a finite-dimensional vector space $\mathcal{X}$. Show that if $\dim(\sum_{i=1}^{k} \mathcal{M}_i) = \sum_{i=1}^{k} \dim(\mathcal{M}_i)$, then $\mathcal{M}_1 + \cdots + \mathcal{M}_k$ is a direct sum.
3. Let $\mathcal{V}$ be a finite-dimensional vector space over $\mathbb{F}$. Let f be a bijective map on $\mathcal{V}$. Define operations by

$$v_1 * v_2 = f^{-1}(f(v_1) + f(v_2)),$$
$$\alpha \nabla v = f^{-1}(\alpha f(v)).$$

Show that with these operations, $\mathcal{V}$ is a vector space over $\mathbb{F}$.
4. Let $\mathcal{V} = \{p_{n-1}z^{n-1} + \cdots + p_1 z + p_0 \in \mathbb{F}[z] \,|\, p_{n-1} + \cdots + p_1 + p_0 = 0\}$. Show that $\mathcal{V}$ is a finite-dimensional subspace of $\mathbb{F}[z]$ and find a basis for it.
5. Let $q(z)$ be a monic polynomial with distinct zeros $\lambda_1, \ldots, \lambda_n$. Let $p(z)$ be a polynomial of degree $n-1$. Show $\sum_{i=1}^{n} \frac{g(\lambda_j)}{f'(\lambda_j)} = 1$.
6. Let $f(z), g(z) \in \mathbb{F}[z]$ with $g(z)$ nonzero. Then $f(z)$ has a unique representation of the form $f(z) = \sum_i a_i(z)g(z)^i$ with $\deg a_i < \deg g$.
7. Let $C_{\mathbb{R}}(X)$ be the space of all real-valued, continuous functions on a topological space X. Let $\mathcal{V}_{\pm} = \{f \in C_{\mathbb{R}}(X) \,|\, f(x) = \pm f(-x)\}$. Show that $C_{\mathbb{R}}(X) = \mathcal{V}_+ \oplus \mathcal{V}_-$.

2.14   Notes and Remarks

Linear algebra is geometric in origin. It traces its development to the work of Fermat and Descartes on analytic geometry. In that context points in the plane are identified with ordered pairs of real numbers. This was extended in the work of Hamilton, who discovered quaternions. In its early history linear algebra focused on the solution of systems of linear equations, and the use of determinants for that purpose was predominant. Matrices came rather late in the game, formalized by Cayley and Sylvester, both of whom made important contributions. This ushered in the era of matrix theory, which also served as the title of the classic book Gantmacher (1959). For a short discussion of the contributions of Cayley and Sylvester to matrix theory, see Higham (2008).

The major step in the development of linear algebra as a field of mathematics is due to Grassmann (1844). In this remarkable book, Grassmann gives a fairly abstract definition of a linear vector space and develops the theory of linear independence very much in the spirit of modern linear algebra expositions. He defines the notions of subspace, independence, span, dimension, sum and intersection of subspaces, as well as projections onto subspaces. He also proves Theorem 2.13, usually attributed to Steinitz. He shows that any finite set has an independent subset with the same span and that any independent set extends to a basis, and he proves the important identity $\dim(\mathscr{U} + \mathscr{V}) = \dim \mathscr{U} + \dim \mathscr{V} - \dim(\mathscr{U} \cap \mathscr{V})$. He obtains the formula for change of coordinates under change of basis, defines elementary transformations of bases, and shows that every change of basis is a product of elementary matrices. Since Grassmann was ahead of his time and did not excel at exposition, his work had little immediate impact.

The first formal, modern definition of a vector space appears in Peano (1888), where it is called a linear system. Peano treats also the set of all linear transformations between two vector spaces. However, his most innovative example of a vector space was that of the polynomial functions of a real variable. He noted that if the polynomial functions were restricted to those of degree at most n, then they would form a vector space of dimension $n + 1$. But if one considered all such polynomial functions, then the vector space would have an infinite dimension. This is a precursor of functional analysis. Thirty years after the publication of Peano's book, an independent axiomatic approach to vector spaces was given by Weyl.

Extensions of the vector space axioms to include topological considerations are due to Schmidt, Banach, Wiener, and Von Neumann.

The classic book van der Waerden (1931) seems to be the first algebra book having a chapter devoted to linear algebra. Adopting Noether's point of view, modules are defined first and vector spaces appear as a special case.

Chapter 3
Determinants

3.1 Introduction

Determinants used to be an important tool for the study of systems of linear equations. With the development of the abstract approach, based on the axioms of vector spaces, determinants lost their central role. Still, we feel that taking them out altogether from a text on linear algebra would be a mistake. Throughout the book we shall return to the use of special determinants for showing linear independence of vectors, invertibility of matrices, and coprimeness of polynomials. Since we shall need to compute determinants with polynomial entries, it is natural to develop the theory of determinants over a commutative ring with an identity.

3.2 Basic Properties

Let R be a commutative ring with identity. Let $X \in R^{n \times n}$ and denote by $x_1, \ldots, x_n$ its columns.

Definition 3.1. A **determinant** is a function $D : R^{n \times n} \longrightarrow R$ that as a function of the columns of a matrix in $R^{n \times n}$ satisfies

1. $D(x_1, \ldots, x_n)$ is **multilinear**, that is, it is a linear function in each of its columns.
2. $D(x_1, \ldots, x_n)$ is **alternating**, by which we mean that it is zero whenever two adjacent columns coincide.
3. It is **normalized** so that if $e_i, \ldots, e_n$ are the columns of the identity matrix, then

$$D(e_1, \ldots, e_n) = 1.$$

Proposition 3.2. *If two adjacent columns in a matrix are interchanged, then the determinant changes sign.*

P.A. Fuhrmann, *A Polynomial Approach to Linear Algebra*, Universitext,
DOI 10.1007/978-1-4614-0338-8_3, © Springer Science+Business Media, LLC 2012

Proof. Let x and y be the the ith and $(i+1)$th columns respectively. Then

$$
\begin{aligned}
0 &= D(\ldots,x+y,x+y,\ldots) \\
&= D(\ldots,x,x+y,\ldots)+D(\ldots,y,x+y,\ldots) \\[6pt]
&= D(\ldots,x,x,\ldots)+D(\ldots,x,y,\ldots)+D(\ldots,y,x,\ldots)+D(\ldots,y,y,\ldots) \\[6pt]
&= D(\ldots,x,y,\ldots)+D(\ldots,y,x,\ldots).
\end{aligned}
$$
∎

This property of determinants explains the usage of *alternating*.

Corollary 3.3. *If any two columns in a matrix are interchanged then the determinant changes sign.*

Proof. Suppose the ith and jth columns of the matrix A are interchanged. Without loss of generality, assume $i < j$. By $j - i$ transpositions of adjacent columns, the jth column can be brought to the ith place. Now the ith column is brought to the jth place by $j - i - 1$ transpositions of adjacent columns. This changes the value of the determinant by only a factor $(-1)^{2j-2i-1} = -1$. ∎

Corollary 3.4. *If in a matrix any two columns are equal, then the determinant vanishes.*

Corollary 3.5. *If the jth column of a matrix, multiplied by c, is added to the ith columnn, the value of the determinant does not change.*

Proof. We use the linearity in the ith variable to compute

$$
\begin{aligned}
D(x_1,&\ldots,x_i+cx_j,\ldots,x_j,\ldots,x_n) \\
&= D(x_1,\ldots,x_i,\ldots,x_j,\ldots,x_n)+D(x_1,\ldots,x_j,\ldots,x_j,\ldots,x_n) \\
&= D(x_1,\ldots,x_i,\ldots,x_j,\ldots,x_n).
\end{aligned}
$$
∎

So far, we have proved elementary properties of the determinant function, but we do not know yet whether such a function exists. Our aim now is to prove the existence of a determinant function, which we will do by a direct, inductive, construction.

Definition 3.6. Let A be an $n \times n$ matrix over R. We define the (i,j)th **minor** of A, which we denote by M_{ij}, as the determinant of the matrix of order $n - 1$ obtained from A by eliminating the ith row and jth column. We define the (i,j)-**cofactor**, which we denote by A_{ij}, by

$$
A_{ij} = (-1)^{i+j} M_{ij}.
$$

The matrix whose i,j entry is A_{ji} will be called the **classical adjoint** of A and denoted by $\mathrm{adj}\,(A)$.

Theorem 3.7. *For each integer $n \geq 1$, a determinant function exists, and is given by the following formula to which we refer as the expansion by the ith row:*

$$D(A) = \sum_{j=1}^{n} a_{ij} A_{ij}. \qquad (3.1)$$

Proof. The construction is by an inductive process. For $n = 1$ we define $D(a) = a$, and this clearly satisfies the required three properties of the determinant function.

Assume that we have constructed determinant functions for integers $\leq n - 1$. Now we define a determinant of A via the expansion by rows formula (3.1). We show now that the properties of a determinant are satisfied.

1. Multilinearity: Fix an index k. We will show that $D(A)$ is linear in the kth column. For $j \neq k$, a_{ij} is not a function of the entries of the kth column, but A_{ij} is linear in the entries of all columns appearing in it, including the kth. For $j = k$, the cofactor A_{ik} is not a function of the elements of the kth column, but the coefficient a_{ik} is a linear function. So $D(A)$ is linear as a function of the kth column.
2. Alternacy: Assume that the kth and $(k+1)$th columns are equal. Then, for each index $j \neq k, k+1$ we clearly have $A_{ij} = (-1)^{i+j} M_{ij}$, and M_{ij} contains two adjacent and equal columns, hence is zero by the induction hypothesis. Therefore

$$D(A) = a_{ik} A_{ik} + a_{i(k+1)} A_{i(k+1)}$$

$$= a_{ik}(-1)^{i+k} M_{ik} + a_{i(k+1)}(-1)^{i+k+1} M_{i(k+1)}$$

$$= a_{ik}(-1)^{i+k} M_{ik} + a_{ik}(-1)^{i+k+1} M_{i(k+1)} = 0,$$

for $a_{ik} = a_{i(k+1)}$ and $M_{ik} = M_{i(k+1)}$.
3. Normalization: We compute

$$D(I) = \sum_{j=1}^{n} \delta_{ij}(-1)^{i+j} I_{ij} = (-1)^{2i} I_{ii} = 1,$$

for I_{ii} is the determinant of the identity matrix of order $n - 1$, hence equal to 1 by the induction hypothesis. ∎

We use now the availability of the determinant function to discuss permutations. By a permutation σ of n elements we will mean a bijective map of the set $\{1, \ldots, n\}$ onto itself. The set of all such permutations S_n is a group and is called the **symmetric group**. We use alternatively the notation $(\sigma_1, \ldots, \sigma_n)$ for the permutation.

Assume now that $(\sigma_1, \ldots, \sigma_n)$ is a permutation. Let $e_1, \ldots, e_n$ be the standard unit vectors in R^n. Since a determinant function exists, $D(e_{\sigma_1}, \ldots, e_{\sigma_n}) = \pm 1$. We can bring the permutation matrix, whose columns are the e_{σ_i}, to the identity matrix in a finite number of column exchanges. Clearly, the sign depends on the permutation only. Thus we define

$$(-1)^\sigma = \text{sign}(\sigma) = D(e_{\sigma_1}, \ldots, e_{\sigma_n}). \tag{3.2}$$

We will call $(-1)^\sigma$ the sign of the permutation. We say that the permutation is even if $(-1)^\sigma = 1$ and odd if $(-1)^\sigma = -1$.

Given a matrix A, we denote by $A^{(i)}$ its ith column. Thus we have

$$A^{(j)} = \sum_{i=1}^{n} a_{ij} e_i.$$

Therefore, we can compute

$$\begin{aligned}
D(A) &= D(A^{(1)}, \ldots, A^{(n)}) \\
&= D(\textstyle\sum_{i_1=1}^{n} a_{i_1 1} e_{i_1}, \ldots, \sum_{i_n=1}^{n} a_{i_n 1} e_{i_n}) \\
&= \textstyle\sum_{i_1=1}^{n} \cdots \sum_{i_n=1}^{n} D(a_{i_1 1} e_{i_1}, \ldots, a_{i_n n} e_{i_n}) \\
&= \textstyle\sum_{\sigma \in S_n} a_{\sigma_1 1} \cdots a_{\sigma_n n} D(e_{\sigma_1}, \ldots, e_{\sigma_n}) \\
&= \textstyle\sum_{\sigma \in S_n} (-1)^\sigma a_{\sigma_1 1} \cdots a_{\sigma_n n} D(e_1, \ldots, e_n) \\
&= \textstyle\sum_{\sigma \in S_n} (-1)^\sigma a_{\sigma_1 1} \cdots a_{\sigma_n n}.
\end{aligned} \tag{3.3}$$

This gives us another explicit expression for the determinant function. In fact, this expression for the determinant function could be an alternative starting point for the development of the theory, but we chose to do it differently.

From now on we will use the notation $\det(A)$ for the determinant of A, i.e., we have

$$\det(A) = \sum_{\sigma \in S_n} (-1)^\sigma a_{\sigma_1 1} \cdots a_{\sigma_n n}. \tag{3.4}$$

The existence of this expression leads immediately to the following.

Theorem 3.8. *Let R be a commutative ring with an identity. Then, for any integer n, there exists a unique determinant function, and it is given by (3.4).*

Proof. Let D be any function defined on $R^{n \times n}$ and satisfying the determinantal axioms. Then clearly, using the computation (3.3), we get $D(A) = \det(A)$. ∎

There was a basic asymmetry in our development of determinants, since we focused on the columns. The next result is a step toward putting the theory of determinants into a more symmetric state.

Theorem 3.9. *Given an $n \times n$ matrix A over the commutative ring R, we have for its transpose $\tilde{A}$,*

$$\det(\tilde{A}) = \det(A). \tag{3.5}$$

Proof. We use the fact that $\tilde{a}_{ij} = a_{ji}$, where $\tilde{a}_{ij}$ denotes the ij entry of the transposed matrix $\tilde{A}$. Thus we have

$$\begin{aligned} \det(\tilde{A}) &= \sum_{\sigma \in S_n}(-1)^\sigma \tilde{a}_{\sigma(1)1}\cdots \tilde{a}_{\sigma(n)n} \\ &= \sum_{\sigma \in S_n}(-1)^\sigma a_{1\sigma(1)}\cdots a_{n\sigma(n)} \\ &= \sum_{\sigma^{-1} \in S_n}(-1)^{\sigma^{-1}} a_{\sigma^{-1}(1)1}\cdots a_{\sigma^{-1}(n)n} \\ &= \det(A). \end{aligned}$$ ∎

Corollary 3.10. *For each integer $n \geq 1$, a determinant function exists, and is given by the following formula, to which we refer as the expansion by the jth column:*

$$D(A) = \sum_{i=1}^{n} a_{ij}A_{ij}. \tag{3.6}$$

Proof. We use the expansion by the ith column for the classical adjoint matrix. We use the fact that $\tilde{a}_{ji} = a_{ij}$ and $\tilde{A}_{ji} = A_{ij}$ to obtain

$$\det(A) = D(\tilde{A}) = \sum_{j=1}^{n} \tilde{a}_{ji}\tilde{A}_{ji} = \sum_{j=1}^{n} a_{ij}A_{ij}.$$ ∎

Corollary 3.11. *The determinant, as a function of the rows of a matrix, is multilinear, alternating, and satisfies $\det(I) = 1$.*

We can prove now the important multiplicative rule of determinants.

Theorem 3.12. *Let R be a commutative ring and let A and B be matrices in $R^{n \times n}$. Then*

$$\det(AB) = \det(A)\det(B).$$

Proof. Let $C = AB$. Let $A^{(j)}$ and $C^{(j)}$ be the jth columns of A and C respectively. Clearly $C^{(j)} = \sum_{i=1}^{n} b_{ij}A^{(i)}$. Hence

$$\begin{aligned} \det(C) &= \det(C^{(1)},\ldots,C^{(n)}) \\ &= \det(\textstyle\sum_{i_1=1}^{n} b_{i_11}A^{(i_1)},\ldots,\sum_{i_n=1}^{n} b_{i_nn}A^{(i_n)}) \\ &= \textstyle\sum_{i_1=1}^{n}\cdots\sum_{i_n=1}^{n} b_{i_11}\cdots b_{i_nn}\det(A^{(i_1)},\ldots,A^{(i_n)}) \\ &= \textstyle\sum_{\sigma \in S_n} b_{\sigma(1)1}\cdots b_{\sigma(n)n}(-1)^\sigma \det(A^{(1)},\ldots,A^{(n)}) \\ &= \det(A)\textstyle\sum_{\sigma \in S_n}(-1)^\sigma b_{\sigma(1)1}\cdots b_{\sigma(n)n} \\ &= \det(A)\det(B). \end{aligned}$$ ∎

We can use the expansion by rows and columns to obtain the following result.

Corollary 3.13. *Let A be an $n \times n$ matrix over R. Then*

$$\begin{aligned} \sum_{k=1}^{n} a_{ik}A_{jk} &= \delta_{ij}\det(A), \\ \sum_{k=1}^{n} a_{ki}A_{kj} &= \delta_{ij}\det(A). \end{aligned} \tag{3.7}$$

Proof. The first equation is the expansion of the determinant by the jth row if $i = j$, whereas if $i \neq j$ it is the expansion by the jth row of a matrix derived from A by replacing the jth row with the ith. Thus, it is the determinant of a matrix with two equal rows; hence its value is zero. The other formula is proved analogously. ∎

The determinant expansion rules can be written in a matrix form by defining a matrix adj$A$, called the **classical adjoint** of A, via

$$(\mathrm{adj}\,A)_{ij} = A_{ji}.$$

Clearly, the expansion formulas, given in Corollary 3.13, can now be written concisely as

$$(\mathrm{adj}\,A)A = A(\mathrm{adj}\,A) = \det A \cdot I. \tag{3.8}$$

This leads to the important determinantal criterion for the nonsingularity of matrices.

Corollary 3.14. *Let A be a square matrix over the field $\mathbb{F}$. Then A is invertible if and only if $\det A \neq 0$.*

Proof. Assume that A is invertible. Then there exists a matrix B such that $AB = I$. By the multiplication rule of determinants, we have

$$\det(AB) = \det(A) \cdot \det(B) = \det(I) = 1,$$

and hence $\det(A) \neq 0$.

Conversely, assume $\det(A) \neq 0$. Then

$$A^{-1} = \frac{1}{\det A}\,\mathrm{adj}\,A. \tag{3.9}$$

∎

3.3   Cramer's Rule

We can use determinants to give a closed-form representation of the solution of a system of linear equations, provided that the coefficient matrix of the system is nonsingular.

Theorem 3.15. *Given the system of linear equations $Ax = b$, with A a nonsingular, $n \times n$ matrix. Denote by $A^{(i)}$ the columns of the matrix A and let $x = \begin{pmatrix} x_1 \\ \vdots \\ x_n \end{pmatrix}$ be the solution vector, that is, $b = \sum_{i=1}^{n} x_i A^{(i)}$. Then*

$$x_i = \frac{\det(A^{(1)}, \ldots, b, \ldots, A^{(n)})}{\det(A)}, \tag{3.10}$$

or equivalently,

$$x_i = \frac{1}{\det A} \begin{vmatrix} a_{11} & . & b_1 & . & . & a_{1n} \\ . & . & . & . & . & . \\ . & . & . & . & . & . \\ . & . & . & . & . & . \\ . & . & . & . & . & . \\ a_{n1} & . & b_n & . & . & a_{nn} \end{vmatrix}. \tag{3.11}$$

Proof. We compute

$$\begin{aligned} \det(A^{(1)}, \ldots, b, \ldots, A^{(n)}) &= \det(A^{(1)}, \ldots, b, \sum_{j=1}^{n} x_j A^{(j)}, A^{(n)}) \\ &= \sum_{j=1}^{n} x_j \det(A^{(1)}, \ldots, b, A^{(j)}, A^{(n)}) \\ &= x_i \det(A^{(1)}, \ldots, b, A^{(j)}, A^{(n)}) \\ &= x_i \det(A). \end{aligned}$$

Another way to see this is to note that in this case, the unique solution is given by $x = A^{-1}b$. We will compute the ith component, x_i:

$$\begin{aligned} x_i = (A^{-1}b)_i = \sum_{j=1}^{n} (A^{-1})_{ij} b_j \\ = \frac{1}{\det A} \sum_{j=1}^{n} (\text{adj} A)_{ij} b_j = \frac{1}{\det A} \sum_{j=1}^{n} b_j A_{ji}. \end{aligned}$$

However, this is just the expansion by the ith column of the matrix A, where the ith column has been replaced by the elements of the vector b. The representation (3.11) of the solution to the system of equations $Ax = b$ goes by the name **Cramer's rule**. ∎

We prove now an elementary, but useful, result about the computation of determinants.

Proposition 3.16. *Let A and B be square matrices, not necessarily of the same size. Then*

$$\det \begin{pmatrix} A & C \\ 0 & B \end{pmatrix} = \det A \cdot \det B. \tag{3.12}$$

Proof. We note that $\det \begin{pmatrix} A & C \\ 0 & B \end{pmatrix}$ is multilinear and alternating in the columns of A and in the rows of B. Therefore

$$\det \begin{pmatrix} A & C \\ 0 & B \end{pmatrix} = \det A \cdot \det B \cdot \det \begin{pmatrix} I & C \\ 0 & I \end{pmatrix}.$$

But a consideration of elementary operations in rows and columns leads to

$$\det \begin{pmatrix} I & C \\ 0 & I \end{pmatrix} = \det \begin{pmatrix} I & 0 \\ 0 & I \end{pmatrix} = 1.$$

∎

Lemma 3.17. *Let $A, B, C,$ and D be matrices of appropriate size, with A, D square, such that C and D commute. Then*

$$\det \begin{pmatrix} A & B \\ C & D \end{pmatrix} = \det(AD - BC). \tag{3.13}$$

Proof. Without loss of generality, let $\det D \neq 0$. From the equality

$$\begin{pmatrix} A & B \\ C & D \end{pmatrix} = \begin{pmatrix} A - BD^{-1}C & B \\ 0 & D \end{pmatrix} \begin{pmatrix} I & 0 \\ D^{-1}C & I \end{pmatrix},$$

we conclude that

$$\begin{aligned} \det \begin{pmatrix} A & B \\ C & D \end{pmatrix} &= \det(A - BD^{-1}C) \cdot \det D \\ &= \det(A - BCD^{-1}) \cdot \det D = \det(AD - BC). \end{aligned} \tag{3.14}$$

∎

3.4   The Sylvester Resultant

We give now a determinantal condition for the coprimeness of two polynomials. Our approach is geometric in nature.

   Let $p(z), q(z) \in \mathbb{F}[z]$. We know, from Corollary 1.38, that $p(z)$ and $q(z)$ are coprime if and only if the Bezout equation $a(z)p(z) + b(z)q(z) = 1$ has a polynomial solution. The following gives a matrix condition for coprimeness.

Theorem 3.18. *Let $p(z), q(z) \in \mathbb{F}[z]$ with $p(z) = p_0 + \cdots + p_m z^m$ and $q(z) = q_0 + q_1 z + \cdots + q_n z^n$. Then the following conditions are equivalent:*

1. *The polynomials $p(z)$ and $q(z)$ are coprime.*
2. *We have the direct sum representation*

$$\mathbb{F}_{m+n}[z] = p\mathbb{F}_n[z] \oplus q\mathbb{F}_m[z]. \tag{3.15}$$

3. *The* **resultant matrix**

$$\text{Res}\,(p,q) = \begin{pmatrix} p_0 & & & & q_0 & & \\ \cdot & \cdot & & & \cdot & \cdot & \\ \cdot & \cdot & \cdot & & \cdot & \cdot & \cdot \\ \cdot & \cdot & \cdot & & \cdot & \cdot & \cdot & \cdot \\ p_m & \cdot & \cdot & \cdot & & \cdot & \cdot & \cdot & q_0 \\ & \cdot & \cdot & \cdot & \cdot & p_0 & q_n & \cdot & \cdot & \cdot & \cdot \\ & & \cdot & \cdot & \cdot & \cdot & & \cdot & \cdot & \cdot & \cdot \\ & & & \cdot & \cdot & \cdot & & & \cdot & \cdot & \cdot \\ & & & & \cdot & \cdot & & & & \cdot & \cdot \\ & & & & & p_m & & & & & q_n \end{pmatrix} \tag{3.16}$$

is nonsingular.

4. *The* **resultant**, *defined as the determinant of the resultant matrix, is nonzero.*

Proof. (1) $\Rightarrow$ (2) Follows from the Bezout equation directly or from Proposition 2.21 as a special case.

(2) $\Rightarrow$ (1). Since $1 \in \mathbb{F}_{m+n}[z]$, it follows that a solution exists to the Bezout equation $a(z)p(z) + b(z)q(z) = 1$. Moreover, with the conditions $\deg a < n, \deg b < m$ satisfied, it is unique.

(2) $\Rightarrow$ (3) By Proposition 2.21, the equality $\mathbb{F}_{m+n}[z] = p\mathbb{F}_n[z] + q\mathbb{F}_m[z]$ is equivalent to this sum being a direct sum. In turn, this implies that a union of bases for $p\mathbb{F}_n[z]$ and $q\mathbb{F}_m[z]$ is a basis for $\mathbb{F}_{m+n}[z]$. Now $\{z^i p(z)|i = 0, \ldots, n-1\}$ is a basis for $p\mathbb{F}_n[z]$, and $\{z^i q(z)|i = 0, \ldots, m-1\}$ is a basis for $q\mathbb{F}_m[z]$. Thus $\mathscr{B}_{res} = \{p(z), zp(z), \ldots, z^{n-1}p(z), q(z), zq(z), \ldots, z^{m-1}q(z)\}$ forms a basis for $\mathbb{F}_{m+n}[z]$. The last space has also the standard basis $\mathscr{B}_{st} = \{1, z, \ldots, z^{m+n-1}\}$. It is easy to check that $\text{Res}\,(p,q) = [I]_{res}^{st}$, i.e., it is a change of basis transformation, hence necessarily invertible.

(3) $\Rightarrow$ (2) If $\text{Res}\,(p,q)$ is nonsingular, this shows that $\mathscr{B}_{res}$ is a basis for $\mathbb{F}_{m+n}[z]$. This implies the equality (3.15).

(3) $\Leftrightarrow$ (4) Follows from Corollary 3.14. ∎

We conclude this short chapter by computing the determinant of the **Vandermonde matrix**, derived in (2.12).

Proposition 3.19. *Let*

$$V(\lambda_1, \ldots, \lambda_n) = \begin{pmatrix} 1 & \lambda_1 & \cdot\cdot & \lambda_1^{n-1} \\ \cdot & \cdot & \cdot\cdot & \cdot \\ \cdot & \cdot & \cdot\cdot & \cdot \\ \cdot & \cdot & \cdot\cdot & \cdot \\ 1 & \lambda_n & \cdot\cdot & \lambda_n^{n-1} \end{pmatrix}$$

be the Vandermonde matrix. Then we have

$$\det V(\lambda_1,\ldots,\lambda_n) = \begin{vmatrix} 1 & \lambda_1 & .. & \lambda_1^{n-1} \\ . & . & ... & . \\ . & . & ... & . \\ . & . & ... & . \\ 1 & \lambda_n & .. & \lambda_n^{n-1} \end{vmatrix} = \Pi_{1 \le j < i \le n}(\lambda_i - \lambda_j). \qquad (3.17)$$

Proof. By induction on n. For $n = 1$ the equality is trivially satisfied. Assume that this holds for all integers $< n$. We consider the following determinant

$$f(z) = \begin{vmatrix} 1 & \lambda_1 & .. & \lambda_1^{n-1} \\ . & . & ... & . \\ . & . & ... & . \\ 1 & \lambda_{n-1} & .. & \lambda_{n-1}^{n-1} \\ 1 & z & .. & z^{n-1} \end{vmatrix}.$$

Obviously, $f(z)$ is a polynomial of degree $n - 1$ that vanishes at the points $\lambda_1,\ldots,\lambda_{n-1}$. Thus $f(z) = C_{n-1}(z - \lambda_1)\cdots(z - \lambda_{n-1})$. Expanding by the last row, and using the induction hypothesis, we get

$$C_{n-1} = \begin{vmatrix} 1 & \lambda_1 & .. & \lambda_1^{n-2} \\ . & . & ... & . \\ . & . & ... & . \\ . & . & ... & . \\ 1 & \lambda_{n-1} & .. & \lambda_{n-1}^{n-2} \end{vmatrix} = \Pi_{1 \le j < i \le n-1}(\lambda_i - \lambda_j).$$

Evaluating $f(z)$ at $z = \lambda_n$ implies (3.17).   ∎

We note that for (3.17) to hold, we do not have to assume the λ_i to be distinct. However, clearly, the determinant of a Vandermonde matrix is nonzero if and only if the λ_i are distinct.

3.5 Exercises

1. Let A, B be $n \times n$ matrices. For the classical adjoint show the following:

    a. $\mathrm{adj}\,(AB) = \mathrm{adj}\,A \cdot \mathrm{adj}\,B$.
    b. $\det \mathrm{adj}\,A = (\det A)^{n-1}$.
    c. $\mathrm{adj}\,(\mathrm{adj}\,A) = (\det A)^{n-2} \cdot A$.
    d. $\det \mathrm{adj}\,(\mathrm{adj}\,A) = (\det A)^{(n-1)^2}$.

2. Let a square $n \times n$ matrix A be defined by $a_{ij} = 1 - \delta_{ij}$, where δ_{ij} is the Kronecker delta. Show that $\det A = (n-1)(-1)^{n-1}$.

3. Let A, B be square matrices. Show that

$$\det \begin{pmatrix} A & B \\ B & A \end{pmatrix} = \det(A+B) \cdot \det(A-B).$$

4. Given the block matrix

$$\begin{pmatrix} A & B \\ C & D \end{pmatrix},$$

with A, D square and D nonsingular. Show that

$$\det \begin{pmatrix} A & B \\ C & D \end{pmatrix} = \det D \cdot \det(A - BD^{-1}C).$$

5. Let A be a square $n \times n$ matrix, $x, y \in \mathbb{F}^n$, and $0 \neq \alpha \in \mathbb{F}$. Show that

$$\det \begin{pmatrix} A & x \\ \tilde{y} & \alpha \end{pmatrix} = \det(\alpha A - x\tilde{y}).$$

Show that

$$\operatorname{adj}(I - x\tilde{y}) = x\tilde{y} - (1 - \tilde{y}x)I.$$

6. Let $p(z) = \sum_{i=0}^{n-1} p_i z^i$ be a polynomial of degree $\leq n-1$. Show that $p(z)$ and $z^n - 1$ are coprime if and only if

$$\det \begin{pmatrix} p_0 & p_{n-1} & \cdot & \cdot & p_1 \\ p_1 & p_0 & \cdot & \cdot & \\ \cdot & & \cdot & \cdot & \cdot \\ \cdot & & \cdot & \cdot & p_{n-1} \\ p_{n-1} & & \cdot & p_1 & p_0 \end{pmatrix} \neq 0.$$

7. Let $V(\lambda_1, \ldots, \lambda_n)$ be the Vandermonde determinant. Show that

$$\begin{vmatrix} 1 & \lambda_1 & \cdot\cdot & \lambda_1^{n-2} & \lambda_2 \cdots \lambda_n \\ \cdot & \cdot & \cdot\cdot & \cdot & \cdot \\ \cdot & \cdot & \cdot\cdot & \cdot & \cdot \\ \cdot & \cdot & \cdot\cdot & \cdot & \cdot \\ 1 & \lambda_n & \cdot\cdot & \lambda_n^{n-2} & \lambda_1 \cdots \lambda_{n-1} \end{vmatrix} = (-1)^{n-1} V(\lambda_1, \ldots, \lambda_n).$$

8. Let $V(\lambda_1, \ldots, \lambda_n)$ be the Vandermonde determinant. Show that

$$
\begin{vmatrix}
1 & \lambda_1 & .. & \lambda_1^{n-2} & (\lambda_2 + \cdots + \lambda_n)^{n-1} \\
. & . & . & . & . \\
. & . & . & . & . \\
. & . & . & . & . \\
. & . & . & . & . \\
1 & \lambda_n & .. & \lambda_n^{n-2} & (\lambda_1 + \cdots + \lambda_{n-1})^{n-1}
\end{vmatrix}
= (-1)^{n-1} V(\lambda_1, \ldots, \lambda_n).
$$

9. Given $\lambda_1, \ldots, \lambda_n$, define $s_k = p_1 \lambda_1^k + \cdots + p_n \lambda_n^k$. Set, for $i, j = 0, \ldots, n-1$, $a_{ij} = s_{i+j}$. Show that

$$
\det(a_{ij}) =
\begin{vmatrix}
s_0 & \cdots & s_{n-1} \\
. & . & . \\
. & . & . \\
. & . & . \\
s_{n-1} & \cdots & s_{2n-2}
\end{vmatrix}
= p_1 \cdots p_n \prod_{i>j} (\lambda_i - \lambda_j)^2.
$$

10. Show that

$$
\begin{vmatrix}
\dfrac{1}{x_1 + y_1} & \cdots & \dfrac{1}{x_1 + y_n} \\
. & . & . \\
. & . & . \\
. & . & . \\
\dfrac{1}{x_n + y_1} & \cdots & \dfrac{1}{x_n + y_n}
\end{vmatrix}
= \frac{\prod_{i>j}(x_i - x_j)(y_i - y_j)}{\prod_{i,j}(x_i + y_j)}.
$$

This determinant is generally known as the **Cauchy determinant**.

3.6   Notes and Remarks

The theory of determinants predates the development of linear algebra, and it reached its high point in the nineteenth century. Thomas Muir, in his monumental *The Theory of Determinants in the Historical Order of Development*, Macmillan, 4 volumes, 1890-1923, (reprinted by Dover Books 1960), begins with Leibniz in 1693. However, a decade earlier, the Japanese mathematician Seki Kowa developed a theory of determinants; see Joseph (2000).

The first systematic study is due to Cauchy, who apparently coined the term *determinant* in the sense we use today, and made important contributions, including the multiplication rule and the expansion rules that are named after him. The axiomatic treatment presented here is due to Kronecker and Weierstrass, probably in the 1860s. Their work became known in 1903, when Weierstrass' *On Determinant Theory* and Kronecker's *Lectures on Determinant Theory* were published posthumously.

Chapter 4
Linear Transformations

4.1 Introduction

The study of linear transformations, and their structure, provides the core of linear algebra. We shall study matrix representations of linear transformations, linear functionals, and duality and the adjoint transformation. To conclude, we show how a linear transformation in a vector space induces a module structure over the corresponding ring of polynomials. This simple construction opens the possibility of using general algebraic results as a tool for the study of linear transformations.

4.2 Linear Transformations

Definition 4.1. Let $\mathcal{V}$ and $\mathcal{W}$ be two vector spaces over the field $\mathbb{F}$. A mapping T from $\mathcal{V}$ to $\mathcal{W}$, denoted by $T : \mathcal{V} \longrightarrow \mathcal{W}$, is a **linear transformation**, or a **linear operator**, if

$$T(\alpha x + \beta y) = \alpha(Tx) + \beta(Ty) \tag{4.1}$$

holds for all $x, y \in \mathcal{V}$ and all $\alpha, \beta \in \mathbb{F}$.

For an arbitrary vector space $\mathcal{V}$, we define the zero and identity transformations by $0x = 0$ and $I_{\mathcal{V}} x = x$, for all $x \in \mathcal{V}$, respectively. Clearly, both are linear transformations. If there is no confusion about the space, we may drop the subscript and write I.

The property of linearity of a transformation T is equivalent to the following two properties:

- Additivity: $T(x + y) = Tx + Ty$.
- Homogeneity: $T(\alpha x) = \alpha(Tx)$.

The structure of a linear transformation is very rigid. It suffices to know its action on basis vectors in order to determine the transformation uniquely.

P.A. Fuhrmann, *A Polynomial Approach to Linear Algebra*, Universitext,
DOI 10.1007/978-1-4614-0338-8_4, © Springer Science+Business Media, LLC 2012

Theorem 4.2. *Let $\mathcal{V}, \mathcal{W}$ be two vector spaces over the same field $\mathbb{F}$. Let $\{e_1, \ldots, e_n\}$ be a basis for $\mathcal{V}$ and let $\{f_1, \ldots, f_n\}$ be arbitrary vectors in $\mathcal{W}$. Then there exists a unique linear transformation $T : \mathcal{V} \longrightarrow \mathcal{W}$ for which $Te_i = f_i$, for all $i = 1, \ldots, n$.*

Proof. Every $x \in \mathcal{V}$ has a unique representation $x = \sum_{i=1}^{n} \alpha_i e_i$. We define $T : \mathcal{V} \longrightarrow \mathcal{W}$ by

$$Tx = \sum_{i=1}^{n} \alpha_i f_i.$$

Clearly $Te_i = f_i$, for all $i = 1, \ldots, n$. Next we show that T is linear. Let $y \in \mathcal{V}$ have the representation $y = \sum_{i=1}^{n} \beta_i e_i$. Then

$$T(\alpha x + \beta y) = T \sum_{i=1}^{n} (\alpha \alpha_i + \beta \beta_i) e_i = \sum_{i=1}^{n} (\alpha \alpha_i + \beta \beta_i) f_i$$

$$= \sum_{i=1}^{n} (\alpha \alpha_i) f_i + \sum_{i=1}^{n} (\beta \beta_i) f_i = \alpha \sum_{i=1}^{n} \alpha_i f_i + \beta \sum_{i=1}^{n} \beta_i f_i$$

$$= \alpha(Tx) + \beta(Ty).$$

To prove uniqueness, let $S : \mathcal{V} \longrightarrow \mathcal{W}$ be a linear transformation satisfying $Se_i = f_i$ for all i. Then, for every $x \in \mathcal{V}$,

$$Sx = S \sum_{i=1}^{n} \alpha_i e_i = \sum_{i=1}^{n} \alpha_i Se_i = \sum_{i=1}^{n} \alpha_i f_i = Tx.$$

This implies $T = S$.   ∎

Given a linear transformation $T : \mathcal{V} \longrightarrow \mathcal{W}$, there are two important subspaces that are determined by it, namely

1. $\operatorname{Ker} T \subset \mathcal{V}$, the **kernel** of T, defined by

$$\operatorname{Ker} T = \{x \in \mathcal{V} | Tx = 0\}.$$

2. $\operatorname{Im} T \subset \mathcal{W}$, the **image** of T, defined by

$$\operatorname{Im} T = \{Tx | x \in \mathcal{V}\}.$$

Theorem 4.3. *Let $T : \mathcal{V} \longrightarrow \mathcal{W}$ be a linear transformation. Then $\operatorname{Ker} T$ and $\operatorname{Im} T$ are subspaces of $\mathcal{V}$ and $\mathcal{W}$ respectively.*

Proof. 1. It is clear that $\operatorname{Ker} T \subset \mathcal{V}$ and $0 \in \operatorname{Ker} T$. Let $x, y \in \operatorname{Ker} T$ and $\alpha, \beta \in \mathbb{F}$. Since

$$T(\alpha x + \beta y) = \alpha(Tx) + \beta(Ty) = 0,$$

it follows that $\alpha x + \beta y \in \operatorname{Ker} T$.
2. By definition $\operatorname{Im} T \subset \mathcal{W}$. Let $y_1, y_2 \in \operatorname{Im} T$. Thus there exist vectors $x_1, x_2 \in \mathcal{V}$ for which $y_i = Tx_i$. Let $\alpha_1, \alpha_2 \in \mathbb{F}$. Then

$$\alpha_1 y_1 + \alpha_2 y_2 = \alpha_1 T x_1 + \alpha_2 T x_2 = T(\alpha_1 x_1 + \alpha_2 x_2),$$

and $\alpha_1 y_1 + \alpha_2 y_2 \in \operatorname{Im} T$.

∎

We define the **rank** of a linear transformation T, denoted by $\operatorname{rank}(T)$, by

$$\operatorname{rank}(T) = \dim \operatorname{Im} T, \tag{4.2}$$

and the **nullity** of T, denoted $\operatorname{null}(T)$, by

$$\operatorname{null}(T) = \dim \operatorname{Ker} T. \tag{4.3}$$

The rank and nullity of a linear transformation are connected via the following theorem.

Theorem 4.4. *Let $\mathscr{V}$ and $\mathscr{W}$ be finite-dimensional vector spaces over $\mathbb{F}$ and let $T : \mathscr{V} \longrightarrow \mathscr{W}$ be a linear transformation. Then*

$$\operatorname{rank}(T) + \operatorname{null}(T) = \dim \mathscr{V}, \tag{4.4}$$

or equivalently,

$$\dim \operatorname{Im} T + \dim \operatorname{Ker} T = \dim \mathscr{V}. \tag{4.5}$$

Proof. Let $\{e_1, \ldots, e_n\}$ be a basis of $\mathscr{V}$ for which $\{e_1, \ldots, e_p\}$ is a basis of $\operatorname{Ker} T$. Let $x \in \mathscr{V}$, then x has a unique representation of the form $x = \sum_{i=1}^{n} \alpha_i e_i$. Since T is linear, we have

$$T x = T \sum_{i=1}^{n} \alpha_i e_i = \sum_{i=1}^{n} \alpha_i T e_i = \sum_{i=p+1}^{n} \alpha_i T e_i,$$

since $T e_i = 0$ for $i = 1, \ldots, p$. Therefore it follows that $\{T e_{p+1}, \ldots, T e_n\}$ is a spanning set of $\operatorname{Im} T$. We will show that they are also linearly independent. Assume that there exist $\{\alpha_{p+1}, \ldots, \alpha_n\}$ such that $\sum_{i=p+1}^{n} \alpha_i (T e_i) = 0$. Now $\sum_{i=p+1}^{n} \alpha_i (T e_i) = T \sum_{i=p+1}^{n} \alpha_i e_i$, which implies that $\sum_{i=p+1}^{n} \alpha_i e_i \in \operatorname{Ker} T$. Hence, it can be represented as a linear combination of the form $\sum_{i=1}^{p} \alpha_i e_i$. So we get

$$\sum_{i=1}^{p} \alpha_i e_i - \sum_{i=p+1}^{n} \alpha_i e_i = 0.$$

Using the fact that $\{e_1, \ldots, e_n\}$ is a basis of $\mathscr{V}$, we conclude that $\alpha_i = 0$ for all $i = 1, \ldots, n$. Thus we have proved that $\{T e_{p+1}, \ldots, T e_n\}$ is a basis of $\operatorname{Im} T$. Therefore $\operatorname{null}(T) = p$, $\operatorname{rank}(T) = n - p$, and $\operatorname{null}(T) + \operatorname{rank}(T) = p + (n - p) = n = \dim \mathscr{V}$.

∎

Let $\mathscr{U}$ and $\mathscr{V}$ be vector spaces over the field $\mathbb{F}$. We denote by $L(\mathscr{U}, \mathscr{V})$, or $\operatorname{Hom}_{\mathbb{F}}(\mathscr{U}, \mathscr{V})$, the space of all linear transformations from $\mathscr{U}$ to $\mathscr{V}$. Given two transformations $T, S \in L(\mathscr{U}, \mathscr{V})$, we define their sum by

$$(T+S)x = Tx + Sx,$$

and the product with any scalar $\alpha \in \mathbb{F}$ by

$$(\alpha T)x = \alpha(Tx).$$

Theorem 4.5. *With the previous definitions, $L(\mathcal{U}, \mathcal{V})$ is a vector space.*

Proof. We show that $T + S$ and αT are indeed linear transformations. Let $x, y \in \mathcal{V}$, $\alpha \in \mathbb{F}$. We have

$$\begin{aligned}
(T+S)(x+y) &= T(x+y) + S(x+y) = (Tx + Ty) + (Sx + Sy) \\
&= (Tx + Sx) + (Ty + Sy) = (T+S)x + (T+S)y.
\end{aligned}$$

Similarly,

$$\begin{aligned}
(T+S)(\alpha x) &= T(\alpha x) + S(\alpha x) \\
&= \alpha(Tx) + \alpha(Sx) = \alpha(Tx + Sx) \\
&= \alpha((T+S)x).
\end{aligned}$$

This shows that $T + S$ is a linear transformation. The proof for αT goes along similar lines. We leave it to the reader to verify that all the axioms of a linear space hold for $L(\mathcal{U}, \mathcal{V})$. ∎

It is clear that the dimension of $L(\mathcal{U}, \mathcal{V})$ is determined by the spaces $\mathcal{U}$ and $\mathcal{V}$. The next theorem characterizes it.

Theorem 4.6. *Let $\mathcal{U}$ and $\mathcal{V}$ be vector spaces over the field $\mathbb{F}$. Then*

$$\dim L(\mathcal{U}, \mathcal{V}) = \dim \mathcal{U} \cdot \dim \mathcal{V}. \tag{4.6}$$

Proof. Let $\{e_1, \ldots, e_n\}$ be a basis for $\mathcal{U}$ and $\{f_1, \ldots, f_m\}$ a basis for $\mathcal{V}$. We proceed to construct a basis for $L(\mathcal{U}, \mathcal{V})$. By Theorem 4.2 a linear transformation is determined by its values on basis elements. We define a set of mn linear transformations E_{ij}, $i = 1, \ldots, m, j = 1, \ldots, n$, by letting

$$E_{ij}e_k = \delta_{jk} f_i, \qquad k = 1, \ldots, n. \tag{4.7}$$

Here δ_{ij} is the Kronecker delta function, defined by (2.4). On the vector $x = \sum_{k=1}^{n} \alpha_k e_k$ the transformation E_{ij} acts as

$$E_{ij}x = E_{ij} \sum_{k=1}^{n} \alpha_k e_k = \sum_{k=1}^{n} \alpha_k E_{ij}e_k = \sum_{k=1}^{n} \alpha_k \delta_{jk} f_i = \alpha_j f_i,$$

or

$$E_{ij}x = \alpha_j f_i.$$

We now show that the set of transformations E_{ij} forms a basis for $L(\mathscr{U},\mathscr{V})$. To prove this we have to show that this set is linearly independent and spans $L(\mathscr{U},\mathscr{V})$. We begin by showing linear independence. Assume there exist $\alpha_{ij} \in \mathbb{F}$ for which

$$\sum_{i=1}^{m}\sum_{j=1}^{n}\alpha_{ij}E_{ij} = 0.$$

We operate with the zero transformation, in both its representations, on an arbitrary basis element e_k in $\mathscr{V}$:

$$0 = 0 \cdot e_k = \sum_{i=1}^{m}\sum_{j=1}^{n}\alpha_{ij}E_{ij}e_k = \sum_{i=1}^{m}\sum_{j=1}^{n}\alpha_{ij}\delta_{jk}f_i = \sum_{i=1}^{m}\alpha_{ik}f_i.$$

Since the f_i are linearly independent, we get $\alpha_{ik} = 0$ for all $i = 1,\ldots,m$. The index k was chosen arbitrarily; hence we conclude that all the α_{ij} are zero. This completes the proof of linear independence.

Let T be an arbitrary linear transformation in $L(\mathscr{U},\mathscr{V})$. Since $Te_k \in \mathscr{V}$, it has a unique representation of the form $Te_k = \sum_{i=1}^{m} b_{ik}f_i$, $k = 1,\ldots,n$. We show now that $T = \sum_{i=1}^{m}\sum_{j=1}^{n}b_{ij}E_{ij}$. Of course, by Theorem 4.2, it suffices to prove the equality on the basis elements in $\mathscr{U}$. We compute

$$(\sum_{i=1}^{m}\sum_{j=1}^{n}b_{ij}E_{ij})e_k = \sum_{i=1}^{m}\sum_{j=1}^{n}b_{ij}(E_{ij}e_k) = \sum_{i=1}^{m}\sum_{j=1}^{n}b_{ij}\delta_{jk}f_i$$
$$= \sum_{i=1}^{m}b_{ik}f_i = Te_k.$$

Since k can be chosen arbitrarily, the proof is complete. ∎

In some cases linear transformations can be composed, or multiplied. Let $\mathscr{U},\mathscr{V},\mathscr{W}$ be vector spaces over $\mathbb{F}$. Let $S \in L(\mathscr{V},\mathscr{W})$ and $T \in L(\mathscr{U},\mathscr{V})$, i.e., $\mathscr{U} \xrightarrow{T} \mathscr{V} \xrightarrow{S} \mathscr{W}$.

We define the transformation $ST : \mathscr{U} \longrightarrow \mathscr{W}$ by

$$(ST)x = S(Tx), \qquad \forall x \in \mathscr{U}. \tag{4.8}$$

We call the transformation ST the **composition**, or **product**, of S and T.

Theorem 4.7. *The transformation $ST : \mathscr{U} \longrightarrow \mathscr{W}$ is linear, i.e., $ST \in L(\mathscr{U},\mathscr{W})$.*

Proof. Let $x,y \in \mathscr{U}$ and $\alpha,\beta \in \mathbb{F}$. By the linearity of S and T we get

$$(ST)(\alpha x + \beta y) = S(T(\alpha x + \beta y))$$
$$= S(\alpha(Tx) + \beta(Ty)) = \alpha S(Tx) + \beta S(Ty)$$
$$= \alpha(ST)x + \beta(ST)y.$$

∎

Composition of transformations is associative; that is, if

$$\mathscr{U} \xrightarrow{T} \mathscr{V} \xrightarrow{S} \mathscr{W} \xrightarrow{R} Y,$$

then it is easily checked that with the composition of linear transformations defined by (4.8), the following associative rule holds:

$$(RS)T = R(ST). \tag{4.9}$$

We saw already the existence of the identity map I in each linear space $\mathscr{U}$. Clearly, we have $IT = TI = T$ for every linear transformation T.

Definition 4.8. Let $T \in L(\mathscr{V}, \mathscr{W})$. We say that T is **right invertible** if there exists a linear transformation $S \in L(\mathscr{W}, \mathscr{V})$ for which

$$TS = I_{\mathscr{W}}.$$

Similarly, we say that T is **left invertible** if there exists a linear transformation $S \in L(\mathscr{W}, \mathscr{V})$ for which

$$ST = I_{\mathscr{V}}.$$

We say that T is **invertible** if it is both right and left invertible.

Proposition 4.9. *If T is both right and left invertible, then the right inverse and the left inverse are equal and will be denoted by T^{-1}.*

Proof. Let $TS = I_{\mathscr{W}}$ and $RT = I_{\mathscr{V}}$. Then $R(TS) = (RT)S$ implies

$$R = RI_{\mathscr{W}} = I_{\mathscr{V}}S = S.$$

■

Left and right invertibility are intrinsic properties of the transformation T, as follows from the next theorem.

Theorem 4.10. *Let $T \in L(\mathscr{V}, \mathscr{W})$. Then*

1. *T is right invertible if and only if T is surjective.*
2. *T is left invertible if and only if T is injective.*

Proof. 1. Assume T is right invertible. Thus, there exists $S \in L(\mathscr{W}, \mathscr{V})$ for which $TS = I_{\mathscr{W}}$. Therefore, for all $x \in \mathscr{W}$, we have

$$T(Sx) = (TS)x = I_{\mathscr{W}}x = x,$$

or $x \in \operatorname{Im}T$, and so $\operatorname{Im}T = \mathscr{W}$ and T is surjective.

Conversely, assume T is surjective. Let $\{f_1, \ldots, f_m\}$ be a basis in $\mathscr{W}$ and let $\{e_1, \ldots, e_m\}$ be vectors in $\mathscr{V}$ satisfying $Te_i = f_i$, for $i = 1, \ldots, m$. Clearly, the vectors $\{e_1,\ldots,e_m\}$ are linearly independent. We define a linear transformation $S \in L(\mathcal{W},\mathcal{V})$ by

$$Sf_i = \begin{cases} e_i, & i = 1,\ldots,m, \\ 0, & i = m+1,\ldots,n. \end{cases}$$

Then, for all i, $(TS)f_i = T(Sf_i) = Te_i = f_i$, i.e., $TS = I_{\mathcal{W}}$.

2. Assume T is left invertible. Let $S \in L(\mathcal{W},\mathcal{V})$ be a left inverse, that is, $ST = I_{\mathcal{V}}$. Let $x \in \operatorname{Ker} T$. Then

$$x = I_{\mathcal{V}}x = (ST)x = S(Tx) = S(0) = 0,$$

that is, $\operatorname{Ker} T = \{0\}$.

Conversely, assume $\operatorname{Ker} T = \{0\}$. Let $\{e_1,\ldots,e_n\}$ be a basis in $\mathcal{V}$. It follows from our assumption that $\{Te_1,\ldots,Te_n\}$ are linearly independent vectors in $\mathcal{W}$. We complete this set of vectors to a basis $\{Te_1,\ldots,Te_m,f_{m+1},\ldots,f_n\}$. We now define a linear transformation $S \in L(\mathcal{W},\mathcal{V})$ by

$$S \begin{cases} Te_i = e_i, & i = 1,\ldots,m, \\ Tf_i = 0, & i = m+1,\ldots,n. \end{cases}$$

Obviously, $STe_i = e_i$ for all i, and hence $ST = I_{\mathcal{V}}$.

∎

It should be noted that the left inverse is generally not unique. In our construction, we had the freedom to define S differently on the vectors $\{f_{m+1},\ldots,f_n\}$, but we opted for the simplest definition.

The previous characterizations are special cases of the more general result.

Theorem 4.11. *Let $\mathcal{U},\mathcal{V},\mathcal{W}$ be vector spaces over $\mathbb{F}$.*

1. *Assume $A \in L(\mathcal{U},\mathcal{W})$ and $B \in L(\mathcal{U},\mathcal{V})$. Then there exists $C \in L(\mathcal{V},\mathcal{W})$ such that*

$$A = CB$$

if and only if

$$\operatorname{Ker} A \supset \operatorname{Ker} B. \tag{4.10}$$

2. *Assume $A \in L(\mathcal{U},\mathcal{W})$ and $B \in L(\mathcal{V},\mathcal{W})$. Then there exists $C \in L(\mathcal{U},\mathcal{V})$ such that*

$$A = BC$$

if and only if

$$\operatorname{Im} A \subset \operatorname{Im} B. \tag{4.11}$$

Proof. 1. If $A = CB$, then for $x \in \operatorname{Ker} B$ we have

$$Ax = C(Bx) = C0 = 0,$$

i.e., $x \in \operatorname{Ker} A$ and (4.10) follows.

Conversely, assume $\operatorname{Ker}A \supset \operatorname{Ker}B$. Let $\{e_1,\ldots,e_r,e_{r+1},\ldots,e_p,e_{p+1},\ldots,e_n\}$ be a basis in $\mathscr{U}$ such that $\{e_1,\ldots,e_p\}$ is a basis for $\operatorname{Ker}A$ and $\{e_1,\ldots,e_r\}$ is a basis for $\operatorname{Ker}B$. The vectors $\{Be_{r+1},\ldots,Be_n\}$ are linearly independent in $\mathscr{V}$. We complete them to a basis of $\mathscr{V}$ and define a linear transformation $C : \mathscr{V} \longrightarrow \mathscr{W}$ by

$$CBe_i = \begin{cases} Ae_i, & i = p+1,\ldots,n, \\ 0, & i = r+1,\ldots,p, \end{cases}$$

and arbitrarily on the other basis elements of $\mathscr{V}$. Then, for every $x = \sum_{i=1}^n \alpha_i e_i \in \mathscr{U}$ we have

$$Ax = A\sum_{i=1}^n \alpha_i e_i = \sum_{i=1}^n \alpha_i Ae_i = \sum_{i=p+1}^n \alpha_i(Ae_i)$$

$$= \sum_{i=p+1}^n \alpha_i(CB)e_i = CB\sum_{i=p+1}^n \alpha_i e_i = CB\sum_{i=1}^n \alpha_i e_i = CBx,$$

and so $A = CB$.

2. If $A = BC$, then for every $x \in \mathscr{U}$ we have

$$Ax = (BC)x = B(Cx),$$

or $\operatorname{Im}A \subset \operatorname{Im}B$.

Conversely, assume $\operatorname{Im}A \subset \operatorname{Im}B$. Then we let $\{e_1,\ldots,e_r,e_{r+1},\ldots,e_n\}$ be a basis for $\mathscr{U}$ such that $\{e_{r+1},\ldots,e_n\}$ is a basis for $\operatorname{Ker}A$. Then $\{Ae_1,\ldots,Ae_r\}$ are linearly independent vectors in $\mathscr{W}$. By our assumption, $\operatorname{Im}A \subset \operatorname{Im}B$, there exist vectors $\{g_1,\ldots,g_r\}$ in $\mathscr{V}$, necessarily linearly independent, such that

$$Ae_i = Bg_i, \qquad i = 1,\ldots,r.$$

We complete them to a basis $\{g_1,\ldots,g_r,g_{r+1},\ldots,g_m\}$ of $\mathscr{W}$. We define now $C \in L(\mathscr{U},\mathscr{V})$ by

$$Ce_i = \begin{cases} g_i, & i = 1,\ldots,r, \\ 0, & i = r+1,\ldots,n. \end{cases}$$

For $x = \sum_{i=1}^n \alpha_i e_i \in \mathscr{U}$, we get

$$A\sum_{i=1}^n \alpha_i e_i = \sum_{i=1}^r \alpha_i Ae_i = \sum_{i=1}^r \alpha_i Bg_i$$

$$= B\sum_{i=1}^r \alpha_i g_i = B\sum_{i=1}^r \alpha_i Ce_i = B\sum_{i=1}^n \alpha_i Ce_i = BC\sum_{i=1}^n \alpha_i e_i = BCx,$$

and $A = BC$. ∎

Theorem 4.12. *Let $T \in L(\mathscr{U},\mathscr{V})$ be invertible. Then $T^{-1} \in L(\mathscr{V},\mathscr{U})$.*

Proof. Let $y_1,y_2 \in \mathscr{V}$ and $\alpha_1,\alpha_2 \in \mathbb{F}$. There exist unique vectors $x_1,x_2 \in \mathscr{U}$ such that $Tx_i = y_i$, or equivalently, $x_i = T^{-1}y_i$. Since

$$T(\alpha_1 x_1 + \alpha_2 x_2) = \alpha_1 T x_1 + \alpha_2 T x_2 = \alpha_1 y_1 + \alpha_2 y_2,$$

therefore

$$T^{-1}(\alpha_1 y_1 + \alpha_2 y_2) = \alpha_1 x_1 + \alpha_2 x_2 = \alpha_1 T^{-1} y_1 + \alpha_2 T^{-1} y_2.$$ ∎

A linear transformation $T : \mathcal{U} \longrightarrow \mathcal{V}$ is called an **isomorphism** if it is invertible. In this case we will say that $\mathcal{U}$ and $\mathcal{V}$ are isomorphic spaces. Clearly, isomorphism of vector spaces is an equivalence relation.

Theorem 4.13. *Every n-dimensional vector space $\mathcal{U}$ over $\mathbb{F}$ is isomorphic to $\mathbb{F}^n$.*

Proof. Let $\mathcal{B} = \{e_1, \ldots, e_n\}$ be a basis for $\mathcal{U}$. The map $x \mapsto [x]^{\mathcal{B}}$ is an isomorphism. ∎

Theorem 4.14. *Two finite-dimensional vector spaces $\mathcal{V}$ and $\mathcal{W}$ over $\mathbb{F}$ are isomorphic if and only if they have the same dimension.*

Proof. If $\mathcal{U}$ and $\mathcal{V}$ have the same dimension, then we can construct a map that maps the basis elements in $\mathcal{U}$ onto the basis elements in $\mathcal{V}$. This map is necessarily an isomorphism.

Conversely, if $\mathcal{U}$ and $\mathcal{V}$ are isomorphic, the image of a basis in $\mathcal{U}$ is necessarily a basis in $\mathcal{V}$. Thus dimensions are equal. ∎

4.3 Matrix Representations

Let $\mathcal{U}, \mathcal{V}$ be vector spaces over a field $\mathbb{F}$. Let $B_{\mathcal{U}} = \{e_1, \ldots, e_n\}$ and $B_{\mathcal{V}} = \{f_1, \ldots, f_m\}$ be bases of $\mathcal{U}$ and $\mathcal{V}$ respectively. We saw before that the set of linear transformations $\{E_{ij}\}$, $i = 1, \ldots, m$, $j = 1, \ldots, n$, defined by

$$E_{ij} e_k = \delta_{jk} f_i, \tag{4.12}$$

is a basis for $L(\mathcal{U}, \mathcal{V})$. We denote this basis by $\mathcal{B}_{\mathcal{U}} \times \mathcal{B}_{\mathcal{V}}$.

A natural problem presents itself, namely finding the coordinate vector of a linear transformation $T \in L(\mathcal{U}, \mathcal{V})$ with respect to this basis. We observed before, in Theorem 4.6, that T can be written as

$$T = \sum_{i=1}^{m} \sum_{j=1}^{n} t_{ij} E_{ij}. \tag{4.13}$$

Hence

$$Te_k = \sum_{i=1}^{m} \sum_{j=1}^{n} t_{ij} E_{ij} e_k = \sum_{i=1}^{m} \sum_{j=1}^{n} t_{ij} \delta_{jk} f_i = \sum_{i=1}^{m} t_{ik} f_i, \tag{4.14}$$

i.e.,

$$Te_k = \sum_{i=1}^{m} t_{ik} f_i. \tag{4.15}$$

Therefore, the coordinate vector of T with respect to the basis $\mathscr{B}_{\mathscr{U}} \times \mathscr{B}_{\mathscr{V}}$ will have the entries t_{ik} arranged in the same order in which the basis elements are ordered.

In contrast to the case of an abstract vector space, we will arrange the coordinates in this case in an $m \times n$ matrix (t_{ij}) and call it the **matrix representation** of T with respect to the bases $\mathscr{B}_{\mathscr{U}}$ in $\mathscr{U}$ and $\mathscr{B}_{\mathscr{V}}$ in $\mathscr{V}$. We will use the following notation:

$$[T]_{\mathscr{B}_{\mathscr{U}}}^{\mathscr{B}_{\mathscr{V}}} = (t_{ik}). \tag{4.16}$$

The importance of the matrix representation stems froms the following theorem, which reduces the application of an arbitrary linear transformation to matrix multiplication.

Theorem 4.15. *Let $\mathscr{U}$ and $\mathscr{V}$ be n-dimensional and m-dimensional vector spaces with respective bases $\mathscr{B}_{\mathscr{U}}$ and $\mathscr{B}_{\mathscr{V}}$, and let $T : \mathscr{U} \longrightarrow \mathscr{V}$ be a linear transformation. Then the following diagram is commutative:*

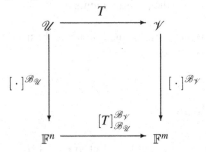

In other words, we have

$$[Tx]^{\mathscr{B}_{\mathscr{V}}} = [T]_{\mathscr{B}_{\mathscr{U}}}^{\mathscr{B}_{\mathscr{V}}} [x]^{\mathscr{B}_{\mathscr{U}}}.$$

Proof. Assume $x = \sum_{j=1}^{n} \alpha_j e_j$ and $Te_j = \sum_{i=1}^{m} t_{ij} f_i$. Then

$$Tx = T \sum_{j=1}^{n} \alpha_j e_j = \sum_{j=1}^{n} \alpha_j \sum_{i=1}^{m} t_{ij} f_i = \sum_{i=1}^{m} \left(\sum_{j=1}^{n} t_{ij} \alpha_j \right) f_i.$$

Thus

$$[Tx]^{\mathscr{B}_{\mathscr{V}}} = \begin{pmatrix} \sum_{j=1}^{n} t_{1j} \alpha_j \\ \cdot \\ \cdot \\ \cdot \\ \sum_{j=1}^{n} t_{mj} \alpha_j \end{pmatrix} = [T]_{\mathscr{B}_{\mathscr{U}}}^{\mathscr{B}_{\mathscr{V}}} [x]^{\mathscr{B}_{\mathscr{U}}}.$$

∎

The previous theorem shows that by choosing bases, we can pass from an abstract representation of spaces and transformations to a concrete one in the form of column vectors and matrices. In the latter form, computations are easily mechanized.

The next theorem deals with the matrix representation of the product of two linear transformations.

Theorem 4.16. *Let $\mathscr{U}, \mathscr{V}, \mathscr{W}$ be vector spaces over a field $\mathbb{F}$ of dimensions n, m, and p respectively. Let $T \in L(\mathscr{U}, \mathscr{V})$ and $S \in L(\mathscr{V}, \mathscr{W})$. Let $\mathscr{B}_U = \{e_1, \ldots, e_n\}$, $\mathscr{B}_V = \{f_1, \ldots, f_m\}$, and $\mathscr{B}_W = \{g_1, \ldots, g_p\}$ be bases in $\mathscr{U}, \mathscr{V}$, and $\mathscr{W}$ respectively. Then*

$$[ST]_{\mathscr{B}_\mathscr{U}}^{\mathscr{B}_\mathscr{W}} = [S]_{\mathscr{B}_\mathscr{V}}^{\mathscr{B}_\mathscr{W}} [T]_{\mathscr{B}_\mathscr{U}}^{\mathscr{B}_\mathscr{V}}. \tag{4.17}$$

Proof. Let $Te_j = \sum_{k=1}^{m} t_{kj} f_k$, $Sf_k = \sum_{i=1}^{p} s_{ik} g_i$ and $(ST)e_j = \sum_{i=1}^{p} r_{ij} g_i$. Therefore

$$(ST)e_j = S(Te_j) = S\sum_{k=1}^{m} t_{kj} f_k$$

$$= \sum_{k=1}^{m} t_{kj} S f_k = \sum_{k=1}^{m} t_{kj} \sum_{i=1}^{p} s_{ik} g_i$$

$$= \sum_{i=1}^{p} [\sum_{k=1}^{m} s_{ik} t_{kj}] g_i.$$

Since $\{g_1, \ldots, g_p\}$ is a basis for $\mathscr{W}$, we get

$$r_{ij} = \sum_{k=1}^{m} s_{ik} t_{kj},$$

and (4.17) follows. ∎

Clearly, we have proved the commutativity of the following diagram:

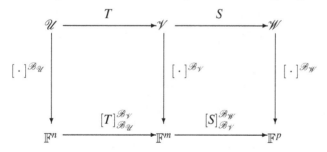

Corollary 4.17. *Let the vector space $\mathscr{V}$ have the two bases $\mathscr{B}_1$ and $\mathscr{B}_2$. Then we have*

$$[I]_{\mathscr{B}_2}^{\mathscr{B}_1} [I]_{\mathscr{B}_1}^{\mathscr{B}_2} = [I]_{\mathscr{B}_1}^{\mathscr{B}_1} = I, \tag{4.18}$$

or alternatively,

$$[I]_{\mathscr{B}_2}^{\mathscr{B}_1} = \left([I]_{\mathscr{B}_1}^{\mathscr{B}_2}\right)^{-1}. \tag{4.19}$$

An important special case of the previous result is that of the relation between two matrix representations of a given linear transformation, taken with respect to two different bases. We begin by noting that the following diagram commutes.

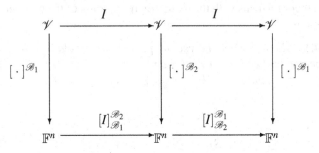

Theorem 4.18. *Let T be a linear transformation in a vector space $\mathscr{V}$. Let $\mathscr{B}_1$ and $\mathscr{B}_2$ be two bases of $\mathscr{V}$. Then we have*

$$[T]_{\mathscr{B}_2}^{\mathscr{B}_2} = [I]_{\mathscr{B}_1}^{\mathscr{B}_2}[T]_{\mathscr{B}_1}^{\mathscr{B}_1}[I]_{\mathscr{B}_2}^{\mathscr{B}_1}, \tag{4.20}$$

or

$$[T]_{\mathscr{B}_2}^{\mathscr{B}_2} = \left([I]_{\mathscr{B}_2}^{\mathscr{B}_1}\right)^{-1}[T]_{\mathscr{B}_1}^{\mathscr{B}_1}[I]_{\mathscr{B}_2}^{\mathscr{B}_1}. \tag{4.21}$$

Proof. We observe the following commutative diagram, from which the result immediately follows:

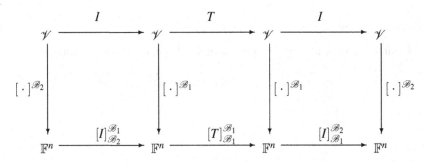

$\blacksquare$

Equation (4.21) leads us to the following definition.

Definition 4.19. Given matrices $A, B \in \mathbb{F}^{n \times n}$, we say that A and B are **similar**, and write $A \simeq B$, if there exists an invertible matrix $R \in \mathbb{F}^{n \times n}$ for which

$$A = R^{-1}BR. \tag{4.22}$$

Clearly, similarity is an equivalence relation.

Theorem 4.18 plays a key role in the study of linear transformations. The structure of a linear transformation is determined through the search for simple matrix representations, which is equivalent to finding appropriate bases. The previous theorem's content is that through a change of basis the matrix representation of a given linear transformation undergoes a similarity transformation.

4.4 Linear Functionals and Duality

We note that a field $\mathbb{F}$ is at the same time a one-dimensional vector space over itself. Given a vector space $\mathscr{V}$ over a field $\mathbb{F}$, a linear transformation from $\mathscr{V}$ to $\mathbb{F}$ is called a **linear functional**. The set of all linear functionals on $\mathscr{V}$, i.e., $L(\mathscr{V}, \mathbb{F})$, is called the **dual space** of $\mathscr{V}$ and will be denoted by $\mathscr{V}^*$. Thus

$$\mathscr{V}^* = L(\mathscr{V}, \mathbb{F}). \tag{4.23}$$

Given elements $x \in \mathscr{V}$ and $x^* \in \mathscr{V}^*$, we will write also

$$x^*(x) = \langle x, x^* \rangle \tag{4.24}$$

and call $\langle x, x^* \rangle$ a duality pairing of $\mathscr{V}$ and $\mathscr{V}^*$. Note that $\langle x, x^* \rangle$ is linear in both variables. More generally, given $\mathbb{F}$-vector spaces $\mathscr{V}, \mathscr{W}$, we say that a map $\langle v, w \rangle$: $\mathscr{V} \times \mathscr{W} \longrightarrow \mathbb{F}$ is a **bilinear pairing** or a **bilinear form** if $\langle v, w \rangle$ is linear in both variables, i.e.,

$$\begin{aligned} \langle \alpha_1 v_1 + \alpha_2 v_2, w \rangle &= \alpha_1 \langle v_1, w \rangle + \alpha_2 \langle v_2, w \rangle, \\ \langle v, \alpha_1 w_1 + \alpha_2 w_2 \rangle &= \alpha_1 \langle v, w_1 \rangle + \alpha_2 \langle v, w_2 \rangle, \end{aligned} \tag{4.25}$$

for all $\alpha_i \in \mathbb{F}, v_i \in \mathscr{V}, w_i \in \mathscr{W}$.

Examples:

- Let $\mathscr{V} = \mathbb{F}^n$. Then, for $\alpha_1, \ldots, \alpha_n \in \mathbb{F}$, the map $f : \mathbb{F}^n \longrightarrow \mathbb{F}$ defined by

$$f \begin{pmatrix} x_1 \\ \cdot \\ \cdot \\ \cdot \\ x_n \end{pmatrix} = \sum_{i=1}^n \alpha_i x_i$$

defines a linear functional.
- Let $\mathbb{F}^{n \times n}$ be the space of all square matrices of order n. Given a linear transformation T acting in a finite-dimensional vector space $\mathscr{U}$, we define the **trace** of T, $\mathrm{Trace}(T)$, by

$$\mathrm{Trace}(T) = \sum_{i=1}^n t_{ii}, \tag{4.26}$$

where (t_{ij}) is a matrix representation of T with respect to an arbitrary basis of $\mathscr{U}$. It can be shown directly that the trace is independent of the particular choice of basis. Let $X, A \in \mathbb{F}^{n \times n}$. Then f defined by

$$f(X) = \text{Trace}\,(AX) \tag{4.27}$$

is a linear functional.

Let $\mathscr{V}$ be an n-dimensional vector space, and let $\mathscr{B} = \{e_1, \ldots, e_n\}$ be a basis for $\mathscr{V}$. Thus, each vector $x \in \mathscr{V}$ has a unique representation of the form $x = \sum_{j=1}^{n} \alpha_j e_j$. We define functionals f_i by

$$f_i(x) = \alpha_i, \qquad i = 1, \ldots, n.$$

It is easily checked that the f_i so defined are linear functionals. To see this, let $y = \sum_{j=1}^{n} \beta_j e_j$. Then

$$f_i(x+y) = f_i(\textstyle\sum_{j=1}^{n} \alpha_j e_j + \sum_{j=1}^{n} \beta_j e_j) = f_i(\textstyle\sum_{j=1}^{n} (\alpha_j + \beta_j)e_j)$$

$$= \alpha_i + \beta_i = f_i(x) + f_i(y).$$

In the same way,

$$f_i(\alpha x) = f_i(\alpha \textstyle\sum_{j=1}^{n} \alpha_j e_j) = f_i(\textstyle\sum_{j=1}^{n} (\alpha \alpha_j)e_j) = \alpha \alpha_i = \alpha f_i(x).$$

Since clearly, $e_i = \sum_{j=1}^{n} \delta_{ij} e_j$, we have $f_i(e_j) = \delta_{ij}$ for $1 \le i, j \le n$.

Theorem 4.20. *Let $\mathscr{V}$ be an n-dimensional vector space, and let $\mathscr{B} = \{e_1, \ldots, e_n\}$ be a basis for $\mathscr{V}$. Then*

1. We have

$$\dim \mathscr{V}^* = \dim \mathscr{V}. \tag{4.28}$$

2. The set of linear functionals $\{f_1, \ldots, f_n\}$ defined, through linear extensions, by $f_i(e_j) = \delta_{ij}$ is a basis for $\mathscr{V}^*$.

Proof. 1. We apply Theorem 4.6, i.e., the fact that $\dim L(\mathscr{U}, \mathscr{V}) = \dim \mathscr{U} \dim \mathscr{V}$ and the fact that $\dim \mathbb{F} = 1$.

2. The functionals $\{f_1, \ldots, f_n\}$ are linearly independent. For let $\sum_{i=1}^{n} \alpha_i f_i = 0$. Then

$$0 = 0 \cdot e_j = \left(\sum_{i=1}^{n} \alpha_i f_i \right) e_j = \sum_{i=1}^{n} \alpha_i f_i(e_j) = \sum_{i=1}^{n} \alpha_i \delta_{ij} = \alpha_j.$$

Therefore $\alpha_j = 0$ for all j, and linear independence is proved. Since $\dim \mathscr{V}^* = n$, it follows that $\{f_1, \ldots, f_n\}$ is a basis for $\mathscr{V}^*$. ∎

Definition 4.21. Let $\mathscr{V}$ be an n-dimensional vector space, and let $\mathscr{B} = \{e_1, \ldots, e_n\}$ be a basis for $\mathscr{V}$. Then the basis $\{f_1, \ldots, f_n\}$ defined by $f_i(e_j) = \delta_{ij}$ is called the **dual basis** to $\mathscr{B}$ and will be denoted by $\mathscr{B}^*$.

Definition 4.22. Let $\mathscr{S}$ be a subset of a vector space $\mathscr{V}$. We denote by $\mathscr{S}^{\perp}$ the subset of $\mathscr{V}^*$ defined by

$$\mathscr{S}^{\perp} = \{f \in \mathscr{V}^* | f(s) = 0, \text{ for all } s \in \mathscr{S}\}.$$

The set $\mathscr{S}^{\perp}$ is called the **annihilator** of $\mathscr{S}$.

Proposition 4.23. *Let $\mathscr{S}$ be a subset of the vector space $\mathscr{V}$. Then*

1. *The set $\mathscr{S}^{\perp}$ is a subspace of $\mathscr{V}^*$.*
2. *If $\mathscr{Q} \subset \mathscr{S}$ then $\mathscr{S}^{\perp} \subset \mathscr{Q}^{\perp}$.*
3. *We have $\mathscr{S}^{\perp} = (\text{span}\,(\mathscr{S}))^{\perp}$.*

Proof. 1. Let $f, g \in \mathscr{S}^{\perp}$ and $\alpha, \beta \in \mathbb{F}$. Then for an arbitrary $x \in \mathscr{S}$,

$$(\alpha f + \beta g)x = \alpha f(x) + \beta g(x) = \alpha \cdot 0 + \beta \cdot 0 = 0,$$

that is, $\alpha f + \beta g \in \mathscr{S}^{\perp}$.

2. Let $f \in \mathscr{S}^{\perp}$. Then $f(x) = 0$ for all $x \in \mathscr{S}$ and particularly, by the inclusion $\mathscr{Q} \subset \mathscr{S}$, for all $x \in \mathscr{Q}$. So $f \in \mathscr{Q}^{\perp}$.

3. It is clear that $\mathscr{S} \subset \text{span}\,(\mathscr{S})$, and therefore $\text{span}\,(\mathscr{S})^{\perp} \subset \mathscr{S}^{\perp}$. On the other hand, if $f \in \mathscr{S}^{\perp}$, $x_i \in \mathscr{S}$ and $\alpha_i \in \mathbb{F}$, then

$$f\left(\sum_{i=1}^n \alpha_i x_i\right) = \sum_{i=1}^n \alpha_i f(x_i) = \sum_{i=1}^n \alpha_i \cdot 0 = 0,$$

which means that f annihilates all linear combinations of elements of $\mathscr{S}$. So $f \in \text{span}\,(\mathscr{S})^{\perp}$ or $\mathscr{S}^{\perp} \subset \text{span}\,(\mathscr{S})^{\perp}$. From these two inclusions the equality $\mathscr{S}^{\perp} = \text{span}\,(\mathscr{S})^{\perp}$ follows. ∎

We will find the following proposition useful in the study of duality.

Proposition 4.24. *Let $\mathscr{V}$ be a vector space and $\mathscr{M}$ its subspace. Then the dual space to the quotient space $\mathscr{V}/\mathscr{M}$ is isomorphic to $\mathscr{M}^{\perp}$.*

Proof. Let $\phi \in \mathscr{M}^{\perp}$. Then ϕ induces a linear functional Φ on $\mathscr{V}/\mathscr{M}$ by $\Phi([x]) = \phi(x)$. The functional Φ is well defined, since $[x_1] = [x_2]$ if and only if $x_1 - x_2 \in \mathscr{M}$, and this implies $\phi(x_1) - \phi(x_2) = \phi(x_1 - x_2) = 0$. The linearity of Φ follows from the linearity of ϕ.

Conversely, given $\Phi \in (\mathscr{V}/\mathscr{M})^*$, we define ϕ by $\phi(x) = \Phi([x])$. Clearly $\phi(x) = 0$ for $x \in \mathscr{M}$, i.e., $\phi \in \mathscr{M}^{\perp}$.

Finally, we note that the map $\phi \mapsto \Phi$ is linear, which proves the required isomorphsm. ∎

Theorem 4.25. *Let $\mathscr{V}$ be an n-dimensional vector space. Let $\mathscr{M}$ be a subspace of $\mathscr{V}$ and $\mathscr{M}^\perp$ its annihilator. Then*

$$\dim \mathscr{V} = \dim \mathscr{M} + \dim \mathscr{M}^\perp.$$

Proof. Let $\{e_1,\ldots,e_n\}$ be a basis for $\mathscr{V}$, with $\{e_1,\ldots,e_m\}$ a basis for $\mathscr{M}$. Let $\{f_1,\ldots,f_n\}$ be the dual basis. We will show that $\{f_{m+1},\ldots,f_n\}$ is a basis for $\mathscr{M}^\perp$. The linear independence of the vectors $\{f_{m+1},\ldots,f_n\}$ is clear, since they are a subset of a basis. Moreover, they are clearly in $\mathscr{M}^\perp$. It remains to show that they actually span $\mathscr{M}^\perp$. So let $f \in \mathscr{M}^\perp$. It can be written as $f = \sum_{j=1}^n \alpha_j f_j$. But since $f \in \mathscr{M}^\perp$, we have $f(e_1) = \cdots = f(e_m) = 0$. Since $f(e_i) = \sum_{j=1}^n \alpha_j f_j(e_i) = \sum_{j=1}^n \alpha_j \delta_{ji} = \alpha_i$, we conclude that $\alpha_1 = \cdots = \alpha_m = 0$ and therefore $f = \sum_{j=m+1}^n \alpha_j f_j$. The proof is complete, since $\dim \mathscr{M} = m$ and $\dim \mathscr{M}^\perp = n - m$. ∎

Let $\mathscr{U}$ be a vector space over $\mathbb{F}$. We define a **hyperspace** $\mathscr{M}$ to be a maximal nontrivial subspace of $\mathscr{U}$.

Proposition 4.26. *Let $\dim \mathscr{U} = n$. The following statements are equivalent.*

1. *$\mathscr{M}$ is a hyperspace.*
2. *$\dim \mathscr{M} = n - 1$.*
3. *$\mathscr{M}$ is the kernel of a nonzero functional ϕ.*

Proof. $(1) \Rightarrow (2)$. Assume $\mathscr{M}$ is a hyperspace. Fix a nonzero vector $x \notin \mathscr{M}$. Then $\text{span}(x, \mathscr{M}) = \{\alpha x + m \mid \alpha \in \mathbb{F}, m \in \mathscr{M}\}$ is a subspace that properly contains $\mathscr{M}$. Thus necessarily $\text{span}(x, \mathscr{M}) = \mathscr{U}$. Let $\{e_1,\ldots,e_k\}$ be a basis for $\mathscr{M}$. Then $\{e_1,\ldots,e_k,x\}$ are linearly independent and span $\mathscr{U}$, so they are a basis for $\mathscr{U}$. Necessarily $k = \dim \mathscr{M} = n - 1$.

$(2) \Rightarrow (3)$. Let $\mathscr{M}$ be an $(n-1)$ dimensional subspace of $\mathscr{U}$. Let $x \notin \mathscr{M}$. Define a linear functional ϕ by setting

$$\phi(x) = 1,$$
$$\phi|_{\mathscr{M}} = 0.$$

For each $u \in \mathscr{U}$ there exist unique $\gamma \in \mathbb{F}$ and $m \in \mathscr{M}$ such that $u = \gamma f + m$. It follows that

$$\phi(u) = \phi(\gamma f + m) = \gamma \phi(f) + \phi(m) = \gamma.$$

Therefore, $u \in \text{Ker}\,\phi$ if and only if $u \in \mathscr{M}$.

$(3) \Rightarrow (1)$. Assume $\mathscr{M} = \text{Ker}\,\phi$ with ϕ a nontrivial functional. Fix $f \notin \mathscr{M}$, which exists by the nontriviality of ϕ. We show now that every $u \in \mathscr{U}$ has a unique representation of the form $u = \gamma f + m$ with $m \in \mathscr{M}$. In fact, it suffices to take $\gamma = \frac{\phi(u)}{\phi(f)}$. Thus $\mathscr{M}$ is a hyperspace. ∎

Recalling Definition 2.25, where the codimension of a space was introduced, we can extend Proposition 4.26 in the following way.

Proposition 4.27. *Let $\mathscr{U}$ be an n-dimensional vector space over $\mathbb{F}$, and let $\mathscr{M}$ be a subspace. Then the following statements are equivalent:*

1. $\mathscr{M}$ has codimension k.
2. We have $\dim \mathscr{M} = n - k$.
3. $\mathscr{M}$ is the intersection of the kernels of k linearly independent functionals.

Proof. (1) $\Rightarrow$ (2). Let $f_1, \ldots, f_k$ be linearly independent vectors over $\mathscr{M}$ and let $L(\mathscr{M}, f_1, \ldots, f_k) = \mathscr{U}$. Let $e_1, \ldots, e_p$ be a basis for $\mathscr{M}$. We claim that $\{e_1, \ldots, e_p, f_1, \ldots, f_k\}$ is a basis for $\mathscr{U}$. The spanning property is obvious. To see linear independence, assume there exist $\alpha_i, \beta_i \in \mathbb{F}$ such that $\sum_{i=1}^{p} \alpha_i e_i + \sum_{i=1}^{k} \beta_i f_i = 0$. This implies $\sum_{i=1}^{k} \beta_i f_i = -\sum_{i=1}^{p} \alpha_i e_i \in \mathscr{M}$. This implies in turn that all β_i are zero. From this we conclude that all α_i are zero. Thus we must have that $p + k = n$, or $\dim \mathscr{M} = n - k$.

(2) $\Rightarrow$ (3). Assume $\dim \mathscr{M} = n - k$. Let $\{e_1, \ldots, e_{n-k}\}$ be a basis for $\mathscr{M}$. We extend it to a basis $\{e_1, \ldots, e_n\}$. Define, using linear extensions, k linear functionals $\phi_1, \ldots, \phi_k$ such that

$$\phi_i(e_j) = \delta_{ij}, \qquad j = n - k + 1, \ldots, n.$$

Obviously, $\phi_1, \ldots, \phi_k$ are linearly independent and $\cap_i \mathrm{Ker}\, \phi_i = \mathscr{M}$.

(3) $\Rightarrow$ (1). Assume $\phi_1, \ldots, \phi_k$ are linearly independent and $\cap_i \mathrm{Ker}\, \phi_i = \mathscr{M}$. We choose a sequence of vectors x_i inductively. We choose a vector x_1 such that $\phi_1(x_1) = 1$. Suppose we have chosen already $x_1, \ldots, x_{i-1}$. Then we choose x_i such that $x_i \in \cap_{j=1}^{i-1} \mathrm{Ker}\, \phi_j$ and $\phi_i(x_i) = 1$. To see that this is possible, we note that if this is not the case, then

$$\mathrm{Ker}\, \phi_i \supset \cap_{j=1}^{i-1} \mathrm{Ker}\, \phi_j = \mathrm{Ker} \begin{pmatrix} \phi_1 \\ \cdot \\ \cdot \\ \cdot \\ \phi_{i-1} \end{pmatrix}.$$

Using Theorem 4.11, it follows that there exist α_j such that $\phi_i = \sum_{j=1}^{i-1} \alpha_j \phi_j$, contrary to the assumption of linear independence. Clearly, the vectors $x_1, \ldots, x_k$ are linearly independent. Moreover, it is easy to check that $x - \sum_{j=1}^{k} \alpha_j x_j \in \cap_{j=1}^{k} \mathrm{Ker}\, \phi_j$. This implies $\dim \mathscr{M} = n - k$. $\blacksquare$

Let $\mathscr{V}$ be a vector space over $\mathbb{F}$ and let $\mathscr{V}^*$ be its dual space. Thus, given $x \in \mathscr{V}$ and $f \in \mathscr{V}^*$ we have $f(x) \in \mathbb{F}$. But, fixing a vector $x \in \mathbb{F}$, we can view the expression $f(x)$ as a $\mathbb{F}$-valued function defined on the dual space $\mathscr{V}^*$. We denote this function by $\hat{x}$, and it is defined by

$$\hat{x}(f) = f(x). \tag{4.29}$$

Theorem 4.28. *The function $\hat{x}: \mathscr{V}^* \longrightarrow \mathbb{F}$, defined by (4.29), is a linear map.*

Proof. Let $f, g \in \mathscr{V}^*$ and $\alpha, \beta \in \mathbb{F}$. Then

$$\hat{x}(\alpha f + \beta g) = (\alpha f + \beta g)(x) = (\alpha f)(x) + (\beta g)(x)$$
$$= \alpha f(x) + \beta g(x) = \alpha \hat{x}(f) + \beta \hat{x}(g). \qquad \blacksquare$$

Together with the function $\hat{x} \in L(\mathscr{V}^*, \mathbb{F}) = \mathscr{V}^{**}$, we have the function $\phi : \mathscr{V} \longrightarrow \mathscr{V}^{**}$ defined by

$$\phi(x) = \hat{x}. \tag{4.30}$$

The function ϕ so defined is referred to as the **canonical embedding** of $\mathscr{V}$ in $\mathscr{V}^{**}$.

Theorem 4.29. *Let $\mathscr{V}$ be a finite-dimensional vector space over $\mathbb{F}$. The function $\phi : \mathscr{V} \longrightarrow \mathscr{V}^{**}$ defined by (4.30) is injective and surjective, i.e., it is an invertible linear map.*

Proof. We begin by showing the linearity of ϕ. Let $x, y \in \mathscr{V}$, $\alpha, \beta \in \mathbb{F}$, and $f \in \mathscr{V}^*$. Then

$$
\begin{aligned}
(\phi(\alpha x + \beta y))f &= f(\alpha x + \beta y) = \alpha f(x) + \beta f(y) \\
&= \alpha \hat{x}(f) + \beta \hat{y}(f) = (\alpha \hat{x} + \beta \hat{y})(f) \\
&= (\alpha \phi(x) + \beta \phi(y))f.
\end{aligned}
$$

Since this is true for all $f \in \mathscr{V}^*$, we get

$$\phi(\alpha x + \beta y) = \alpha \phi(x) + \beta \phi(y),$$

i.e., the canonical map ϕ is linear.

It remains to show that ϕ is invertible. Since $\dim \mathscr{V}^* = \dim \mathscr{V}$, we get also $\dim \mathscr{V}^{**} = \dim \mathscr{V}$. So, to show that ϕ is invertible, it suffices to show injectivity. To this end, let $x \in \operatorname{Ker} \phi$. Then for each $f \in \mathscr{V}^*$ we have

$$(\phi(x))(f) = 0 = \hat{x}(f) = f(x).$$

This implies $x = 0$. For if $x \neq 0$, there exists a functional f for which $f(x) \neq 0$. The easiest way to see this is to complete x to a basis and take the dual basis. ∎

Corollary 4.30. *Let $L \in \mathscr{V}^{**}$. Then there exists a $x \in \mathscr{V}$ such that $L = \hat{x}$.*

Corollary 4.31. *Let $\mathscr{V}$ be an n-dimensional vector space and let $\mathscr{V}^*$ be its dual space. Let $\{f_1, \ldots, f_n\}$ be a basis for $\mathscr{V}^*$. Then there exists a basis $\{e_1, \ldots, e_n\}$ for $\mathscr{V}$ for which*

$$f_i(e_j) = \delta_{ij},$$

i.e., every basis in $\mathscr{V}^*$ is the dual of a basis in $\mathscr{V}$.

Proof. Let $\{E_1, \ldots, E_n\}$ be the basis in $\mathscr{V}^{**}$ that is dual to the basis $\{f_1, \ldots, f_n\}$ of $\mathscr{V}^*$. Now, there exist unique vectors $\{e_1, \ldots, e_n\}$ in $\mathscr{V}$ for which $E_i = \hat{e}_i$. So, for each $f \in \mathscr{V}^*$, we have

$$E_i(f) = \hat{e}_i(f) = f(e_i),$$

and in particular,

$$f_j(e_i) = E_i(f_j) = \delta_{ij}. \qquad ∎$$

Let now $\mathcal{M} \subset \mathcal{V}^*$ be a subspace. We define

$$^\perp\mathcal{M} = \cap_{f \in \mathcal{M}} \operatorname{Ker} f = \{x \in \mathcal{V} | f(x) = 0, \forall f \in \mathcal{M}\}.$$

Theorem 4.32. *We have*

$$\phi(^\perp\mathcal{M}) = \mathcal{M}^\perp.$$

Proof. We have $x \in \cap_{f \in \mathcal{M}} \operatorname{Ker} f$ if and only if for each $f \in \mathcal{M}$ we have $\hat{x}(f) = f(x) = 0$. However, the last condition is equivalent to $\hat{x} \in \mathcal{M}^\perp$. ∎

Corollary 4.33. *Let $\mathcal{V}$ be an n-dimensional vector space and $\mathcal{M}$ an $(n-k)$-dimensional subspace of $\mathcal{V}^*$. Let $\phi(^\perp\mathcal{M}) = \cap_{f \in \mathcal{M}} \operatorname{Ker} f$. Then $\dim \phi(^\perp\mathcal{M}) = k$.*

Proof. We have

$$\dim \mathcal{V}^* = n = \dim \mathcal{M} + \dim \mathcal{M}^\perp = n - k + \dim \mathcal{M}^\perp,$$

that is, $\dim \mathcal{M}^\perp = k$. We conclude by observing that $\dim \phi(^\perp\mathcal{M}) = \dim \mathcal{M}^\perp$. ∎

Theorem 4.34. *Let $f, g_1, \ldots, g_p \in \mathcal{V}^*$. Then $\operatorname{Ker} f \supset \cap_{i=1}^p \operatorname{Ker} g_i$ if and only if $f \in \operatorname{span}(g_1, \ldots, g_p)$.*

Proof. Define the map $g : \mathcal{V} \longrightarrow \mathbb{F}^p$ by

$$g(x) = \begin{pmatrix} g_1(x) \\ \cdot \\ \cdot \\ \cdot \\ g_p(x) \end{pmatrix}. \tag{4.31}$$

Obviously, we have the equality $\operatorname{Ker} g = \cap_{i=1}^p \operatorname{Ker} g_i$. We apply Theorem 4.11 to conclude that there exists a map $\alpha = \begin{pmatrix} \alpha_1 & \ldots & \alpha_p \end{pmatrix} : \mathbb{F}^p \longrightarrow \mathbb{F}$ such that $f = \alpha g = \sum_{i=1}^p \alpha_i g_i$. ∎

4.5 The Adjoint Transformation

Let $\mathcal{U}$ and $\mathcal{V}$ be two vector spaces over the field $\mathbb{F}$. Assume $T \in L(\mathcal{U}, \mathcal{V})$ and $f \in \mathcal{V}^*$. Let us consider the composition of maps

$$\mathcal{U} \xrightarrow{T} \mathcal{V} \xrightarrow{f} \mathbb{F}.$$

It is clear that the product, or composition, of f and T, i.e., fT, is a linear transformation from $\mathcal{U}$ to $\mathbb{F}$. This means that $fT \in \mathcal{U}^*$ and we denote this functional by T^*f. Therefore $T^* : \mathscr{V}^* \longrightarrow \mathscr{U}^*$ is defined by

$$T^*f = fT, \tag{4.32}$$

or

$$(T^*f)u = f(Tu), \qquad u \in \mathscr{U}. \tag{4.33}$$

The transformation T^* is called the **adjoint transformation** of T. In terms of the duality pairing $f(v) = \langle v, f \rangle$, we can write (4.33) as

$$\langle Tu, f \rangle = \langle u, T^*f \rangle. \tag{4.34}$$

Theorem 4.35. *The transformation T^* is linear, i.e., $T^* \in L(\mathscr{V}^*, \mathscr{U}^*)$.*

Proof. Let $f, g \in \mathscr{V}^*$ and $\alpha, \beta \in \mathbb{F}$. Then, for every $x \in \mathscr{U}$,

$$
\begin{aligned}
(T^*(\alpha f + \beta g))x &= (\alpha f + \beta g)(Tx) \\
&= \alpha f(Tx) + \beta g(Tx) = \alpha(T^*f)x + \beta(T^*g)x \\
&= (\alpha T^*f + \beta T^*g)x.
\end{aligned}
$$

Therefore $T^*(\alpha f + \beta g) = \alpha T^*f + \beta T^*g$, which proves linearity.

Theorem 4.36. *Let $T \in L(\mathscr{U}, \mathscr{V})$. Then*

$$(\operatorname{Im}T)^\perp = \operatorname{Ker}T^*. \tag{4.35}$$

Proof. For every $x \in \mathscr{U}$ and $f \in \mathscr{V}^*$ we have $(T^*f)x = f(Tx)$, so $f \in (\operatorname{Im}T)^\perp$ if and only if for all $x \in \mathscr{U}$, we have $0 = f(Tx) = (T^*f)x$, which is equivalent to $f \in \operatorname{Ker}T^*$. ∎

Corollary 4.37. *Let $T \in L(\mathscr{U}, \mathscr{V})$. Then*

$$\operatorname{rank}(T) = \operatorname{rank}(T^*). \tag{4.36}$$

Proof. Let $\dim\mathscr{U} = n$ and $\dim\mathscr{V} = m$. Assume $\operatorname{rank}T = \dim\operatorname{Im}T = p$. Now

$$
\begin{aligned}
m = \dim\mathscr{V} &= \dim\mathscr{V}^* = \dim\operatorname{Im}T^* + \dim\operatorname{Ker}T^* \\
&= \operatorname{rank}T^* + \dim\operatorname{Ker}T^*.
\end{aligned}
$$

Since we have

$$
\begin{aligned}
\dim\operatorname{Ker}T^* &= \dim(\operatorname{Im}T)^\perp = m - \dim(\operatorname{Im}T) \\
&= m - \operatorname{rank}T = m - p,
\end{aligned}
$$

it follows that $\operatorname{rank}T^* = p$. ∎

Theorem 4.38. *Let* $T \in L(\mathcal{U}, \mathcal{V})$ *and let* $\mathcal{B}_{\mathcal{U}} = \{e_1, \ldots, e_n\}$ *be a basis in* $\mathcal{U}$ *and let* $\mathcal{B}_{\mathcal{V}} = \{f_1, \ldots, f_m\}$ *be a basis in* $\mathcal{V}$. *Let* $\mathcal{B}^*_{\mathcal{U}} = \{\phi_1, \ldots, \phi_n\}$ *and* $\mathcal{B}^*_{\mathcal{V}} = \{\psi_1, \ldots, \psi_m\}$ *be the dual bases in* $\mathcal{U}^*$ *and* $\mathcal{V}^*$ *respectively. Then*

$$[T^*]^{\mathcal{B}^*_{\mathcal{U}}}_{\mathcal{B}^*_{\mathcal{V}}} = \widetilde{[T]^{\mathcal{B}_{\mathcal{V}}}_{\mathcal{B}_{\mathcal{U}}}}.$$

Proof. We recall that

$$([T]^{\mathcal{B}_{\mathcal{V}}}_{\mathcal{B}_{\mathcal{U}}})_{ij} = \psi_i(Te_j),$$

so in order to compute, we have to find the dual basis to $\mathcal{B}^*_{\mathcal{U}}$. Now we know that $B^{**}_{\mathcal{U}} = \{\hat{e}_1, \ldots, \hat{e}_n\}$, so

$$([T^*]^{\mathcal{B}^*_{\mathcal{U}}}_{\mathcal{B}^*_{\mathcal{V}}})_{ij} = \hat{e}_i(T^*\psi_j) = (T^*\psi_j)e_i = \psi_j(Te_i) = ([T]^{\mathcal{B}_{\mathcal{V}}}_{\mathcal{B}_{\mathcal{U}}})_{ji}.$$

∎

Corollary 4.39. *Let* A *be an* $m \times n$ *matrix. Then the row and column ranks of* A *are equal.*

Proof. Follows from (4.36). ∎

We consider now a special class of linear transformations T that satisfy $T^* = T$. Assuming $T \in L(\mathcal{U}, \mathcal{V})$, it follows that $T^* \in L(\mathcal{V}^*, \mathcal{U}^*)$. For the equality $T^* = T$ to hold, it is therefore necessary that $\mathcal{V} = \mathcal{U}^*$ and $\mathcal{V}^* = \mathcal{U}$. Of course $\mathcal{V} = \mathcal{U}^*$ implies $\mathcal{V}^* = \mathcal{U}^{**}$, and therefore $\mathcal{V}^* = \mathcal{U}$ will hold only if we identify $\mathcal{U}^{**}$ with $\mathcal{U}$, which we can do using the canonical embedding.

Let now $x, y \in \mathcal{U}$. Then we have

$$(Tx)(y) = \hat{y}(Tx) = (T^*\hat{y})(x) = \hat{x}(T^*\hat{y}). \tag{4.37}$$

If we rewrite now the action of a functional x^* on a vector x by

$$x^*(x) = \langle x, x^* \rangle, \tag{4.38}$$

we can rewrite (4.37) as

$$\langle Tx, y \rangle = \langle x, T^*y \rangle. \tag{4.39}$$

We say that a linear transformation $T \in L(\mathcal{U}, \mathcal{U}^*)$ is **self-dual** if $T^* = T$, or equivalently, for all $x, y \in \mathcal{U}$ we have $\langle Tx, y \rangle = \langle x, Ty \rangle$.

Proposition 4.40. *Let* $\mathcal{U}$ *be a vector space and* $\mathcal{U}^*$ *its dual,* $\mathcal{B}$ *a basis in* $\mathcal{U}$ *and* $\mathcal{B}^*$ *its dual basis. Let* $T \in L(\mathcal{U}, \mathcal{U}^*)$ *be a self-dual linear transformation. Then* $[T]^{\mathcal{B}^*}_{\mathcal{B}}$ *is a symmetric matrix.*

Proof. We use the fact that $\mathcal{U}^{**} = \mathcal{U}$ and $\mathcal{B}^{**} = \mathcal{B}$. Then

$$[T]_{\mathscr{B}}^{\mathscr{B}^*} = [T^*]_{\mathscr{B}^{**}}^{\mathscr{B}^*} = \widetilde{[T]_{\mathscr{B}}^{\mathscr{B}^*}}.$$

■

If $\mathscr{B} = \{e_1, \ldots, e_n\}$, then every vector $x \in \mathscr{U}$ has an expansion $x = \sum_{i=1}^{n} \xi_i e_i$. This leads to $\langle Tx, x \rangle = \sum_{i=1}^{n} \sum_{j=1}^{n} T_{ij} \xi_i \xi_j$, where $(T_{ij}) = [T]_{\mathscr{B}}^{\mathscr{B}^*}$. The expression $\sum_{i=1}^{n} \sum_{j=1}^{n} T_{ij} \xi_i \xi_j$ is called a **quadratic form**. We will return to this topic in much greater detail in Chapter 8.

4.6   Polynomial Module Structure on Vector Spaces

Let $\mathscr{U}$ be a vector space over the field $\mathbb{F}$. Given a linear transformation T in $\mathscr{U}$, there is in $\mathscr{U}$ a naturally induced module structure over the ring $\mathbb{F}[z]$. This module structure, which we proceed to define, will be central to our study of linear transformations.

Given a polynomial $p(z) \in \mathbb{F}[z]$, $p(z) = \sum_{j=0}^{k} p_j z^j$, and a linear transformation T in a vector space $\mathscr{U}$ over $\mathbb{F}$, we define

$$p(T) = \sum_{j=0}^{k} p_j T^j, \qquad (4.40)$$

with $T^0 = I$. The action of a polynomial $p(z)$ on a vector $x \in \mathscr{U}$ is defined by

$$p \cdot x = p(T)x. \qquad (4.41)$$

Proposition 4.41. *Given a linear transformation T in a vector space $\mathscr{U}$ over $\mathbb{F}$, the map $p(z) \mapsto p(T)$ is an algebra homomorphism of $\mathbb{F}[z]$ into $L(\mathscr{U})$.*

Proof. Clearly $(\alpha p + \beta q)(T) = \alpha p(T) + \beta q(T)$ and $(pq)(T) = p(T)q(T)$.   ■

Given two spaces $\mathscr{U}_1, \mathscr{U}_2$ and linear transformations $T_i \in L(\mathscr{U}_i)$, a linear transformation $X : \mathscr{U}_1 \longrightarrow \mathscr{U}_2$ is said to **intertwine** T_1 and T_2 if

$$XT_1 = T_2 X. \qquad (4.42)$$

Obviously (4.42) implies, for an arbitrary polynomial $p(z)$, that $X p(T_1) = p(T_2)X$. Thus, for $x \in \mathscr{U}_1$, we have $X(p \cdot x) = p \cdot Xx$. This shows that intertwining maps are $\mathbb{F}[z]$ module homomorphisms from $\mathscr{U}_1$ to $\mathscr{U}_2$. Two operators T_i are said to be **similar**, and we write $T_1 \simeq T_2$, if there exists an invertible map intertwining them. Equivalently, $T_2 = RT_1R^{-1}$, where $R : \mathscr{U}_1 \longrightarrow \mathscr{U}_2$ is invertible. A natural strategy for studying the similarity of two transformations is to characterize intertwining maps and find conditions guaranteeing their invertibility.

Definition 4.42. Let $\mathcal{U}$ be an n-dimensional vector space over the field $\mathbb{F}$, and $T : \mathcal{U} \longrightarrow \mathcal{U}$ a linear transformation.

1. A subspace $\mathcal{M} \subset \mathcal{U}$ is called an **invariant subspace** of T if for all $x \in \mathcal{M}$ we have also $Tx \in \mathcal{M}$.
2. A subspace $\mathcal{M} \subset \mathcal{U}$ is called a **reducing subspace** of T if it is an invariant subspace of T and there exists another invariant subspace $N \subset \mathcal{U}$ such that

$$\mathcal{U} = \mathcal{M} \oplus N.$$

We note that given a linear transformation T in $\mathcal{U}$, T-invariant subspaces are just $\mathbb{F}[z]$-submodules of $\mathcal{U}$ relative to the $\mathbb{F}[z]$-module structure defined in $\mathcal{U}$ by (4.41). Similarly, reducing subspaces of T are equivalent to module direct summands of the module $\mathcal{U}$.

Proposition 4.43. *1. Let $\mathcal{M}$ be a subspace of $\mathcal{U}$ invariant under T. Let $\mathcal{B} = \{e_1, \ldots, e_n\}$ be a basis for $\mathcal{U}$ such that $\mathcal{B}_1 = \{e_1, \ldots, e_m\}$ is a basis for $\mathcal{M}$. Then, with respect to this basis, T has the block triangular matrix representation*

$$[T]_{\mathcal{B}}^{\mathcal{B}} = \begin{pmatrix} T_{11} & T_{12} \\ 0 & T_{22} \end{pmatrix}.$$

Moreover, $T_{11} = [T|_M]_{\mathcal{B}_1}^{\mathcal{B}_1}$.

2. Let $\mathcal{M}$ be a reducing subspace for T and let $\mathcal{N}$ be a complementary invariant subspace. Let $\mathcal{B} = \{e_1, \ldots, e_n\}$ be a basis for $\mathcal{U}$ such that $\mathcal{B}_1 = \{e_1, \ldots, e_m\}$ is a basis for $\mathcal{M}$ and $\mathcal{B}_2 = \{e_{m+1}, \ldots, e_n\}$ is a basis for $\mathcal{N}$. Then the matrix representation of T with respect to this basis is block diagonal. Specifically,

$$[T]_{\mathcal{B}}^{\mathcal{B}} = \begin{pmatrix} T_1 & 0 \\ 0 & T_2 \end{pmatrix},$$

where $T_1 = [T|_M]_{\mathcal{B}_1}^{\mathcal{B}_1}$ and $T_2 = [T|_N]_{\mathcal{B}_2}^{\mathcal{B}_2}$.

Corollary 4.44. *1. Let T be a linear transformation in $\mathcal{U}$. Let $\{0\} \subset \mathcal{M}_1 \subset \cdots \subset \mathcal{M}_k \subset \mathcal{U}$ be invariant subspaces. Let $\mathcal{B} = \{e_1, \ldots, e_n\}$ be a basis for $\mathcal{U}$ such that $\mathcal{B}_i = \{e_{n_1 + \cdots + n_{i-1} + 1}, \ldots, e_{n_1 + \cdots + n_i}\}$ is a basis for $\mathcal{M}_i$. Then*

$$[T]_{\mathcal{B}}^{\mathcal{B}} = \begin{pmatrix} T_{11} & T_{12} & . & . & T_{1n} \\ 0 & T_{22} & & & . \\ . & . & . & & . \\ . & & . & . & . \\ . & & . & 0 & T_{kk} \end{pmatrix}.$$

2. *Let $\mathcal{U} = \mathcal{M}_1 \oplus \cdots \oplus \mathcal{M}_k$, with $\mathcal{M}_i$ being T-invariant subspaces. Let $\mathcal{B}_i$ be a basis for $\mathcal{M}_i$ and let B be the union of the bases B_i. Then*

$$[T]_{\mathcal{B}}^{\mathcal{B}} = \begin{pmatrix} T_{11} & & & \\ & T_{22} & & \\ & & \ddots & \\ & & & T_{kk} \end{pmatrix}.$$

We leave the proof of the two previous results to the reader.

An invariant subspace of a linear transformation T in $\mathcal{U}$ induces two other transformations, namely the restriction $T|_{\mathcal{M}}$ of T to the invariant subspace $\mathcal{M}$ and the induced transformation $T|_{\mathcal{U}/\mathcal{M}}$ in the quotient space $\mathcal{U}/\mathcal{M}$, which is defined by

$$T|_{\mathcal{U}/\mathcal{M}}[x]_{\mathcal{M}} = [Tx]_{\mathcal{M}}. \tag{4.43}$$

Proposition 4.45. *Let $T \in L(\mathcal{U})$, let $\mathcal{M} \subset \mathcal{U}$ be a T-invariant subspace, and let $\pi : \mathcal{U} \longrightarrow \mathcal{U}/\mathcal{M}$ be the canonical projection, i.e., the map defined by*

$$\pi(x) = [x]_{\mathcal{M}}. \tag{4.44}$$

Then

1. *There exists a unique linear transformation $\overline{T} : \mathcal{U}/\mathcal{M} \longrightarrow \mathcal{U}/\mathcal{M}$ that makes the following diagram commutative:*

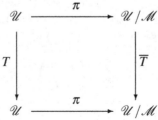

i.e., we have

$$\overline{T}\pi x = \pi T x. \tag{4.45}$$

2. *Let $\mathcal{B} = \{e_1, \ldots, e_n\}$ be a basis for $\mathcal{U}$ such that $\mathcal{B}_1 = \{e_1, \ldots, e_m\}$ is a basis for $\mathcal{M}$. Then $\overline{\mathcal{B}} = \{\pi e_{m+1}, \ldots, \pi e_n\}$ is a basis for $\mathcal{U}/\mathcal{M}$. If*

$$[T]_{\mathcal{B}}^{\mathcal{B}} = \begin{pmatrix} T_{11} & T_{12} \\ 0 & T_{22} \end{pmatrix}, \tag{4.46}$$

then

$$[\overline{T}]_{\overline{\mathcal{B}}}^{\overline{\mathcal{B}}} = T_{22}. \tag{4.47}$$

Proof. 1. In terms of equivalence classes modulo $\mathcal{M}$, we have $\overline{T}[x] = [Tx]$. To show that $\overline{T}$ is well defined, we have to show that $[x] = [y]$ implies $[Tx] = [Ty]$. This is the direct consequence of the invariance of $\mathcal{M}$ under T.

2. Follows from the invariance of $\mathcal{M}$ under T. ∎

Definition 4.46. 1. Let $\mathcal{M} \subset \mathcal{U}$ be a T-invariant subspace. The **restriction** of T to $\mathcal{M}$ is the unique linear transformation $T|_M : \mathcal{M} \longrightarrow \mathcal{M}$ defined by

$$T|_M x = Tx, \qquad x \in \mathcal{M}.$$

2. The **induced map**, namely the map induced by T on the quotient space $\mathcal{U}/\mathcal{M}$, is the unique map defined by (4.45). We will use the notation $T|_{\mathcal{U}/\mathcal{M}}$ for the induced map.

Invariance and reducibility of linear transformations are conveniently described in terms of projection operators. We turn to that.

Assume the vector space $\mathcal{U}$ admits a direct sum decomposition $\mathcal{U} = \mathcal{M} \oplus \mathcal{N}$. Thus every vector $x \in \mathcal{U}$ can be written, in a unique way, as $x = m + n$ with $m \in \mathcal{M}$ and $n \in \mathcal{N}$. We define a transformation $P_{\mathcal{N}} : \mathcal{U} \longrightarrow \mathcal{U}$ by $P_{\mathcal{N}} x = m$. We call $P_{\mathcal{N}}$ the **projection** on $\mathcal{M}$ in the direction of $\mathcal{N}$. Clearly $P_{\mathcal{N}}$ is a linear transformation and satisfies $\operatorname{Ker} P_{\mathcal{N}} = \mathcal{N}$ and $\operatorname{Im} P_{\mathcal{N}} = \mathcal{M}$.

Proposition 4.47. *A linear transformation P is a projection if and only if $P^2 = P$.*

Proof. Assume P_N is the projection on $\mathcal{M}$ in the direction of $\mathcal{N}$. If $x = m + n$, it is clear that $P_N m = m$, so $P_N^2 x = P_N m = m = P_N x$.

Conversely, assume $P^2 = P$. Let $\mathcal{M} = \operatorname{Im} P$ and $\mathcal{N} = \operatorname{Ker} P$. Since for every $x \in \mathcal{U}$, we have $x = Px + (I - P)x$, with $Px \in \operatorname{Im} P$ and $(I - P)x \in \operatorname{Ker} P$, it follows that $\mathcal{U} = \mathcal{M} + \mathcal{N}$. To show that this is a direct sum decomposition, assume $x \in \mathcal{M} \cap \mathcal{N}$. This implies $x = Px = (I - P)y$. This implies in turn that $x = Px = P(I - P)y = 0$. ∎

Proposition 4.48. *A linear transformation P in $\mathcal{U}$ is a projection if and only if $I - P$ is a projection.*

Proof. It suffices to show that $(I - P)^2 = I - P$, and this is immediate. ∎

If $\mathcal{U} = \mathcal{M} \oplus \mathcal{N}$, and P_N is the projection on $\mathcal{M}$ in the direction of $\mathcal{N}$, then $P_M = I - P_N$ is the projection on $\mathcal{N}$ in the direction of $\mathcal{M}$.

The next proposition is central. It expresses the geometric conditions of invariance and reducibility in terms of arithmetic conditions involving projection operators.

Proposition 4.49. *Let P be a projection operator in a vector space $\mathcal{X}$ and let T be a linear transformation in $\mathcal{X}$. Then*

1. A subspace $\mathcal{M} \subset \mathcal{X}$ is invariant under T if and only if for any projection P on $\mathcal{M}$ we have

$$TP = PTP. \tag{4.48}$$

2. *Let $\mathscr{X} = \mathscr{M} \oplus \mathscr{N}$ be a direct sum decomposition of $\mathscr{X}$ and let P be the projection on $\mathscr{M}$ in the direction of $\mathscr{N}$. Then the direct sum decomposition reduces T if and only if*

$$TP = PT. \tag{4.49}$$

Proof. 1. Assume (4.48) holds, and $\operatorname{Im} P = \mathscr{M}$. Then, for every $m \in \mathscr{M}$, we have

$$Tm = TPm = PTPm = PTm \in \mathscr{M}.$$

Thus $\mathscr{M}$ is invariant. The converse is immediate.

2. The direct sum reduces T if and only if both $\mathscr{M}$ and $\mathscr{N}$ are invariant subspaces. By part 1 this is equivalent to the two conditions $TP = PTP$ and $T(I - P) = (I - P)T(I - P)$. These two conditions are equivalent to (4.49). ∎

We clearly have, for a projection P, that $I = P + (I - P)$, which corresponds to the direct sum decomposition $\mathscr{X} = \operatorname{Im} P \oplus \operatorname{Im}(I - P)$. This generalizes in the following way.

Theorem 4.50. *Given the direct sum decomposition $\mathscr{X} = \mathscr{M}_1 \oplus \cdots \oplus \mathscr{M}_k$, there exist k projections P_i on $\mathscr{X}$ such that*

1. $P_i P_j = \delta_{ij} P_j$.
2. $I = P_1 + \cdots + P_k$.
3. $\operatorname{Im} P_i = \mathscr{M}_i$, $i = 1, \ldots, k$.

Conversely, given k operators P_i satisfying (1)-(3), then with $\mathscr{M}_i = \operatorname{Im} P_i$, we have $\mathscr{X} = \mathscr{M}_1 \oplus \cdots \oplus \mathscr{M}_k$.

Proof. Assume we are given the direct sum decomposition $\mathscr{X} = \mathscr{M}_1 \oplus \cdots \oplus \mathscr{M}_k$. Thus, each $x \in \mathscr{X}$ has a unique representation in the form $x = m_1 + \cdots + m_k$ with $m_i \in \mathscr{M}_i$. We define $P_i x = m_i$. The operators P_i are clearly linear, projections with $\operatorname{Im} P_i = \mathscr{M}_i$, and satisfy (1)-(3). Moreover, we have $\operatorname{Ker} P_i = \sum_{j \neq i} \mathscr{M}_j$.

Conversely, assume P_i satisfy conditions (1)-(3), and let $\mathscr{M}_i = \operatorname{Im} P_i$. From $I = P_1 + \cdots + P_k$ it follows that $x = P_1 x + \cdots + P_k x$ and hence $\mathscr{X} = \mathscr{M}_1 + \cdots + \mathscr{M}_k$. This representation of x is unique, for if $x = m_1 + \cdots + m_k$, with $m_i \in \mathscr{M}_i$ another such representation, then since $m_j \in \mathscr{M}_j$, we have $m_j = P_j y_j$ for some y_j. Therefore,

$$P_i x = P_i \sum_{j=1}^{k} m_j = \sum_{j=1}^{k} P_i m_j = \sum_{j=1}^{k} P_i P_j y_j = \sum_{j=1}^{k} \delta_{ij} P_j y_j = P_i y_i = m_i,$$

which shows that the sum is a direct sum. ∎

In search for nontrivial invariant subspaces we begin by looking for those that are 1-dimensional. If $\mathscr{M}$ is a 1-dimensional invariant subspace and x a nonzero vector in $\mathscr{M}$, then every other vector in $\mathscr{M}$ is of the form αx. Thus the invariance condition is $Tx = \alpha x$, and this leads us to the following definition.

Definition 4.51. 1. Let T be a linear transformation in a vector space $\mathscr{U}$ over the field $\mathbb{F}$. A nonzero vector $x \in \mathscr{U}$ will be called an **eigenvector**, or **characteristic vector**, of T if there exists $\alpha \in \mathbb{F}$ such that

$$Tx = \alpha x.$$

Such an α will be called an **eigenvalue**, or **characteristic value**, of T.
2. The **characteristic polynomial** of T, $d_T(z)$, is defined by

$$d_T(z) = \det(zI - T).$$

Clearly,

$$\deg d_T = \dim \mathscr{U}.$$

Proposition 4.52. *Let T be a linear transformation in a finite-dimensional vector space $\mathscr{U}$ over the field $\mathbb{F}$. An element $\alpha \in \mathbb{F}$ is an eigenvalue of T if and only if it is a zero of the characteristic polynomial of T.*

Proof. The homogeneous system $(\alpha I - T)x = 0$ has a nontrivial solution if and only if $\alpha I - T$ is singular, i.e., if and only if $d_T(\alpha) = \det(\alpha I - T) = 0$. ∎

Let T be a linear transformation in an n-dimensional vector space $\mathscr{U}$. We say that a polynomial $p(z) \in \mathbb{F}[z]$ annihilates T if $p(T) = 0$, where $p(T)$ is defined by (4.40). Clearly, every linear transformation is annihilated by the zero polynomial. A priori, it is not clear that an arbitrary linear transformation has a nontrivial annihilator. This is proved next.

Proposition 4.53. *For every linear transformation T in an n-dimensional vector space $\mathscr{U}$ there exists a nontrivial annihilating polynomial.*

Proof. $L(\mathscr{U})$, the space of all linear transformations in $\mathscr{U}$, is n^2-dimensional. Therefore the set of linear transformations $\{I, T, \ldots, T^{n^2}\}$ is linearly dependent. Hence, there exist coefficients $p_i \in \mathbb{F}$, $i = 0, \ldots, n^2$, not all zero, for which $\sum_{i=0}^{n^2} p_i T^i = 0$, or $p(T) = 0$, where $p(z) = \sum_{i=0}^{n^2} p_i z^i$. ∎

Theorem 4.54. *Let T be a linear transformation in an n-dimensional vector space $\mathscr{U}$. There exists a unique monic polynomial of minimal degree that annihilates T.*

Proof. Let

$$J = \{p(z) \in \mathbb{F}[z] \mid p(T) = 0\}.$$

Clearly, J is a nontrivial ideal in $\mathbb{F}[z]$. Since $\mathbb{F}[z]$ is a principal ideal domain, there exists a unique monic polynomial $m_T(z)$ for which $J = m_T \mathbb{F}[z]$. Obviously if $0 \neq p(z) \in J$, then $\deg p \geq \deg m$. ∎

The polynomial $m(z)$ whose existence is established in the previous theorem is called the **minimal polynomial** of T.

Both the characteristic and the minimal polynomial are similarity invariants.

Proposition 4.55. *Let T_1 and T_2 be similar linear transformations. Let $d_{T_1}(z), d_{T_2}(z)$ be their characteristic polynomials and $m_{T_1}(z), m_{T_2}(z)$ their minimal polynomials. Then $d_{T_1}(z) = d_{T_2}(z)$ and $m_{T_1}(z) = m_{T_2}(z)$.*

Proof. If $T_2 \simeq T_1$, then for some invertible R, $zI_2 - T_2 = R(zI_1 - T_1)R^{-1}$. Using the multiplication rule of determinants, it follows that $d_{T_2}(z) = \det(zI_2 - T_2) = \det(zI_1 - T_1) = d_{T_1}(z)$.

Note that $T_2 = RT_1R^{-1}$ implies that for any $p(z) \in \mathbb{F}[z]$, we have $p(T_2) = Rp(T_1)R^{-1}$. In particular, we have $m_{T_1}(T_2) = Rm_{T_1}(T_1)R^{-1} = 0$, which shows that $m_{T_2}(z) \mid m_{T_1}(z)$. By symmetry, we have $m_{T_1}(z) \mid m_{T_2}(z)$, and the equality of the minimal polynomials follows. ∎

Since the characteristic polynomial of a linear transformation T is a similarity invariant, all coefficients of $d_T(z)$ are similarity invariants too. We single out two. If $d_T(z) = z^n + t_{n-1}z^{n-1} + \cdots + t_0$, then it is easily checked that with Trace T defined by (4.26), we have $t_{n-1} = -\text{Trace}\,T = -\sum_{i=1}^{n} t_{ii}$, for any matrix representation of T. In the same way, we check that $t_0 = (-1)^n \det T$, so $\det T$ is also a similarity invariant.

4.7   Exercises

1. Let T be an $m \times n$ matrix over the field $\mathbb{F}$. Define the **determinant rank** of T to be the order of the largest nonvanishing minor of T. Show that the rank of T is equal to its determinantal rank.

2. Let $\mathcal{M}$ be a subspace of a finite-dimensional vector space $\mathcal{V}$. Show that $\mathcal{M}^* \simeq \mathcal{V}^* / \mathcal{M}^\perp$.

3. Let $\mathcal{X}$ be an n-dimensional complex vector space. Let $T : \mathcal{X} \longrightarrow \mathcal{X}$ be such that for every $x \in \mathcal{X}$, the vectors $x, Tx, \ldots, T^m x$ are linearly dependent. Show that $I, T, \ldots, T^m$ are linearly dependent.

4. Let A, B, C be linear transformations. Show that there exists a linear transformation Z such that $C = AZB$ if and only if

$$\text{Im}\,C \subset \text{Im}\,A,$$
$$\text{Ker}\,C \supset \text{Ker}\,B.$$

5. Let $\mathcal{X}, \mathcal{Y}$ be finite-dimensional vector spaces, $A \in L(\mathcal{X})$, and $W, Z \in L(\mathcal{X}, \mathcal{Y})$. Show that the following statements are equivalent:

   a. We have $A\text{Ker}\,W \subset \text{Ker}\,Z$.
   b. We have $\text{Ker}\,W \subset \text{Ker}\,ZA$.
   c. There exists a map $B \in L(\mathcal{Y})$ for which $ZA = BW$.

6. Let $\mathcal{V}$ be a finite-dimensional vector space and let $A_i \in L(\mathcal{V})$, $i = 1, \ldots, s$. Show that if $\mathcal{M} = \sum_{i=1}^{s} \text{Im}\,A_i$ $(\mathcal{M} = \cap_{i=1}^{s} \text{Ker}\,A_i)$, then there exist $B_i \in L(\mathcal{V})$ such that $\text{Im}\, \sum_{i=1}^{s} A_i B_i = \mathcal{M}$ $(\text{Ker}\, \sum_{i=1}^{s} B_i A_i = \mathcal{M})$.

7. Let $\mathscr{V}$ be a finite-dimensional vector space. Show that there is a bijective correspondence between left (right) ideals in $L(\mathscr{V})$ and subspaces of $\mathscr{V}$. The correspondence is given by $J \leftrightarrow \cap_{A \in J} \mathrm{Ker}\, A$ ($J \leftrightarrow \sum_{A \in J} \mathrm{Im}\, A$).

8. Show that $\mathrm{Ker}\, A^2 \supset \mathrm{Ker}\, A$. Show also that $\mathrm{Ker}\, A^2 = \mathrm{Ker}\, A$ implies $\mathrm{Ker}\, A^p = \mathrm{Ker}\, A$ for all $p > 0$.

9. Let T be an injective linear transformation on a (not necessarily finite-dimensional) vector space $\mathscr{V}$. Show that if for some integer k, we have $T^k = T$, then T is also surjective.

10. Let A be an $n \times n$ complex matrix with eigenvalues $\lambda_1, \ldots, \lambda_n$. Show that $\det A = \prod_{i=1}^n \lambda_i$. Show that given a polynomial $p(z)$, the eigenvalues of $p(A)$ are $p(\lambda_1), \ldots, p(\lambda_n)$ (Spectral mapping theorem).

11. Let A be an $n \times n$ complex matrix with eigenvalues $\lambda_1, \ldots, \lambda_n$. Show that the eigenvalues of $\mathrm{adj}\, A$ are $\prod_{j \neq 1} \lambda_j, \ldots, \prod_{j \neq n} \lambda_j$.

12. Let A be invertible and let $d_A(z)$ be its characteristic polynomial. Show that the characteristic polynomial of A^{-1} is $d_{A^{-1}}(z) = d(0)^{-1} z^n d_A(z^{-1})$.

13. Let the minimal polynomial of a linear transformation A be $\prod (z - \lambda_j)^{\nu_j}$. Show that the minimal polynomial of $\begin{pmatrix} A & I \\ 0 & A \end{pmatrix}$ is $\prod (z - \lambda_j)^{\nu_j + 1}$.

14. Let A be a linear transformation in a finite-dimensional vector space $\mathscr{V}$ over the field $\mathbb{F}$. Let $m_A(z)$ be its minimal polynomial. Prove that given a polynomial $p(z)$, $p(A)$ is invertible if and only if $p(z)$ and $m_A(z)$ are coprime. Show that the minimal polynomial can be replaced by the characteristic polynomial and the result still holds.

4.8 Notes and Remarks

An excellent source for linear algebra is Hoffman and Kunze (1961). The classic treatise of Gantmacher (1959), though strongly matrix oriented, is still a rich source for results and ideas and is highly recommended as a general reference. So is Malcev (1963), which is close in spirit to the present book.

We already mentioned, in Section 2.14, the contributions of Grassmann and Peano to the study of linear transformations. The conceptual rigor of this part of linear algebra owes much to Weierstrass and his students Kronecker and Frobenius.

Chapter 5
The Shift Operator

5.1 Introduction

We now turn our attention to the study of a special class of transformations, namely shift operators. These will turn out later to serve as models for all linear transformations, in the sense that every linear transformation is similar to a shift operator.

5.2 Basic Properties

We introduce now an extremely important class of linear transformation that will play a central role in the analysis of the structure of linear transformations. Recall that for a nonzero polynomial $q(z)$, we denote by $\pi_q f$ the remainder of the polynomial $f(z)$ after division by $q(z)$. Clearly, π_q is a projection operator in $\mathbb{F}[z]$. We note that given a nonzero polynomial $q(z)$, any $f(z) \in \mathbb{F}[z]$ has a unique representation of the form

$$f(z) = a(z)q(z) + r(z), \tag{5.1}$$

with $\deg r < \deg q$. The remainder, $r = \pi_q f$, can be written in another form based on the direct sum representation (2.7), namely $\mathbb{F}((z^{-1})) = \mathbb{F}[z] \oplus z^{-1}\mathbb{F}[[z^{-1}]]$. Let π_+, π_- be defined by (1.23), i.e., the projections of $\mathbb{F}((z^{-1}))$ on $\mathbb{F}[z]$ and $z^{-1}\mathbb{F}[[z^{-1}]]$ respectively, which correspond to the above direct sum decomposition. From the representation (5.1) of $f(z)$, we have $q(z)^{-1}f(z) = a(z) + q(z)^{-1}r(z)$. Applying the projection π_-, we have $\pi_- q^{-1} f = \pi_- q^{-1} r = q^{-1} r$, which implies an important alternative representation for the remainder, namely

$$\pi_q f = q\pi_- q^{-1} f. \tag{5.2}$$

P.A. Fuhrmann, *A Polynomial Approach to Linear Algebra*, Universitext, DOI 10.1007/978-1-4614-0338-8_5, © Springer Science+Business Media, LLC 2012

The importance of equation (5.2) stems from the fact that it easily extends to the case of polynomial vectors.

Proposition 5.1. *Let $q(z)$ be a monic polynomial in $\mathbb{F}[z]$ and let $\pi_q : \mathbb{F}[z] \longrightarrow \mathbb{F}[z]$ be the projection map defined in (5.2). Then we have*

$$\operatorname{Ker} \pi_q = q\mathbb{F}[z] \tag{5.3}$$

and the direct sum

$$\mathbb{F}[z] = X_q \oplus q\mathbb{F}[z]. \tag{5.4}$$

Defining the set X_q by

$$X_q = \operatorname{Im} \pi_q = \{\pi_q f | f(z) \in \mathbb{F}[z]\}, \tag{5.5}$$

we have the isomorphism

$$X_q \simeq \mathbb{F}[z]/q\mathbb{F}[z]. \tag{5.6}$$

Proof. Clearly, for each nonzero $q(z) \in \mathbb{F}[z]$, the map π_q is a projection map. Equation (5.3) follows by a simple computation. Since π_q is a projection, so is $I - \pi_q$. The identity $I = \pi_q + (I - \pi_q)$, taken together with (5.3), implies the direct sum (5.4). Finally, the isomorphism (5.6) follows from (5.5) and (5.3). ∎

The isomorphism (5.6) allows us to lift the $\mathbb{F}[z]$-module structure to X_q. This module structure is the one induced by the polynomial z.

Definition 5.2. Let $q(z)$ be a monic polynomial in $\mathbb{F}[z]$. We define a linear transformation $S_q : X_q \longrightarrow X_q$ by

$$S_q f = z \cdot f = \pi_q z f. \tag{5.7}$$

We call S_q the **shift operator** in X_q. We note that for the shift operator S_q, we have for all $k \geq 0$, that $S_q^k f = \pi_q z^k f$. This implies that for any $p(z) \in \mathbb{F}[z]$, we have

$$p \cdot f = p(S_q)f = \pi_q(pf), \qquad f(z) \in X_q. \tag{5.8}$$

We refer to X_q, with the $\mathbb{F}[z]$-module structure induced by S_q, as a **polynomial model**.

The next proposition characterizes elements of a polynomial model and studies some important bases for it.

Proposition 5.3. *Let $q(z) = z^n + q_{n-1}z^{n-1} + \cdots + q_0$. Then*

1. *A polynomial $f(z)$ belongs to X_q if and only if $q(z)^{-1}f(z)$ is strictly proper.*
2. *We have $\dim X_q = \deg q = n$.*
3. *The following sets are bases for X_q:*

   a. *The **standard basis**, namely $\mathscr{B}_{st} = \{1, z, \ldots, z^{n-1}\}$.*

b. The **control basis**, namely $\mathscr{B}_{co} = \{e_1(z), \ldots, e_n(z)\}$, where

$$e_i(z) = z^{n-i} + q_{n-1}z^{n-i-1} + \cdots + q_i.$$

c. In case $\alpha_1, \ldots, \alpha_n$, the zeros of $q(z)$, are distinct then the polynomials $p_i(z) = \Pi_{j \neq i}(z - \alpha_j)$, $i = 1, \ldots, n$ form a basis for X_q. We refer to this as the **spectral basis** $\mathscr{B}_{sp}$ of X_q.

d. Under the same assumption, the Lagrange interpolation polynomials, given by $l_i(z) = \frac{p_i(z)}{p_i(\alpha_i)}$ are a basis for X_q, naturally called the **interpolation basis**.

Proof. 1. Clearly, $f(z) \in X_q$ is equivalent to $\pi_q f = f$. In turn, this can be rewritten as $q^{-1}f = \pi_- q^{-1}f$, and this equality holds if and only if $q^{-1}f$ is strictly proper.
2. Clearly, the elements of X_q are all polynomials of degree $< n = \deg q$. Obviously, this is an n-dimensional space.
3. Each of the sets has n linearly independent elements, hence is a basis for X_q. The linear independence of the Lagrange interpolation polynomials was proved in Chapter 2, and the polynomials $p_i(z)$ are, up to a multiplicative constant, equal to the Lagrange interpolation polynomials. ∎

We proceed by studying the matrix representations of S_q with respect to these bases of X_q.

Proposition 5.4. *Let $S_q : X_q \longrightarrow X_q$ be defined by (5.7).*

1. With respect to the standard basis, S_q has the matrix representation

$$C_q^\sharp = [S_q]_{st}^{st} = \begin{pmatrix} 0 & & & -q_0 \\ 1 & & & \cdot \\ & \cdot & & \cdot \\ & & \cdot & \cdot \\ & & 1 & -q_{n-1} \end{pmatrix}. \tag{5.9}$$

2. With respect to the control basis, S_q has the diagonal matrix representation

$$C_q^\flat = [S_q]_{co}^{co} = \begin{pmatrix} 0 & 1 & & \\ & \cdot & & \cdot \\ & & \cdot & 1 \\ -q_0 & \cdots & & -q_{n-1} \end{pmatrix}. \tag{5.10}$$

3. With respect to the spectral basis, S_q has the matrix representation

$$[S_q]_{sp}^{sp} = \begin{pmatrix} \alpha_1 & & & \\ & \cdot & & \\ & & \cdot & \\ & & & \alpha_n \end{pmatrix}. \tag{5.11}$$

Proof. 1. Clearly, we have

$$S_q z^i = \begin{cases} z^{i+1}, & i = 0, \ldots, n-2, \\ -\sum_{i=0}^{n-1} q_i z^i, & i = n-1. \end{cases}$$

2. We compute, defining $e_0(z) = 0$,

$$\begin{aligned} S_q e_i &= \pi_q z e_i(z) = \pi_q z(z^{n-i} + q_{n-1}z^{n-i-1} + \cdots + q_i) \\ &= \pi_q(z^{n-i+1} + q_{n-1}z^{n-i} + \cdots + q_i z) \\ &= \pi_q(z^{n-i+1} + q_{n-1}z^{n-i} + \cdots + q_i z + q_{i-1}) - q_{i-1}e_n(z). \end{aligned}$$

So we get

$$S_q e_i = e_{i-1} - q_{i-1}e_n. \tag{5.12}$$

3. Noting that $q(z) = (z - \alpha_i)p_i(z)$, we compute

$$S_q p_i(z) = \pi_q(z - \alpha_i + \alpha_i)p_i = \pi_q(q + \alpha_i p_i) = \alpha_i p_i. \qquad\blacksquare$$

The matrices $C_q^{\#}, C_q^{\flat}$ are called the **companion matrices** of the polynomial $q(z)$.

We note that the change of basis transformation from the control to the standard basis has a particularly nice form. In fact, we have

$$[I]_{co}^{st} = \begin{pmatrix} q_1 & \cdot\,\cdot & q_{n-1} & 1 \\ \cdot & & \cdot & \\ \cdot & & \cdot & \\ q_{n-1} & \cdot & & \\ 1 & & & \end{pmatrix}.$$

We relate now the invariant subspaces of the shift operators S_q, or equivalently the submodules of X_q, to factorizations of the polynomial $q(z)$. This is an example of the interplay between algebra and geometry that is one of the salient characteristics of the use of functional models.

Theorem 5.5. *Given a monic polynomial $q(z)$, $M \subset X_q$ is an S_q-invariant subspace if and only if*

$$M = q_1 X_{q_2},$$

for some factorization

$$q(z) = q_1(z)q_2(z).$$

Proof. Assume $q(z) = q_1(z)q_2(z)$ and $M = q_1 X_{q_2}$. Thus $f(z) \in M$ implies $f(z) = q_1(z)f_1(z)$ with $\deg f_1 < \deg q_2$. Using Lemma 1.22, we compute

$$S_q f = \pi_q zf = \pi_{q_1 q_2} z q_1 f_1 = q_1 \pi_{q_2} z f_1 = q_1 S_{q_2} f_1 \in M.$$

Conversely, let M be an S_q-invariant subspace. Now, for each $f(z) \in X_q$ there exists a scalar α that depends on $f(z)$ for which

$$S_q f = zf - \alpha q. \tag{5.13}$$

Consider now the set $N = M + q\mathbb{F}[z]$. Obviously N is closed under addition, and using (5.13), $z\{M + q\mathbb{F}[z]\} \subset \{M + q\mathbb{F}[z]\}$. Thus N is an ideal in $\mathbb{F}[z]$, hence of the form $q_1 \mathbb{F}[z]$. Since, obviously, $q\mathbb{F}[z] \subset q_1 \mathbb{F}[z]$, it follows from Proposition 1.46 that $q_1(z)$ is a divisor of $q(z)$, that is, we have a factorization $q(z) = q_1(z)q_2(z)$. It is clear that

$$M = \pi_q \{M + q\mathbb{F}[z]\} = \pi_q q_1 \mathbb{F}[z] = \pi_{q_1 q_2} q_1 \mathbb{F}[z] = q_1 X_{q_2}. \qquad \blacksquare$$

Proper subspaces of a vector space may have many complementary subspaces and invariant subspaces of polynomial models are no exception. The following proposition exhibits a particular complementary subspace to such a proper invariant subspace.

Proposition 5.6. *Let $q(z) \in \mathbb{F}[z]$ be nonzero and let $q(z) = q_1(z)q_2(z)$ be a factorization. Then, as vector spaces, we have the following direct sum representation:*

$$X_q = X_{q_1} \oplus q_1 X_{q_2}. \tag{5.14}$$

Proof. That $X_{q_1} \subset X_q$ follows from the fact that $\deg q_1 \leq \deg q$. Next, we note that $X_{q_1} \cap q_1 X_{q_2} = \{0\}$ and every nonzero polynomial in X_{q_1} has degree $< \deg q_1$, whereas a nonzero polynomial in $q_1 X_{q_2}$ has degree $\geq \deg q_1$. Finally, the equality in (5.14) follows from

$$\dim X_q = \deg q = \deg q_1 + \deg q_2 = \dim X_{q_1} + \dim X_{q_2} = \dim X_{q_1} \oplus q_1 X_{q_2}.$$

$$\blacksquare$$

Note that $X_{q_1} \subset X_q$ is generally not an invariant subspace for S_q.

The following proposition sums up the basic arithmetic properties of invariant subspaces of the shift operator. This can be viewed as the counterpart of Proposition 1.46.

Proposition 5.7. *Given a monic polynomial $q(z) \in \mathbb{F}[z]$, the following hold:*

1. Let $q(z) = q_1(z)q_2(z) = p_1(z)p_2(z)$ be two factorizations. Then we have the inclusion

$$q_1 X_{q_2} \subset p_1 X_{p_2} \tag{5.15}$$

if and only if $p_1(z) \mid q_1(z)$, or equivalently $q_2(z) \mid p_2(z)$.

2. *Given factorizations $q(z) = p_i(z)q_i(z)$, $i = 1,\ldots,s$, then $\cap_{i=1}^{s} p_i X_{q_i} = p X_q$ with $p(z)$ the l.c.m of the $p_i(z)$ and $q(z)$ the g.c.d. of the $q_i(z)$.*
3. *Given factorizations $q(z) = p_i(z)q_i(z)$, $i = 1,\ldots,s$, then $\sum_{i=1}^{s} p_i X_{q_i} = p X_q$ with $q(z)$ the l.c.m of the $q_i(z)$ and $p(z)$ the g.c.d. of the $p_i(z)$.*

Proof. 1. Assume $p_1(z)|q_1(z)$, i.e., $q_1(z) = p_1(z)r(z)$ for some polynomial $r(z)$. Then $q(z) = q_1(z)q_2(z) = (p_1(z)r(z))q_2(z) = p_1(z)(r(z)q_2(z)) = p_1(z)p_2(z)$, and in particular, $p_2(z) = r(z)q_2(z)$. This implies $q_1 X_{q_2} = p_1 r X_{q_2} \subset p_1 X_{rq_2} = p_1 X_{p_2}$.
Conversely, assume the inclusion (5.15) holds. From this we have

$$q_1 X_{q_2} + q\mathbb{F}[z] = q_1 X_{q_2} + q_1 q_2 \mathbb{F}[z] = q_1[X_{q_2} + q_2 \mathbb{F}[z]] = q_1 \mathbb{F}[z].$$

So (5.15) implies the inclusion $q_1\mathbb{F}[z] \subset p_1\mathbb{F}[z]$. By Proposition 1.46 it follows that $p_1(z)|q_1(z)$.
2. Let $t(z) = p_i(z)q_i(z)$, $i = 1,\ldots,s$. Since $\cap_{i=1}^{s} p_i X_{q_i}$ is a submodule of X_q, it is of the form $p X_q$ with $t(z) = p(z)q(z)$. Now the inclusion $p X_q \subset p_i X_{q_i}$ implies $p_i(z)|p(z)$ and $q(z)|q_i(z)$, so $p(z)$ is a common multiple of the $p_i(z)$, and $q(z)$ a common divisor of the $q_i(z)$. Let now $q'(z)$ be any common divisor of the $q_i(z)$. Since necessarily $q'(z) \mid t(z)$, we can write $t(z) = p'(z)q'(z)$. Now applying Proposition 1.46, we have $p' X_{q'} \subset p_i X_{q_i}$ and hence $p' X_{q'} \subset \cap_{i=1}^{s} p_i X_{q_i} = p X_q$. This implies $q'(z) \mid q(z)$, and hence $q(z)$ is a g.c.d. of the $q_i(z)$. By the same token, we conclude that $p(z)$ is the l.c.m. of the $p_i(z)$.
3. Since $p_1 X_{q_1} + \cdots + p_s X_{q_s}$ is an invariant subspace of X_t it is of the form $p X_q$ with $t(z) = p(z)q(z)$. Now the inclusions $p_i X_{q_i} \subset p X_q$ imply the division relations $q_i(z) \mid q(z)$ and $p(z) \mid p_i(z)$. So $q(z)$ is a common multiple of the $q_i(z)$, and $p(z)$ a common divisor of the $p_i(z)$. Let $p'(z)$ be any other common divisor of the $p_i(z)$. Then $p_i(z) = p'(z)e_i(z)$ for some polynomials $e_i(z)$. Now $t(z) = p_i(z)q_i(z) = p'(z)e_i(z)q_i(z) = p'(z)q'(z)$, so $e_i(z)q_i(z) = q'(z)$, and $q'(z)$ is a common multiple of the $q_i(z)$. Now $p_i(z) = p'(z)e_i(z)$ implies $p_i X_{q_i} = p' e_i X_{q_i} \subset p' X_{e_i q_i} = p' X_{q'}$ and hence $p X_q = p_1 X_{q_1} + \cdots + p_s X_{q_s} = p' X_{q'}$. This shows that $q(z) \mid q'(z)$, and so $q(z)$ is the l.c.m. of the $q_i(z)$.   ∎

Corollary 5.8. *Given the factorizations $q(z) = p_i(z)q_i(z)$, $i = 1,\ldots,s$, then*

1.

$$X_q = p_1 X_{q_1} + \cdots + p_s X_{q_s}$$

if and only if the $p_i(z)$ are coprime.
2. *The sum $p_1 X_{q_1} + \cdots + p_s X_{q_s}$ is a direct sum if and only if $q_1(z),\ldots,q_s(z)$ are mutually coprime.*
3. *We have the direct sum decomposition*

$$X_q = p_1 X_{q_1} \oplus \cdots \oplus p_s X_{q_s}$$

if and only if the $p_i(z)$ are coprime and the $q_i(z)$ are mutually coprime.

4. We have the direct sum decomposition

$$X_q = p_1 X_{q_1} \oplus \cdots \oplus p_s X_{q_s}$$

if and only if the $q_i(z)$ are mutually coprime and $q(z) = q_1(z) \cdots q_s(z)$. In this case $p_i(z) = \Pi_{j \neq i} q_j(z)$.

Proof. 1. Let the invariant subspace $p_1 X_{q_1} + \cdots + p_s X_{q_s}$ have the representation $p_\nu X_{q_\nu}$ with $p_\nu(z)$ the g.c.d. of the $p_i(z)$ and $q_\nu(z)$ the l.c.m. of the $q_i(z)$. Therefore $p_\nu X_{q_\nu} = X_q$ if and only if $p_\nu(z) = 1$ or equivalently $q_\nu(z) = q(z)$.
2. The sum $p_1 X_{q_1} + \cdots + p_s X_{q_s}$ is a direct sum if and only if for each index i, we have

$$p_i X_{q_i} \cap \sum_{j \neq i} p_j X_{q_j} = \{0\}.$$

Now $\sum_{j \neq i} p_j X_{q_j}$ is an invariant subspace and hence of the form $\pi_i X_{\sigma_i}$ for some factorization $q(z) = \pi_i(z) \sigma_i(z)$. Here $\pi_i(z)$ is the g.c.d. of the $p_j(z), j \neq i$ and $\sigma_i(z)$ is the l.c.m. of the $q_j(z)$, $j \neq i$. Now $p_i X_{q_i} \cap \pi_i X_{\sigma_i} = \{0\}$ if and only if $\sigma_i(z)$ and $q_i(z)$ are coprime. This, however, is equivalent to $q_i(z)$ being coprime with each of the $q_j(z)$, $j \neq i$, i.e., to the mutual coprimeness of the $q_i(z)$, $i = 1, \ldots, s$.
3. Follows from the previous two parts.
4. Clearly, the $p_i(z)$ are coprime. ∎

Corollary 5.9. *Let $p(z) = p_1(z)^{v_1} \cdots p_k(z)^{v_k}$ be the primary decomposition of the polynomial $p(z)$. Define $\pi_i(z) = \Pi_{j \neq i} p_j(z)^{v_j}$, for $i = 1, \ldots, k$. Then*

$$X_p = \pi_1 X_{p_1^{v_1}} \oplus \cdots \oplus \pi_k X_{p_k^{v_k}}. \qquad (5.16)$$

Proof. Clearly, the g.c.d. of the $\pi_i(z)$ is 1, whereas the l.c.m. of the $p_i^{v_i}(z)$ is $p(z)$. ∎

The structure of the shift operator restricted to an invariant subspace can be easily deduced from the corresponding factorization.

Proposition 5.10. *Let $q(z) = q_1(z) q_2(z)$. Then we have the similarity*

$$S_q | q_1 X_{q_2} \simeq S_{q_2}. \qquad (5.17)$$

Proof. Let $\phi : X_{q_2} \longrightarrow q_1 X_{q_2}$ be the map defined by

$$\phi(f) = q_1 f,$$

which is clearly an isomorphism of the two spaces. Next we compute, for $f(z) \in X_{q_2}$,

$$\phi S_{q_2} f = q_1 S_{q_2} f = q \pi_q q_1 z f = \pi_q z q_1 f = S_q \phi f.$$

Therefore, the following diagram is commutative:

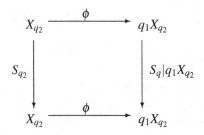

This is equivalent to (5.17).   ∎

Since eigenvectors span 1-dimensional invariant subspaces, we expect a characterization of eigenvectors of the shift S_q in terms of the polynomial $q(z)$.

Proposition 5.11. *Let $q(z)$ be a nonzero polynomial.*

1. *The eigenvalues of S_q coincide with the zeros of $q(z)$.*
2. *$f(z) \in X_q$ is an eigenvector of S_q corresponding to the eigenvalue α if and only*
   if it has the representation

$$f(z) = \frac{cq(z)}{z - \alpha}. \qquad (5.18)$$

Proof. Let $f(z)$ be an eigenvector of S_q corresponding to the eigenvalue α, i.e.,
$S_q f = \alpha f$. By (5.13), there exists a scalar c for which $(S_q f)(z) = zf(z) - cq(z)$.
Thus $zf(z) - cq(z) = \alpha f(z)$, which implies (5.18). Since $f(z)$ is a polynomial, we
must have $q(\alpha) = 0$.

Conversely, if $q(\alpha) = 0$, $q(z)$ is divisible by $z - \alpha$ and hence $f(z)$, defined by
(5.18), is in X_q. We compute

$$(S_q - \alpha I)f = \pi_q(z - \alpha)\frac{cq(z)}{z - \alpha} = \pi_q cq(z) = 0,$$

which shows that α is an eigenvalue of S_q, and a corresponding eigenvector is given
by (5.18).   ∎

The previous proposition suggests that the characteristic polynomial of S_q is $q(z)$
itself. Indeed, this is true and is proved next.

Proposition 5.12. *Let $q(z) = z^n + q_{n-1}z^{n-1} + \cdots + q_0$ be a monic polynomial of
degree n, and let S_q be the shift operator defined by (5.7). Then the characteristic
polynomial of S_q is $q(z)$.*

Proof. It suffices to compute $\det(zI - C)$ for an arbitrary matrix representation of
S_q. We find it convenient to do the computation in the standard basis. In this case
the matrix representation is given by the companion matrix $C_q^{\#}$ of (5.9). We prove
the result by induction on n. For $n = 1$, we have $C = (-q_0)$ and $\det(zI - C) = z + q_0$.

Assume the statement holds up to $n-1$. We proceed to compute, using the induction hypothesis and expanding the determinant by the first row,

$$
\det(zI - C_q^\sharp) = \begin{vmatrix} z & & & q_0 \\ -1 & & & \cdot \\ & \cdot & & \cdot \\ & & \cdot & \\ & & -1 & z+q_{n-1} \end{vmatrix}
$$

$$
= z \begin{vmatrix} z & & & q_1 \\ -1 & & & \cdot \\ & \cdot & & \cdot \\ & & \cdot & \\ & & -1 & z+q_{n-1} \end{vmatrix} + (-1)^{n+1} q_0 \begin{vmatrix} -1 & z & & \\ & \cdot & \cdot & \\ & & \cdot & z \\ & & & -1 \end{vmatrix}
$$

$$
= z(z^{n-1} + q_{n-1}z^{n-2} + \cdots + q_1) + (-1)^{n+1} q_0 (-1)^{n-1}
$$

$$
= q(z). \qquad\qquad\qquad \blacksquare
$$

We introduce now the class of cyclic transformations. Their importance lies in that they turn out to be the building blocks of general linear transformations.

Definition 5.13. Let $\mathscr{U}$ be an n-dimensional vector space over the field $\mathbb{F}$. A map $A : \mathscr{U} \longrightarrow \mathscr{U}$ is called a **cyclic transformation** if there exists a vector $b \in \mathscr{U}$ for which $\{b, Ab, \dots, A^{n-1}b\}$ is a basis for $\mathscr{U}$. Such a vector b will be called a **cyclic vector** for A.

Lemma 5.14. *1. Given a nonzero polynomial $q(z)$ and $f(z) \in X_q$, the smallest S_q-invariant invariant subspace of X_q containing $f(z)$ is $q_1 X_{q_2}$, where $q_1(z) = q(z) \wedge f(z)$ and $q(z) = q_1(z)q_2(z)$.*
2. S_q is a cyclic transformation in X_q.
3. A polynomial $f(z) \in X_q$ is a cyclic vector of S_q if and only if $f(z)$ and $q(z)$ are coprime.

Proof. 1. Let M be the subspace of X_q spanned by the vectors $\{S_q^i f \mid i \geq 0\}$. This is the smallest S_q invariant subspace containing $f(z)$. Therefore it has the representation $M = q_1 X_{q_2}$ for a factorization $q(z) = q_1(z)q_2(z)$. Since $f(z) \in M$, there exists a polynomial $f_1(z) \in X_{q_2}$ for which $f(z) = q_1(z)f_1(z)$. This shows that $q_1(z)$ is a common divisor of $q(z)$ and $f(z)$.

To show that it is the greatest common divisor, let us assume that $q'(z)$ is an arbitrary common divisor of $q(z)$ and $f(z)$. Thus we have $q(z) = q'(z)q''(z)$ and $f(z) = q'(z)f'(z)$. Using Lemma 1.24, we compute

$$
S_q^k f = \pi_q x^k f = \pi_q x^k q' f' = q' \pi_{q''} x^k f' = q' S_{q''}^k f'.
$$

Thus we have $M \subset q'X_{q''}$, or equivalently $q_1X_{q_2} \subset q'X_{q''}$. This implies $q'(z)|q_1(z)$; hence $q_1(z)$ is the g.c.d. of $q(z)$ and $f(z)$.

2. Obviously $1 \in X_q$ and $1 \wedge q(z) = 1$. So 1 is a cyclic vector for S_q.

3. Obviously, since $\dim q_1X_{q_2} = \dim X_{q_2} = \deg q_2$, $X = q_1X_{q_2}$ if and only if $\deg q_1 = 0$, that is, $f(z)$ and $q(z)$ are coprime.   ∎

The availability of eigenvectors of the shift allows us to study under what conditions the shift S_q is diagonalizable, i.e., has a diagonal matrix representation.

Proposition 5.15. *Let $q(z)$ be a monic polynomial of degree n. Then S_q is diagonalizable if and only if $q(z)$ splits into the product of n distinct linear factors, or equivalently, it has n distinct zeros.*

Proof. Assume $\alpha_1, \ldots, \alpha_n$ are the distinct zeros of $q(z)$, i.e., $q(z) = \Pi_{i=1}^n(z - \alpha_i)$. Let $p_i(z) = \frac{q(z)}{z-\alpha_i} = \Pi_{j \neq i}(z - \alpha_j)$. Then $\mathscr{B}_{sp} = \{p_1, \ldots, p_n\}$ is the spectral basis for X_q, differing from the Lagrange interpolation basis by constant factors only. It is easily checked that $(S_q - \alpha_i)p_i = 0$. So

$$[S_q]_{sp}^{sp} = \begin{pmatrix} \alpha_1 & & & \\ & \cdot & & \\ & & \cdot & \\ & & & \alpha_n \end{pmatrix}, \tag{5.19}$$

and S_q is diagonalizable.

Conversely, assume S_q is diagonalizable. Then with respect to some basis it has the representation (5.19). Since S_q is cyclic, its minimal and characteristic polynomials coincide. Necessarily all the α_i are distinct.   ∎

Proposition 5.16. *Let $q(z)$ be a monic polynomial and S_q the shift operator in X_q defined by (5.7). Then*

$$p(S_q)f = \pi_q(pf), \qquad f(z) \in X_q. \tag{5.20}$$

Proof. Using linearity, it suffices to show that

$$S_q^k f = \pi_q z^k f, \qquad f(z) \in X_q.$$

We prove this by induction. For $k = 1$ this is the definition. Assume we proved it up to an integer k. Now, using the fact that $z \operatorname{Ker} \pi_q \subset \operatorname{Ker} \pi_q$, we compute

$$S_q^{k+1} f = S_q S_q^k f = \pi_q z \pi_q z^k f = \pi_q z^{k+1} f.$$ ∎

Clearly, the operators $p(S_q)$ all commute with the shift S_q. We proceed to state the simplest version of the commutant lifting theorem. It characterizes operators commuting with the shift S_q in X_q via operators commuting with the shift S_+ in $\mathbb{F}[z]$. The last class of operators consists of multiplication operators by polynomials.

Theorem 5.17. *1. Let $q(z)$ be a monic polynomial and S_q the shift operator in X_q defined by (5.7). Let Z be any operator in X_q that commutes with S_q. Then there exists an operator $\overline{Z}$ that commutes with S_+ and such that*

$$Z = \pi_q \overline{Z}|_{X_q}. \tag{5.21}$$

Equivalently, the following diagram is commutative:

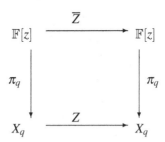

2. An operator Z in X_q commutes with S_q if and only if there exists a polynomial $p(z) \in \mathbb{F}[z]$ for which

$$Zf = \pi_q pf = p(S_q)f, \qquad f(z) \in X_q. \tag{5.22}$$

Proof. 1. By Proposition 6.2, there exists a polynomial $p(z)$ for which $Z = p(S_q)$. We define $\overline{Z} : \mathbb{F}[z] \longrightarrow \mathbb{F}[z]$ by $\overline{Z}f = pf$. It is easily checked that (5.21) holds.
2. If Z is represented by (5.22), then

$$S_q Zf = \pi_q z \pi_q pf = \pi_q zpf = \pi_q pzf = \pi_q p \pi_q zf = ZS_q f.$$

Conversely, if Z commutes with the shift, it has a representation (5.21). Now, any map $\overline{Z}$ commuting with S_+ is a multiplication by a polynomial $p(z)$, which proves (5.22). ∎

The proof of the previous theorem is slightly misleading, inasmuch as it uses the fact that S_q is cyclic. In Chapter 8, we will return to this subject using tensor products and the Bezout map.

Since any operator of the form $p(S_q)$ is completely determined by the two polynomials $p(z)$ and $q(z)$, all its properties should be derivable from these data. The next proposition focuses on the invertibility properties of $p(S_q)$.

Theorem 5.18. *Given the polynomials $p(z), q(z)$, with $q(z)$ monic, let $r(z)$ and $s(z)$ be the g.c.d and the l.c.m. of $p(z)$ and $q(z)$ respectively. Then*

1. We have the factorizations

$$q(z) = r(z)q_1(z),$$
$$p(z) = r(z)p_1(z),$$

(5.23)

with $p_1(z), q_1(z)$ coprime, as well as the factorizations

$$s(z) = r(z)p_1(z)q_1(z) = p(z)q_1(z) = q(z)p_1(z).$$

(5.24)

Moreover, we have

$$\text{Ker}\, p(S_q) = q_1 X_r,$$
$$\text{Im}\, p(S_q) = r X_{q_1}.$$

(5.25)

2. Let $p(z), q(z) \in \mathbb{F}[z]$, with $q(z)$ nonzero. Then the linear transformation $p(S_q)$ is invertible if and only if $p(z)$ and $q(z)$ are coprime. Moreover, we have

$$p(S_q)^{-1} = a(S_q),$$

(5.26)

where the polynomial $a(z)$ arises out of any solution of the Bezout equation

$$a(z)p(z) + b(z)q(z) = 1.$$

(5.27)

Proof. 1. With $r(z)$ and $s(z)$ defined as above, (5.24) is an immediate consequence of (5.23). Applying Theorem 5.5, it follows that $q_1 X_r$ and $r X_{q_1}$ are invariant subspaces of X_q. If $f(z) \in q_1 X_r$, then $f = q_1 g$, with $g \in X_r$. We compute, using (5.23),

$$p(S_q)f = \pi_q p f = q_1 r \pi_- r^{-1} q_1^{-1} r p_1 q_1 g = q_1 r \pi_- g = 0,$$

which shows that $q_1 X_r \subset \text{Ker}\, p(S_q)$.

Conversely, assume $f(z) \in \text{Ker}\, p(S_q)$. Then $\pi_q p f = 0$ or there exists $g(z)$ such that $p(z)f(z) = q(z)g(z)$, which implies $p_1(z)f(z) = q_1(z)g(z)$. Since $p_1(z)$ and $q_1(z)$ are coprime, we have $f(z) = q_1(z)f_1(z)$ for some polynomial $f_1(z)$. As $f(z) \in X_q$, we must have $f_1(z) \in X_r$. So $\text{Ker}\, p(S_q) \subset q_1 X_r$, and the first equality in (5.25) follows.

Next, assume $g(z) \in \text{Im}\, p(S_q)$, i.e., there exists an $f(z) \in X_q$ such that $g = \pi_q p f$. We compute, using (5.23) and (5.24),

$$g = \pi_q p f = q_1 r \pi_- r^{-1} q_1^{-1} r p_1 f = r \pi_{q_1} p_1 f \in r X_{q_1},$$

i.e., we have the inclusion $\text{Im}\, p(S_q) \subset r X_{q_1}$.

Conversely, assume $g(z) \in r X_{q_1}$, i.e., $g(z) = r(z)g_1(z)$ with $g_1(z) \in X_{q_1}$. By the coprimeness of $p_1(z)$ and $q_1(z)$, the map $f_1 \mapsto \pi_{q_1} p_1 f_1$, acting in X_{q_1}, is an invertible map. Hence, there exists $f_1(z) \in X_{q_1}$ for which $g_1 = \pi_{q_1} p_1 f_1$. This implies

$$rg_1 = rq_1\pi_-r^{-1}q_1^{-1}p_1rf_1 = \pi_q prf_1.$$

This shows that $rX_{q_1} \subset \operatorname{Im} p(S_q)$, and the second equality in (5.25) follows.

2. From the characterization (5.25) of $\operatorname{Ker} p(S_q)$ and $\operatorname{Im} p(S_q)$, it follows that the injectivity of $p(S_q)$ is equivalent to the coprimeness of $p(z)$ and $q(z)$, and the same holds for surjectivity.

In order to actually invert $p(S_q)$, we use the coprimeness of $p(z)$ and $q(z)$. This implies the existence of polynomials $a(z), b(z)$ that solve the Bezout equation $a(z)p(z) + b(z)q(z) = 1$. Applying the functional calculus to the shift S_q, and noting that $q(S_q) = 0$, we get

$$a(S_q)p(S_q) + b(S_q)q(S_q) = a(S_q)p(S_q) = I,$$

and (5.26) follows.   ∎

5.3 Circulant Matrices

In this section we give a short account of a special class of structured matrices, called circulant matrices. We do this for its own sake, and as an illustration of the power of polynomial algebra.

Definition 5.19. An $n \times n$ matrix over a field $\mathbb{F}$ is called a **circulant matrix**, or simply **circulant** for short, if it has the form

$$C = \operatorname{circ}(c_0, \dots, c_{n-1}) = \begin{pmatrix} c_0 & c_{n-1} & \cdot & \cdot & c_1 \\ c_1 & c_0 & \cdot & \cdot & \cdot \\ \cdot & & \cdot & \cdot & \cdot \\ \cdot & & \cdot & \cdot & c_{n-1} \\ c_{n-1} & \cdot & & \cdot & c_1 & c_0 \end{pmatrix},$$

i.e., $c_{ij} = c_{(i-j \bmod n)}$. We define a polynomial $c(z)$ by $c(z) = c_0 + c_1 z + \cdots + c_{n-1}z^{n-1}$. The polynomial c is called the **representer** of $\operatorname{circ}(c_0, \dots, c_{n-1})$.

Theorem 5.20. *For circulant matrices the following properties hold:*

1. *The circulant matrix* $\operatorname{circ}(c_0, \dots, c_{n-1})$ *is the matrix representation of* $c(S_{z^n-1})$ *with respect to the standard basis of* X_{z^n-1}.
2. *The sum of circulant matrices is a circulant matrix. Specifically,*

$$\operatorname{circ}(a_1, \dots, a_n) + \operatorname{circ}(b_1, \dots, b_n) = \operatorname{circ}(a_1 + b_1, \dots, a_n + b_n).$$

3. *We have*

$$\alpha \operatorname{circ}(a_1, \dots, a_n) = \operatorname{circ}(\alpha a_1, \dots, \alpha a_n).$$

4. *Define the special circulant matrix* Π *by*

$$\Pi = \text{circ}\,(0,1,0,\ldots,0) = \begin{pmatrix} 0 & & 1 \\ 1 & \cdot\cdot & \cdot \\ & \cdot\cdot\cdot & \\ & & \cdot\cdot \\ & & 1\ 0 \end{pmatrix}.$$

Then $C \in \mathbb{F}^{n \times n}$ *is a circulant if and only if* $C\Pi = \Pi C$.
5. *The product of circulants is commutative.*
6. *The product of circulants is a circulant.*
7. *The inverse of a circulant is a circulant. Moreover, the inverse of a circulant with representer* $c(z)$ *is a circulant with representer* $a(z)$, *where* $a(z)$ *comes from a solution of the Bezout equation* $a(z)c(z) + b(z)(z^n - 1) = 1$.
8. *Over an algebraically closed field, circulants are diagonalizable.*

Proof. 1. We compute

$$[S_{z^n-1}]_{st}^{st} = \begin{pmatrix} 0 & \ldots & 1 \\ 1 & & \cdot \\ & \cdot & \cdot \\ & & \cdot \\ & 1 & 0 \end{pmatrix}.$$

This is a special case of the companion matrix in (5.9). Now

$$S_{z^n-1}^i z^j = \begin{cases} z^{i+j}, & i+j \le n-1, \\ z^{i+j-n}, & i+j \ge n. \end{cases}$$

So

$$c(S_{z^n-1})z^j = \sum_{i=0}^{n-1} c_i S_{z^n-1}^i z^j = \sum_{i=0}^{n-1-j} c_i z^{i+j} + \sum_{i=n-j}^{n-1} c_i z^{i+j-n},$$

and this implies the equality

$$[c(S_{z^n-1})]_{st}^{st} = \text{circ}\,(c_0,\ldots,c_{n-1}).$$

2. Given polynomials $a(z), b(z) \in \mathbb{F}[z]$, we have

$$(a+b)(S_{z^n-1}) = a(S_{z^n-1}) + b(S_{z^n-1}),$$

and hence

$$\begin{aligned} \text{circ}\,(a_0+b_0,\ldots,a_{n-1}+b_{n-1}) &= [(a+b)(S_{z^n-1})]_{st}^{st} \\ &= [a(S_{z^n-1})]_{st}^{st} + [b(S_{z^n-1})]_{st}^{st} \\ &= \text{circ}\,(a_0,\ldots,a_{n-1}) + \text{circ}\,(b_0,\ldots,b_{n-1}). \end{aligned}$$

3. We compute
$$\mathrm{circ}\,(\alpha a_0,\ldots,\alpha a_{n-1}) = [\alpha a(S_{z^n-1})]_{st}^{st} = \alpha[a(S_{z^n-1})]_{st}^{st} = \alpha\,\mathrm{circ}\,(a_0,\ldots,a_{n-1}).$$

4. Clearly, $\Pi = \mathrm{circ}\,(0,1,0,\ldots,0) = [S_{z^n-1}]_{st}^{st}$. Obviously, S_{z^n-1} is cyclic. Hence, a linear transformation K commutes with S_{z^n-1} if and only if $K = c(S_{z^n-1})$ for some polynomial $c(z)$. Thus, assume $C\Pi = \Pi C$. Then there exists a linear transformation K in X_{z^n-1} satisfying $[K]_{st}^{st} = C$, and K commutes with S_{z^n-1}. Therefore $K = c(S_{z^n-1})$ and $C = \mathrm{circ}\,(c_0,\ldots,c_{n-1})$.
   Conversely, if $C = \mathrm{circ}\,(c_0,\ldots,c_{n-1})$, we have

$$C\Pi = [c(S_{z^n-1})]_{st}^{st}[S_{z^n-1}]_{st}^{st} = [S_{z^n-1}]_{st}^{st}[c(S_{z^n-1})]_{st}^{st} = \Pi C.$$

5. Follows from
$$c(S_{z^n-1})d(S_{z^n-1}) = d(S_{z^n-1})c(S_{z^n-1}).$$

6. Follows from
$$c(S_{z^n-1})d(S_{z^n-1}) = (cd)(S_{z^n-1}).$$

7. Let $C = \mathrm{circ}\,(c_0,\ldots,c_{n-1}) = c(S_{z^n-1})$, where $c(z) = c_0 + c_1 z + \cdots + c_{n-1}z^{n-1}$. By Theorem 5.18, $c(S_{z^n-1})$ is invertible if and only if $c(z)$ and $z^n - 1$ are coprime. In this case there exist polynomials $a(z), b(z)$ satisfying the Bezout identity $a(z)c(z) + b(z)(z^n - 1) = 1$. We may assume without loss of generality that $\deg a < n$. From the Bezout identity we conclude that $c(S_{z^n-1})c(S_{z^n-1}) = I$ and hence
$$\mathrm{circ}\,(c_0,\ldots,c_{n-1})^{-1} = [a(S_{z^n-1})]_{st}^{st} = \mathrm{circ}\,(a_0,\ldots,a_{n-1}).$$

8. The polynomial $z^n - 1$ has a multiple zero if and only if $z^n - 1$ and nz^{n-1} have a common zero. Clearly, this cannot occur. Since all the roots of $z^n - 1$ are distinct, it follows from Proposition 5.15 that S_{z^n-1} is diagonalizable. This implies the diagonalizability of $c(S_{z^n-1})$. ∎

5.4 Rational Models

Given a field $\mathbb{F}$, we saw that the ring of polynomials $\mathbb{F}[z]$ is an entire ring. Hence, by Theorem 1.49, it is embeddable in its field of quotients. We call the field of quotients of $\mathbb{F}[z]$ the field of **rational functions** and denote it by $\mathbb{F}(z)$. Strictly speaking, the elements of $\mathbb{F}(z)$ are equivalence classes of pairs of polynomials $(p(z), q(z))$, with $q(z)$ nonzero. However, in each nonzero equivalence class, there is a unique pair with $p(z), q(z)$ coprime and $q(z)$ monic. The corresponding equivalence class will be denoted by $\frac{p(z)}{q(z)}$. Given such a pair of polynomials $p(z), q(z)$, there is a unique representation of $p(z)$ in the form $p(z) = a(z)q(z) + r(z)$, with $\deg r < \deg q$. This allows us to write

$$\frac{p(z)}{q(z)} = a(z) + \frac{r(z)}{q(z)}. \tag{5.28}$$

A rational function $r(z)/q(z)$ with $\deg r \leq \deg q$ will be called **proper**, and if $\deg r < \deg q$ is satisfied, **strictly proper**. Thus, any rational function $g(z) = \frac{p(z)}{q(z)}$ has a unique representation as a sum of a polynomial and a strictly proper rational function. We denote by $\mathbb{F}_-(z)$ the space of strictly proper rational functions and observe that it is an infinite-dimensional linear space. Equation (5.28) means that for $\mathbb{F}(z)$ we have the following direct sum decomposition:

$$\mathbb{F}(z) = \mathbb{F}[z] \oplus \mathbb{F}_-(z). \tag{5.29}$$

With this direct sum decomposition we associate two projection operators in $\mathbb{F}(z)$, π_+ and π_-, with images $\mathbb{F}[z]$ and $\mathbb{F}_-(z)$ respectively. To be precise, given the representation (5.28), we have

$$\pi_+ \left(\frac{p}{q} \right) = a$$

$$\pi_- \left(\frac{p}{q} \right) = \frac{r}{q}. \tag{5.30}$$

Proper rational functions have an expansion as formal power series in the variable z^{-1}, i.e., in the form $g(z) = \sum_{i=0}^{\infty} \frac{g_i}{z^i}$. Assume $p(z) = \sum_{k=0}^{n} p_k z^k, q(z) = \sum_{i=0}^{n} q_i z^i$, with $q_n \neq 0$. We compute

$$\sum_{k=0}^{n} p_k z^k = \sum_{i=0}^{n} q_i z^i \sum_{j=0}^{\infty} \frac{g_j}{z^j} = \sum_{k=-\infty}^{n} \left\{ \sum_{i=k}^{n} q_i g_{i-k} \right\} z^k.$$

By comparing coefficients we get the infinite system of linear equations

$$p_k = \sum_{i=k}^{n} q_i g_{i-k}, \quad -\infty < k \leq n.$$

This system has a unique solution, which can be found by solving it recursively, starting with $k = n$. An alternative way of finding this expansion is via the process of long division of $p(z)$ by $q(z)$.

This generalizes easily to the case of the field of rational functions. We can consider the field $\mathbb{F}(z)$ of rational functions as a subfield of $\mathbb{F}((z^{-1}))$, the field of truncated Laurent series. The space $\mathbb{F}_-(z)$ of strictly proper rational functions can be viewed as a subspace of $z^{-1}\mathbb{F}[[z^{-1}]]$. In the same way that we have the isomorphism $z^{-1}\mathbb{F}[[z^{-1}]] \simeq \mathbb{F}((z^{-1}))/\mathbb{F}[z]$, we have also $\mathbb{F}_-(z) \simeq \mathbb{F}(z)/\mathbb{F}[z]$. In fact, the projection $\pi_- : \mathbb{F}(z) \longrightarrow \mathbb{F}_-(z)$ is a surjective linear map with kernel equal to $\mathbb{F}[z]$. However, the spaces $\mathbb{F}((z^{-1}))$, $\mathbb{F}(z)$, and $\mathbb{F}[z]$ all carry also a natural $\mathbb{F}[z]$-module structure, with polynomials acting by multiplication. This structure induces an $\mathbb{F}[z]$-module structure in the quotient spaces. This module structure is transferred to $\mathbb{F}_-(z)$ by defining, for $p(z) \in \mathbb{F}[z]$,

$$p \cdot g = \pi_-(pg), \qquad g(z) \in \mathbb{F}_-(z). \tag{5.31}$$

In particular, we define the **backward shift operator** S_-, acting in $\mathbb{F}_-(z)$, by

$$S_- g = \pi_-(zg), \qquad g \in \mathbb{F}_-(z). \tag{5.32}$$

We shall use the same notation when $z^{-1}\mathbb{F}[[z^{-1}]]$ replaces $\mathbb{F}_-(z)$. We point out that in the behavioral literature, S_- is denoted by σ. Using this, equation (5.31) can be rewritten, for $p(z) \in \mathbb{F}[z]$, as a Toeplitz-like operator

$$p(\sigma)g = \pi_-(pg), \qquad g(z) \in z^{-1}\mathbb{F}[[z^{-1}]]. \tag{5.33}$$

In terms of the expansion of $g(z) \in \mathbb{F}_-(z)$ around infinity, i.e., in terms of the representation $g(z) = \sum_{i=1}^{\infty} \frac{g_i}{z^i}$ we have

$$S_- \sum_{i=1}^{\infty} \frac{g_i}{z^i} = \sum_{i=1}^{\infty} \frac{g_{i+1}}{z^i}.$$

This explains the use of the term backward shift for this operator.

We will show now that for the backward shift S_- in $z^{-1}\mathbb{F}[[z^{-1}]]$, any $\alpha \in \mathbb{F}$ is an eigenvalue and it has associated eigenfunctions of any order. Indeed, let $\alpha \in \mathbb{F}$. Then it is easy to check that $(z - \alpha)^{-1} = \sum_{i=1}^{\infty} \frac{\alpha^{i-1}}{z^i}$ is an eigenvector of S_- correspomding to the eigenvalue α. Analogously, $(z - \alpha)^{-k}$ is a generalized eigenvector of S_- of order k. Moreover, any other such eigenvector is necessarily of the form $c(z - \alpha)^{-k}$. Therefore, we have, for $k \geq 1$, $\dim \text{Ker}\,(\alpha I - S_-)^k = k$.

In contrast to this richness of eigenfunctions of the backward shift S_-, the forward shift S_+ does not have any eigenfunctions. This asymmetry will be further explored in the study of duality.

The previous discussion indicates that the spectral structure of the shift S_-, with finite multiplicity, might be rich enough to model all finite-dimensional linear transformations, up to similarity transformations, as the backward shift S_- restricted to a finite-dimensional backward-shift-invariant subspace. This turns out to be the case, and this result lies at the heart of the matter, explaining the effectiveness of the use of shifts in linear algebra and system theory. In fact, the whole book can be considered, to a certain extent, to be an elaboration of this idea. We will study a special case of this in Chapter 6.

It is easy to construct many finite-dimensional S_--invariant subspaces of $\mathbb{F}_-(z)$. In fact, given any nonzero polynomial $d(z)$, we let

$$X^d = \left\{ \frac{r}{d} \mid \deg r < \deg d \right\}.$$

It is easily checked that X^d is indeed an S_--invariant subspace, and its dimension equals the degree of $d(z)$.

It is natural to consider the restriction of the operator S_- to X^d. Thus we define a linear transformation $S^d : X^d \longrightarrow X^d$ by $S^d = S_-|X^d$, or equivalently, for $h \in X^d$,

$$S^d h = S_- h = \pi_- z h. \tag{5.34}$$

The modules X_d and X^d have the same dimension and are defined by the same polynomial. Just as the polynomial model X_d has been defined in terms of the projection π_d, so can the rational model X^d be characterized as the image of a projection.

Definition 5.21. Let the map $\pi^d : z^{-1}\mathbb{F}[[z^{-1}]] \longrightarrow z^{-1}\mathbb{F}[[z^{-1}]]$ be defined by

$$\pi^d h = \pi_- d^{-1} \pi_+ dh, \qquad h(z) \in \mathbb{F}[[z^{-1}]]. \tag{5.35}$$

Proposition 5.22. *Let π^d be defined by (5.35). Then*

1. π^d is a projection in $z^{-1}\mathbb{F}[[z^{-1}]]$.
2. We have
$$X^d = \operatorname{Im}\pi^d. \tag{5.36}$$

3. With $d(\sigma)$ defined by (5.33), we have

$$X^d = \operatorname{Ker} d(\sigma). \tag{5.37}$$

Proof. 1. For $h(z) \in z^{-1}\mathbb{F}[[z^{-1}]]$, we compute

$$
\begin{aligned}
(\pi^d)^2 h &= \pi^d(\pi^d h) = \pi_- d^{-1}\pi_+ d\pi_- d^{-1}\pi_+ dh \\
&= \pi_- d^{-1}\pi_+ dd^{-1}\pi_+ dh = \pi_- d^{-1}\pi_+ \pi_+ dh \\
&= \pi_- d^{-1}\pi_+ dh = \pi^d h,
\end{aligned}
$$

i.e., π^d is a projection.
2. Assume $h(z) = d^{-1}r$, with $\deg r < \deg d$. Then

$$\pi^d(d^{-1}r) = \pi_- d^{-1}\pi_+ dd^{-1}r = \pi_- d^{-1}r = d^{-1}r,$$

which shows that $X^d \subset \operatorname{Im}\pi^d$.
Conversely, assume $h(z) \in \operatorname{Im}\pi^d$. Thus there exists $g(z) \in z^{-1}\mathbb{F}[[z^{-1}]]$ for which $h(z) = \pi^d g$. Hence

$$\pi^d h = (\pi^d)^2 g = \pi^d g = h.$$

From this it follows that

$$\pi_- dh = \pi_- d\pi^d h = \pi_- d\pi_- d^{-1}\pi_+ dh = \pi_- dd^{-1}\pi_+ dh = \pi_- \pi_+ dh = 0.$$

This implies that there exists $r(z) \in \mathbb{F}[z]$ for which $d(z)h(z) = r(z)$ and $\deg r < \deg d$. In turn, we conclude that $h(z) = d(z)^{-1}r(z) \in X^d$, which implies the inclusion $\operatorname{Im} \pi^d \subset X^d$.

3. Assume $h \in \operatorname{Ker} d(\sigma)$, which implies $\pi_+ dh = dh$. We compute

$$\pi^d h = \pi_- d^{-1}\pi_+ dh = \pi_- d^{-1}dh = \pi_- h = h,$$

i.e., $\operatorname{Ker} d(\sigma) \subset X^d$.

Conversely, assume $h \in X^d$. This implies $h = \pi_- d^{-1}\pi_+ dh$. We compute

$$d(\sigma)h = \pi_- dh = \pi_- d\pi_- d^{-1}\pi_+ dh = \pi_- dd^{-1}\pi_+ dh = \pi_- \pi_+ dh = 0,$$

i.e., $X^d \subset \operatorname{Ker} d(\sigma)$. The two inclusions imply (5.37). ∎

It is natural to conjecture that the polynomial model X_d and the rational model X^d must be isomorphic, and this is indeed the case, as the next theorem shows.

Theorem 5.23. *Let $d(z)$ be a nonzero polynomial. Then the operators S_d and S^d are isomorphic. The isomorphism is given by the map $\rho_d : X^d \longrightarrow X_d$ defined by*

$$\rho_d h = dh. \tag{5.38}$$

Proof. We compute

$$\rho_d S^d h = \rho_d \pi_- zh = d\pi_- zh = d\pi_- d^{-1}dzh = \pi_d z(dh) = S_d(\rho_d h),$$

i.e.,

$$\rho_d S^d = S_d \rho_d, \tag{5.39}$$

which, by the invertibility of ρ_d, proves the isomorphism. ∎

The polynomial and rational models that have been introduced are isomorphic, yet they represent two fundamentally different points of view. In the case of polynomial models, all spaces of the form X_q, with $\deg q = n$, contain the same elements, but the associated shifts S_q act differently. On the other hand, the operators S^q in the spaces X^q act in the same way, since they are all restrictions of the backward shift S_-. However, the spaces are different. Thus the polynomial models represent an arithmetic perspective, whereas rational models represent a geometric one.

Our next result is the characterization of all finite-dimensional S_--invariant subspaces.

Proposition 5.24. *A subset M of $\mathbb{F}_-(z)$ is a finite-dimensional S_--invariant subspace if and only if we have $M = X^d$.*

Proof. Assume that for some polynomial $d(z)$, $M = X^d$. Then

$$S_- \frac{r}{d} = \pi_- z \frac{r}{d} = d^{-1} d\pi_- d^{-1} zr = d^{-1} \pi_d zr \in M.$$

Conversely, let M be a finite-dimensional S_--invariant subspace. By Theorem 4.54, there exists a nonzero polynomial $p(z)$ of minimal degree such that $\pi_- ph = 0$, for all $h \in M$. Thus, we get $M \subset X^p$. It follows that pM is a submodule of X_p, hence of the form $p_1 X_{p_2}$ for some factorization $p(z) = p_1(z) p_2(z)$. We conclude that $M = p^{-1} p_1 X_{p_2} = p_2^{-1} p_1^{-1} p_1 X_{p_2} = X^{p_2}$. The minimality of $p(z)$ implies $p(z) = p_2(z)$, up to a constant nonzero factor. ∎

The spaces of rational functions of the form X^d will be referred to as **rational models**. Note that in the theory of differential equations, these spaces appear as the Laplace transforms of the spaces of solutions of a homogeneous linear differential equation with constant coefficients.

The following sums up the basic arithmetic properties of rational models. It is the counterpart of Proposition 5.7.

Proposition 5.25. *1. Given polynomials $p(z), q(z) \in \mathbb{F}[z]$, we have the inclusion $X^p \subset X^q$ if and only if $p(z) \mid q(z)$.*
2. Given polynomials $p_i(z) \in \mathbb{F}[z]$, $i = 1, \ldots, s$, then $\cap_{i=1}^s X^{p_i} = X^p$ with $p(z)$ the g.c.d. of the $p_i(z)$.
3. Given polynomials $p_i \in \mathbb{F}[z]$, then $\sum_{i=1}^s X^{p_i} = X^q$ with $q(z)$ the l.c.m. of the $p_i(z)$.

Proof. Follows from Proposition 5.7, using the isomorphism of polynomial and rational models given by Theorem 5.23. ∎

The primary decomposition theorem and the direct sum representation (5.16) have a direct implication toward the partial fraction decomposition of rational functions.

Theorem 5.26. *Let $p(z) = \Pi_{i=1}^s p_i(z)^{v_i}$ be the primary decomposition of the nonzero polynomial $p(z)$. Then*

1. We have

$$X^p = X^{p_1^{v_1}} \oplus \cdots \oplus X^{p_s^{v_s}}. \tag{5.40}$$

2. Each rational function $g(z) \in X^p$ has a unique representation of the form

$$g(z) = \frac{r(z)}{p(z)} = \sum_{i=1}^s \sum_{j=1}^{v_i} \frac{r_{ij}}{p_i^j},$$

with $\deg r_{ij} < \deg p_i$, $j = 1, \ldots, v_i$.

Proof. 1. Given the primary decomposition of $p(z)$, we define $\pi_i(z) = \Pi_{j \neq i} p_i^{v_i}$. Clearly, by Corollary 5.9, we have

$$X_p = \pi_1 X_{p_1^{v_1}} \oplus \cdots \oplus \pi_s X_{p_s^{v_s}}. \tag{5.41}$$

We use the isomorphism of the modules X_p and X^p and the fact that $p(z) = \pi_i(z)p_i(z)^{v_i}$, which implies $p^{-1}\pi_i X_{p_i^{v_i}} = p_i^{-v_i}\pi_i - 1\pi_i X_{p_i^{v_i}} = X^{p_i^{v_i}}$, to get the direct sum decomposition (5.40).

2. For any $r_i(z) \in X_{p_i^{v_i}}$ we have $r_i = \sum_{j=0}^{v_i-1} r_{i(v_i-j)}p_i^j$. With $\frac{r}{p} = \sum_{i=1}^{s} \frac{r_i}{p_i^{v_i}}$, using (5.41) and taking $r_i \in X_{p_i^{v_i}}$, we have

$$\frac{r_i}{p_i^{v_i}} = \sum_{j=1}^{v_i} \frac{r_{ij}}{p_i^j}. \tag{5.42}$$

∎

The isomorphism (5.39) between the shifts S_q and S^q can be used to transform Theorems 5.17 and 5.18 into the context of rational models. Thus we have the following Theorem.

Theorem 5.27. *1. Let $q(z)$ be a monic polynomial and S^q the shift operator in X^q defined by (5.34). Let W be any operator in X^q that commutes with S^q. Then there exists an operator Z that commutes with S_q and such that*

$$Z = \rho_q W \rho_q^{-1}. \tag{5.43}$$

Equivalently, the following diagram is commutative:

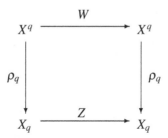

2. An operator W in X^q commutes with S^q if and only if there exists a polynomial $p(z) \in \mathbb{F}[z]$ for which

$$Wh = \pi_- ph = p(S^q)h, \qquad h(z) \in X^q. \tag{5.44}$$

Proof. 1. By the isomorphism (5.39), we have $\rho_q S^q = S_q \rho_q$. Since $WS^q = S^q W$, it follows that $W\rho_q^{-1}S_q\rho_q = \rho_q^{-1}S_q\rho_q W$, or $(\rho_q W \rho_q^{-1})S_q = S_q(\rho_q W \rho_q^{-1})$. Defining $Z = \rho_q W \rho_q^{-1}$, we have $ZS_q = S_q Z$.

2. If W is represented by (5.44), then

$$S^q Wh = \pi^q z\pi_- ph = \pi_- q^{-1}\pi_+ qz\pi_- ph = \pi_- pq^{-1}q^{-1}z\pi_- ph$$
$$= \pi_- z\pi_- ph = WS^q h.$$

Conversely, assume $WS^q = S^qW$, and define Z by (5.43). Then we have $ZS_q = S_qZ$. Applying Theorem 5.17, there exists a polynomial $p(z) \in \mathbb{F}[z]$ for which $Z = p(S_q)$. Since $W = \rho_q^{-1}Z\rho_q$, this implies

$$Wh = \rho_q^{-1}Z\rho_q h = q^{-1}q\pi_- q^{-1}pqh = \pi_- ph. \qquad\blacksquare$$

Theorem 5.28. *Given the polynomials $p(z), q(z)$, with $q(z)$ monic. Let $r(z)$ and $s(z)$ be the g.c.d. and the l.c.m. of $p(z)$ and $q(z)$ respectively. Then*

1. We have the factorizations

$$q(z) = r(z)q_1(z),$$

$$p(z) = r(z)p_1(z), \qquad (5.45)$$

with $p_1(z), q_1(z)$ coprime, as well as the factorizations

$$s(z) = r(z)p_1(z)q_1(z) = p(z)q_1(z) = q(z)p_1(z). \qquad (5.46)$$

We have

$$\operatorname{Ker} p(S^q) = X^r,$$

$$\operatorname{Im} p(S^q) = X^{q_1}. \qquad (5.47)$$

2. Let $p(z), q(z) \in \mathbb{F}[z]$, with $q(z)$ nonzero. Then the linear transformation $p(S^q)$ is invertible if and only if $p(z)$ and $q(z)$ are coprime. Moreover, we have

$$p(S^q)^{-1} = a(S^q), \qquad (5.48)$$

where the polynomial $a(z)$ arises out of any solution of the Bezout equation

$$a(z)p(z) + b(z)q(z) = 1. \qquad (5.49)$$

$$\blacksquare$$

5.5   The Chinese Remainder Theorem and Interpolation

The roots of the Chinese remainder theorem are in number theory. However, we interpret it, the underlying ring taken to be $\mathbb{F}[z]$, as an interpolation result.

Theorem 5.29 (Chinese remainder theorem). *Let $q_i(z) \in \mathbb{F}[z]$ be mutually coprime polynomials and let $q(z) = q_1(z) \cdots q_s(z)$. Then, given polynomials $a_i(z)$ such that $\deg a_i < \deg q_i$, there exists a unique polynomial $f(z) \in X_q$, i.e., such that $\deg f < \deg q$, and for which $\pi_{q_i} f = a_i$.*

Proof. The interesting thing about the proof of the Chinese remainder theorem is its use of coprimeness in two distinct ways. One is geometric, the other one is spectral. Let us define $d_j(z) = \prod_{i \neq j} q_i(z)$.

The mutual coprimeness of the $q_i(z)$ implies the direct sum decomposition

$$X_q = d_1 X_{q_1} \oplus \cdots \oplus d_s X_{q_s}. \tag{5.50}$$

This is the geometric use of the coprimeness assumption. The condition $\deg f < \deg q$ is equivalent to $f(z) \in X_q$. Let $f(z) = \sum_{j=1}^{s} d_j(z) f_j(z)$ with $f_j(z) \in X_{q_j}$. Since for $i \neq j$, $q_i(z) \mid d_j(z)$, it follows that in this case $\pi_{q_i} d_j f_j = 0$. Hence

$$\pi_{q_i} f = \pi_{q_i} \sum_{j=1}^{s} d_j f_j = \pi_{q_i} d_i f_i = d_i(S_{q_i}) f_i.$$

For the spectral use of coprimeness, we observe that the pairwise coprimeness of the $q_i(z)$ implies the coprimeness of $d_i(z)$ and $q_i(z)$. In turn, this shows, applying Theorem 5.18, that the module homomorphism $d_i(S_{q_i})$ in X_{q_i} is actually an isomorphism. Hence, there exists a unique $f_i(z)$ in X_{q_i} such that $a_i = d_i(S_{q_i}) f_i$ and $f_i = d_i(S_{q_i})^{-1} a_i$. So $f = \sum_{j=1}^{s} d_j d_i(S_{q_i})^{-1} a_i$ is the required polynomial. Note that the inversion of $d_i(S_{q_i})$ can be done easily, as in Theorem 5.18, using the Euclidean algorithm.

The uniqueness of $f(z)$, under the condition $f(z) \in X_q$, follows from the fact that (5.50) is a direct sum representation. This completes the proof. ∎

5.5.1 Lagrange Interpolation Revisited

Another solution to the Lagrange interpolation problem, introduced in Chapter 2, can be easily derived from the Chinese remainder theorem. Indeed, given distinct numbers $\alpha_i \in \mathbb{F}$, $i = 1, \ldots, n$, the polynomials $q_i(z) = (z - \alpha_i)$ are mutually coprime. We define $q(z) = \prod_{j=1}^{n} q_j(z)$ and $q_i(z) = \prod_{j \neq i} q_j(z)$, which leads to the factorizations $q(z) = d_i(z) q_i(z)$. The coprimeness assumption implies the direct sum representation $X_q = d_1 X_{q_1} \oplus \cdots \oplus d_n X_{q_n}$. Thus, any polynomial $f(z) \in X_q$ has a unique representation of the form

$$f(z) = \sum_{j=1}^{n} c_j d_j(z). \tag{5.51}$$

We want to find a polynomial $f(z)$ that satisfies the interpolation conditions $f(\lambda_i) = a_i$, $i = 1, \ldots, n$. Applying the projection π_{q_i} to the expansion (5.51), and noting that $\pi_{q_i} d_j = 0$ for $j \neq i$, we obtain

$$a_i = f(\lambda_i) = \pi_{q_i} f = \pi_{q_i} \sum_{j=1}^{n} c_j d_j(z) = c_i d_i(\lambda_i).$$

Defining the Lagrange interpolation polynomials by $l_i(z) = d_i(z)/d_i(\lambda_i)$, we get $f(z) = \sum_{j=1}^{n} a_j l_j(z)$ for the unique solution in X_q of the Lagrange interpolation problem. Any other solution differs by a polynomial that has a zero at all the points λ_i, hence is divisible by $q(z)$.

5.5.2 Hermite Interpolation

We apply now the Chinese remainder theorem to the problem of higher-order interpolation, or **Hermite interpolation**. In Hermite interpolation, which is a generalization of the Lagrange interpolation problem, we prescribe not only the value of the interpolating polynomial at given points, but also the value of a certain number of derivatives, the number of which may differ from point to point.

Specifying the first ν derivatives, counting from zero, of a polynomial $p(z)$ at a point α means that we are given a representation

$$f(z) = \sum_{i=0}^{\nu-1} f_{i,\alpha}(z - \alpha_i)^i + (z - \alpha)^\nu g(z).$$

Of course, since $\deg \sum_{i=0}^{\nu-1} f_{i,\alpha}(z - \alpha_i)^i < \nu$, this means that

$$\deg \sum_{i=0}^{\nu-1} f_{i,\alpha}(z - \alpha)^i = \deg \pi_{(z-\alpha)^\nu} f.$$

Hence we can formulate the **Hermite interpolation problem:** Given distinct $\alpha_1, \ldots, \alpha_k \in \mathbb{F}$, positive integers $\nu_1, \ldots, \nu_k$, and polynomials $f_i(z) = \sum_{j=0}^{\nu_i-1} f_{j,\alpha_i}(z - \alpha_i)^j$, find a polynomial $f(z)$ such that

$$\pi_{(z-\alpha_i)^{\nu_i}} f = f_i, \qquad i = 1, \ldots, k. \tag{5.52}$$

Proposition 5.30. *There exists a unique solution $f(z)$, of degree $< n = \sum_{i=1}^{k} \nu_i$, to the Hermite interpolation problem. Any other solution of the Hermite interpolation problem is of the form $f(z) + p(z)g(z)$, where $g(z)$ is an arbitrary polynomial and $p(z)$ is given by*

$$p(z) = \Pi_{i=1}^{k}(z - \alpha_i)^{\nu_i}. \tag{5.53}$$

Proof. We apply the Chinese remainder theorem. Obviously, the polynomials $(z - \alpha_i)^{v_i}$, $i = 1, \ldots, k$ are mutually coprime. Then, with $p(z)$ defined by (5.53), there exists a unique $f(z)$ with $\deg f < n$ for which (5.52) holds.

If $\hat{f}(z)$ is any other solution, then $h(z) = (f(z) - \hat{f}(z))$ satisfies $\pi_{(z - \alpha_i)^{v_i}} h = 0$, that is $h(z)$ is divisible by $(z - \alpha_i)^{v_i}$. Since these polynomials are mutually coprime, it follows that $p(z) \mid h(z)$, or $h(z) = p(z)g(z)$ for some polynomial $g(z)$. ∎

5.5.3 Newton Interpolation

There are situations in which the interpolation data are given to us sequentially. At each stage we solve the corresponding interpolation problem. Newton interpolation is recursive in the sense that the solution at time k is the basis for the solution at time $k + 1$.

Definition 5.31. Given $\lambda_i, a_i \in \mathbb{F}$, $i = 0, 1, \ldots$, we define the **Newton interpolation problem** NIP(i):
Find polynomials $f_i(z) \in X_{d_i}$, $i \geq 1$, satisfying the interpolation conditions

$$\mathbf{NIP}(i) : f_i(\lambda_j) = a_j, \qquad 0 \leq j \leq i - 1. \tag{5.54}$$

Theorem 5.32. *Given $\lambda_i, a_i \in \mathbb{F}$, $i = 0, 1, \ldots$, with the λ_i distinct, we define polynomials by*

$$q_i(z) = z - \lambda_i,$$
$$d_i(z) = \Pi_{j=0}^{i-1}(z - \lambda_j). \tag{5.55}$$

Then

1. We have the factorizations

$$d_{i+1}(z) = d_i(z)q_i(z), \tag{5.56}$$

with $d_i(z), q_i(z)$ coprime.
2. We have the following direct sum decomposition:

$$X_{d_{i+1}} = X_{d_i} \oplus d_i X_{q_i}. \tag{5.57}$$

3. Every $f(z) \in X_{d_{i+1}}$ has a unique representation of the form

$$f(z) = g(z) + cd_i(z), \tag{5.58}$$

for some $g(z) \in X_{d_i}$ and $c \in \mathbb{F}$.

4. Let $f_i(z)$ be the solution to the Newton interpolation problem **NIP**(i), then there exists a constant for which

$$f_{i+1}(z) = f_i(z) + d_i(z)c_i \qquad (5.59)$$

is the solution to the Newton interpolation problem **NIP**$(i+1)$ and it has the representation

$$f_{i+1}(z) = \sum_{j=0}^{i} c_j d_j(z), \qquad (5.60)$$

where

$$c_j = \frac{f_{j+1}(\lambda_j) - f_j(\lambda_j)}{d_j(\lambda_j)}. \qquad (5.61)$$

Proof. 1. Follows from (5.55). The coprimeness of $d_i(z)$ and $q_i(z)$ is a consequence of our assumption that the λ_j are distinct.
2. Follows from the factorization (5.56) by applying Proposition 5.6.
3. Follows from the direct sum representation (5.57).
4. In equation (5.59), we substitute the corresponding expression for $f_i(z)$ and proceed inductively to get (5.60).

   Alternatively, we note that $\deg d_i = i$, for $i \geq 0$; hence $\{d_0(z), \ldots, d_i(z)\}$ is a basis for $X_{d_{i+1}}$ and an expansion (5.60) exists.

   Let $f_{i+1}(z)$ be defined by (5.59). Since $d_i(\lambda_j) = 0$ for $j = 0, \ldots, i-1$, it follows that $f_{i+1}(\lambda_j) = a_j$ for $j = 0, \ldots, i-1$. All we need for $f_{i+1}(z)$ to be a solution of **NIP**$(i+1)$ is to choose c_i such that $a_{i+1} = f_{i+1}(\lambda_i) = f_i(\lambda_i) + d_i(\lambda_i)c_i$ or, since $d_i(\lambda_i) \neq 0$, we get (5.61). ∎

5.6 Duality

The availability of both polynomial and rational models allows us to proceed with a deeper study of duality. Our aim is to obtain an identification of the dual space to a polynomial model in terms of a polynomial model.

On $\mathbb{F}(z)$, and more generally on $\mathbb{F}((z^{-1}))$, we introduce a bilinear form as follows. Given $f(z) = \sum_{j=-\infty}^{\bar{n}_f} f_j z^j$ and $g(z) = \sum_{j=-\infty}^{n_g} g_j z^j$, let

$$[f, g] = \sum_{j=-\infty}^{\infty} f_j g_{-j-1}. \qquad (5.62)$$

Clearly, the sum in (5.62) is well defined, since only a finite number of summands are nonzero. Given a subspace $M \subset \mathbb{F}(z)$, we let $M^{\perp} = \{f \in \mathbb{F}(z) \,|\, [m, f] = 0, \forall m \in M\}$. It is easy to check that $\mathbb{F}[z]^{\perp} = \mathbb{F}[z]$.

We will need the following simple computational rule.

Proposition 5.33. *Let $\phi(z), f(z), g(z)$ be rational functions. Then*

$$[\phi f, g] = [f, \phi g]. \tag{5.63}$$

Proof. With the obvious notation we compute

$$[\phi f, g] = \sum_{j=-\infty}^{\infty} (\phi f)_j h_{-j-1} = \sum_{j=-\infty}^{\infty} \left(\sum_{i=-\infty}^{\infty} \phi_i f_{j-i} \right) h_{-j-1}$$

$$= \sum_{i=-\infty}^{\infty} f_{j-i} \sum_{j=-\infty}^{\infty} \phi_i h_{-j-1} = \sum_{k=-\infty}^{\infty} f_k \sum_{j=-\infty}^{\infty} \phi_{j-k} h_{-j-1}$$

$$= \sum_{k=-\infty}^{\infty} f_k \sum_{i=-\infty}^{\infty} \phi_i h_{-i-k-1} = \sum_{k=-\infty}^{\infty} f_k (\phi h)_{-k-1} = [f, \phi h].$$

∎

Multiplication operators in $\mathbb{F}(z)$ of the form $L_\phi h = \phi h$ are called **Laurent operators**. The function ϕ is called the **symbol** of the Laurent operator.

Before getting the representation of the dual space to a polynomial model, we derive a representation of the dual space to $\mathbb{F}[z]$.

Theorem 5.34. *The dual space of $\mathbb{F}[z]$ can be identified with $z^{-1}\mathbb{F}[[z^{-1}]]$.*

Proof. Clearly, every element $h(z) \in z^{-1}\mathbb{F}[[z^{-1}]]$ defines, by way of the pairing (5.62), a linear functional on $\mathbb{F}[z]$. Conversely, given a linear functional Φ on $\mathbb{F}[z]$, it induces linear functionals ϕ_i on $\mathbb{F}$ by defining, for $\xi \in \mathbb{F}$,

$$\phi_i(\xi) = [z^i \xi, \phi] = \Phi(z^i \xi),$$

and an element $h \in z^{-1}\mathbb{F}[[z^{-1}]]$ is defined by letting $h(z) = \sum_{j=0}^{\infty} \phi_j z^{-j-1}$. It follows that $\Phi(f) = [f, h]$. ∎

Point evaluations are clearly linear functionals in $\mathbb{F}[z]$. It is easy to identify the representing functions. In fact, this is an algebraic version of Cauchy's theorem.

Proposition 5.35. *Let $\alpha \in \mathbb{F}$ and $f(z) \in \mathbb{F}[z]$. Then*

$$f(\alpha) = [f, (z-\alpha)^{-1}].$$

Proof. We have $(z-\alpha)^{-1} = \sum_{i=1}^{\infty} \alpha^{i-1} z^{-i}$. So this follows from (5.62), since

$$\left[f, \frac{1}{z-\alpha} \right] = \sum_{i=0}^{n} f_i \alpha^i = f(\alpha).$$

∎

Theorem 5.36. *Let $M = d\mathbb{F}[z]$ with $d(z) \in \mathbb{F}[z]$. Then $M^\perp = X^d$.*

Proof. Let $f(z) \in \mathbb{F}[z]$ and $h(z) \in M$. Then

$$0 = [df, h] = [f, dh] = [f, \pi_- dh].$$

But this implies $d(z)h(z) \in X_d$, or $h(z) \in X^d$.   ∎

Next, we compute the adjoint of the projection $\pi_d : \mathbb{F}[z] \longrightarrow \mathbb{F}[z]$. Clearly π_d^* is a transformation acting in $z^{-1}\mathbb{F}[[z^{-1}]]$.

Theorem 5.37. *The adjoint of π_d is π^d.*

Proof. Let $f(z) \in \mathbb{F}[z]$ and $h(z) \in z^{-1}\mathbb{F}[[z^{-1}]]$. Then

$$
\begin{aligned}
[\pi_d f, h] &= [d\pi_- d^{-1}f, h] = [\pi_- d^{-1}f, \tilde{d}h] = [d^{-1}f, \pi_+ \tilde{d}h] \\
&= [f, \tilde{d}^{-1}\pi_+ \tilde{d}h] = [\pi_+ f, \tilde{d}^{-1}\pi_+ \tilde{d}h] = [f, \pi_- \tilde{d}^{-1}\pi_+ \tilde{d}h] \\
&= [f, \pi^d h].
\end{aligned}
$$
                                                                                ∎

Not only are we interested in the study of duality on the level of $\mathbb{F}[z]$ and its dual space $z^{-1}\mathbb{F}[[z^{-1}]]$, but also we would like to study it on the level of the modules X_d and X^d. The key to this study is the fact that if X is a vector space and M a subspace, then $(X/M)^* \simeq M^\perp$.

Theorem 5.38. *Let $d(z) \in \mathbb{F}[z]$ be nonsingular. Then X_d^* is isomorphic to X^d and $S_d^* = S^d$.*

Proof. Since X_d is isomorphic to $\mathbb{F}[z]/d\mathbb{F}[z]$, then X_d^* is isomorphic to $(\mathbb{F}[z]/d\mathbb{F}[z])^*$, which, in turn, is isomorphic to $(d\mathbb{F}[z])^\perp$. However, this last module is X^d. It is clear that under the duality pairing we introduced, we actually have $X_d^* = X^d$. Finally, let $f(z) \in X_d$ and let $h(z) \in X^d$. Then

$$
\begin{aligned}
[S_d f, h] &= [\pi_d z f, h] = [zf, \pi^d h], \\
[zf, h] &= [f, zh] = [\pi_+ f, zh], \\
[f, \pi_- zh] &= [f, S_- h] = [f, S^d h].
\end{aligned}
$$

Hence, we can identify $X_{\tilde{d}}$ with X_d^* by defining a new pairing

$$\langle f, g \rangle = [d^{-1}f, g] = [f, d^{-1}g] \tag{5.64}$$

for all $f(z), g(z) \in X_d$.   ∎

As a direct corollary of Theorem 5.38 we have the following.

Theorem 5.39. *The dual space of X_d under the pairing $\langle \, , \, \rangle$ introduced in (5.64) is X_d, and moreover, $S_d^* = S_d$.*   ∎

With the identification of the polynomial model X_d with its dual space, we can identify some pairs of dual bases.

Proposition 5.40. *1. Let $d(z) = z^n + d_{n-1}z^{n-1} + \cdots + d_0$, and let $\mathcal{B}_{st} = \{1, z, \ldots, z^{n-1}\}$ be the standard and $\mathcal{B}_{co} = \{e_1(z), \ldots, e_n(z)\}$ the control bases respectively of X_d. Then $\mathcal{B}_{co} = \mathcal{B}_{st}^*$, that is, the control and standard bases are dual to each other.*

2. Let $d(z) = \Pi_{i=1}^{n}(z - \lambda_i)$, with the λ_i distinct. Let $\mathcal{B}_{in} = \{\pi_1(z), \ldots, \pi_n(z)\}$, with $\pi_i(z)$ the Lagrange interpolation polynomials, be the interpolation basis in X_d. Let $\mathcal{B}_{sp} = \{p_1(z), \ldots, p_n(z)\}$ be the spectral basis, with $p_i(z) = \Pi_{j \neq i}(z - \lambda_j)$. Then $\mathcal{B}_{sp}^* = \mathcal{B}_{in}$.

Proof. 1. We use (5.64) to compute

$$\langle z^{i-1}, e_j \rangle = [d^{-1}z^{i-1}, \pi_+ z^{-j} d] = [d^{-1}z^{i-1}, z^{-j}d]$$
$$= [z^{i-j-1}, 1] = \delta_{ij}.$$

2. We use Proposition 5.35 and note that for every $f(z) \in X_d$ we have

$$\langle f, p_i \rangle = [d^{-1}f, p_i] = \left[f, \frac{p_i}{d} \right] = \left[f, \frac{1}{z - \lambda_i} \right] = f(\lambda_i).$$

In particular, $\langle \pi_i, p_j \rangle = \pi_i(\lambda_j) = \delta_{ij}$. ∎

This result explains the connection between the two companion matrices given in Proposition 5.4. Indeed,

$$C_q^\sharp = [S_q]_{st}^{st} = \widetilde{[S_q^*]_{co}^{co}} = \widetilde{[S_q]_{co}^{co}} = \widetilde{C_q^\flat}.$$

Next we compute the change of basis transformations.

Proposition 5.41. *1. Let $q(z) = z^n + q_{n-1}z^{n-1} + \cdots + q_0,$. Then*

$$[I]_{co}^{st} = \begin{pmatrix} q_1 & \cdots & q_{n-1} & 1 \\ \cdot & & \cdot & \\ \cdot & & \cdot & \\ q_{n-1} & \cdot & & \\ 1 & & & \end{pmatrix} \tag{5.65}$$

and

$$[I]_{st}^{co} = \begin{pmatrix} & & & 1 \\ & & \cdot & \psi_1 \\ & \cdots & & \\ \cdot & \cdots & & \\ 1 & \psi_1 & \cdots & \psi_{n-1} \end{pmatrix}, \tag{5.66}$$

where for $q^{\sharp}(z) = z^n q(z^{-1})$, $\psi(z) = \psi_0 + \cdots + \psi_{n-1} z^{n-1}$ is the unique solution, of degree $< n$, of the Bezout equation

$$q^{\sharp}(z)\psi(z) + z^n \sigma(z) = 1. \tag{5.67}$$

2. The matrices in (5.65) and (5.66) are inverses of each other.
3. Let $d(z) = \Pi_{j=1}^n (z - \alpha_j)$ with $\alpha_1, \ldots, \alpha_n$ distinct and let $\mathscr{B}_{sp}, \mathscr{B}_{in}$ be the corresponding spectral and interpolation bases of X_d. Then we have the following change of basis transformations:

$$[I]_{st}^{in} = \begin{pmatrix} 1 & \alpha_1 & .. & \alpha_1^{n-1} \\ . & . & . & . \\ . & . & . & . \\ . & . & . & . \\ 1 & \alpha_n & .. & \alpha_n^{n-1} \end{pmatrix}, \tag{5.68}$$

$$[I]_{sp}^{co} = \begin{pmatrix} 1 & \cdots & 1 \\ \alpha_1 & \cdots & \alpha_n \\ . & \cdots & . \\ . & \cdots & . \\ \alpha_1^{n-1} & \cdots & \alpha_n^{n-1} \end{pmatrix}, \tag{5.69}$$

$$[I]_{sp}^{in} = \begin{pmatrix} p_1(\alpha_1) & & \\ & . & \\ & & . \\ & & & p_n(\alpha_n) \end{pmatrix}. \tag{5.70}$$

Proof. 1. Note that if $\psi(z)$ is a solution of (5.67), then necessarily $\psi_0 = 1$. With J the transposition matrix defined in (8.39), we compute

$$J[I]_{st}^{co} = ([I]_{co}^{st} J)^{-1} = (q^{\sharp}(S_{z^n}))^{-1}.$$

However, if we consider the map S_{z^n}, then

$$\begin{pmatrix} 1 & q_{n-1} & \cdots & q_1 \\ . & . & . & . \\ & & . & . \\ & & & . \\ & & & . & q_{n-1} \\ & & & & 1 \end{pmatrix} = q^{\sharp}(S_{z^n}),$$

and its inverse is given by $\psi(S_{z^n})$, where $\psi(z)$ solves the Bezout equation (5.67). This completes the proof.

2. Follows from the fact that $[I]_{co}^{st}[I]_{st}^{co} = I$.
3. The matrix representation for $[I]_{st}^{in}$ has been derived in Corollary 2.36.

The matrix representation for $[I]_{sp}^{co}$ follows from (5.68) by applying duality theory, in particular Theorem 4.38. We can also derive this matrix representation directly, which we proceed to do. For this, we define polynomials $s_1(z), \ldots, s_n(z)$ by

$$s_i(z) = e_1(z) + \alpha_i e_2(z) + \cdots + \alpha_i^{n-1} e_n(z).$$

We claim that $s_i(z)$ are eigenfunctions of S_d corresponding to the eigenvalues α_i. Indeed, using equation (5.12) and the fact that $0 = q(\alpha_i) = q_0 + q_1\alpha_i + \cdots + \alpha_i^n$, we have

$$
\begin{aligned}
S_q s_i &= -q_0 e_n + \alpha_i(e_1 - q_1 e_n) + \cdots + \alpha_i(e_{n-1} - q_{n-1} e_n) \\
&= \alpha_i e_1 + \cdots + \alpha_i^{n-1} e_{n-1} - (q_0 + \cdots + q_{n-1}\alpha_i^{n-1}) e_n \\
&= \alpha_i e_1 + \cdots + \alpha_i^n e_n = \alpha_i s_i(z).
\end{aligned}
$$

This implies that there exist constants γ_i such that $s_i(z) = \gamma_i p_i(z)$. Since both $s_i(z)$ and $e_i(z)$ are obviously monic, it follows that necessarily $\gamma_i = 1$ and $s_i(z) = p_i(z)$. The equations

$$p_i(z) = e_1(z) + \alpha_i e_2(z) + \cdots + \alpha_i^{n-1} e_n(z)$$

imply the matrix representation (5.69).

Finally, the matrix representation in (5.70) follows from the trivial identities

$$p_i(z) = p_i(\alpha_i)\pi_i(z).$$

■

5.7 Universality of Shifts

We end this chapter by explaining the reason that the polynomial and rational models we introduced can be used so effectively in linear algebra and its applications. The main reason for this is the universality property of shifts. For our purposes, we shall show that any linear transformation in a finite-dimensional vector space $\mathcal{V}$ over the field $\mathbb{F}$ is isomorphic to the compression of the forward shift $S_+ : \mathbb{F}[z]^n \longrightarrow \mathbb{F}[z]^n$, defined in (1.31), to a quotient space, or alternatively, to the restriction of the backward shift $S_- : z^{-1}\mathbb{F}[[z^{-1}]]^n \longrightarrow z^{-1}\mathbb{F}[[z^{-1}]]^n$, defined in (1.32), to an invariant subspace. For greater generality, we will have to stray away from scalar-valued polynomial and rational functions.

Assume we are given a linear transformation A in a finite-dimensional vector space $\mathcal{V}$, over the field $\mathbb{F}$. Without loss of generality, by choosing a matrix representation, we may as well assume that $\mathcal{V} = \mathbb{F}^n$ and that A is a square matrix.

By $\mathbb{F}^n[z]$ we denote the space of vector polynomials with coefficients in $\mathbb{F}^n$, whereas $\mathbb{F}[z]^n$ will denote the space of vectors with polynomial entries. We will use freely the isomorphism between $\mathbb{F}^n[z]$ and $\mathbb{F}[z]^n$ and we will identify the two spaces.

With the linear polynomial matrix $zI - A$ we associate a map $\pi_{zI-A} : \mathbb{F}[z]^n \longrightarrow \mathbb{F}^n$, given by

$$\pi_{zI-A} \sum_{j=0}^{k} \xi_j z^j = \sum_{j=0}^{k} A^j \xi_j. \tag{5.71}$$

The operation defined above can be considered as taking the remainder of a polynomial vector after left division by the polynomial matrix $zI - A$.

Proposition 5.42. *For the map π_{zI-A} defined by (5.71)*

1. *We have π_{zI-A} is surjective.*
2. *We have*

$$\operatorname{Ker} \pi_{zI-A} = (zI - A)\mathbb{F}[z]^n. \tag{5.72}$$

3. *For the map $S_+ : \mathbb{F}[z]^n \longrightarrow \mathbb{F}[z]^n$ defined by*

$$(S_+ f)(z) = z f(z), \tag{5.73}$$

the following diagram is commutative:

$$
\begin{array}{ccc}
\mathbb{F}[z]^n & \xrightarrow{\ \pi_{zI-A}\ } & \mathbb{F}^n \\[2em]
{\scriptstyle S_+}\Big\downarrow & & \Big\downarrow {\scriptstyle A} \\[2em]
\mathbb{F}[z]^n & \xrightarrow{\ \pi_{zI-A}\ } & \mathbb{F}^n
\end{array}
$$

This implies that

$$Ax = \pi_{zI-A} z \cdot x. \tag{5.74}$$

4. *Given a polynomial $p(z) \in \mathbb{F}[z]$, we have*

$$p(A)x = \pi_{zI-A} p(z)x. \tag{5.75}$$

Proof. 1. For each constant polynomial $x \in \mathbb{F}[z]^n$, we have $\pi_{zI-A} x = x$. The surjectivity follows.
2. Assume $f(z) = (zI - A)g(z)$ with $g(z) = \sum_{i=0}^{k} g_i z^i$. Then

$$f(z) = (zI - A) \sum_{i=0}^{k} g_i z^i = \sum_{i=0}^{k} g_i z^{i+1} - \sum_{i=0}^{k} A g_i z^i.$$

Therefore

$$\pi_{zI-A} f = \sum_{i=0}^{k} A^{i+1} g_i - \sum_{i=0}^{k} A^i A g_i = 0,$$

i.e., $(zI - A)\mathbb{F}[z]^n \subset \operatorname{Ker} \pi_{zI-A}$.

Conversely, assume $f(z) = \sum_{i=0}^{k} f_i z^i \in \text{Ker}\,\pi_{zI-A}$, that is, $\sum_{i=0}^{k} A^i f_i = 0$. Recalling that $z^i I - A^i = (zI - A)\sum_{j=0}^{i-1} z^{i-1-j} A^j$, we compute

$$
\begin{aligned}
f(z) &= \sum_{i=0}^{k} f_i z^i = \sum_{i=0}^{k} f_i z^i - \sum_{i=0}^{k} A^i f_i \\
&= \sum_{i=0}^{k}(z^i I - A^i) f_i = \sum_{i=0}^{k}(zI - A)(\sum_{j=0}^{i-1} z^{i-1-j} A^j) f_i \\
&= (zI - A)\sum_{i=0}^{k}(\sum_{j=0}^{i-1} z^{i-1-j} A^j) f_i = (zI - A)g,
\end{aligned}
$$

so $\text{Ker}\,\pi_{zI-A} \subset (zI - A)\mathbb{F}[z]^n$; Hence the equality (5.72) follows.
3. To prove the commutativity of the diagram, we compute, with $f(z) = \sum_{i=0}^{k} f_i z^i$,

$$
\begin{aligned}
\pi_{zI-A} S_+ f &= \pi_{zI-A} z \sum_{i=0}^{k} f_i z^i = \pi_{zI-A} \sum_{i=0}^{k} f_i z^{i+1} \\
&= \sum_{i=0}^{k} A^{i+1} f_i = A \sum_{i=0}^{k} A^i f_i \\
&= A\pi_{zI-A} f.
\end{aligned}
$$

4. By linearity, it suffices to prove this for polynomials of the form z^k. We do this by induction. For $k = 1$ this holds by equation (5.74). Assume it holds up to $k - 1$. Using the fact that

$$
z\text{Ker}\,(zI - A) = z(zI - A)\mathbb{F}[z]^n \subset (zI - A)\mathbb{F}[z]^n = \text{Ker}\,(zI - A),
$$

we compute

$$
\pi_{zI-A} z^k x = \pi_{zI-A} z \pi_{zI-A} z^{k-1} x = \pi_{zI-A} z A^{k-1} x = A A^{k-1} x = A^k x. \qquad \blacksquare
$$

Clearly, the equality $f(z) = (zI - A)g(z) + \pi_{zI-A} f$ can be interpreted as $\pi_{zI-A} f$ being the remainder of $f(z)$ after division by $zI - A$.

As a corollary, we obtain the celebrated Cayley–Hamilton theorem.

Theorem 5.43 (Cayley–Hamilton). *Let A be a linear transformation in an n-dimensional vector space $\mathcal{U}$ over $\mathbb{F}$ and let $d_A(z)$ be its characteristic polynomial. Then*

$$
d_A(A) = 0.
$$

Proof. By Cramer's rule, we have $d_A(z)I = (zI - A)\text{adj}\,(zI - A)$; hence we have the inclusion

$$
d_A(z)\mathbb{F}[z]^n \subset (zI - A)\mathbb{F}[z]^n.
$$

This implies, for each $x \in \mathcal{U}$, that $d_A(A)x = \pi_{zI-A} d(z)x = 0$, so $d_A(A) = 0$. $\qquad \blacksquare$

Corollary 5.44. *Let A be a linear transformation in an n-dimensional vector space $\mathcal{U}$ over $\mathbb{F}$. Then its minimal polynomial $m_A(z)$ divides its characteristic polynomial $d_A(z)$.*

Theorem 5.45. *Let A be a linear transformation in $\mathbb{F}^n$. Then A is isomorphic to S_- restricted to a finite-dimensional S_--invariant subspace of $z^{-1}\mathbb{F}[[z^{-1}]]^n$. Specifically, let $\Phi : \mathbb{F}^n \longrightarrow z^{-1}\mathbb{F}[[z^{-1}]]^n$ be defined by*

$$\Phi\xi = (zI - A)^{-1}\xi, \tag{5.76}$$

and let

$$\mathscr{L} = \operatorname{Im}\Phi. \tag{5.77}$$

Then the following diagram is commutative:

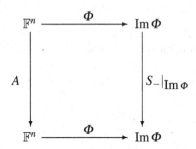

which implies the isomorphism

$$A \simeq S_-|_{\operatorname{Im}\Phi}. \tag{5.78}$$

Proof. Note that the map Φ is injective, hence invertible as a map from $\mathbb{F}^n$ onto $\mathscr{L} = \operatorname{Im}\Phi = \{(zI - A)^{-1}\xi \mid \xi \in \mathbb{F}^n\}$. Since

$$(zI - A)^{-1}\xi = \sum_{j=0}^{\infty} A^j\xi z^{-(j+1)}$$

$\mathscr{L}$ is a subspace of $z^{-1}\mathbb{F}[[z^{-1}]]^n$. Since

$$\begin{aligned}
S_-\Phi\xi &= S_-(zI - A)^{-1}\xi = \pi_- z(zI - A)^{-1}\xi \\
&= \pi_-(zI - A + A)(zI - A)^{-1}\xi = (zI - A)^{-1}A\xi \\
&= \Phi A\xi,
\end{aligned}$$

the S_--invariance of $\mathscr{L}$ follows as well as the isomorphism (5.78).   ∎

Remark 5.46. 1. The underlying idea is that the subspace $\operatorname{Im}\Phi$ contains all the information about A. The study of $\mathscr{L} = \operatorname{Im}\Phi$ as an $\mathbb{F}[z]$-module is a tool for the study of the transformation A.

2. The subspace $\mathscr{L} = \operatorname{Im}\Phi$ inherits an $\mathbb{F}[z]$-module structure from $z^{-1}\mathbb{F}[[z^{-1}]]^m$. With $\mathbb{F}^n$ having the $\mathbb{F}[z]$-module structure induced by A, then Φ is clearly an $\mathbb{F}[z]$-module homomorphism.

5.8 Exercises

1. Assume $q(z) = z^n + q_{n-1}z^{n-1} + \cdots + q_0$ is a real or complex polynomial. Show that the solutions of the linear, homogeneous differential equation

$$y^{(n)} + q_{n-1}y^{(n-1)} + \cdots + q_0 y = 0$$

   form an n-dimensional space. Use the Laplace transform $\mathscr{L}$ to show that y is a solution if and only if $\mathscr{L}(y) \in X^q$.

2. Let T be a cyclic transformation with minimal polynomial $m_T(z)$ of degree n, and let $p(z) \in \mathbb{F}[z]$. Show that the following statements are equivalent:

   a. The operator $p(T)$ is cyclic.
   b. There exists a polynomial $q(z) \in \mathbb{F}[z]$ such that $(q \circ p)(z) = z \mod (m_T)$.
   c. The map $\Lambda_p : \mathbb{F}[z] \longrightarrow X_{m_T}$ defined by $\Lambda_p(q) = \pi_{m_T}(q \circ p)$ is surjective.
   d. We have $\det(\pi_{kj}) \neq 0$, where $\pi_k = \pi_{m_T}(p^k)$ and $\pi_k(z) = \sum_{j=0}^{n-1} \pi_{kj}z^j$.

3. Assume the minimal polynomial of a cyclic operator T factors into linear factors, i.e., $m_T(z) = \prod(z - \lambda_i)^{v_i}$ with the λ_i distinct. Show that $p(T)$ is cyclic if and only if $\lambda_i \neq \lambda_j$ implies $p(\lambda_i) \neq p(\lambda_j)$ and $p'(\lambda_i) \neq 0$ whenever $v_i > 1$.

4. Show that if $\pi(z) \in \mathbb{F}[z]$ is irreducible and $\deg p$ is a prime number, then $p(S_\pi)$ is either scalar or a cyclic transformation.

5. Let $q(z) = z^n + q_{n-1}z^{n-1} + \cdots + q_0$ and let $C_q^\sharp$ be its companion matrix. Let $f(z) \in X_q$ with $f(z) = f_0 + \cdots + f_{n-1}z^{n-1}$. We put

$$\mathbf{f} = \begin{pmatrix} f_0 \\ \cdot \\ \cdot \\ \cdot \\ f_{n-1} \end{pmatrix} \in \mathbb{F}^n.$$

   Show that $f(C_q) = \left(\mathbf{f}, C_q\mathbf{f}, \dots, C_q^{n-1}\mathbf{f}\right)$.

6. Given a linear transformation A in $\mathscr{V}$, a vector $x \in \mathscr{V}$, and a polynomial $p(z)$, define the **Jacobson chain matrix** by

$$C_m(p, x, A) = (x, Ax, \dots, A^{r-1}x, \dots, p(A)^{m-1}x, p(A)^{m-1}Ax, \dots, p(A)^{m-1}A^{r-1}x).$$

   Prove the following

   a. Let $E = C_m(p, 1, C_{p^m})$ then E is a solution of $C_{p^m}E = EH(p^m)$.
   b. Every solution X of $C_{p^m}X = XH(p^m)$ is of the form $X = f(C_{p^m})E$ for some polynomial $f(z)$ of degree $< m \deg p$.

7. Let $p(z) = z^m + p_{m-1}z^{m-1} + \cdots + p_0$ and $q(z) = z^n + q_{n-1}z^{n-1} + \cdots + q_0$. Let $H(p,q) = \begin{pmatrix} C_p^\sharp & 0 \\ N & C_q^\sharp \end{pmatrix}$, where N is the $n \times m$ matrix whose only nonzero element is $N_{1m} = 1$.

Show that the general solution to the equation $C_{pq}^\sharp X = XH(pq)$ is of the form $X = f(C_{pq}^\sharp)K$, where $\deg f < \deg p + \deg q$ and

$$
K = \begin{pmatrix}
1 & & & & p_0 & & & & \\
 & \cdot & & & & \cdot & \cdot & & \\
 & & \cdot & & & & \cdot & \cdot & \\
 & & & \cdot & & & & \cdot & \\
 & & & & 1 & p_{m-1} & & \cdot & \\
 & & & & & 1 & & p_0 & \cdot \\
 & & & & & & \cdot & & \cdot \\
 & & & & & & & \cdot & \\
 & & & & & & & & \cdot \cdot \\
 & & & & & & & & 1
\end{pmatrix}
$$

8. Let C be the circulant matrix

$$
C = \mathrm{circ}\,(c_0, \ldots, c_{n-1}) = \begin{pmatrix}
c_0 & c_{n-1} & \cdot & \cdot & c_1 \\
c_1 & c_0 & \cdot & \cdot & \\
\cdot & & \cdot & \cdot & \cdot \\
\cdot & & \cdot & \cdot & c_{n-1} \\
c_{n-1} & \cdot & & \cdot & c_1 & c_0
\end{pmatrix} .
$$

Show that $\det(C) = \prod_{i=1}^{n}(c_0 + c_1\zeta_i + \cdots + c_{n-1}\zeta_i^{n-1})$, where $1 = \zeta_1, \ldots, \zeta_n$ are the distinct nth roots of unity, that is, the zeros of $z^n - 1$.

9. Prove the following **barycentric representation** for the minimal-degree polynomial satisfying the interpolation constraints $f(\lambda_i) = a_i$, $i = 0, \ldots, n-1$, namely

$$
f(z) = \begin{cases}
\dfrac{\sum_{i=0}^{n-1} \dfrac{w_j}{z - \lambda_j} a_i}{\sum_{i=0}^{n-1} \dfrac{w_j}{z - \lambda_j}}, & z \neq \lambda_j, j = 0, \ldots, n-1, \\[2em]
a_i, & z = \lambda_j, j = 0, \ldots, n-1.
\end{cases}
$$

Here $w_i = \dfrac{1}{\prod_{j \neq i}(\lambda_i - \lambda_j)}$.

5.9   Notes and Remarks

Shift operators are a cornerstone of modern operator theory. As we explained in Section 5.7, their interest lies in their universality properties, an extremely important fact, first pointed out in Rota (1960). This observation is the key insight to the use of shift operators in modeling linear transformations as well as, more generally, linear systems. The fact that any linear transformation T acting in a finite-dimensional vector space is isomorphic to either a compression of S_+ or to a restriction of S_- has far-reaching implications. This leads to the use of functional models, in our case polynomial and rational models, in the study of linear transformations. The reason for the great effectiveness of the use of functional models can be traced back to the compactness of the polynomial notation as well as that the polynomial setting provides a richer algebraic language, with terms as zeros, factorizations, ideals, homomorphisms, and modules conveniently at hand. Of the two classes of models we employ, the polynomial models emphasize the arithmetic properties of the transformation, whereas the module structure of rational models is completely determined by the geometry of the model space. In other words, in the case of polynomial models, the space is simple but the tranformation is complicated. On the other hand, for rational models, the space is complicated but the transformation, i.e., the restricted backward shift, is simple.

The material in this chapter is mostly standard. The definitions given in Equations (5.71) and (5.75) have far reaching implications. They can be generalized, replacing $zI - A$ by an arbitrary nonsingular polynomial matrix, thus leading to the theory of polynomial models, initiated in Fuhrmann (1976). This is the algebraic counterpart of the functional models used so effectively in operator theory, e.g., Sz.-Nagy and Foias (1970).

The results and methods presented in this chapter originate in Fuhrmann (1976) and a long series of follow-up articles. In an analytic, infinite-dimensional setting, Nikolskii (1985) is a very detailed study of shift operators. The central results of this chapter are two theorems. Theorem 5.17 characterizes the maps intertwining the shifts of the form S_q, while Theorem 5.18 studies their invertibility properties in terms of factorizations and polynomial coprimeness. These results generalize to the use of shifts of higher multiplicity, a generalization that requires the use of the algebra of polynomial matrices, which is beyond the scope of the present book.

The analytic analogues of these result are the commutant lifting theorem, see Sarason (1967) and Sz.-Nagy and Foias (1970), and a corresponding spectral mapping theorem; see Fuhrmann (1968a,b). We shall return to these topics in Chapter 11.

The kernel representation (5.37) of rational models has far-reaching generalizations. With the scalar polynomial replaced by a rectangular polynomial matrix, it is the basic characterizations of behaviors in the behavioral approach to linear systems initiated in Willems (1986, 1989, 1991). In this connection, see also Fuhrmann (2002).

Theorem 5.17 can be extended to the characterization of intertwining maps between two polynomial, or rational, models. We will return to this in Chapter 8, where we give a different proof using tensor products and Bezoutians.

Theorem 5.18 highlights the importance of coprimenes and the Bezout equation. Since the Bezout equation can be solved using the Euclidean algorithm, this brings up the possibility of a recursive approach to inversion algorithms for structured matrices. This will be further explored in Chapter 8. The Bezout equation reappears, this time over the ring $\mathbf{RH}_+^\infty$, in Chapters 11 and 12.

Companion matrices of the polynomial $q(z)$ appear in Krull's thesis. The particularly musical notation for the matrices $C_q^\sharp, C_q^\flat$, defined in Proposition 5.4, was introduced by Kalman.

Section 5.3 on circulant matrices is based on results taken from Davis (1979).

The Chinese remainder theorem has its roots in the problem of solving simultaneous congruences arising in modular arithmetic. It appears in a third-century mathematical treatise by Sun Tzu. For a very interesting account of the history of early non-European mathematics, see Joseph (2000). In working over the ring of polynomials $\mathbb{F}[z]$ rather than over the integes $\mathbb{Z}$, it can be applied to interpolation problems. The possibility of applying the Chinese remainder theorem to interpolation problems is not mentioned in many of the classic algebra texts such as van der Waerden, Lang, Mac Lane and Birkhoff or even Gantmacher (1959). Of these monographs, van der Waerden (1931) stands out inasmuch as it discusses both the Lagrange and the Newton interpolation formulas. In connection to interpolation, the following quotation, from Schoenberg (1987), may be of interest: "Sometime in the 1950's the late Hungarian-Swedish mathematican Marcel Riesz visited the University of Pennsylvania and told us informally that the Chinese remainder theorem (1) can be thought of as an analogue of the interpolation by polynomials."

Cayley, in a letter to Sylvester, actually stated a more general version of the Cayley–Hamilton theorem. That version says that if the square matrices A and B commute and $f(x,y) = \det(xA - yB)$, then $f(B,A) = 0$. This was generalized in Livsic (1983).

Chapter 6
Structure Theory of Linear Transformations

6.1 Introduction

In this chapter, we study the structure of linear transformations. Our aim is to understand the structure of a linear transformation in terms of its most elementary components. This we do by representing the transformation by its matrix representation with respect to particularly appropriate bases. The ultimate goal, not always achievable, is to represent a linear transformation in diagonal form.

6.2 Cyclic Transformations

We begin our study of linear transformations in a finite-dimensional linear space by studying a special subclass, namely the class of cyclic linear transformations, introduced in Definition 5.13. Later on, we will show that every linear transformation is isomorphic to a direct sum of cyclic transformations. Thus the structure theory will be complete.

Proposition 6.1. *Let A be a cyclic linear transformation in an n-dimensional vector space $\mathcal{U}$. Then its characteristic and minimal polynomials coincide.*

Proof. We prove the statement by contradiction. Suppose $m(z) = z^k + m_{k-1}z^{k-1} + \cdots + m_0$ is the minimal polynomial of A, and assume $k < n$. Then $A^k = -m_0 I - \cdots - +m_{k-1}A^{k-1}$. This shows that given an arbitrary vector $x \in \mathcal{U}$, we have $A^k x \in \text{span}\,(x, Ax, \ldots, A^{k-1}x)$. By induction, it follows that for any integer p we have $\text{span}\,(x, Ax, \ldots, A^p x) \subset \text{span}\,(x, Ax, \ldots, A^{k-1}x)$. Since $\dim \text{span}\,(x, Ax, \ldots, A^{k-1}x) \leq k < n$, it follows that A is not cyclic. ∎

Later on, we will prove that the coincidence of the characteristic and minimal polynomials implies cyclicity.

P.A. Fuhrmann, *A Polynomial Approach to Linear Algebra*, Universitext,
DOI 10.1007/978-1-4614-0338-8_6, © Springer Science+Business Media, LLC 2012

Given a linear transformation T in a vector space $\mathcal{U}$, its **commutant** is defined as the set $\mathscr{C}(T) = \{X \in L(\mathcal{U}) | XT = TX\}$. The next result is a characterization of the commutant of a cyclic transformation.

Proposition 6.2. *Let $T : \mathcal{U} \longrightarrow \mathcal{U}$ be a cyclic linear transformation. Then $X : \mathcal{U} \longrightarrow \mathcal{U}$ commutes with T if and only if $X = p(T)$ for some polynomial $p(z)$.*

Proof. Let $p(z) \in \mathbb{F}[z]$. Then clearly, $Tp(T) = p(T)T$.

Conversely, let X commute with T, where T is cyclic. Let $x \in \mathcal{U}$ be a cyclic vector, i.e., $x, Tx, \ldots, T^{n-1}x$ form a basis for $\mathcal{U}$. Thus there exist $p_i \in \mathbb{F}$ for which

$$Xx = \sum_{i=0}^{n-1} p_i T^i x.$$

This implies

$$XTx = TXx = T \sum_{i=0}^{n-1} p_i T^i x = \sum_{i=0}^{n-1} p_i T^i Tx,$$

and by induction, $XT^k x = \sum_{i=0}^{n-1} p_i T^i T^k x$. Since the $T^k x$ span $\mathcal{U}$, this implies $X = p(T)$, where $p(z) = \sum_{i=0}^{n-1} p_i z^i$. ∎

Next, we show that the class of shift transformations S_m is not that special. In fact, every cyclic linear transformation in a finite-dimensional vector space is isomorphic to a unique transformation of this class.

Theorem 6.3. *Let $\mathcal{V}$ be a finite-dimensional linear space and let $A : \mathcal{V} \longrightarrow \mathcal{V}$ be a cyclic transformation, with $b \in \mathcal{V}$ a cyclic vector. We consider $\mathcal{V}$ an $\mathbb{F}[z]$-module, with the module structure induced by A. Then*

1. Define a map $\Phi : \mathbb{F}[z] \longrightarrow \mathcal{V}$, by

$$\Phi \sum_{i=0}^{k} f_i z^i = \sum_{i=0}^{k} f_i A^i b. \tag{6.1}$$

Then the following diagram commutes:

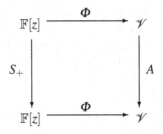

i.e., we have

$$\Phi S_+ = A\Phi, \tag{6.2}$$

which shows that Φ is an $\mathbb{F}[z]$-homomorphism.

2. *The map Φ is surjective and $\mathrm{Ker}\,\Phi$ is a submodule of $\mathbb{F}[z]$, i.e., an S_+-invariant subspace.*
3. *There exists a unique monic polynomial $m(z) \in \mathbb{F}[z]$, with $\deg m = \dim \mathscr{V}$, for which*

$$\mathrm{Ker}\,\Phi = m\mathbb{F}[z]. \tag{6.3}$$

4. *Let $\phi : \mathbb{F}[z]/m\mathbb{F}[z] \longrightarrow \mathscr{V}$ be the map induced by Φ on the quotient module, i.e.,*

$$\phi[f]_m = \Phi f. \tag{6.4}$$

Let $\pi : \mathbb{F}[z] \longrightarrow \mathbb{F}[z]/m\mathbb{F}[z]$ be the canonical projection map defined by

$$\pi f = [f]_m, \tag{6.5}$$

where $[f]_m = f(z) + m(z)\mathbb{F}[z]$. Then $\phi\pi = \Phi$, and the map $\phi : \mathbb{F}[z]/m\mathbb{F}[z] \longrightarrow \mathscr{V}$ is an isomorphism.

  Identifying the quotient module $\mathbb{F}[z]/m\mathbb{F}[z]$ with the polynomial model X_m, we have the $\mathbb{F}[z]$-module isomorphism $\phi : X_m \longrightarrow \mathscr{V}$ given by

$$\phi f = \Phi f, \qquad f(z) \in X_m, \tag{6.6}$$

i.e., $\phi = \Phi_{X_m}$.

5. *We have*

$$\phi S_m = A\phi, \tag{6.7}$$

i.e., the following diagram is commutative:

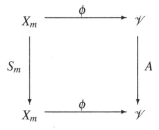

Proof. 1. We compute

$$\Phi(zf) = \Phi \sum_{i=0}^{k} f_i z^{i+1} = \sum_{i=0}^{k} f_i A^{i+1} b = A \sum_{i=0}^{k} f_i A^i b = A\Phi f.$$

This proves (6.2).

2. Since $\{b, Ab, \ldots, A^{n-1}b\}$ is a basis of $\mathscr{V}$, the map Φ is clearly surjective. We show that $\operatorname{Ker}\Phi$ is an ideal, or equivalently, a submodule in $\mathbb{F}[z]$. Assume $f(z), g(z) \in \operatorname{Ker}\Phi$. Then

$$\Phi(\alpha f + \beta g) = \alpha\Phi(f) + \beta\Phi(g) = 0,$$

so $\alpha f + \beta g \in \operatorname{Ker}\Phi$. Similarly, if $f \in \operatorname{Ker}\Phi$, then

$$\Phi(zf) = A\Phi(f) = A0 = 0,$$

which shows that $\operatorname{Ker}\phi$ is indeed an ideal. In fact, as is easily seen, a non-trivial ideal.

That $\operatorname{Ker}\Phi$ is a submodule of $\mathbb{F}[z]$ follows from (6.2).

3. Follows from the fact that $\mathbb{F}[z]$ is a principal ideal domain; hence there exists a unique monic polynomial for which $\operatorname{Ker}\Phi = m\mathbb{F}[z]$.
4. Follows from Proposition 1.55.
5. We compute

$$A\phi f = A\phi\pi_m f = A\Phi f = \Phi zf = \phi\pi_m zf = \phi S_m f. \qquad \blacksquare$$

From this theorem it follows that the study of an arbitrary cyclic transformation reduces to the study of the class of model transformations of the form S_m.

By Theorem 5.23, the polynomial and rational models that correspond to the same polynomial are isomorphic. This indicates that Theorem 6.3 can be dualized in order to represent linear transformations by S_- restricted to finite-dimensional backward-shift-invariant subspaces. This leads to the following direct proof of Rota's theorem restricted to cyclic transformations.

Theorem 6.4. *Let A be a cyclic linear transformation in $\mathscr{V}$ and let $b \in \mathscr{V}$ be a cyclic vector for A. Then*

1. Let Φ, π and ϕ be defined by (6.1), (6.5) and (6.4) respectively. Then

   a. The adjoint map $\Phi^* : \mathscr{V}^* \longrightarrow z^{-1}\mathbb{F}[[z^{-1}]]$ is given by

$$\Phi^*\eta = \sum_{i=1}^{\infty} \frac{(b, A^{*(i-1)}\eta)}{z^i} \tag{6.8}$$

   and is injective.

   b. The adjoint map $\pi^* : X^m \longrightarrow z^{-1}\mathbb{F}[[z^{-1}]]$ is given by

$$\pi^*h = h, \tag{6.9}$$

   i.e., it is the injection of X^m into $\mathbb{F}[[z^{-1}]]$.

   c. The adjoint map $\phi^* : \mathscr{V}^* \longrightarrow X^m$ is given, for $\eta \in \mathscr{V}^*$, by

$$\phi^*\eta = \sum_{i=1}^{\infty} \frac{(b, A^{*(i-1)}\eta)}{z^i}. \tag{6.10}$$

d. *The dual of the diagram in Theorem 6.3 is given by*

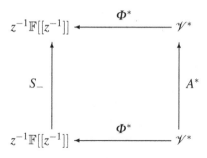

i.e., we have

$$\Phi^* A^* = S_- \Phi^*. \tag{6.11}$$

e. Im Φ^* *is a finite-dimensional, S_--invariant subspace of $z^{-1}\mathbb{F}[[z^{-1}]]$.*

f. *We have the following isomorphism:*

$$S_- \phi^* = \phi^* A^*. \tag{6.12}$$

2. *Let $\mathcal{V}$ be a finite-dimensional linear space and let $A : \mathcal{V} \longrightarrow \mathcal{V}$ be a cyclic transformation, with $b \in \mathcal{V}$ a cyclic vector. Then A is isomorphic to S_- restricted to a finite-dimensional S_--invariant subspace of $z^{-1}\mathbb{F}[[z^{-1}]]$. Specifically, let $\Psi : \mathcal{V} \longrightarrow z^{-1}\mathbb{F}[[z^{-1}]]$ be defined by*

$$\Psi \xi = \sum_{i=1}^{\infty} \frac{(b, A^{i-1}\xi)}{z^i}. \tag{6.13}$$

Then the following diagram is commutative:

$$
\begin{array}{ccc}
\mathcal{V} & \xrightarrow{\ \Psi\ } & \mathrm{Im}\,\Psi \\
\Big\downarrow{\scriptstyle A} & & \Big\downarrow{\scriptstyle S_-|\mathrm{Im}\,\Psi} \\
\mathcal{V} & \xrightarrow{\ \Psi\ } & \mathrm{Im}\,\Psi
\end{array}
$$

This implies the isomorphism

$$A \simeq S_- | \mathrm{Im}\,\Psi. \tag{6.14}$$

Proof. 1. a. We compute

$$
\begin{aligned}
(\Phi f, \eta) &= (\Phi \textstyle\sum_{i=0}^{k} f_i z^i, \eta) = (\textstyle\sum_{i=0}^{k} f_i A^i b, \eta) \\
&= \textstyle\sum_{i=0}^{k} f_i (A^i b, \eta) = \textstyle\sum_{i=0}^{k} f_i (b, (A^*)^i \eta) \\
&= [f, \Phi^* \eta],
\end{aligned}
$$

which implies (6.8).

To show injectivity of Φ^*, assume $\eta \in \operatorname{Ker} \Phi^*$. This means that for all $i \geq 0$, we have $(b, (A^*)^i \eta) = (A^i b, \eta) = 0$. Since the $A^i b$ span $\mathcal{V}$, necessarily $\eta = 0$, which proves injectivity.

b. We have made the identification $(\mathbb{F}[z]/m\mathbb{F}[z])^* = (m\mathbb{F}[z])^\perp = X^m$. For $f \in \mathbb{F}[z]$ and $h \in X^m$, we compute

$$
[\pi f, h] = [[f]_m, h] = [f, h] = [f, \pi^* h],
$$

from which (6.9) follows.

c. For $f \in \mathbb{F}[z]$ and $\eta \in \mathcal{V}^*$, we compute

$$
(\phi [f]_m, \eta) = (\Phi f, \eta) = (f, \Phi^*) = ([f]_m, \phi^*),
$$

i.e., $\phi^* = \Phi^*$. Using (6.8), (6.10) follows.

d. Follows by dualizing (6.7) and using the representation (6.8).

e. Follows from the fact that $\mathcal{V}^*$ is finite-dimensional.

f. Follows from (6.11) and the equality $\phi^* = \Phi^*$.

2. Follows from the representation of the adjoints, computed in the previous part, taken together with $\mathbb{F}[z]^* \simeq z^{-1}\mathbb{F}[[z^{-1}]]$. Identity (6.11) is obtained by dualizing (6.2). ∎

6.2.1 Canonical Forms for Cyclic Transformations

In this section we exhibit canonical forms for cyclic transformations.

Theorem 6.5. *Let $p(z) \in \mathbb{F}[z]$ and let $p(z) = p_1(z)^{m_1} \cdots p_k(z)^{m_k}$ be the primary decomposition of $p(z)$. We define*

$$
\pi_i(z) = \prod_{j \neq i} p_j(z)^{m_j}. \tag{6.15}
$$

Then we have the direct sum decomposition

$$
X_p = \pi_1(z) X_{p_1^{m_1}} \oplus \cdots \oplus \pi_k(z) X_{p_k^{m_k}} \tag{6.16}
$$

and the associated **spectral representation**

$$S_p \simeq S_{p_1^{m_1}} \oplus \cdots \oplus S_{p_k^{m_k}}. \tag{6.17}$$

Proof. The direct sum decomposition (6.16) was proved in Theorem 2.23.
We define a map

$$Z : X_{p_1^{m_1}} \oplus \cdots \oplus X_{p_k^{m_k}} \longrightarrow X_p = \pi_1(z) X_{p_1^{m_1}} \oplus \cdots \oplus \pi_k(z) X_{p_k^{m_k}}$$

by

$$Z(f_1, \ldots, f_k) = \sum_{i=1}^{k} \pi_i f_i. \tag{6.18}$$

Clearly, Z is invertible, and it is easily checked that the following diagram is commutative:

$$
\begin{array}{ccc}
X_{p_1^{m_1}} \oplus \cdots \oplus X_{p_k^{m_k}} & \xrightarrow{\ \ Z\ \ } & X_p \\[4pt]
\Big\downarrow{\scriptstyle S_{p_1^{m_1}} \oplus \cdots \oplus S_{p_k^{m_k}}} & & \Big\downarrow{\scriptstyle S_p} \\[4pt]
X_{p_1^{m_1}} \oplus \cdots \oplus X_{p_k^{m_k}} & \xrightarrow{\ \ Z\ \ } & X_p
\end{array}
$$

which shows that Z is an $\mathbb{F}[z]$-isomorphism.   ∎

Thus, by choosing appropriate bases in the spaces $X_{p_i^{m_i}}$, we get a block matrix representation for S_p. It suffices to consider the matrix representation of a typical operator S_{p^m}, where $p(z)$ is an irreducible polynomial. This is the content of the next proposition.

Proposition 6.6. *Let* $p(z) = z^r + p_{r-1} z^{r-1} + \cdots + p_0$. *Then*

1.

$$\dim X_{p^m} = mr. \tag{6.19}$$

2. The set of polynomials

$$\mathscr{B}_{jo} = \{1, z, \ldots, z^{r-1}, p, zp, \ldots, z^{r-1}p, \ldots, p^{m-1}, z p^{m-1}, \ldots, z^{r-1} p^{m-1}\} \tag{6.20}$$

forms a basis for X_{p^m}. *We call this the* **Jordan basis**.

3. The matrix representation of S_{p^m} with respect to this basis is of the form

$$
[S_{p^m}]_{jo}^{jo} = \begin{pmatrix} C_p^\sharp & & & & \\ N & C_p^\sharp & & & \\ & N & \cdot & & \\ & & & \ddots & \\ & & & & \ddots \\ & & & & N & C_p^\sharp \end{pmatrix},
\tag{6.21}
$$

where $C_p^\sharp$ is the companion matrix of $p(z)$, defined by (5.9), and

$$
N = \begin{pmatrix} 0 \ldots 0 \ 1 \\ & & 0 \\ \cdot & & \\ \cdot & & \cdot \\ \cdot & & \cdot \\ 0 \ldots \ldots 0 \end{pmatrix}.
\tag{6.22}
$$

We refer to (6.21) as the **Jordan canonical form** *of S_{p^m}.*

Proof. 1. We have
$$
\dim X_{p^m} = \deg p^m = m \deg p = mr.
$$

2. In $\mathscr{B}$ there is one polynomial of each degree and mr polynomials altogether, so necessarily they form a basis.

3. We prove this by simple computations. Notice that for $j < m - 1$,

$$
\begin{aligned}
S_{p^m} z^{r-1} p^j &= \pi_{p^m} z \cdot z^{r-1} p^j = \pi_{p^m} z^r p^j \\
&= \pi_{p^m} (z^r + p_{r-1} z^{r-1} + \cdots + p_0 - p_{r-1} z^{r-1} - \cdots - p_0) p^j \\
&= \pi_{p^m} p^{j+1} - \sum_{i=0}^{r-1} p_i (z^i p^j).
\end{aligned}
$$

On the other hand, for $j = m - 1$, we have

$$
\begin{aligned}
S_{p^m} z^{r-1} p^{m-1} &= \pi_{p^m} z^r p^{m-1} \\
&= \pi_{p^m} (z^r + p_{r-1} z^{r-1} + \cdots + p_0 - p_{r-1} z^{r-1} - \cdots - p_0) p^{m-1} \\
&= \pi_{p^m} p^m - \sum_{i=0}^{r-1} p_i (z^i p^{m-1}) \\
&= - \sum_{i=0}^{r-1} p_i (z^i p^{m-1}).
\end{aligned}
$$

∎

It is of interest to compute the dual basis to the Jordan basis. This leads us to the Jordan form in the block upper triangular form.

Proposition 6.7. *Given the monic polynomial $p(z)$ of degree r, let $\{e_1(z), \ldots, e_r(z)\}$ be the control basis of X_p. Then*

1. *The basis of X_{p^m} dual to the Jordan basis of (6.20) is given by*

$$\widetilde{\mathscr{B}}_{jo} = \{p^{m-1}(z)e_1(z),\ldots,p^{m-1}(z)e_r(z),\ldots,e_1(z),\ldots,e_r(z)\}. \tag{6.23}$$

*Here $e_1(z),\ldots,e_r(z)$ are the elements of the control basis of X_p. We call this basis the **dual Jordan basis**.*

2. *The matrix representation of S_{q^m} with respect to the dual Jordan basis is given by*

$$[S_{q^m}]^{\widetilde{jo}}_{\widetilde{jo}} = \begin{pmatrix} C_p^\flat & \tilde{N} & & & \\ & C_p^\flat & \cdot & & \\ & & \cdot & \cdot & \\ & & & \cdot & \cdot \\ & & & & \cdot & \tilde{N} \\ & & & & & C_p^\flat \end{pmatrix}. \tag{6.24}$$

Note that $C_p^\flat$ is defined in (5.10).

3. *The change of basis transformation from the dual Jordan basis to the Jordan basis is given by*

$$[I]^{jo}_{\widetilde{jo}} = \begin{pmatrix} & & & K \\ & & \cdot & \\ & \cdot & & \\ K & & & \end{pmatrix},$$

where $K = [I]^{st}_{co} = \begin{pmatrix} p_1 & \cdot\cdot & p_{r-1} & 1 \\ \cdot & & \cdot & \\ \cdot & & & \\ p_{r-1} & \cdot & & \\ 1 & & & \end{pmatrix}.$

4. *The Jordan matrices in (6.21) and (6.24) are similar. A similarity is given by $R = [I]^{jo}_{\widetilde{jo}}.$*

Proof. 1. This is an extension of Proposition 5.40. We compute, with $0 \le \alpha, \beta < m$ and $i, j < r$,

$$\langle p^\alpha e_i, p^\beta z^{j-1}\rangle = [p^{-m}p^\alpha \pi_+ z^{-i}p, p^\beta z^{j-1}] = [\pi_+ z^{-i}p, p^{\alpha+\beta-m}z^{j-1}].$$

The last expression is clearly zero whenever $\alpha + \beta \ne m - 1$. When $\alpha + \beta = m - 1$, we have

$$[\pi_+ z^{-i}p, p^{\alpha+\beta-m}z^{j-1}] = [\pi_+ z^{-i}p, p^{-1}z^{j-1}]$$

$$= [z^{-i}p, p^{-1}z^{j-1}] = [z^{j-i-1}, 1] = \delta_{ij}.$$

2. Follows either by direct computation or the fact that $S_q^* = S_q$ coupled with an application of Theorem 4.38.
3. This results from a simple computation.
4. From $S_{p^m}I = IS_{p^m}$ we get $[S_{p^m}]_{jo}^{jo}[I]_{\bar{jo}}^{jo} = [I]_{jo}^{jo}[S_{p^m}]_{\bar{jo}}^{\bar{jo}}$.   ■

Note that if the polynomial $p(z)$ is monic of degree 1, i.e., $p(z) = z - \alpha$, then with respect to the basis $\mathcal{B} = \{1, z - \alpha, \ldots, (z - \alpha)^{n-1}\}$ the matrix representation has the following form:

$$[S_{(z-\alpha)^n}]_{\mathcal{B}}^{\mathcal{B}} = \begin{pmatrix} \alpha & & & & \\ 1 & \alpha & & & \\ & 1 & . & & \\ & & . & . & \\ & & & . & . \\ & & & & 1 & \alpha \end{pmatrix}.$$

If our field $\mathbb{F}$ is algebraically closed, as is the field of complex numbers, then every irreducible polynomial is of degree 1. In case we work over the real field, the irreducible monic polynomials are either linear, or of the form $(z - \alpha)^2 + \beta^2$, with $\alpha, \beta \in \mathbb{R}$. In this case, we make a variation on the canonical form obtained in Proposition 6.6. This leads to the **real Jordan form**.

Proposition 6.8. *Let $p(z)$ be the real polynomial $p(z) = (z - \alpha)^2 + \beta^2$. Then*

1. We have

$$\dim X_{p^n} = 2n. \tag{6.25}$$

2. The set of polynomials

$$\mathcal{B} = \{\beta, z - \alpha, \beta p(z), (z - \alpha)p(z), \ldots, \beta p^{n-1}(z), (z - \alpha)p^{n-1}(z)\}$$

forms a basis for X_{p^n}.
3. The matrix representation of S_{p^n} with respect to this basis is given by

$$[S_{p^n}]_{jo}^{jo} = \begin{pmatrix} A & & & & \\ N & A & & & \\ & N & . & & \\ & & . & . & \\ & & & . & . \\ & & & & N & A \end{pmatrix}, \tag{6.26}$$

where

$$A = \begin{pmatrix} \alpha & -\beta \\ \beta & \alpha \end{pmatrix}, \qquad N = \begin{pmatrix} 0 & \beta^{-1} \\ 0 & 0 \end{pmatrix}.$$

Proof. The proof is analogous to that of Proposition 6.6 and is omitted.   ■

6.3 The Invariant Factor Algorithm

The **invariant factor algorithm** is the tool that allows us to move from the special case of cyclic operators to the general case.

We begin by introducing an equivalence relation in the space $\mathbb{F}[z]^{m \times n}$, i.e., the space of all $m \times n$ matrices over the ring $\mathbb{F}[z]$. We will use the identification of $\mathbb{F}[z]^{m \times n}$ with $\mathbb{F}^{m \times n}[z]$, the space of matrix polynomials with coefficients in $\mathbb{F}^{m \times n}$. Thus given a matrix $A \in \mathbb{F}^{m \times m}$, then $zI - A$ is a matrix polynomial of degree one, but usually we will consider it also as a polynomial matrix.

Definition 6.9. A polynomial matrix $U(z) \in \mathbb{F}[z]^{m \times m}$ is called **unimodular** if it is invertible in $\mathbb{F}[z]^{m \times m}$, that is, there exists a matrix $V(z) \in \mathbb{F}[z]^{m \times m}$ for which

$$U(z)V(z) = V(z)U(z) = I. \tag{6.27}$$

Lemma 6.10. $U(z) \in \mathbb{F}[z]^{m \times m}$ is unimodular if and only if $\det U(z) = \alpha \in \mathbb{F}$ and $\alpha \neq 0$.

Proof. Assume $U(z)$ is unimodular. Then there exists $V(z) \in \mathbb{F}[z]^{m \times m}$ for which $U(z)V(z) = V(z)U(z) = I$. By the multiplicative rule of determinants we have $\det U(z) \det V(z) = 1$. This implies that both $\det U(z)$ and $\det V(z)$ are nonzero scalars.

Conversely, if $\det U(z)$ is a nonzero scalar, then, using (3.9), we conclude that $U(z)^{-1}$ is a polynomial matrix. ∎

Definition 6.11. Two matrices $A, B \in \mathbb{F}[z]^{m \times n}$ are called **equivalent** if there exist unimodular matrices $U(z) \in \mathbb{F}[z]^{m \times m}$ and $V(z) \in \mathbb{F}[z]^{n \times n}$ such that

$$U(z)A(z) = B(z)V(z) = I.$$

It is easily checked that this definition indeed yields an equivalence relation. The next theorem leads to a canonical form for this equivalence. However, we find it convenient to prove first the following lemma.

Lemma 6.12. Let $A(z) \in \mathbb{F}[z]^{m \times n}$. Then $A(z)$ is equivalent to a block matrix of the form

$$\begin{pmatrix} a(z) & 0 & . & . & . & 0 \\ . & & & & & \\ . & & \mathbf{B}(z) & & & \\ . & & & & & \\ 0 & & & & & \end{pmatrix},$$

with $a(z) \mid b_{ij}(z)$.

Proof. If $A(z)$ is the zero matrix, there is nothing to prove. Otherwise, interchanging rows and columns as necessary, we can bring a nonzero entry of least degree to the upper left-hand corner. We use the division rule of polynomials to write each element of the first row in the form $a_{1j}(z) = c_{1j}(z)a_{11}(z) + a'_{1j}(z)$, with $\deg a'_{1j} < \deg a_{11}$. Next, we subtract the first column, multiplied by $c_{1j}(z)$, from the jth column. We repeat the process with the first column. If all the remainders are zero we check whether $a_{11}(z)$ divides all other entries in the matrix. If it does, we are through. Otherwise, we bring, using column and row interchanges, a lowest-degree nonzero element to the upper left-hand corner and repeat the process. ∎

Theorem 6.13 (The invariant factor algorithm). *Let* $A(z) \in \mathbb{F}[z]^{m \times n}$. *Then*

1. $A(z)$ *is equivalent to a diagonal polynomial matrix with the diagonal elements* $d_i(z)$ *monic polynomials satisfying* $d_i(z) \mid d_{i-1}(z)$.
2. *The polynomials* $d_i(z)$ *are uniquely determined and are called the* **invariant factors** *of* $A(z)$.

Proof. We use the previous lemma inductively. Thus, by elementary operations, A is reducible to diagonal form with the diagonal entries $d_i(z)$. If $d_1(z)$ does not divide $d_i(z)$, then we add the ith row to the first. Then, by an elementary column operation, reduce $d_i(z)$ modulo $d_1(z)$ and reapply Lemma 6.12. We repeat the process as needed, till $d_1(z) \mid d_i(z)$ for $i > 1$. We proceed by induction to get $d_i(z) \mid d_{i+1}(z)$. This is the opposite ordering from that in the statement of the theorem. We can get the right ordering by column and row interchanges. ∎

The polynomial matrix $D(z)$ having nontrivial invariant factors on the diagonal will be called the **Smith form** of $A(z)$.

Proposition 6.14. *A polynomial matrix* $U(z)$ *is unimodular if and only if it is the product of a finite number of elementary unimodular matrices.*

Proof. The product of unimodular matrices is, by the multiplicative rule for determinants, also unimodular.

Conversely, assume $U(z)$ is unimodular. By elementary row and column operations, it can be brought to its Smith form $D(z)$, with the polynomials $d_i(z)$ on the diagonal. Since $D(z)$ is also unimodular, $D(z) = I$. Let now $U_i(z)$ and $V_j(z)$ be the elementary unimodular matrices representing the elementary operations in the diagonalization process. Thus $U_k(z) \cdots U_1(z)U(z)V_1(z) \cdots V_l(z) = I$, which implies $U(z) = U_1(z)^{-1} \cdots U_k(z)^{-1} V_l(z)^{-1} \cdots V_1(z)^{-1}$. This is a representation of $U(z)$ as the product of elementary unimodular matrices. ∎

To prove the uniqueness of the invariant factors, we introduce the determinantal divisors.

Definition 6.15. *Let* $A(z) \in \mathbb{F}[z]^{m \times n}$. *Then we define* $D_0(A) = 1$ *and* $D_k(A)$ *to be the g.c.d., taken to be monic, of all* $k \times k$ *minors of* $A(z)$. *The* $D_i(A)$ *are called the* **determinantal divisors** *of* $A(z)$.

Proposition 6.16. *Let* $A(z), B(z) \in \mathbb{F}[z]^{m \times n}$. *If* $B(z)$ *is obtained from* $A(z)$ *by a series of elementary row or column operations, then* $D_k(A) = D_k(B)$.

Proof. We consider the set of all $k \times k$ minors of $A(z)$. We consider the effect of applying elementary transformations on $A(z)$ on this set. Multiplication of a row or column by a nonzero scalar α can change a minor by at most a factor α. This does not change the g.c.d. Interchanging two rows or columns leaves the set of minors unchanged, except for a possible sign. The only elementary operation that has a nontrivial effect is adding the jth row, multiplied by a polynomial $p(z)$, to the ith row. In a minor that does not contain the ith row, or contains elements of both the ith and the jth rows, this has no effect. If a minor contains elements of the ith row but not of the jth, we expand by the ith row. We get the original minor added to the product of another minor by the polynomial $p(z)$. Again, this does not change the g.c.d. $\blacksquare$

Theorem 6.17. *Let $A(z), B(z) \in \mathbb{F}[z]^{m \times n}$. Then $A(z)$ and $B(z)$ are equivalent if and only if $D_k(A) = D_k(B)$.*

Proof. It is clear from the previous proposition that if a single elementary operation does not change the set of elementary divisors, then also a finite sequence of elementary operation leaves the determinantal divisors unchanged. Since any unimodular matrix is the product of elementary unimodular matrices, we conclude that the equivalence of $A(z)$ and $B(z)$ implies $D_k(A) = D_k(B)$.

Assume now that $D_k(A) = D_k(B)$ for all k. Using the invariant factor algorithm, we reduce $A(z)$ and $B(z)$ to their Smith form with the invariant factors $d_1(z), \ldots, d_r(z)$, ordered by $d_{i-1}(z) \mid d_i(z)$, and $e_1(z), \ldots, e_s(z)$, similarly ordered, respectively. Clearly $D_k(A) = d_1(z) \cdots d_k(z)$ and $D_k(B) = e_1(z) \cdots e_k(z)$. This implies $r = s$ and, assuming the invariant factors are all monic, $e_i(z) = d_i(z)$. By transitivity of equivalence we get $A(z) \simeq B(z)$. $\blacksquare$

Corollary 6.18. *The invariant factors of $A(z) \in \mathbb{F}[z]^{m \times n}$ are uniquely determined.*

We can view the invariant factor algorithm as a far-reaching generalization of the Euclidean algorithm, for the Euclidean algorithm is the invariant factor algorithm as applied to a polynomial matrix of the form $(p(z) \quad q(z))$.

6.4 Noncyclic Transformations

We continue now with the study of general, i.e., not necessarily cyclic, transformations in finite-dimensional vector spaces. Our aim is to reduce their study to that of the cyclic ones, and we will do so using the invariant factor algorithm. However, to use it effectively, we will have to extend the modular polynomial arithmetic to the case of vector polynomials. We will not treat the case of remainders with respect to general nonsingular polynomial matrices, since for our purposes, it will suffice to consider two special cases. The first case is the operator π_{zI-A}, defined in (5.71) and discussed in Proposition 5.42. The second case is that of nonsingular diagonal polynomial matrices.

There is a projection map, similar to the projection π_d, defined in (5.2), that we shall need in the following, and which we proceed to introduce. Assume $D(z) = \begin{pmatrix} d_1(z) \\ & \ddots \\ & & d_n(z) \end{pmatrix}$, with $d_i(z)$ nonzero polynomials. Given a vector polynomial $f(z) = \begin{pmatrix} f_1(z) \\ \vdots \\ f_n(z) \end{pmatrix}$, we define

$$\pi_D f = \begin{pmatrix} \pi_{d_1} f_1 \\ \vdots \\ \pi_{d_n} f_n \end{pmatrix}, \tag{6.28}$$

where $\pi_{d_i} f_i$ are the remainders of $f_i(z)$ after division by $d_i(z)$.

Proposition 6.19. *Given a nonsingular diagonal polynomial matrix $D(z)$, then*

1. We have

$$\operatorname{Ker} \pi_D = D\mathbb{F}^n[z]. \tag{6.29}$$

2.

$$X_D = \operatorname{Im} \pi_D = \left\{ \begin{pmatrix} f_1 \\ \vdots \\ f_n \end{pmatrix} \mid f_i(z) \in \operatorname{Im} \pi_{d_i} \right\}. \tag{6.30}$$

In particular, $X_D \simeq X_{d_1} \oplus \cdots \oplus X_{d_n}$.

3. Defining the map $S_D : X_D \longrightarrow X_D$ by $S_D f = \pi_D z f$ for $f(z) \in X_D$, we have $S_D \simeq S_{d_1} \oplus \cdots \oplus S_{d_n}$.

Proof. 1. Clearly, $f(z) \in \operatorname{Ker} \pi_D$ if and only if for all i, $f_i(z)$ is divisible by $d_i(z)$. Thus

$$\operatorname{Ker} \pi_D = \left\{ \begin{pmatrix} d_1 f_1 \\ \vdots \\ d_n f_n \end{pmatrix} \mid f_i(z) \in \mathbb{F}[z] \right\} = D\mathbb{F}^n[z].$$

2. Follows from the definition of π_D.

3. We compute

$$S_D \begin{pmatrix} f_1 \\ \vdots \\ f_n \end{pmatrix} = \pi_D z \begin{pmatrix} f_1 \\ \vdots \\ f_n \end{pmatrix} = \begin{pmatrix} \pi_{d_1} f_1 \\ \vdots \\ \pi_{d_n} f_n \end{pmatrix} = \begin{pmatrix} S_{d_1} f_1 \\ \vdots \\ S_{d_n} f_n \end{pmatrix}. \qquad \blacksquare$$

With this background material, we can proceed to the structure theorem we have been after. It shows that an arbitrary linear transformation is isomorphic to a direct sum of cyclic transformations corresponding to the invariant factors of A.

Theorem 6.20. *Let A be a linear transformation in $\mathbb{F}^n$. Let $d_1(z),\dots,d_n(z)$ be the invariant factors of $zI - A$. Then A is isomorphic to $S_{d_1} \oplus \cdots \oplus S_{d_n}$.*

Proof. Since $d_1(z),\dots,d_n(z)$ are the invariant factors of $zI - A$, there exist unimodular polynomial matrices $U(z)$ and $V(z)$ satisfying

$$U(z)(zI - A) = D(z)V(z). \qquad (6.31)$$

This equation implies

$$U \operatorname{Ker} \pi_{zI-A} \subset \operatorname{Ker} \pi_D. \qquad (6.32)$$

Since $U(z)$ and $V(z)$ are invertible as polynomial matrices, we have also

$$U(z)^{-1}D(z) = (zI - A)V(z)^{-1}, \qquad (6.33)$$

which implies

$$U^{-1} \operatorname{Ker} \pi_D \subset \operatorname{Ker} \pi_{zI-A}. \qquad (6.34)$$

We define now two maps, $\Phi : \mathbb{F}^n \longrightarrow X_D$ and $\Psi : X_D \longrightarrow \mathbb{F}^n$, by

$$\Phi x = \pi_D U x, \qquad x \in \mathbb{F}^n, \qquad (6.35)$$

and

$$\Psi f = \pi_{zI-A} U^{-1} f, \qquad f(z) \in X_D. \qquad (6.36)$$

We claim now that Φ is invertible and its inverse is Ψ. Indeed, using (6.34)-(6.36), we compute

$$\Psi \Phi x = \pi_{zI-A} U^{-1} \pi_D U x = \pi_{zI-A} U^{-1} U x = \pi_{zI-A} x = x.$$

Similarly, using (6.32), we have, for $f \in X_D$,

$$\Phi \Psi f = \pi_D U \pi_{zI-A} U^{-1} f = \pi_D U U^{-1} f = \pi_D f = f.$$

Next, we show that the following diagram is commutative:

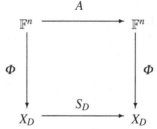

Indeed, for $x \in \mathbb{F}^n$, using (6.32) and the fact that $z \mathrm{Ker}\, \pi_D \subset \mathrm{Ker}\, \pi_D$, we have

$$
\begin{aligned}
S_D \Phi x &= \pi_D z \cdot \pi_D U x = \pi_D z U x = \pi_D U z x \\
&= \pi_D U \pi_{zI-A} z x = \pi_D U (Ax) = \Phi A x.
\end{aligned}
$$

Since Φ is an isomorphism, we have $A \simeq S_D$. But $S_D \simeq S_{d_1} \oplus \cdots \oplus S_{d_n}$, and the result follows.   ∎

Corollary 6.21. *Let A be a linear transformation in an n-dimensional vector space V. Let $d_1(z), \ldots, d_n(z)$ be the invariant factors of A. Then A is similar to the diagonal block matrix composed of the companion matrices corresponding to the invariant factors, i.e.,*

$$
A \simeq \begin{pmatrix} C_{d_1}^{\sharp} & & & \\ & \cdot & & \\ & & \cdot & \\ & & & C_{d_n}^{\sharp} \end{pmatrix}. \tag{6.37}
$$

Proof. By Theorem 6.20, A is isomorphic to $S_{d_1} \oplus \cdots \oplus S_{d_n}$ acting in $X_{d_1} \oplus \cdots \oplus X_{d_n}$. Choosing the standard basis in each X_{d_i} and taking the union of these basis elements, we have

$$
[S_{d_1} \oplus \cdots \oplus S_{d_n}]_{st}^{st} = \mathrm{diag}\,([S_{d_1}]_{st}^{st}, \ldots, [S_{d_n}]_{st}^{st}).
$$

We conclude by recalling that $[S_{d_i}]_{st}^{st} = C_{d_i}^{\sharp}$.   ∎

Theorem 6.22. *Let A and B be linear transformations in n-dimensional vector spaces $\mathscr{U}$ and $\mathscr{V}$ respectively. Then the following statements are equivalent.*

1. A and B are similar.
2. The polynomial matrices $zI - A$ and $zI - B$ are equivalent.
3. The invariant factors of A and B coincide.

Proof. 1. Assume A and B are similar. Let X be an invertible matrix for which $XA = BX$. This implies $X(zI - A) = (zI - B)X$ and so the equivalence of $zI - A$ and $zI - B$.
2. The equivalence of the polynomial matrices $zI - A$ and $zI - B$ shows that their invariant factors coincide.
3. Assuming the invariant factors coincide, both A and B are similar to $S_{d_1} \oplus \cdots \oplus S_{d_n}$, and the claim follows from the transitivity of similarity.   ∎

Corollary 6.23. *Let A be a linear transformation in an n-dimensional vector space $\mathscr{U}$ over $\mathbb{F}$. Let $d_1(z), \ldots, d_n(z)$ be the invariant factors of A, ordered so that $d_i \mid d_{i-1}$. Then*

1. *The minimal polynomial of A, $m_A(z)$, satisfies*

$$m_A(z) = d_1(z).$$

2. *$d_A(z)$, the characteristic polynomial of A, satisfies*

$$d_A(z) = \prod_{i=1}^{n} d_i(z). \tag{6.38}$$

Proof. 1. Since $A \simeq S_{d_1} \oplus \cdots \oplus S_{d_n}$, for an arbitrary polynomial $p(z)$ we have $p(A) \simeq p(S_{d_1}) \oplus \cdots \oplus p(S_{d_n})$. So $p(A) = 0$ if and only if $p(S_{d_i}) = 0$ for all i, or equivalently, $d_i \mid p$. In particular, $d_1 \mid p$. But $d_1(z)$ is the minimal polynomial of S_{d_1}, so it is also the minimal polynomial of $S_{d_1} \oplus \cdots \oplus S_{d_n}$ and, by similarity, also of A.

2. Using Proposition 6.20 and the matrix representation (6.37), (6.38) follows. ∎

Next we give a characterization of cyclicity in terms of the characteristic and minimal polynomials.

Proposition 6.24. *A transformation T in $\mathscr{U}$ is cyclic if and only if its characteristic and minimal polynomials coincide.*

Proof. It is obvious that if T is cyclic, its characteristic and minimal polynomials have to coincide.

Conversely, if the characteristic and minimal polynomials coincide, then the only nontrivial invariant factor $d(z)$ is equal to the characteristic polynomial. By Theorem 6.20 we have $T \simeq S_d$, and since S_d is cyclic, so is T. ∎

6.5 Diagonalization

The maximal simplification one can hope to obtain for the structure of an arbitrary linear transformation in a finite-dimensional vector space is diagonalizability. Unhappily, that is too much to ask. The following theorem gives a complete characterization of diagonalizability.

Theorem 6.25. *Let T be a linear transformation in a finite-dimensional vector space $\mathscr{U}$. Then T is diagonalizable if and only if $m_T(z)$, the minimal polynomial of T, splits into distinct linear factors.*

Proof. This follows from Proposition 5.15, taken in conjunction with Theorem 6.20. However, we find it of interest to give a direct proof.

That the existence of a diagonal representation implies that $m_T(z)$, the minimal polynomial of T, splits into distinct linear factors is obvious.

Assume now that the minimal polynomial splits into distinct linear factors. So $m_T(z) = \Pi_{i=1}^k (z - \lambda_i)$, where the λ_i are distinct. Let $\pi_i(z)$ be the corresponding Lagrange interpolation polynomials. Clearly $(z - \lambda_i)\pi_i(z) = m_T(z)$. Since for $i \neq j$, $(z - \lambda_i) \mid \pi_j(z)$, we get

$$m_T(z) \mid \pi_i(z)\pi_j(z). \tag{6.39}$$

Similarly, the polynomial $\pi_i(z)^2 - \pi_i(z)$ vanishes at all the points λ_i, $i = 1, \ldots, k$. So

$$m_T(z) \mid \pi_i(z)^2 - \pi_i(z). \tag{6.40}$$

The two division relations imply now

$$\pi_i(T)\pi_j(T) = 0, \qquad i \neq j, \tag{6.41}$$

and

$$\pi_i(T)^2 = \pi_i(T). \tag{6.42}$$

It follows that the $\pi(T)$ are projections on independent subspaces. We set $\mathcal{U}_i = \operatorname{Im} \pi_i(T)$. Then, since the equality $1 = \sum_{i=1}^k \pi_i(z)$ implies $I = \sum_{i=1}^k \pi_i(T)$, we get the direct sum decomposition $\mathcal{U} = \mathcal{U}_1 \oplus \cdots \oplus \mathcal{U}_k$. Moreover, since $T\pi_i(T) = \pi_i(T)T$, the subspaces $\mathcal{U}_i$ are T-invariant. From the equality $(z - \lambda_i)\pi_i(z) = m_T(z)$ we get $(T - \lambda_i I)\pi_i(T) = 0$, or $\mathcal{U}_i \subset \operatorname{Ker}(\lambda_i I - T)$. This can be restated as $T|\mathcal{U}_i = \lambda_i I_{\mathcal{U}_i}$. Choosing bases in the subspaces $\mathcal{U}_i$ and taking their union leads to a basis of $\mathcal{U}$ made up of eigenvectors. With respect to this basis, T has a diagonal matrix representation.   ■

The following is the basic structure theorem, yielding a spectral decomposition for a linear transformation. The result follows from our previous considerations and can be read off from the Jordan form. Still, we find it of interest to give also another proof.

Theorem 6.26. *Let T be a linear transformation in a finite-dimensional vector space $\mathcal{U}$. Let $m_T(z)$ be the minimal polynomial of T and let $m_T(z) = p_1(z)^{v_1} \cdots p_k(z)^{v_k}$ be the primary decomposition of $m_T(z)$. We set $\mathcal{U}_i = \operatorname{Ker} p_i(T)^{v_i}$, $i = 1, \ldots, k$, and define polynomials by $\pi_i(z) = \Pi_{j \neq i} p_i^{v_i}(z)$. Then*

1. There exists a unique representation

$$1 = \sum_{j=1}^k \pi_j(z)a_j(z), \tag{6.43}$$

with $a_j \in X_{p_j^{v_j}}$.

2. We have

$$m_T \mid \pi_i \pi_j, \quad for \ i \neq j, \tag{6.44}$$

and

$$m_T \mid \pi_i^2 a_i^2 - \pi_i a_i. \tag{6.45}$$

3. *The operators $P_i = \pi_i(T)a_i(T)$, $i = 1,\ldots,k$, are projection operators satisfying*

$$P_i P_j = \delta_{ij} P_j,$$
$$I = \Sigma_{i=1}^k P_i,$$
$$T P_i = P_i T P_i,$$
$$\operatorname{Im} P_i = \operatorname{Ker} p_i^{v_i}(T).$$

4. *Define $\mathscr{U}_i = \operatorname{Im} P_i$. Then $\mathscr{U} = \mathscr{U}_1 \oplus \cdots \oplus \mathscr{U}_k$.*
5. *The subspaces $\mathscr{U}_i$ are T-invariant.*
6. *Let $T_i = T|\mathscr{U}_i$. Then the minimal polynomial of T_i is $p_i(z)^{v_i}$.*

Proof. 1. Clearly, the $\pi_i(z)$ are relatively prime and $p_i(z)^{v_i}\pi_i(z) = m_T(z)$. By the Chinese remainder theorem, there exists a unique representation of the form $1 = \Sigma_{j=1}^k \pi_j(z)a_j(z)$, with $a_j(z) \in X_{p_j^{v_j}}$. The $a_j(z)$ can be computed as follows. Since for $i \neq j$ we have $\pi_{p_i^{v_i}}\pi_j f = 0$, we have

$$1 = \pi_{p_i^{v_i}} \sum_{j=1}^k \pi_j a_j = \pi_{p_i^{v_i}} \pi_i a_i = \pi_i(S_{p_i^{v_i}})a_i,$$

and hence $a_i = \pi_i(S_{p_i^{v_i}})^{-1}1$. That the operator $\pi_i(S_{p_i^{v_i}}) : X_{p_i^{v_i}} \longrightarrow X_{p_i^{v_i}}$ is invertible follows, by Theorem 5.18, from the coprimeness of $\pi_i(z)$ and $p_i^{v_i}(z)$.

2. To see (6.44), note that if $j \neq i$, $p_i^{v_i} \mid \pi_j$, and hence $m_T(z)$ is a factor of $\pi_i(z)\pi_j(z)$. Naturally, we get also $m_T \mid \pi_i a_i \pi_j a_j$.

   To see (6.45), it suffices to show that for all j, $p_i^{v_i} \mid \pi_j^2 a_j^2 - \pi_j a_j$, or equivalently, that $\pi_j^2 a_j^2$ and $\pi_j a_j$ have the same remainder after division by $p_i^{v_i}$. For $j \neq i$ this is trivial, since $p_i^{v_i}|\pi_j$. So we assume $j = i$ and compute

$$\pi_{p_i^{v_i}}(\pi_i^2 a_i^2) = \pi_{p_i^{v_i}}(\pi_i a_i)\pi_{p_i^{v_i}}\pi_i a_i = 1,$$

for $\pi_{p_i^{v_i}}\pi_i a_i = \pi_i(S_{p_i^{v_i}})a_i = 1$.

3. Properties (6.44) and (6.45) have now the following consequences. For $i \neq j$ we have $\pi_i(T)a_i(T)\pi_j(T)a_j(T) = 0$, and $(\pi_i(T)a_i(T))^2 = \pi_i(T)a_i(T)$. This implies that $P_i = \pi_i(T)a_i(T)$ are commuting projections on independent subspaces.

4. Set now $\mathscr{U}_i = \operatorname{Im} P_i = \operatorname{Im} \pi_i(T)a_i(T)$. Equation (6.43) implies $I = \Sigma_{i=1}^k P_i$, and hence we get the direct sum decomposition $\mathscr{U} = \mathscr{U}_1 \oplus \cdots \oplus \mathscr{U}_k$.

5. Since P_i commutes with T, the subspaces $\mathscr{U}_i$ actually reduce T.

6. Let $T_i = T|\mathscr{U}_i$. Since $p_i^{v_i}\pi_i = m_T$, it follows that for all $x \in \mathscr{U}$,

$$p_i^{v_i}(T)P_i x = p_i^{v_i}(T)\pi_i(T)a_i(T)x = 0,$$

i.e., the minimal polynomial of T_i divides $p_i^{v_i}$. To see the converse, let $q(z)$ be any polynomial such that $q(T_i) = 0$. Then $q(T)\pi_i(T) = 0$. It follows that $q(z)\pi_i(z)$ is divisible by the minimal polynomial of T. But that means that $p_i^{v_i}\pi_i \mid q\pi_i$ or $p_i^{v_i} \mid q$. We conclude that the minimal polynomial of T_i is $p_i(z)^{v_i}$. ∎

6.6   Exercises

1. Let $T(z)$ be an $n \times n$ nonsingular polynomial matrix. Show that there exists a unimodular matrix $P(z)$ such that $P(z)T(z)$ has the upper triangular form

$$\begin{pmatrix} t_{11}(z) & \ldots & t_{1n}(z) \\ 0 & \ldots & \cdot \\ \cdot & \ldots & \cdot \\ \cdot & \ldots & \cdot \\ 0 & \ldots 0 & t_{nn}(z) \end{pmatrix}$$

   with $t_{ii}(z) \neq 0$ and $\deg(t_{ji}) < \deg(t_{ii})$.

2. Let T be a linear operator on $\mathbb{F}^2$. Prove that any nonzero vector that is not an eigenvector of T is a cyclic vector for T. Prove that T is either cyclic or a multiple of the identity.

3. Let T be a diagonalizable operator in an n-dimensional vector space. Show that if T has a cyclic vector, then T has n distinct eigenvalues. If T has n distinct eigenvalues, and if $x_1, \ldots, x_n$ is a basis of eigenvectors, show that $x = x_1 + \cdots + x_n$ is a cyclic vector for T.

4. Let T be a linear transformation in the finite-dimensional vector space $\mathcal{V}$. Prove that $\operatorname{Im} T$ has a complementary invariant subspace if and only if $\operatorname{Im} T$ and $\operatorname{Ker} T$ are independent subspaces. Show that if $\operatorname{Im} T$ and $\operatorname{Ker} T$ are independent subspaces, then $\operatorname{Ker} T$ is the unique complementary invariant subspace for $\operatorname{Im} T$.

5. How many possible Jordan forms are there for a 6×6 complex matrix with characteristic polynomial $(z+2)^4(z-1)^2$.

6. Given a linear transformation A in an n-dimensional space $\mathcal{V}$, let $d(z)$ and $m(z)$ be its characteristic and minimal polynomials respectively. Show that $d(z)$ divides $m^n(z)$.

7. Let $p(z) = p_0 + p_1 z + \cdots + p_{k-1} z^{k-1}$. Solve the equation

$$\pi_{z^k} q p = 1.$$

8. Let $A = \begin{pmatrix} a & c \\ b & d \end{pmatrix}$ be a complex matrix. Apply the Gram-Schmidt process to the columns of A. Show that this results in a basis for $\mathbb{C}^2$ if and only if $\det A \neq 0$.

9. Let A be an $n \times n$ matrix over the field $\mathbb{F}$. We say that A is **diagonalizable** if there exists a matrix representation of A that is diagonal. Show that *diagonalizable* and *cyclic* are independent concepts, i.e., show the existence of matrices that are $DC, \overline{D}C, D\overline{C}, \overline{D}\overline{C}$. (Here $\overline{D}C$ means nondiagonalizable and cyclic.)

10. Let A be an $n \times n$ Jordan block matrix, i.e.,

$$A = \begin{pmatrix} \lambda & 1 & & & \\ & \ddots & & & \\ & & \ddots & & \\ & & & \ddots & 1 \\ & & & & \lambda \end{pmatrix},$$

and

$$N = \begin{pmatrix} 0 & 1 & & & \\ & \ddots & & & \\ & & \ddots & & \\ & & & \ddots & 1 \\ & & & & 0 \end{pmatrix},$$

i.e., $A = \lambda I + N$. Let p be a polynomial of degree r. Show that

$$p(A) = \sum_{k=0}^{n-1} \frac{1}{k!} p^{(k)}(\lambda) N^k.$$

11. Show that if A is diagonalizable and $p(z) \in \mathbb{F}[z]$, then $p(A)$ is diagonalizable.
12. Let $p(z) \in \mathbb{F}[z]$ be of degree n.

   a. Show that the dimension of the smallest invariant subspace of X_p that contains $f(z)$ is $n - r$, where $r = \deg(f \wedge p)$.
   b. Conclude that $f(z)$ is a cyclic vector of $S_p(z)$ if and only if $f(z)$ and $p(z)$ are coprime.

13. Let $A : \mathcal{V} \longrightarrow \mathcal{V}$ and $A_1 : \mathcal{V}_1 \longrightarrow \mathcal{V}_1$ be linear transformations with minimal polynomials m and m_1 respectively.

   a. Show that if there exists a surjective (onto) map $Z : \mathcal{V} \longrightarrow \mathcal{V}_1$ satisfying $ZA = A_1 Z$, then $m_1 \mid m$. (Is the converse true? Under what conditions?)
   b. Show that if there exists an injective (1 to 1) map $Z : \mathcal{V}_1 \longrightarrow \mathcal{V}$ satisfying $ZA_1 = AZ$, then $m_1 \mid m$. (Is the converse true? Under what conditions?)
   c. Show that the same is true with the minimal polynomials replaced by the characteristic polynomials.

14. Let $A : \mathcal{V} \longrightarrow \mathcal{V}$ be a linear transformation and $\mathcal{M} \subset \mathcal{V}$ an invariant subspace of A. Show that the minimal and characteristic polynomials of $A|\mathcal{M}$ divide the minimal and characteristic polynomials of A respectively.

15. Let $A : \mathcal{V} \longrightarrow \mathcal{V}$ be a linear transformation and $\mathcal{M} \subset \mathcal{V}$ an invariant subspace of A. Let $d_1, \ldots, d_n$ be the invariant factors of A and $e_1, \ldots, e_m$ be the invariant factors of $A|\mathcal{M}$.

   a. Show that $e_i \mid d_i$ for $i = 1, \ldots, m$.
   b. Show that given arbitrary vectors $v_1, \ldots, v_k$ in $\mathcal{V}$, we have

$$\dim \mathrm{span}\,\{A^i v_j \mid j = 1, \ldots, k, \quad i = 0, \ldots, n-1\} \leq \sum_{j=1}^{k} \deg d_j.$$

16. Let $A : \mathscr{V} \longrightarrow \mathscr{V}$ and assume

$$\mathscr{V} = \mathrm{span}\,\{A^i v_j \mid j = 1, \ldots, k, \quad i = 0, \ldots, n-1\}.$$

Show that the number of nontrivial invariant factors of A is $\leq k$.

17. Let $A(z)$ and $B(z)$ be polynomial matrices and let $C(z) = A(z)B(z)$. Let $a_i(z), b_i(z), c_i(z)$ be the invariant factors of A, B, C respectively. Show that $a_i \mid c_i$ and $b_i \mid c_i$.

18. Let A and B be cyclic linear transformations. Show that if $ZA = BZ$ and Z has rank r, then the degree of the greatest common divisor of the characteristic polynomials of A and B is at least r.

19. Show that if the minimal polynomials of A_1 and A_2 are coprime, then the only solution to $ZA_1 = A_2 Z$ is the zero solution.

20. If there exists a rank r solution to $ZA_1 = A_2 Z$, what can you say about the characteristic polynomials of A_1 and A_2?

21. Let $\mathscr{V}$ be an n-dimensional linear space. Assume $T : \mathscr{V} \longrightarrow \mathscr{V}$ is diagonalizable. Prove

   a. If T is cyclic, then T has n distinct eigenvalues.
   b. If T has n distinct eigenvalues, then T is cyclic.
   c. If $b_1, \ldots, b_n$ are the eigenvectors corresponding to the distinct eigenvalues, then $b = b_1 + \cdots + b_n$ is a cyclic vector for T.

22. Show that if T^2 is cyclic, then T is cyclic. Is the converse true?

23. Let $\mathscr{V}$ be an n-dimensional linear space over the field $\mathbb{F}$. Show that every nonzero vector in $\mathscr{V}$ is a cyclic vector of T if and only if the characteristic polynomial of T is irreducible over $\mathbb{F}$.

24. Show that if $m(z)$ is the minimal polynomial of a linear transformation $A : \mathscr{V} \longrightarrow \mathscr{V}$, then there exists a vector $x \in \mathscr{V}$ such that

$$\{p \in \mathbb{F}[z] \mid p(A)x = 0\} = m\mathbb{F}[z].$$

25. Let $A : \mathscr{V} \longrightarrow \mathscr{V}$ be a linear transformation. Show that:

   a. For each $x \in \mathscr{V}$, there exists a smallest A-invariant subspace M_x containing x.
   b. x is a cyclic vector for $A|M_x$.

26. Show that if A_1, A_2 are cyclic maps with cyclic vectors b_1, b_2 respectively and if the minimal polynomials of A_1 and A_2 are coprime, then $b_1 \oplus b_2$ is a cyclic vector for $A_1 \oplus A_2$.

27. Let A be a cyclic transformation in $\mathscr{V}$ and let $\mathscr{M} \subset \mathscr{V}$ be an invariant subspace of T. Show that M has a complementary invariant subspace if and only if the characteristic polynomials of $T|_{\mathscr{M}}$ and $T_{\mathscr{V}/\mathscr{M}}$ are coprime.

28. Show that if $T^2 - T = 0$, then T is similar to a matrix of the form $\operatorname{diag}(1,\ldots,1, 0,\ldots,0)$.

29. The following series of exercises, based on Axler (1995), provides an alternative approach to the structure theory of a linear transformation, which avoids the use of determinants. We do not assume the underlying field to be algebraically closed.

    Let T be a linear transformation in a finite-dimensional vector space $\mathscr{X}$ over the field $\mathbb{F}$. A nonzero polynomial $p(z) \in \mathbb{F}[z]$ is called a **prime of T** if $p(z)$ is monic, irreducible and there exists a nonzero vector $x \in X$ for which $p(T)x = 0$. Such a vector x is called a p-null vector. Clearly, these notions generalize eigenvalues and eigenvectors.

    a. Show that every linear transformation T in a finite-dimensional vector space $\mathscr{X}$ has at least one prime.
    b. Let $p(z)$ be a prime of T and let x be a p-null vector. If $q(z)$ is any polynomial for which $q(T)x = 0$, then $p(z) \mid q(z)$. In particular, if $\deg q < \deg p$ and $q(T)x = 0$, then necessarily $q(z) = 0$.
    c. Show that nonzero p_i-null vectors corresponding to distinct primes are linearly independent.
    d. Let $p(z)$ be a prime of T with $\deg p = v$. Show that

    $$\{x \mid p(T)^k x = 0, \text{for some } k = 1,2,\ldots\} = \{x \mid p(T)^{\lceil n/v \rceil} x = 0\}.$$

    We call elements of $\operatorname{Ker} p(T)^{\lceil n/v \rceil}$ generalized p-null vectors.
    e. Show that $\mathscr{X}$ is spanned by generalized p-null vectors.
    f. Show that generalized p-null vectors corresponding to distinct primes are linearly independent.
    g. Let $p_1(z),\ldots,p_m(z)$ be the distinct primes of T and let $\mathscr{X}_i = \operatorname{Ker} p_i(T)^{\lceil n/v_i \rceil}$. Prove

       i. $\mathscr{X}$ has the direct sum representation $\mathscr{X} = \mathscr{X}_1 \oplus \cdots \oplus \mathscr{X}_m$.
       ii. $\mathscr{X}_i$ are T-invariant subspaces.
       iii. The operators $p_i(T)|_{\mathscr{X}_i}$ are nilpotent.
       iv. The operator $T|_{\mathscr{X}_i}$ has only one prime, namely $p_i(z)$.

30. Given a nilpotent operator N, show that it is similar to a matrix of the form

$$\begin{pmatrix} N_1 & & & \\ & \cdot & & \\ & & \cdot & \\ & & & \cdot \\ & & & & N_k \end{pmatrix},$$

with N_i of the form

$$\begin{pmatrix} 0 & 1 & & & \\ & & \cdot & \cdot & \\ & & & \cdot & \cdot & \\ & & & & \cdot & 1 \\ & & & & & 0 \end{pmatrix}.$$

31. Let T be a linear operator in an n-dimensional complex vector space $\mathscr{X}$. Let $\lambda_1, \ldots, \lambda_m$ be the distinct eigenvalues of T and let $m(z) = \prod_{i=1}^{m}(z - \lambda_i)^{v_i}$ be its minimal polynomial. Let $H_i(z)$ be the minimal degree solution to the Hermite interpolation problem $H_i^{(k)}(\lambda_j) = \delta_{0k}\delta_{ij}$, $i, j = 1, \ldots, m$, $k = 0, \ldots, v_j - 1$. Define $P_k = H_k(T)$. Show that for any polynomial $p(z)$, we have

$$p(T) = \sum_{k=1}^{m} \sum_{j=0}^{v_k - 1} \frac{p^{(j)}(\lambda_k)}{j!}(T - \lambda_k I)^j P_k.$$

32. Let T be a linear operator in an n-dimensional vector space $\mathscr{X}$. Let $d_i(z)$ be the invariant factors of T and let $\delta_i = \deg d_i$. Show that the commutant of T has dimension equal to $\sum_{i=1}^{n}(2i - 1)\delta_i$. Conclude that the commutant of T is $L(X)$ if and only if T is scalar, i.e., $T = \alpha I$ for some $\alpha \in \mathbb{F}$. Show also that the dimension of the commutant of T is equal to n if and only if T is cyclic.

6.7   Notes and Remarks

The polynomial module structure induced by a linear transformation acting in a vector space is what in functional analysis is called a functional calculus; see Dunford and Schwartz (1958).

The Jordan canonical form originated with C. Jordan and K. Weierstrass, using different techniques. Actually, in 1874, Jordan and Kronecker, himself a student of Weierstrass, were quarreling over priorities regarding pencil equivalence and the reduction to canonical forms. Jordan was stressing spectral theory, i.e., eigenvalues and eigenvectors, whereas Weierstrass, defended by Kronecker, was stressing invariant factors and elementary divisors. For a comprehensive discussion of this controversy, see Brechenmacher (2007).

The study of a linear transformation on a vector space via the study of the polynomial module structure induced by it on that space appears already in van der Waerden (1931). Although it is very natural, it did not become the standard approach in the literature, most notably in books aimed at a broader mathematical audience. This is probably due to the perception that the concept of module is too abstract.

Our approach to the structure theory of linear transformations and the Jordan canonical form, based on polynomial models, rational models, and shift operators, is a concretization of that approach. It was originated in Fuhrmann (1976), which in turn was based on ideas stemming from Hilbert space operator theory. In this connection, the reader is advised to consult Sz.-Nagy and Foias (1970) and Fuhrmann (1981). The transition from the cyclic case to the noncyclic case, described in Section 6.4, uses a special case of general polynomial model techniques introduced in Fuhrmann (1976).

Chapter 7
Inner Product Spaces

7.1 Introduction

In this chapter we focus on the study of vector spaces and linear transformations that relate to notions of distance, angle, and orthogonality. We will restrict ourselves throughout to the case of the real field $\mathbb{R}$ or the complex field $\mathbb{C}$.

7.2 Geometry of Inner Product Spaces

Definition 7.1. Let $\mathcal{U}$ be a vector space over $\mathbb{C}$ (or $\mathbb{R}$). An **inner product** on $\mathcal{U}$ is a function $(\cdot, \cdot) : \mathcal{U} \times \mathcal{U} \longrightarrow \mathbb{C}$ that satisfies the following conditions:

1. $(x,x) \geq 0$, and $(x,x) = 0$ if and only if $x = 0$.
2. For $\alpha_1, \alpha_2 \in \mathbb{C}$ and $x_1, x_2, y \in \mathcal{U}$ we have

$$(\alpha_1 x_1 + \alpha_2 x_2, y) = \alpha_1(x_1, y) + \alpha_2(x_2, y).$$

3. We have

$$(x,y) = \overline{(y,x)}.$$

We note that these axioms imply the antilinearity of the inner product in the right variable, that is,

$$(x, \alpha_1 y_1 + \alpha_2 y_2) = \overline{\alpha_1}(x, y_1) + \overline{\alpha_2}(x, y_2).$$

A function $\phi(x,y)$ that is linear in x and antilinear in y and satisfies $\phi(x,y) = \overline{\phi(y,x)}$ is called a **Hermitian form**. Thus the inner product is a Hermitian form in $\mathcal{U}$.

We define the **norm** of a vector $x \in \mathcal{U}$ by

$$\|x\| = (x,x)^{\frac{1}{2}},$$

P.A. Fuhrmann, *A Polynomial Approach to Linear Algebra*, Universitext,
DOI 10.1007/978-1-4614-0338-8_7, © Springer Science+Business Media, LLC 2012

where we take the nonnegative square root. A vector $x \in \mathscr{U}$ is called normalized, or a unit vector, if $\|x\| = 1$. The existence of the inner product allows us to introduce and utilize the all-important notion of orthogonality.

We say that $x, y \in \mathscr{U}$ are **orthogonal**, and write $x \perp y$, if $(x, y) = 0$. Given a set S, we write $x \perp S$ if $(x, s) = 0$ for all $s \in S$. We define the set $S^{\perp}$ by

$$S^{\perp} = \{x \mid (x, s) = 0, \text{ for all } s \in S\} = \bigcap_{s \in S} \{x \mid (x, s) = 0\}.$$

A set of vectors $\{x_i\}$ is called **orthogonal** if $(x_i, x_j) = 0$ whenever $i \neq j$. A set of vectors $\{x_i\}$ is called **orthonormal** if $(x_i, x_j) = \delta_{ij}$.

Theorem 7.2 (The Pythagorean theorem). *Let $\{x_i\}_{i=1}^{k}$ be an orthogonal set of vectors in $\mathscr{U}$. Then*

$$\left\| \sum_{i=1}^{k} x_i \right\|^2 = \sum_{i=1}^{k} \|x_i\|^2.$$

Proof.

$$\left\| \sum_{i=1}^{k} x_i \right\|^2 = \left(\sum_{i=1}^{k} x_i, \sum_{j=1}^{k} x_j \right) = \sum_{i=1}^{k} \sum_{j=1}^{k} (x_i, x_j) = \sum_{i=1}^{k} \|x_i\|^2. \qquad \blacksquare$$

Theorem 7.3 (The Schwarz inequality). *For all $x, y \in \mathscr{U}$ we have*

$$|(x, y)| \leq \|x\| \cdot \|y\|. \tag{7.1}$$

Proof. If $y = 0$, the equality holds trivially. So, assuming $y \neq 0$, we set $e = y/\|y\|$. We note that $x = (x, e)e + (x - (x, e)e)$ and $(x - (x, e)e) \perp (x, e)e$. Applying the Pythagorean theorem, we get

$$\|x\|^2 = \|(x, e)e\|^2 + \|x - (x, e)e\|^2 \geq \|(x, e)e\|^2 = |(x, e)|^2,$$

or $|(x, e)| \leq \|x\|$. Substituting $y/\|y\|$ for e, inequality (7.1) follows. $\qquad \blacksquare$

Theorem 7.4 (The triangle inequality). *For all $x, y \in \mathscr{U}$ we have*

$$\|x + y\| \leq \|x\| + \|y\|. \tag{7.2}$$

Proof. We compute

$$\begin{aligned}
\|x + y\|^2 &= (x + y, x + y) = (x, x) + (x, y) + (y, x) + (y, y) \\
&= \|x\|^2 + 2\mathrm{Re}(x, y) + \|y\|^2 \\
&\leq \|x\|^2 + 2|(x, y)| + \|y\|^2 \\
&\leq \|x\|^2 + 2\|x\|\|y\| + \|y\|^2 = (\|x\| + \|y\|)^2. \qquad \blacksquare
\end{aligned}$$

Next, we prove two important identities. The proof of both is computational and we omit it.

Proposition 7.5. *For all vectors* $x, y \in \mathscr{U}$ *we have*

1. The polarization identity:

$$(x,y) = \frac{1}{4}\{\|x+y\|^2 - \|x-y\|^2 + i\|x+iy\|^2 - i\|x-iy\|^2\} \tag{7.3}$$

2. The parallelogram identity:

$$\|x+y\|^2 + \|x-y\|^2 = 2\left(\|x\|^2 + \|y\|^2\right). \tag{7.4}$$

Proposition 7.6 (The Bessel inequality). *Let* $\{e_i\}_{i=1}^k$ *be an orthonormal set of vectors in* $\mathscr{U}$. *Then*

$$\|x\|^2 \ge \sum_{i=1}^k |(x,e_i)|^2.$$

Proof. Let x be an arbitrary vector in $\mathscr{U}$. Obviously, we can write $x = \sum_{i=1}^k (x,e_i)e_i + \left(x - \sum_{i=1}^k (x,e_i)e_i\right)$. Since the vector $x - \sum_{i=1}^k (x,e_i)e_i$ is orthogonal to all e_j, $j = 1,\ldots,k$, we can use the Pythagorean theorem to obtain

$$\begin{aligned}
\|x\|^2 &= \|x - \sum_{i=1}^k (x,e_i)e_i\|^2 + \|\sum_{i=1}^k (x,e_i)e_i\|^2 \\
&\ge \|\sum_{i=1}^k (x,e_i)e_i\|^2 = \sum_{i=1}^k |(x,e_i)|^2.
\end{aligned}$$ ∎

Proposition 7.7. *An orthogonal set of nonzero vectors* $\{x_i\}_{i=1}^k$ *is linearly independent.*

Proof. Assume $\sum_{i=1}^k c_i x_i = 0$. Taking the inner product with x_j, we get

$$0 = \sum_{i=1}^k c_i(x_i,x_j) = c_j\|x_j\|^2.$$

This implies $c_j = 0$ for all j. ∎

Since orthogonality implies linear independence, we can search for bases consisting of orthogonal, or even better, orthonormal, vectors. Thus an **orthogonal basis** for $\mathscr{U}$ is an orthogonal set that is also a basis for $\mathscr{U}$. Similarly, an **orthonormal basis** for $\mathscr{U}$ is an orthonormal set that is also a basis for $\mathscr{U}$.

The next theorem, generally known as the **Gram–Schmidt orthonormalization process**, gives a constructive method for computing orthonormal bases.

Theorem 7.8 (Gram–Schmidt orthonormalization). *Let* $\{x_1,\ldots,x_k\}$ *be linearly independent vectors in* $\mathscr{U}$. *Then there exists an orthonormal set* $\{e_1,\ldots,e_k\}$ *such that*

$$\operatorname{span}(e_1,\ldots,e_j) = \operatorname{span}(x_1,\ldots,x_j), \qquad 1 \le j \le k. \tag{7.5}$$

Proof. We prove this by induction. For $j = 1$ we set $e_1 = \frac{x_1}{\|x_1\|}$. Assume we have constructed $e_1, \ldots, e_{j-1}$ as required. We consider next $x'_j = x_j - \sum_{i=1}^{j-1} (x_j, e_i) e_i$. Clearly $x'_j \neq 0$, for otherwise, $x_j \in \mathrm{span}(e_1, \ldots, e_{j-1}) = \mathrm{span}(x_1, \ldots, x_{j-1})$, contrary to the assumption of linear independence of the x_i. We define now

$$e_j = \frac{x'_j}{\|x'_j\|} = \frac{x_j - \sum_{i=1}^{j-1} (x_j, e_i) e_i}{\|x_j - \sum_{i=1}^{j-1} (x_j, e_i) e_i\|}.$$

Since $e_i \in \mathrm{span}(x_1, \ldots, x_j)$ for $i = 1, \ldots, j-1$ and since $e_j \in \mathrm{span}(e_1, \ldots, e_{j-1}, x_j) \subset \mathrm{span}(x_1, \ldots, x_j)$, we obtain the inclusion $\mathrm{span}(e_1, \ldots, e_j) \subset \mathrm{span}(x_1, \ldots, x_j)$.

To get the inverse inclusion, we observe that $\mathrm{span}(x_1, \ldots, x_{j-1}) \subset \mathrm{span}(e_1, \ldots, e_j)$ follows from the

$$x_j = \sum_{i=1}^{j-1} (x_j, e_i) e_i + \left\| x_j - \sum_{i=1}^{j-1} (x_j, e_i) e_i \right\| e_j$$

implies $x_j \in \mathrm{span}(e_1, \ldots, e_j)$. So $\mathrm{span}(x_1, \ldots, x_j) \subset \mathrm{span}(e_1, \ldots, e_j)$, and these two inclusions imply (7.5). ∎

Corollary 7.9. *Every finite-dimensional inner product space has an orthonormal basis.*

Corollary 7.10. *Every orthonormal set in a finite-dimensional inner product space $\mathscr{U}$ can be extended to an orthonormal basis.*

Proof. Let $\{e_1, \ldots, e_m\}$ be an orthonormal set, thus necessarily linearly independent. Hence, by Theorem 2.13, it can be extended to a basis $\{e_1, \ldots, e_m, x_{m+1}, \ldots, x_n\}$ for $\mathscr{U}$. Applying the Gram–Schmidt orthonormalization process, we get an orthonormal basis $\{e_1, \ldots, e_n\}$. The first m vectors remain unchanged. ∎

Orthonormal bases are particularly convenient for vector expansions, or alternatively, for computing coordinates.

Proposition 7.11. *Let $\{e_1, \ldots, e_n\}$ be an orthonormal basis for a finite-dimensional inner product space $\mathscr{U}$. Then every vector $x \in \mathscr{U}$ has a unique representation in the form*

$$x = \sum_{i=1}^{n} (x, e_i) e_i.$$

We consider next an important approximation problem. We define the distance of a vector $x \in \mathscr{U}$ to a subspace $\mathscr{M}$ by

$$\delta(x, \mathscr{M}) = \inf\{\|x - m\| \mid m \in \mathscr{M}\}.$$

A best approximant for x in $\mathscr{M}$ is a vector $m_0 \in \mathscr{M}$ that satisfies $\|x - m_0\| = \delta(x, \mathscr{M})$.

Theorem 7.12. *Let* $\mathscr{U}$ *be a finite-dimensional inner product space and* $\mathscr{M}$ *a subspace. Let* $x \in \mathscr{U}$. *Then*

1. *A best approximant for* x *in* $\mathscr{M}$ *exists.*
2. *A best approximant for* x *in* $\mathscr{M}$ *is unique.*
3. *A vector* $m_0 \in \mathscr{M}$ *is a best approximant for* x *in* $\mathscr{M}$ *if and only if* $x - m_0$ *is orthogonal to* $\mathscr{M}$.

Proof. 1. Let $\{e_1, \ldots, e_m\}$ be an orthonormal basis for $\mathscr{M}$. Extend it to an orthonormal basis $\{e_1, \ldots, e_n\}$ for $\mathscr{U}$. We claim that $m_0 = \sum_{i=1}^{m} (x, e_i) e_i$ is a best approximant. Clearly $m_0 \in \mathscr{M}$. Any other vector $m \in \mathscr{M}$ has a unique representation of the form $m = \sum_{i=1}^{m} c_i e_i$. Therefore

$$x - \sum_{i=1}^{m} c_i e_i = x - \sum_{i=1}^{m} (x, e_i) e_i + \sum_{i=1}^{m} [(x, e_i) - c_i] e_i.$$

Now, the vector $x - \sum_{i=1}^{m} (x, e_i) e_i$ is orthogonal to $L(e_1, \ldots, e_m)$. Hence, applying the Pythagorean theorem, we get

$$\begin{aligned} \|x - \sum_{i=1}^{m} c_i e_i\|^2 &= \|x - \sum_{i=1}^{m} (x, e_i) e_i\|^2 + \|\sum_{i=1}^{m} [(x, e_i) - c_i] e_i\|^2 \\ &\geq \|x - \sum_{i=1}^{m} (x, e_i) e_i\|^2. \end{aligned}$$

2. Let $\sum_{i=1}^{m} c_i e_i$ be another best approximant. By the previous computation, we must have $\|\sum_{i=1}^{m} [(x, e_i) - c_i] e_i\|^2 = \sum_{i=1}^{m} |(x, e_i) - c_i|^2$, and this happens if and only if $c_i = (x, e_i)$ for all i. Thus the two approximants coincide.
3. Assume $m_0 \in \mathscr{M}$ is the best approximant. We saw that $m_0 = \sum_{i=1}^{m} (x, e_i) e_i$ and hence $x - \sum_{i=1}^{m} (x, e_i) e_i$ is orthogonal to $\mathscr{M} = L(e_1, \ldots, e_m)$.

    Conversely, assume $m_0 \in \mathscr{M}$ and $x - m_0 \perp \mathscr{M}$. Then, for any vector $m \in \mathscr{M}$, we have $x - m = (x - m_0) + (m - m_0)$. Since $m - m_0 \in \mathscr{M}$, using the Pythagorean theorem, we have

$$\|x - m\|^2 = \|x - m_0\|^2 + \|m - m_0\|^2 \geq \|x - m_0\|^2.$$

Hence m_0 is the best approximant. ∎

In the special case of inner product spaces, we will reserve the notation $\mathscr{U} = \mathscr{M} \oplus \mathscr{N}$ for the case of orthogonal direct sums, that is, $\mathscr{U} = \mathscr{M} + \mathscr{N}$ and $\mathscr{M} \perp \mathscr{N}$. We note that $\mathscr{M} \perp \mathscr{N}$ implies $\mathscr{M} \cap \mathscr{N} = \{0\}$.

Theorem 7.13. *Let* $\mathscr{U}$ *be a finite-dimensional inner product space and* $\mathscr{M}$ *a subspace. Then we have*

$$\mathscr{U} = \mathscr{M} \oplus \mathscr{M}^{\perp}. \tag{7.6}$$

Proof. The subspaces $\mathscr{M}$ and $\mathscr{M}^{\perp}$ are orthogonal. It suffices to show that they span $\mathscr{U}$. Thus, let $x \in \mathscr{U}$, and let m be its best approximant in $\mathscr{M}$. We can write $x = m + (x - m)$. By Theorem 7.12, we have $x - m \in \mathscr{M}^{\perp}$. ∎

Inner product spaces are a particularly convenient setting for the development of duality theory. The key to this is the identification of the dual space to an inner product space with the space itself. This is done by identifying a linear functional with an inner product with a vector.

Theorem 7.14. *Let $\mathscr{U}$ be a finite-dimensional inner product space. Then f is a linear functional on $\mathscr{U}$ if and only if there exists a vector $\xi_f \in \mathscr{U}$ such that*

$$f(x) = (x, \xi_f). \tag{7.7}$$

Proof. If f is defined by (7.7), then obviously it is a linear functional on $\mathscr{U}$.

To prove the converse, we note first that if f is the zero functional, then we just choose $\xi_f = 0$. Thus, without loss of generality, we can assume that f is a nonzero functional. In this case, $\mathscr{M} = \operatorname{Ker} f$ is a subspace of codimension 1. This implies $\dim \mathscr{M}^\perp = 1$. Let us choose an arbitrary nonzero vector $y \in \mathscr{M}^\perp$ and set $\xi_f = \alpha y$. Since $(y, \xi_f) = (y, \alpha y) = \overline{\alpha} \|y\|^2$, by choosing $\alpha = \frac{\overline{f(y)}}{\|y\|^2}$, we get $f(y) = (y, \xi_f)$, and the same holds, by linearity, for all vectors in $\mathscr{M}^\perp$. For $x \in \mathscr{M}$ it is clear that $f(x) = (x, \xi_f) = 0$. Thus (7.7) follows.   ∎

7.3 Operators in Inner Product Spaces

In this section we study some important classes of operators in inner product spaces. We begin by studying the adjoint transformation. Due to the availability of inner products and the representation of linear functionals, the definition is slightly different from that given for general transformations.

7.3.1 The Adjoint Transformation

Theorem 7.15. *Let $\mathscr{U}, \mathscr{V}$ be inner product spaces and let $T : \mathscr{U} \longrightarrow \mathscr{V}$ be a linear transformation. Then there exists a unique linear transformation $T^* : \mathscr{V} \longrightarrow \mathscr{U}$ such that*

$$(Tx, y) = (x, T^* y) \tag{7.8}$$

for all $x \in \mathscr{U}$ and $y \in \mathscr{V}$.

Proof. Fix a vector $y \in \mathscr{V}$. Then $f(x) = (Tx, y)$ defines a linear functional on $\mathscr{U}$. Thus, by Theorem 7.14, there exists a unique vector $\xi \in \mathscr{U}$ such that $(Tx, y) = (x, \xi)$. We define T^* by $T^* y = \xi$. Therefore, T^* is a map from $\mathscr{V}$ to $\mathscr{U}$. It remains to show that T^* is a linear map. To this end, we compute

$$
\begin{aligned}
(x, T^*(\alpha_1 y_1 + \alpha_2 y_2)) &= (Tx, \alpha_1 y_1 + \alpha_2 y_2) \\
&= \overline{\alpha_1}(Tx, y_1) + \overline{\alpha_2}(Tx, y_2) \\
&= \overline{\alpha_1}(x, T^* y_1) + \overline{\alpha_2}(x, T^* y_2) \\
&= (x, \alpha_1 T^* y_1) + (x, \alpha_2 T^* y_2) \\
&= (x, \alpha_1 T^* y_1 + \alpha_2 T^* y_2).
\end{aligned}
$$

By the uniqueness of the representing vector, we get

$$
T^*(\alpha_1 y_1 + \alpha_2 y_2) = \alpha_1 T^* y_1 + \alpha_2 T^* y_2. \qquad \blacksquare
$$

We will call the transformation T^* the **adjoint** or, if emphasis is needed, the **Hermitian adjoint** of T. Given a complex matrix $A = (a_{ij})$, its adjoint $A^* = (a_{ij}^*)$ is defined by $a_{ij}^* = \overline{a_{ji}}$, or alternatively, by $A^* = \overline{\tilde{A}}$.

Proposition 7.16. *Let $T : \mathscr{U} \longrightarrow \mathscr{V}$ be a linear transformation. Then*

$$
(\operatorname{Im} T)^\perp = \operatorname{Ker} T^*.
$$

Proof. We have for $x \in \mathscr{U}, y \in \mathscr{V}$, $(Tx, y) = (x, T^* y)$. So $y \perp \operatorname{Im} T$ if and only if $T^* y \perp \mathscr{U}$, i.e., if and only if $y \in \operatorname{Ker} T^*$. $\qquad \blacksquare$

Corollary 7.17. *Let T be a linear transformation in an inner product space $\mathscr{U}$. Then we have the following direct sum decompositions:*

$$
\mathscr{U} = \operatorname{Im} T \oplus \operatorname{Ker} T^*,
$$
$$
\mathscr{U} = \operatorname{Im} T^* \oplus \operatorname{Ker} T.
$$

Moreover, $\operatorname{rank} T = \operatorname{rank} T^*$ *and* $\dim \operatorname{Ker} T = \dim \operatorname{Ker} T^*$.

Proof. Follows from Theorem 7.13 and Proposition 7.16. $\qquad \blacksquare$

Proposition 7.18. *Let T be a linear transformation in $\mathscr{U}$. Then $\mathscr{M}$ is a T-invariant subspace of $\mathscr{U}$ if and only if $\mathscr{M}^\perp$ is a T^*-invariant subspace.*

Proof. Follows from equation (7.8). $\qquad \blacksquare$

We proceed to study the matrix representation of a linear transformation T with respect to orthonormal bases. These are particularly easy to compute.

Proposition 7.19. *Let $T : \mathscr{U} \longrightarrow \mathscr{V}$ be a linear transformation and let $\mathscr{B} = \{e_1, \ldots, e_n\}$ and $\mathscr{B}_1 = \{f_1, \ldots, f_m\}$ be orthonormal bases for $\mathscr{U}$ and $\mathscr{V}$ respectively. Then*

1. *For the matrix representation $[T]_{\mathscr{B}}^{\mathscr{B}_1} = (t_{ij})$, we have*

$$
t_{ij} = (Te_j, f_i).
$$

2. We have for the adjoint T^*,

$$[T^*]_{\mathscr{B}_1}^{\mathscr{B}} = ([T]_{\mathscr{B}}^{\mathscr{B}_1})^*.$$

Proof. 1. The matrix representation of T with respect to the given bases is defined by $Te_j = \sum_{k=1}^{m} t_{ki} f_k$. We compute

$$(Te_j, f_i) = (\sum_{k=1}^{m} t_{kj} f_k, f_i) = \sum_{k=1}^{m} t_{kj}(f_k, f_i) = \sum_{k=1}^{m} t_{kj} \delta_{ik} = t_{ij}.$$

2. Let $[T^*]_{\mathscr{B}_1}^{\mathscr{B}} = (t_{ij}^*)$. Then

$$t_{ij}^* = (T^* f_j, e_i) = (f_j, Te_i) = \overline{(Te_i, f_j)} = \overline{t_{ji}}. \qquad \blacksquare$$

Next, we study the properties of the map from $T \mapsto T^*$ as a function from $L(\mathscr{U}, \mathscr{V})$ into $L(\mathscr{V}, \mathscr{U})$.

Proposition 7.20. *Let $\mathscr{U}, \mathscr{V}$ be inner product spaces. Then the adjoint map has the following properties:*

1. *$(T + S)^* = T^* + S^*$.*
2. *$(\alpha T)^* = \overline{\alpha} T^*$.*
3. *$(ST)^* = T^* S^*$.*
4. *$(T^*)^* = T$.*
5. *For the identity map $I_{\mathscr{U}}$ in $\mathscr{U}$, we have $I_{\mathscr{U}}^* = I_{\mathscr{U}}$.*
6. *If $T : \mathscr{U} \longrightarrow \mathscr{V}$ is invertible, then $(T^{-1})^* = (T^*)^{-1}$.*

Proof. 1. Let $x \in \mathscr{U}$ and $y \in \mathscr{V}$. We compute

$$\begin{aligned}
(x, (T + S)^* y) &= ((T + S)x, y) = (Tx + Sx, y) = (Tx, y) + (Sx, y) \\
&= (x, T^* y) + (x, S^* y) = (x, T^* y + S^* y) \\
&= (x, (T^* + S^*)y).
\end{aligned}$$

By uniqueness we get $(T + S)^* = T^* + S^*$.

2. Computing

$$\begin{aligned}
(x, (\alpha T)^* y) &= ((\alpha T)x, y) = (\alpha Tx, y) \\
&= \alpha(Tx, y) = \alpha(x, T^* y) = (x, \overline{\alpha} T^* y),
\end{aligned}$$

we get, using uniqueness, that $(\alpha T)^* = \overline{\alpha} T^*$.

3. We compute

$$(x, (ST)^* y) = (STx, y) = (Tx, S^* y) = (x, T^* S^* y),$$

or $(ST)^* = T^* S^*$.

4.

$$(Tx,y) = (x,T^*y) = \overline{(T^*y,x)} = \overline{(y,T^{**}x)} = (T^{**}x,y),$$

and this implies $T^{**} = T$.

5. Let $I_{\mathscr{U}}$ be the identity map in $\mathscr{U}$. Then

$$(I_{\mathscr{U}}x,y) = (x,y) = (x,I_{\mathscr{U}}y),$$

so $I_{\mathscr{U}}^* = I_{\mathscr{U}}$.

6. Assume $T : \mathscr{U} \longrightarrow \mathscr{V}$ is invertible. Then $T^{-1}T = I_{\mathscr{U}}$ and $TT^{-1} = I_{\mathscr{V}}$. The first equality implies $I_{\mathscr{U}} = I_{\mathscr{U}}^* = T^*(T^{-1})^*$, that is, $(T^{-1})^*$ is a right inverse of T^*. The second equality implies that it is a left inverse, and $(T^{-1})^* = (T^*)^{-1}$ follows. ∎

The availability of the norm function on an inner product space allows us to consider a numerical measure of the size of a linear transformation.

Definition 7.21. Let $\mathscr{U}_1, \mathscr{U}_2$ be two inner product spaces. Given a linear transformation $T : \mathscr{U}_1 \longrightarrow \mathscr{U}_2$, we define its **norm** by

$$\|T\| = \sup_{\|x\| \leq 1} \|Tx\|.$$

Clearly, for an arbitrary linear transformation T, the norm $\|T\|$ is finite. For it is defined as the supremum of a continuous function on the unit ball $\{x \in \mathscr{U}_1 \mid \|x\| \leq 1\}$, which is a compact set. It follows that the supremum is actually attained, and thus there exists a, not necessarily unique, unit vector x for which $\|Tx\| = \|T\|$.

The following proposition, whose standard proof we omit, sums up the basic properties of the operator norm.

Proposition 7.22. *We have the following properties of the operator norm:*

1. $\|T + S\| \leq \|T\| + \|S\|$.
2. $\|\alpha T\| = |\alpha| \cdot \|T\|$.
3. $\|TS\| \leq \|T\| \cdot \|S\|$.
4. $\|T^*\| = \|T\|$.
5. $\|I\| = 1$.
6. $\|T\| = \sup_{\|x\|=1} \|Tx\| = \sup_{\|x\|,\|y\|\leq 1} |(Tx,y)| = \sup_{\|x\|,\|y\|=1} |(Tx,y)|$.

7.3.2 Unitary Operators

The extra structure that inner product spaces have over linear spaces, namely the existence of inner products, leads to a new definition, that of an isomorphism of two inner product spaces.

Definition 7.23. Let $\mathcal{V}$ and $\mathcal{W}$ be two inner product spaces over the same field. A linear transformation $T : \mathcal{V} \longrightarrow \mathcal{W}$ is an **isometry** if it preserves inner products, i.e., if $(Tx, Ty) = (x, y)$, for all $x, y \in \mathcal{V}$. An **isomorphism of inner product spaces** is an isomorphism that is isometric. We will also call it an **isometric isomorphism** or alternatively a **unitary isomorphism**.

Proposition 7.24. *Given a linear transformation* $T : \mathcal{V} \longrightarrow \mathcal{W}$, *then the following properties are equivalent:*

1. *T preserves inner products.*
2. *For all $x \in \mathcal{V}$ we have $\|Tx\| = \|x\|$.*
3. *We have $T^*T = I$.*

Proof. Assume T preserves inner products, i.e., $(Tx, Ty) = (x, y)$, for all $x, y \in \mathcal{V}$. Choosing $y = x$, we get $\|Tx\|^2 = (Tx, Tx) = (x, x) = \|x\|^2$.

If $\|Tx\| = \|x\|$, then using the polarization identity (7.3), we get

$$(Tx, Ty) = \frac{1}{4}\{(\|Tx + Ty\|)^2 - (\|Tx - Ty\|)^2 + i(\|Tx + iTy\|)^2 - i(\|Tx - iTy\|)^2\}$$

$$= \frac{1}{4}\{(\|x + y\|)^2 - (\|x - y\|)^2 + i(\|x + iy\|)^2 - i(\|x - iy\|)^2\} = (x, y).$$

Next observe that if T preserves inner products, then $((I - T^*T)x, y) = 0$, which implies $T^*T = I$. Finally, if the equality $T^*T = I$ holds, then clearly T preserves inner products. ∎

Corollary 7.25. *An isometry $T : \mathcal{V} \longrightarrow \mathcal{W}$ maps orthonormal sets into orthonormal sets.*

The next result is the counterpart, for inner product spaces, of Theorem 4.14.

Corollary 7.26. *Two inner product spaces are isomorphic if and only if they have the same dimension.*

Proof. Assume a unitary isomorphism $U : \mathcal{V} \longrightarrow \mathcal{W}$ exists. Since it maps bases into bases, the dimensions of $\mathcal{V}$ and $\mathcal{W}$ coincide.

Conversely, suppose $\mathcal{V}$ and $\mathcal{W}$ have the same dimension. We choose orthonormal bases $\{e_1, \ldots, e_n\}$ and $\{f_1, \ldots, f_n\}$ in $\mathcal{V}$ and $\mathcal{W}$ respectively. We define a linear map $U : \mathcal{V} \longrightarrow \mathcal{W}$ by $Ue_i = f_i$. It is immediate that the operator U so defined is a unitary isomorphism. ∎

Proposition 7.27. *Let $U : \mathcal{V} \longrightarrow \mathcal{W}$ be a linear transformation. Then U is unitary if and only if $U^* = U^{-1}$.*

Proof. If $U^* = U^{-1}$, we get $U^*U = UU^* = I$, i.e., U is an isometric isomorphism.

Conversely, if U is unitary, we have $U^*U = I$. Since U is an isomorphism and the inverse transformation, if it exists, is unique, we obtain $U^* = U^{-1}$. ∎

Next, we consider matrix representations. We say that a complex matrix is unitary if $A^*A = I$. Here A^* is defined by $a_{ij}^* = \overline{a_{ji}}$. We proceed to consider matrix representations for unitary maps. We say that a complex $n \times n$ matrix A is unitary if $A^*A = I$, where A^* is the Hermitian adjoint of A. We expect that the matrix representation of a unitary matrix, when taken with respect to orthonormal bases, should reflect the unitarity property. We state this in the following proposition, omitting the simple proof.

Proposition 7.28. *Let $U : \mathcal{V} \longrightarrow \mathcal{W}$ be a linear transformation between two inner product spaces. Let $\mathcal{B}_1 = \{e_1, \dots, e_n\}$ and $\mathcal{B}_2 = \{f_1, \dots, f_n\}$, be orthonormal bases in $\mathcal{V}$ and $\mathcal{W}$ respectively. Then U is unitary if and only if its matrix representation $[U]_{\mathcal{B}_1}^{\mathcal{B}_2}$ is a unitary matrix.*

The notion of similarity can be strengthened in the context of inner product spaces.

Definition 7.29. Let $\mathcal{V}$ and $\mathcal{W}$ be two inner product spaces over the same field, and let $T : \mathcal{V} \longrightarrow \mathcal{V}$ and $S : \mathcal{W} \longrightarrow \mathcal{W}$ be linear transformations. We say that T and S are **unitarily equivalent** if there exists a unitary map $U : \mathcal{V} \longrightarrow \mathcal{W}$ for which $UT = SU$, or equivalently, for which the following diagram is commutative:

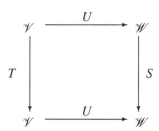

The unitary transformations play, in the algebra of linear transformations on an inner product space, the same role that complex numbers of absolute value one play in the complex number field $\mathbb{C}$. This is reflected in the eigenvalues of unitary transformations as well as in their structure.

Proposition 7.30. *Let U be a unitary operator in an inner product space $\mathcal{V}$. Then*

1. All eigenvalues of U have absolute value one.
2. If $Ux = \lambda x$, then $U^*x = \overline{\lambda}x$.
3. Eigenvectors corresponding to different eigenvalues are orthogonal.
4. We have

$$\operatorname{Ker}(U - \lambda_1 I)^{\nu_1} \cdots (U - \lambda_k I)^{\nu_k} = \operatorname{Ker}(U - \lambda_1 I) \cdots (U - \lambda_k I).$$

5. The minimal polynomial $m_U(z)$ of U splits into distinct linear factors.

Proof. 1. Let λ be an eigenvalue of U and x a corresponding eigenvector. Then $Ux = \lambda x$ implies $(Ux,x) = (\lambda x,x) = \lambda(x,x)$. Now, using the unitarity of U, we have

$$\|x\|^2 = \|Ux\|^2 = \|\lambda x\|^2 = |\lambda|^2 \|x\|^2.$$

This implies $|\lambda| = 1$.

2. Assume $Ux = \lambda x$. Using the fact that $U^*U = UU^* = I$, we compute

$$
\begin{aligned}
\|U^*x - \overline{\lambda}x\|^2 &= (U^*x - \overline{\lambda}x, U^*x - \overline{\lambda}x) \\
&= (U^*x, U^*x) - \overline{\lambda}(x, U^*x) - \lambda(x, Ux) + |\lambda|^2 \|x\|^2 \\
&= (UU^*x, x) - \overline{\lambda}(Ux, x) - \lambda(x, Ux) + \|x\|^2 \\
&= \|x\|^2 - \overline{\lambda}(\lambda x, x) - \lambda(x, \lambda x) + \|x\|^2 \\
&= \|x\|^2 - 2|\lambda|^2 \|x\|^2 + \|x\|^2 = 0.
\end{aligned}
$$

3. Let λ, μ be distinct eigenvalues and x, y the corresponding eigenvectors. Then

$$\lambda(x,y) = (Ux,y) = (x, U^*y) = (x, \overline{\mu}y) = \mu(x,y).$$

Hence, $(\lambda - \mu)(x,y) = 0$. Since $\lambda \neq \mu$, we conclude that $(x,y) = 0$ or $x \perp y$.

4. Since U is unitary, U and U^* commute. Therefore also $U - \lambda I$ and $U^* - \overline{\lambda}I$ commute. Assume now that $(U - \lambda I)^v x = 0$. Without loss of generality we may assume that $(U - \lambda I)^{2^n} x = 0$. This implies

$$0 = (U^* - \overline{\lambda}I)^{2^n}(U - \lambda I)^{2^n}x = [(U^* - \overline{\lambda}I)^{2^{n-1}}(U - \lambda I)^{2^{n-1}}]^2 x.$$

So

$$0 = ([(U^* - \overline{\lambda}I)^{2^{n-1}}(U - \lambda I)^{2^{n-1}}]^2 x, x) = \|(U^* - \overline{\lambda}I)^{2^{n-1}}(U - \lambda I)^{2^{n-1}}x\|^2,$$

or $(U^* - \overline{\lambda}I)^{2^{n-1}}(U - \lambda I)^{2^{n-1}}x = 0$. This implies in turn that $\|(U - \lambda I)^{2^{n-1}}x\| = 0$, and hence $(U - \lambda I)^{2^{n-1}}x = 0$. By repeating the argument we conclude that $(U - \lambda I)x = 0$.

Next, we use the fact that all factors $U - \lambda_i I$ commute. Assuming $(U - \lambda_1 I)^{v_1} \cdots (U - \lambda_k I)^{v_k} x = 0$, we conclude

$$
\begin{aligned}
0 &= (U - \lambda_1 I)(U - \lambda_2 I)^{v_2} \cdots (U - \lambda_k I)^{v_k} x \\
&= (U - \lambda_2 I)^{v_2}(U - \lambda_1 I)(U - \lambda_3 I)^{v_3} \cdots (U - \lambda_k I)^{v_k} x.
\end{aligned}
$$

This implies $(U - \lambda_1 I)(U - \lambda_2 I)(U - \lambda_3 I)^{v_3} \cdots (U - \lambda_k I)^{v_k} x = 0$. Proceeding inductively, we get $(U - \lambda_1 I) \cdots (U - \lambda_k I)x = 0$.

5. Follows from part 4.   ∎

7.3.3 Self-adjoint Operators

Probably the most important class of operators in an inner product space is that of self-adjoint operators. An operator T in an inner product space $\mathscr{U}$ is called **self-adjoint**, or **Hermitian**, if $T^* = T$. Similarly, a complex matrix A is called self-adjoint or Hermitian if it satisfies $A^* = A$, or equivalently $a_{ij} = \overline{a_{ji}}$.

The next proposition shows that self-adjoint operators in an inner product space play in $L(\mathscr{U})$ a role similar to that of the real numbers in the complex field.

Proposition 7.31. *Every linear transformation T in an inner product space can be written in the form*

$$T = T_1 + iT_2$$

with T_1, T_2 self-adjoint. Such a representation is unique.

Proof. Write

$$T = \frac{1}{2}(T + T^*) + i\frac{1}{2i}(T - T^*)$$

and note that $T_1 = \frac{1}{2}(T + T^*)$ and $T_2 = \frac{1}{2i}(T - T^*)$ are both self-adjoint.

Conversely, if $T = T_1 + iT_2$ with T_1, T_2 self-adjoint, then $T^* = T_1 - iT_2$. By elimination we get that T_1, T_2 have necessarily the form given previously. ∎

In preparation for the spectral theorem, we prove a few properties of self-adjoint operators that are of intrinsic interest.

Proposition 7.32. *Let T be a self-adjoint operator in an inner product space $\mathscr{U}$. Then*

1. All eigenvalues of T are real.
2. Eigenvectors corresponding to different eigenvalues are orthogonal.
3. We have

$$\mathrm{Ker}\,(T - \lambda_1 I)^{\nu_1} \cdots (T - \lambda_k I)^{\nu_k} = \mathrm{Ker}\,(T - \lambda_1 I) \cdots (T - \lambda_k I).$$

4. The minimal polynomial $m_T(z)$ of T splits into distinct linear factors.

Proof. 1. Let λ be an eigenvalue of T and x a corresponding eigenvector. Then $Tx = \lambda x$ implies $(Tx, x) = (\lambda x, x) = \lambda(x, x)$. Now, using the self-adjointness of T, we have $\overline{(Tx, x)} = (x, Tx) = (Tx, x)$, that is, (Tx, x) is real. So is $\lambda = \frac{(Tx, x)}{(x, x)}$.

2. Let λ, μ be distinct eigenvalues of T, and x, y corresponding eigenvectors. Using the fact that eigenvalues of T are real, we get $(Tx, y) = (\lambda x, y) = \lambda(x, y)$ and

$$(Tx, y) = (x, Ty) = (x, \mu y) = \overline{\mu}(x, y) = \mu(x, y).$$

By subtraction we get $(\lambda - \mu)(x, y) = 0$, and since $\lambda \neq \mu$, we conclude that $(x, y) = 0$, or $x \perp y$.

3. We prove this by induction on the number of factors. So, let $k = 1$ and assume that $(T - \lambda_1 I)^{v_1} x = 0$. Then, by multiplying this equation by $(T - \lambda_1 I)^\mu$, we may assume without loss of generality that $(T - \lambda_1 I)^{2^n} x = 0$. Then we get

$$0 = ((T - \lambda_1 I)^{2^n} x, x) = ((T - \lambda_1 I)^{2^{n-1}} x, (T - \lambda_1 I)^{2^{n-1}} x) = \|(T - \lambda_1 I)^{2^{n-1}} x\|,$$

which implies $(T - \lambda_1 I)^{2^{n-1}} x = 0$. Repeating this argument, we finally obtain $(T - \lambda_1 I) x = 0$. Assume now the statement holds for up to $k - 1$ factors and that $\Pi_{i=1}^k (T - \lambda_i I)^{v_i} x = 0$. Therefore $(T - \lambda_k I)^{v_k} \Pi_{i=1}^{k-1} (T - \lambda_i I)^{v_i} x = 0$. By the argument for $k = 1$ we conclude that

$$0 = (T - \lambda_i I) \Pi_{i=1}^{k-1} (T - \lambda_i I)^{v_i} x = \Pi_{i=1}^{k-1} (T - \lambda_i I)^{v_i} [(T - \lambda_k I) x].$$

By the induction hypothesis we get

$$0 = \Pi_{i=1}^{k-1} (T - \lambda_i I)[(T - \lambda_k I) x] = \Pi_{i=1}^k (T - \lambda_i I) x,$$

or, since x is arbitrary, $\Pi_{i=1}^k (T - \lambda_i I) = 0$.
4. Follows from the previous part.   ∎

Theorem 7.33. *Let T be self-adjoint. Then there exists an orthonormal basis consisting of eigenvectors of T.*

Proof. Let $m_T(z)$ be the minimal polynomial of T. Then $m_T(z) = \Pi_{i=1}^k (z - \lambda_i)$, with the λ_i distinct. Let $\pi_i(z)$ be the corresponding Lagrange interpolation polynomials. We observe that the $\pi_i(T)$ are orthogonal projections. That they are projections has been proved in Theorem 6.25. That they are orthogonal projections follows from their self-adjointness, noting that, given a real polynomial $p(z)$ and a self-adjoint operator T, necessarily $p(T)$ is self-adjoint. This implies that $U = \mathrm{Ker}\,(\lambda_1 I - T) \oplus \cdots \oplus \mathrm{Ker}\,(\lambda_k I - T)$ is an orthogonal direct sum decomposition. Choosing an orthonormal basis in each subspace $\mathrm{Ker}\,(\lambda_i I - T)$, we get an orthonormal basis made of eigenvectors.   ∎

As a consequence, we obtain the **spectral theorem**.

Theorem 7.34. *Let T be self-adjoint. Then it has a unique representation of the form*

$$T = \sum_{i=1}^s \lambda_i P_i,$$

where the $\lambda_i \in \mathbb{R}$ are distinct and P_i are orthogonal projections satisfying

$$P_i P_j = \delta_{ij} P_j, \sum_{i=1}^s P_i = I.$$

Proof. Let λ_i be the distinct eigenvalues of T. We define the P_i to be the orthogonal projections on $\mathrm{Ker}\,(\lambda_i I - T)$.

Conversely, if such a representation exists, it follows that necessarily, the λ_i are the eigenvalues of T and $\mathrm{Im}\,P_i = \mathrm{Ker}\,(\lambda_i I - T)$. ∎

In the special case of self-adjoint operators, the computation of the norm has a further characterization.

Proposition 7.35. *Let T be a self-adjoint operator in a finite-dimensional inner product space $\mathscr{U}$. Then*

1. *We have $\|T\| = \sup_{\|x\|\leq 1} |(Tx,x)| = \sup_{\|x\|=1} |(Tx,x)|$.*
2. *Let $\lambda_1,\dots,\lambda_n$ be the eigenvalues of T ordered so that $|\lambda_i| \geq |\lambda_{i+1}|$. Then $\|T\| = |\lambda_1|$, i.e., the norm equals the modulus of the largest eigenvalue.*

Proof. 1. Since for x satisfying $\|x\| \leq 1$, we have

$$|(Tx,x)| \leq \|Tx\| \cdot \|x\| \leq \|T\| \cdot \|x\|^2 \leq \|T\|,$$

it follows that $\sup_{\|x\|=1} |(Tx,x)| \leq \sup_{\|x\|\leq 1} |(Tx,x)| \leq \|T\|$.

To prove the converse inequality, we argue as follows. We use the following version of the polarization identity:

$$4\mathrm{Re}\,(Tx,y) = (T(x+y),x+y) - (T(x-y),x-y),$$

which implies

$$|(\mathrm{Re}\,(Tx,y)| \leq \frac{1}{4} \sup_{\|z\|\leq 1} |(Tz,z)|[\|x+y\|^2 + \|x-y\|^2],$$

$$|(\mathrm{Re}\,(Tx,y)| \leq \frac{1}{4} \sup_{\|z\|\leq 1} |(Tz,z)|[2\|x\|^2 + 2\|y\|^2] \leq \sup_{\|z\|\leq 1} |(Tz,z)|.$$

Choosing $y = \frac{Tx}{\|Tx\|}$ it follows that $\|Tx\| \leq \sup_{\|z\|\leq 1}|(Tz,z)|$. The equality $\sup_{\|z\|\leq 1}|(Tz,z)| = \sup_{\|z\|=1}|(Tz,z)|$ is obvious.

2. Let λ_1 be the eigenvalue of T of largest modulus and x_1 a corresponding, normalized, eigenvector. Then

$$|\lambda_1| = |(\lambda_1(x_1,x_1)| = |(\lambda_1 x_1, x_1)| = |(Tx_1,x_1)| \leq \|T\| \cdot \|x_1\|^2 = \|T\|.$$

So the absolute values of all eigenvalues of T are bounded by $\|T\|$. Assume now that x is a vector for which $\|Tx\| = \|T\|$. We will show that it is an eigenvector of T corresponding either to $\|T\|$ or to $-\|T\|$. The equality $\sup|(Tx,x)| = \|T\|$ implies either $\sup(Tx,x) = \|T\|$ or $\inf(Tx,x) = -\|T\|$. We assume the first is satisfied. Thus there exists a unit vector x for which $(Tx,x) = \|T\|$. We compute

$$0 \le \|Tx - \|T\|x\|^2 = \|Tx\|^2 - 2\|T\|(Tx,x) + \|T\|^2\|x\|^2$$
$$\le \|T\|^2\|x\|^2 - 2\|T\|^2 + \|T\|^2\|x\|^2 = 0.$$

So necessarily $Tx = \|T\| \cdot \|x\|$, or $\|T\| = \lambda_1$. If $\inf(Tx,x) = -\|T\|$, the same argument can be used.   ∎

7.3.4   The Minimax Principle

For self-adjoint operators we have a very nice characterization of eigenvalues, known as the **minimax principle**.

Theorem 7.36. *Let T be a self-adjoint operator on an n-dimensional inner product space U. Let $\lambda_1 \ge \cdots \ge \lambda_n$ be the eigenvalues of T. Then*

$$\lambda_k = \min_{\{\mathcal{M} \mid \dim \mathcal{M} = n-k+1\}} \max_{x \in \mathcal{M}} \{(Tx,x) \mid \|x\| = 1\}. \tag{7.9}$$

Proof. Let $\{e_1,\ldots,e_n\}$ be an orthonormal basis of $\mathcal{U}$ consisting of eigenvectors of T. Let now $\mathcal{M}$ be an arbitrary subspace of $\mathcal{U}$ of dimension $n - k + 1$. Let $\mathcal{M}_k = L(e_1,\ldots,e_k)$. Then since $\dim \mathcal{M}_k + \dim \mathcal{N}_k = k + (n - k + 1) = n + 1$, their intersection $\mathcal{M}_k \cap \mathcal{N}_k$ is nontrivial and contains a unit vector x. For this vector we have

$$(Tx,x) = \sum_{i=1}^{k} \lambda_i |(x,e_i)|^2 \ge \lambda_k \sum_{i=1}^{k} |(x,e_i)|^2 = \lambda_k.$$

This shows that $\min_{\{\mathcal{M} \mid \dim \mathcal{M} = n-k+1\}} \max_{x \in \mathcal{M}} \{(Tx,x) \mid \|x\| = 1\} \ge \lambda_k$.

To complete the proof, we need to exhibit at least one subspace on which the reverse inequality holds. To this end, let $\mathcal{M}_k = L(e_k,\ldots,e_n)$. Then for any $x \in \mathcal{M}$ with $\|x\| = 1$, we have

$$(Tx,x) = \sum_{i=k}^{n} \lambda_i |(x,e_i)|^2 \le \lambda_k \sum_{i=k}^{n} |(x,e_i)|^2 \le \lambda_k \|x\|^2 = \lambda_k.$$

Therefore, we have

$$\max_{x \in \mathcal{M}} \{(Tx,x) \mid \|x\| = 1\} = \sum_{i=k}^{n} \lambda_i |(x,e_i)|^2 \le \lambda_k \sum_{i=k}^{n} |(x,e_i)|^2 \le \lambda_k \|x\|^2 = \lambda_k. \quad ∎$$

7.3.5   The Cayley Transform

The classes of unitary and self-adjoint operators generalize the sets of complex numbers of absolute value one and the set of real numbers respectively. Now the fractional linear transformation $w = \frac{z-i}{z+i}$ maps the upper half-plane onto the unit disk, and in particular the real line onto the unit circle. Naturally, one wonders whether this can be extended to a map of self-adjoint operators onto unitary operators. The next theorem focuses on this map.

Theorem 7.37. *Let A be a self-adjoint operator in a finite-dimensional inner product space $\mathcal{V}$. Then*

1. For each vector $x \in \mathcal{V}$, we have

$$\|(A + iI)x\|^2 = \|(A - iI)x\|^2 = \|Ax\|^2 + \|x\|^2.$$

2. The operators $A + iI, A - iI$ are both injective, hence invertible.
3. The operator U defined by

$$U = (A - iI)(A + iI)^{-1} \tag{7.10}$$

is a unitary operator for which 1 is not an eigenvalue. The operator U is called the **Cayley transform** *of A.*
4. Given a unitary map U in a finite-dimensional inner product space such that 1 is not an eigenvalue, the operator A defined by

$$A = i(I + U)(I - U)^{-1} \tag{7.11}$$

is a self-adjoint operator. This map is called the **inverse Cayley transform**.

Proof. 1. We compute

$$\begin{aligned}\|(A + iI)x\|^2 &= \|Ax\|^2 + \|x\|^2 + (ix, Ax) + (Ax, ix) \\ &= \|Ax\|^2 + \|x\|^2 = \|(A - iI)x\|^2.\end{aligned}$$

2. The injectivity of both $A + iI$ and $A - iI$ is an immediate consequence of the previous equality. This shows the invertibility of both operators.
3. Let $x \in \mathcal{V}$ be arbitrary. Then there exists a unique vector z such that $x = (A + iI)z$ or $z = (A + iI)^{-1}x$. Therefore

$$\|(A - iI)(A + iI)^{-1}x\| = \|(A - iI)z\| = \|(A + iI)z\| = \|x\|.$$

This shows that U defined by (7.10) is unitary.
    To see that 1 is not an eigenvalue of U, assume $Ux = x$. Thus $(A - iI)(A + iI)^{-1}x = x = (A + iI)(A + iI)^{-1}x$ and hence $2i(A + iI)^{-1}x = 0$. This implies $x = 0$.
4. Assume U is a unitary map such that 1 is not an eigenvalue. Then $I - U$ is invertible. Defining A by (7.11), and using the fact that $U^* = U^{-1}$, we compute

$$\begin{aligned}A^* &= -i(I + U^*)(I - U^*)^{-1} = -i(I + U^{-1})UU^{-1}(I - U^{-1})^{-1} \\ &= -i(U + I)(U - I)^{-1} = i(I + U)(I - U)^{-1} = A.\end{aligned}$$

So A is self-adjoint. ∎

7.3.6 Normal Operators

Our analysis of the classes of unitary and self-adjoint operators in inner product spaces showed remarkable similarities. In particular, both classes admitted the existence of orthonormal bases made up of eigenvectors. One wonders whether the set of all operators having this property can be characterized. In fact, this can be done, and the corresponding class is that of normal operators, which we proceed to introduce.

Definition 7.38. Let T be a linear transformation in a finite-dimensional inner product space. We say that T is **normal** if it satisfies

$$T^*T = TT^*. \tag{7.12}$$

Proposition 7.39. *The operator T is normal if and only if for every $x \in \mathscr{V}$, we have*

$$\|Tx\| = \|T^*x\|. \tag{7.13}$$

Proof. We compute, assuming T is normal,

$$\|Tx\|^2 = (Tx, Tx) = (T^*Tx, x) = (TT^*x, x) = (T^*x, T^*x) = \|T^*x\|^2,$$

which implies (7.13).

Conversely, assuming the equality (7.13), we have $(TT^*x, x) = (T^*Tx, x)$, and hence, by Proposition 7.35, we get $TT^* = T^*T$, i.e., T is normal. ■

Corollary 7.40. *Let T be a normal operator. Let α be an eigenvalue of T. Then $Tx = \alpha x$ implies $T^*x = \overline{\alpha}x$.*

Proof. If T is normal, so is $T - \alpha I$, and we apply Proposition 7.39. ■

Proposition 7.41. *Let T be a normal operator in a complex inner product space. If for some complex number λ, we have $(T - \lambda I)^{\nu}x = 0$, then $(T - \lambda I)x = 0$.*

Proof. The proof is exactly as in the case of unitary operators and is omitted. ■

Lemma 7.42. *Let T be a normal operator and assume λ, μ are distinct eigenvalues of T. If x, y are corresponding eigenvectors, then $x \perp y$.*

Proof. We compute

$$\lambda(x, y) = (\lambda x, y) = (Tx, y) = (x, T^*y) = (x, \overline{\mu}y) = \mu(x, y).$$

Hence $(\lambda - \mu)(x, y) = 0$, and since $\lambda \neq \mu$, we conclude that $(x, y) = 0$. Thus x, y are orthogonal. ■

The following is known as the spectral theorem for normal operators.

Theorem 7.43. *Let T be a normal operator in a finite-dimensional complex inner product space $\mathscr{V}$. Then there exists an orthonormal basis of $\mathscr{V}$ consisting of eigenvectors of T.*

Proof. Let $m_T(z) = \Pi_{i=1}^{k}(z - \lambda_i)^{v_i}$ be the primary decomposition of the minimal polynomial of T. We set $\mathscr{M}_i = \text{Ker}\,(\lambda_i I - T)^{v_i}$. By Proposition 7.41 we have $\mathscr{M}_i = \text{Ker}\,(\lambda_i I - T)$, i.e., it is the eigenspace corresponding to the eigenvalue λ_i. By Lemma 7.42, the subspaces $\mathscr{M}_i$ are mutually orthogonal. We apply now Theorem 6.26 to conclude that $\mathscr{V} = \mathscr{M}_1 \oplus \cdots \oplus \mathscr{M}_k$, where now this is an orthogonal direct sum decomposition. Choosing an orthonormal basis in each subspace $\mathscr{M}_i$ and taking their union, we get an orthonormal basis for $\mathscr{V}$ made of eigenvectors. ∎

Theorem 7.44. *Let T be a normal operator in a finite-dimensional inner product space $\mathscr{V}$. Then*

1. *A subspace $\mathscr{M} \subset \mathscr{V}$ is invariant under T if and only if it is invariant under T^*.*
2. *Let $\mathscr{M}$ be an invariant subspace for T. Then the direct sum decomposition $\mathscr{V} = \mathscr{M} \oplus \mathscr{M}^{\perp}$ reduces T.*
3. *T reduced to an invariant subspace $\mathscr{M}$, that is, $T|_{\mathscr{M}}$ is a normal operator.*

Proof. 1. Let $\mathscr{M}$ be invariant under T. Since $T|_{\mathscr{M}}$ has at least one eigenvalue and a corresponding eigenvector, say $Tx_1 = \lambda_1 x_1$, then also $T^*x_1 = \overline{\lambda_1}x_1$. Let $\mathscr{M}_1$ be the subspace spanned by x_1. We consider $\mathscr{M} \ominus \mathscr{M}_1 = \mathscr{M} \cap \mathscr{M}_1^{\perp}$. This is also invariant under T. Proceeding by induction, we conclude that $\mathscr{M}$ is spanned by eigenvectors of T hence, by Proposition 7.41, by eigenvectors of T^*. This shows that $\mathscr{M}$ is T^* invariant. The converse follows by symmetry.

2. Since $\mathscr{M}$ is invariant under both T and T^*, so is $\mathscr{M}^{\perp}$.

3. Let $\{e_1, \ldots, e_m\}$ be an orthonormal basis of $\mathscr{M}$ consisting of eigenvectors. Thus $Te_i = \lambda_i e_i$ and $T^*e_i = \overline{\lambda_i}e_i$. Let $x = \sum_{i=1}^{m} \alpha_i e_i \in \mathscr{M}$. Then

$$T^*Tx = T^*T \sum_{i=1}^{m} \alpha_i e_i = T^* \sum_{i=1}^{m} \alpha_i \lambda_i e_i$$
$$= \sum_{i=1}^{m} \alpha_i |\lambda_i|^2 e_i = T \sum_{i=1}^{m} \alpha_i \overline{\lambda_i} e_i$$
$$= TT^* \sum_{i=1}^{m} \alpha_i e_i = TT^*x.$$

This shows that $(T|_{\mathscr{M}})^* = T^*|_{\mathscr{M}}$. ∎

Theorem 7.45. *The operator T is normal if and only if $T^* = p(T)$ for some polynomial p.*

Proof. If $T^* = p(T)$, then obviously T and T^* commute, i.e., T is normal.

Conversely, if T is normal, there exists an orthonormal basis $\{e_1, \ldots, e_n\}$ consisting of eigenvectors corresponding to the eigenvalues $\lambda_1, \ldots, \lambda_n$. Let p be any polynomial that interpolates the values $\overline{\lambda_i}$ at the points λ_i. Then

$$p(T)e_i = p(\lambda_i)e_i = \overline{\lambda_i}e_i = T^*e_i.$$

This shows that $T^* = p(T)$. ∎

7.3.7 Positive Operators

We consider next a subclass of self-adjoint operators, that of positive operators.

Definition 7.46. An operator T on an inner product space $\mathcal{V}$ is called **nonnegative** if for all $x \in \mathcal{V}$, we have $(Tx,x) \geq 0$. The operator T is called **positive** if it is nonnegative and $(Tx,x) = 0$ for $x = 0$ only.

Similarly, a complex Hermitian matrix A is called **nonnegative** if for all $x \in \mathbb{C}^n$, we have $(Ax,x) \geq 0$.

Proposition 7.47. *1. A Hermitian operator T in an inner product space is nonnegative if and only if all its eigenvalues are nonnegative.*

2. A Hermitian operator T in an inner product space is positive if and only if all its eigenvalues are positive.

Proof. 1. Assume T is Hermitian and nonnegative. Let λ be an arbitrary eigenvalue of T and x a corresponding eigenvector. Then

$$\lambda \|x\|^2 = \lambda(x,x) = (\lambda x,x) = (Tx,x) \geq 0,$$

which implies $\lambda \geq 0$.

Conversely, assume T is Hermitian and all its eigenvalues are nonnegative. Let $\{e_1,\ldots,e_n\}$ be an orthonormal basis consisting of eigenvectors corresponding to the eigenvalues $\lambda_i \geq 0$. An arbitrary vector x has the expansion $x = \sum_{i=1}^n (x,e_i)e_i$; hence

$$
\begin{aligned}
(Tx,x) &= (T\sum_{i=1}^n (x,e_i)e_i, \sum_{j=1}^n (x,e_j)e_j) \\
&= \sum_{i=1}^n \sum_{j=1}^n (x,e_i)\overline{(x,e_j)}(Te_i,e_j) \\
&= \sum_{i=1}^n \sum_{j=1}^n (x,e_i)\overline{(x,e_j)}(\lambda_i e_i,e_j) \\
&= \sum_{i=1}^n \lambda_i |(x,e_i)|^2 \geq 0.
\end{aligned}
$$

2. The proof is the same except for the inequalities being strict. ∎

An easy way of producing nonnegative operators is to consider operators of the form A^*A or AA^*. The next result shows that this is the only way.

Proposition 7.48. $T \in L(\mathcal{U})$ *is nonnegative if and only if $T = A^*A$ for some operator A in $\mathcal{U}$.*

Proof. That A^*A is nonnegative is immediate.

Conversely, assume T is nonnegative. By Theorem 7.33, there exists an orthonormal basis $\mathcal{B} = \{e_1,\ldots,e_n\}$ in $\mathcal{U}$ made out of eigenvectors of T corresponding to the eigenvalues $\lambda_1,\ldots,\lambda_n$. Since T is nonnegative, all the λ_i are nonnegative and have nonnegative square roots $\lambda_i^{\frac{1}{2}}$. Define a linear transformation S in $\mathcal{U}$ by letting $Se_i = \lambda_i^{\frac{1}{2}} e_i$. Clearly, $S^2 e_i = \lambda_i e_i = Te_i$. The operator S so defined is self-adjoint, for given $x,y \in \mathcal{U}$, we have

$$(Sx,y) = (S\sum_{i=1}^{n}(x,e_i)e_i, \sum_{j=1}^{n}(y,e_j)e_j)$$
$$= (\sum_{i=1}^{n}\lambda_i^{\frac{1}{2}}(x,e_i)e_i, \sum_{j=1}^{n}(y,e_j)e_j)$$
$$= (\sum_{i=1}^{n}(x,e_i)e_i, \sum_{j=1}^{n}\lambda_i^{\frac{1}{2}}(y,e_j)e_j)$$
$$= (\sum_{i=1}^{n}(x,e_i)e_i, S\sum_{j=1}^{n}(y,e_j)e_j)$$
$$= (x,Sy).$$

So $T = S^2 = S^*S$.   ∎

Uniqueness of a nonnegative square root will be proved in Theorem 7.51.

Definition 7.49. Given vectors $x_1,\ldots,x_k$ in an inner product space $\mathscr{V}$, we define the corresponding **Gram matrix**, or **Gramian**, $G = (g_{ij})$ by $g_{ij} = (x_i,x_j)$.

Clearly, a Gramian is always a Hermitian matrix. The next results connects Gramians with positivity.

Theorem 7.50. *Let $\mathscr{V}$ be a complex n-dimensional inner product space. A necessary and sufficient condition for a $k \times k$, with $k \le n$, Hermitian matrix G to be nonnegative definite is that it be the Gram matrix of k vectors in $\mathscr{V}$.*

Proof. Assume G is a Gramian, i.e., there exist vectors $x_1,\ldots,x_k \in \mathscr{V}$ for which $g_{ij} = (x_i,x_j)$. Let $\begin{pmatrix} \xi_1 \\ \vdots \\ \xi_k \end{pmatrix} \in \mathbb{C}^k$. Then we compute

$$(G\xi,\xi) = \sum_{i=1}^{n}\sum_{j=1}^{n}g_{ij}\xi_j\overline{\xi}_i = \sum_{i=1}^{n}\sum_{j=1}^{n}(x_i,x_j)\xi_j\overline{\xi}_i$$
$$= (\sum_{i=1}^{n}\overline{\xi}_ix_i, \sum_{j=1}^{n}\overline{\xi}_ix_j) \ge 0,$$

which shows that G is nonnegative.

To prove the converse, assume G is nonnegative. By Theorem 7.34, there exists a $k \times k$ unitary matrix U, with columns $u_1,\ldots,u_k$, for which $U^*GU = \mathrm{diag}\,(\delta_1,\ldots,\delta_k)$, with $\delta_i \ge 0$. Let γ_i be the nonnegative square root of δ_i. We define $\Gamma = \mathrm{diag}\,(\gamma_1,\ldots,\gamma_k)$. Thus we have $U^*GU = \Gamma^*\Gamma$, or with $R = \Gamma U^{-1}$, $G = R^*R$. Let $r_1,\ldots,r_k$ be the columns of R. Then it follows that $g_{ij} = (r_j,r_i)$. i.e., G is the Gramian of h vectors in $\mathbb{C}^k$. To show that G is also the Gramian of k vectors in $\mathscr{V}$, we choose an arbitrary orthonormal basis $e_1,\ldots,e_n$ in $\mathscr{V}$. We define a map $S : \mathbb{C}^k \longrightarrow \mathscr{V}$ by $Su_i = e_i$ and extend it linearly. Clearly, S is isometric, and with $x_i = \gamma_ie_i \in \mathscr{V}$, it follows that G is also the Gramian of $x_1,\ldots,x_k$.   ∎

Effectively, in the proof of Proposition 7.48, we have constructed a nonnegative square root. We formalize this.

Theorem 7.51. *A nonnegative operator T has a unique nonnegative square root.*

Proof. The existence of a nonnegative square root has been proved in Proposition 7.48.

To prove uniqueness, let $\lambda_1 \geq \cdots \geq \lambda_n$ be the eigenvalues of T and $\mathscr{B} = \{e_1,\ldots,e_n\}$ an orthonormal basis in U made out of eigenvectors. Let A be an arbitrary nonnegative square root of T, i.e., $T = A^2$. We compute

$$0 = (\lambda_i I - A^2)e_i = (\lambda_i^{\frac{1}{2}} I + A)(\lambda_i^{\frac{1}{2}} I - A)e_i.$$

Now, in case $\lambda_i > 0$, the operators $\lambda_i^{\frac{1}{2}} I + A$ are invertible, and hence $Ae_i = \lambda_i^{\frac{1}{2}} e_i$. In case $\lambda_i = 0$ we compute

$$\|Ae_i\|^2 = (Ae_i, Ae_i) = (A^2 e_i, e_i) = (Te_i, e_i) = 0.$$

So $Ae_i = 0$. Thus a nonnegative square root is completely determined by T, whence uniqueness. ∎

7.3.8   Partial Isometries

Definition 7.52. Let $\mathscr{U}_1, \mathscr{U}_2$ be inner product spaces. An operator $U : \mathscr{U}_1 \longrightarrow \mathscr{U}_2$ is called a **partial isometry** if there exists a subspace $\mathscr{M} \subset \mathscr{U}_1$ such that

$$\|Ux\| = \begin{cases} \|x\|, & x \in \mathscr{M}, \\ 0, & x \perp \mathscr{M}. \end{cases}$$

The space $\mathscr{M}$ is called the **initial space** of U and the image of U, $U\mathscr{M}$, the **final space**.

In the following proposition we collect the basic facts on partial isometries.

Proposition 7.53. *Let $\mathscr{U}_1, \mathscr{U}_2$ be inner product spaces, and $U : \mathscr{U}_1 \longrightarrow \mathscr{U}_2$ a linear transformation. Then*

1. *U is a partial isometry if and only if U^*U is an orthogonal projection onto the initial space of U.*
2. *U is a partial isometry with initial space $\mathscr{M}$ if and only if U^* is a partial isometry with initial space $U\mathscr{M}$.*
3. *UU^* is the orthogonal projection on the final space of U and U^*U is the orthogonal projection on the initial space of U.*

Proof. 1. Assume U is a partial isometry with initial space $\mathscr{M}$. Let P be the orthogonal projection of $\mathscr{U}_1$ on $\mathscr{M}$. Then for $x \in \mathscr{M}$, we have

$$(U^*Ux, x) = \|Ux\|^2 = \|x\|^2 = (x, x) = (Px, x).$$

On the other hand, if $x \perp \mathscr{M}$, then

$$(U^*Ux,x) = \|Ux\|^2 = 0 = (Px,x).$$

So $((U^*U - P)x,x) = 0$ for all x, and hence $U^*U = P$.

Conversely, suppose $P = U^*U$ is a, necessarily orthogonal, projection in $\mathscr{U}_1$. Let $\mathscr{M} = \{x | U^*Ux = x\}$. Then for any $x \in \mathscr{U}_1$,

$$\|Ux\|^2 = (U^*Ux,x) = (Px,x) = \|Px\|^2.$$

This shows that U is a partial isometry with initial space $\mathscr{M}$.

2. Assume U is a partial isometry with initial space $\mathscr{M}$. Note that we have $\mathscr{U}_2 = \operatorname{Im} U \oplus \operatorname{Ker} U^* = U\mathscr{M} \oplus \operatorname{Ker} U^*$. Obviously $U^*|\operatorname{Ker} U^* = 0$, whereas if $x \in \mathscr{M}$, then $U^*Ux = Px = x$. So $\|U^*Ux\| = \|x\| = \|Ux\|$, i.e., U^* is a partial isometry with initial space $U\mathscr{M}$. By the previous part, it follows that UU^* is the orthogonal projection on $U\mathscr{M}$, the final space of U.

3. From Part 1, it follows that UU^* is the orthogonal projection on the initial space of U^*, which is $U\mathscr{M}$, the final space of U. The rest follows by duality. ∎

7.3.9 The Polar Decomposition

In analogy with the polar representation of a complex number in the form $z = re^{i\theta}$, we have a representation of an arbitrary linear transformation in an inner product space as the product of a unitary operator and a nonnegative one.

Theorem 7.54. *Let $\mathscr{U}_1, \mathscr{U}_2$ be two inner product spaces and let $T : \mathscr{U}_1 \longrightarrow \mathscr{U}_2$ be a linear transformation. Then there exist partial isometries $V : \mathscr{U}_1 \longrightarrow \mathscr{U}_2$ and $W : \mathscr{U}_2 \longrightarrow \mathscr{U}_1$ for which*

$$T = V(T^*T)^{\frac{1}{2}} = (TT^*)^{\frac{1}{2}}W. \tag{7.14}$$

The partial isometry V is uniquely defined if we require $\operatorname{Ker} V = \operatorname{Ker} T$. Similarly, the partial isometry W is uniquely defined if we require $\operatorname{Ker} W = \operatorname{Ker} T^*$.

Proof. For $x \in \mathscr{U}_1$, we compute

$$\|(T^*T)^{\frac{1}{2}}x\|^2 = ((T^*T)^{\frac{1}{2}}x,(T^*T)^{\frac{1}{2}}x) = (T^*Tx,x) = (Tx,Tx) = \|Tx\|^2. \tag{7.15}$$

Note that we have

$$\operatorname{Im}(T^*T)^{\frac{1}{2}} = \operatorname{Im} T^*,$$

$$\operatorname{Ker}(T^*T)^{\frac{1}{2}} = \operatorname{Ker} T. \tag{7.16}$$

We define $V : \mathscr{U}_1 \longrightarrow \mathscr{U}_2$ by

$$V(T^*T)^{\frac{1}{2}}x = Tx, \quad x \in \mathscr{U}_1,$$

$$Vz = 0, \quad z \in \mathrm{Ker}\, T. \tag{7.17}$$

Recalling the direct sum representation $\mathscr{U}_1 = \mathrm{Im}\,(T^*T)^{\frac{1}{2}} \oplus \mathrm{Ker}\, T$, it follows from (7.15) that V is a well-defined partial isometry with $\mathrm{Im}\,(T^*T)^{\frac{1}{2}}$ as initial space and $\mathrm{Im}\, T = \mathrm{Im}\,(TT^*)^{\frac{1}{2}}$ as final space. The other representation in (7.14) follows by duality. $\blacksquare$

A remark on uniqueness is in order. It is clear from the construction of V that it is uniquely determined on $\mathrm{Im}\,(T^*T)^{\frac{1}{2}}$. Therefore V is uniquely determined if $\mathrm{Ker}\, T = \mathrm{Ker}\,(T^*T)^{\frac{1}{2}} = \{0\}$. Noting that the definition of V on $\mathrm{Ker}\, T$ was a choice we made, it follows that we could make a different choice, provided $\mathrm{Im}\, T$ has a nontrivial orthogonal complement. Since $(\mathrm{Im}\, T)^{\perp} = \mathrm{Ker}\, T^*$, $\mathrm{Ker}\, T^* = \{0\}$ is also a sufficient condition for the uniqueness of V. In general, if $\dim \mathscr{U}_2 \geq \dim \mathscr{U}_1$, we can assume V to be a, not necessarily unique, isometry. If $\dim \mathscr{U}_1 \geq \dim \mathscr{U}_2$, we can assume W to be a, not necessarily unique, coisometry, i.e., W^* an isometry. In case we have $\dim \mathscr{U}_1 = \dim \mathscr{U}_2$, we can assume both V and W to be unitary. In this case, V and W are uniquely determined if and only if T is invertible.

7.4   Singular Vectors and Singular Values

In this section we study the basic properties of singular vectors and singular values. We use this to give a characterization of singular values as approximation numbers. These results will be applied, in Chapter 12, to Hankel norm approximation problems.

Starting from a polar decomposition of a linear transformation, and noting that the norm of a unitary operator, or for that matter also of isometries and nontrivial partial isometries, is 1, it seems plausible that the operators $(T^*T)^{\frac{1}{2}}$ and $(TT^*)^{\frac{1}{2}}$ provide a measure of the size of T.

Proposition 7.55. *Let $\mathscr{U}_1, \mathscr{U}_2$ be two inner product spaces and let $T : \mathscr{U}_1 \longrightarrow \mathscr{U}_2$ be a linear transformation. Let $(T^*T)^{\frac{1}{2}}$ and $(TT^*)^{\frac{1}{2}}$ be the nonnegative square roots of T^*T and TT^* respectively. Then $(T^*T)^{\frac{1}{2}}$ and $(TT^*)^{\frac{1}{2}}$ have the same nonzero eigenvalues, including multiplicities.*

Proof. Let μ be a nonzero eigenvalue of $(T^*T)^{\frac{1}{2}}$ and $x \neq 0$ a corresponding eigenvector. Clearly, this implies that also $T^*Tx = \mu^2 x$. Define $y = \frac{Tx}{\mu}$. We proceed to compute

$$TT^*y = TT^* \frac{Tx}{\mu} = \frac{T}{\mu}(T^*Tx) = \frac{T}{\mu}(\mu^2 x) = \mu^2 \frac{Tx}{\mu} = \mu^2 y.$$

Note that $x \neq 0$ implies $y \neq 0$, and this shows that μ^2 is an eigenvalue of TT^*. In turn, this implies that μ is an eigenvalue of $(TT^*)^{\frac{1}{2}}$. Let now $\mathscr{V}_i = \{x \in \mathscr{U}_1 | (T^*T)^{\frac{1}{2}} x = \mu_i x\}$ and $\mathscr{W}_i = \{y \in \mathscr{U}_2 | (TT^*)^{\frac{1}{2}} y = \mu_i y\}$. We define maps $\Phi_i : \mathscr{V}_i \longrightarrow \mathscr{W}_i$ and $\Psi_i : \mathscr{W}_i \longrightarrow \mathscr{V}_i$ by

$$\Phi_i x = \frac{Tx}{\mu_i}, \quad x \in \mathscr{V}_i,$$

$$\Psi_i y = \frac{T^* y}{\mu_i}, \quad y \in \mathscr{W}_i. \tag{7.18}$$

We compute

$$\|\Phi_i x\|^2 = \left(\frac{Tx}{\mu_i}, \frac{Tx}{\mu_i} \right) = \frac{1}{\mu_i^2} (T^* Tx, x) = \|x\|^2,$$

i.e., Φ_i is an isometry. The same holds for Ψ_i by symmetry. Finally, it is easy to check that $\Psi_i \Phi_i = I$ and $\Phi_i \Psi_i = I$, i.e., both maps are unitary. Hence, $\mathscr{V}_i, \mathscr{W}_i$ have the same dimension, and so the multiplicity of μ_i as an eigenvalue of $(T^*T)^{\frac{1}{2}}$ and $(TT^*)^{\frac{1}{2}}$ is the same. ∎

This leads us to the following definition.

Definition 7.56. Let $T : \mathscr{U}_1 \longrightarrow \mathscr{U}_2$ be a linear transformation. A pair of vectors $\{\phi, \psi\}$, with $\phi \in \mathscr{U}_1$ and $\psi \in \mathscr{U}_2$, is called a **Schmidt pair** of T, corresponding to the nonzero **singular value** μ, if

$$\begin{aligned} T\phi &= \mu\psi, \\ T^*\psi &= \mu\phi, \end{aligned} \tag{7.19}$$

holds.

We also refer to ϕ, ψ as the **singular vectors** of T and T^* respectively.

The minimax principle leads to the following characterization of singualr values.

Theorem 7.57. *Let $\mathscr{U}_1, \mathscr{U}_2$ be inner product spaces and let $T : \mathscr{U}_1 \longrightarrow \mathscr{U}_2$ be a linear transformation. Let $\mu_1 \geq \cdots \geq \mu_n \geq 0$ be its singular values. Then*

$$\mu_k = \min_{\{\mathscr{M} | \mathrm{codim}\, \mathscr{M} = k-1\}} \max_{x \in \mathscr{M}} \frac{\|Tx\|}{\|x\|}.$$

Proof. The μ_i^2 are the eigenvalues of T^*T. Applying the minimax principle, we have

$$\mu_k^2 = \min_{\{\mathscr{M} | \mathrm{codim}\, \mathscr{M} = k-1\}} \max_{x \in \mathscr{M}} \frac{(T^*Tx, x)}{(x, x)}$$

$$= \min_{\{\mathscr{M} | \mathrm{codim}\, \mathscr{M} = k-1\}} \max_{x \in \mathscr{M}} \frac{\|Tx\|^2}{\|x\|^2},$$

which is equivalent to the statement of the theorem. ∎

With the use of the polar decomposition and the spectral representation of self-adjoint operators, we get a convenient representation of arbitrary operators.

Theorem 7.58. *Let $T : \mathscr{U}_1 \longrightarrow \mathscr{U}_2$ be a linear transformation and let $\mu_1 \geq \cdots \geq \mu_r$ be its nonzero singular values and let $\{\phi_i, \psi_i\}$ be the corresponding orthonormal sets of Schmidt pairs. Then for every pair of vectors $x \in \mathscr{U}_1, y \in \mathscr{U}_2$, we have*

$$Tx = \sum_{i=1}^{r} \mu_i(x, \phi_i)\psi_i \tag{7.20}$$

and

$$T^*y = \sum_{i=1}^{r} \mu_i(y, \psi_i)\phi_i. \tag{7.21}$$

Proof. Let $\mu_1 \geq \cdots \geq \mu_r > 0$ the nonzero singular values and $\{\phi_i, \psi_i\}$ the corresponding Schmidt pairs.

We have seen that with V defined by $V\phi_i = \psi_i$, and of course $V^*\psi_i = \phi_i$, we have $T = V(T^*T)^{\frac{1}{2}}$. Now, for $x \in \mathscr{U}_1$, we have

$$(T^*T)^{\frac{1}{2}}x = \sum_{i=1}^{r} \mu_i(x, \phi_i)\phi_i,$$

and hence

$$Tx = V(T^*T)^{\frac{1}{2}}x = V\sum_{i=1}^{r} \mu_i(x, \phi_i)\phi_i = \sum_{i=1}^{r} \mu_i(x, \phi_i)\psi_i.$$

The other representation for T^* can be derived analogously, or alternatively, can be obtained by computing the adjoint of T using (7.20). ∎

The availability of the previously obtained representations for T and T^* leads directly to an extremely important characterization of singular values in terms of approximation properties. The question we address is the following. Given a linear transformation of rank m between two inner product spaces, how well can it be approximated, in the operator norm, by transformations of lower rank. We have the following characterization.

Theorem 7.59. *Let $T : \mathscr{U}_1 \longrightarrow \mathscr{U}_2$ be a linear transformation of rank m. Let $\mu_1 \geq \cdots \geq \mu_m$ be its nonzero singular values. Then*

$$\mu_k = \inf\{\|T - T'\| \mid \operatorname{rank} T' \leq k - 1\}. \tag{7.22}$$

Proof. By Theorem 7.58, we may assume that $Tx = \sum_{i=1}^{n} \mu_i(x, \phi_i)\psi_i$. We define a linear transformation T' by $T'x = \sum_{i=1}^{k-1} \mu_i(x, \phi_i)\psi_i$. Obviously $\operatorname{rank}(T') = k - 1$ and $(T - T')x = \sum_{i=k}^{n} \mu_i(x, \phi_i)\psi_i$, and therefore we conclude that $\|T - T'\| = \mu_k$.

Conversely, let us assume $\operatorname{rank}(T') = k - 1$, i.e., $\operatorname{codim} \operatorname{Ker} T' = k - 1$. By Theorem 7.57, we have

$$\mu_k \leq \max_{x \in \operatorname{Ker} T'} \frac{\|Tx\|}{\|x\|} = \max_{x \in \operatorname{Ker} T'} \frac{\|(T - T')x\|}{\|x\|} \leq \|T - T'\|.$$ ■

7.5 Unitary Embeddings

The existence of nonnegative square roots is a central tool in modern operator theory. In this section we discuss some related embedding theorems. The basic question we address is, given a linear transformation A in an inner product space U, when can it be embedded in a 2×2 block unitary operator matrix of the form $V = \begin{pmatrix} A & B \\ C & D \end{pmatrix}$.

This block matrix is to be considered initially to be an operator defined on the inner product space $U \oplus U$, by $\begin{pmatrix} A & B \\ C & D \end{pmatrix} \begin{pmatrix} x \\ y \end{pmatrix} = \begin{pmatrix} Ax + By \\ Cx + Dy \end{pmatrix}$. It is clear that if V given before is unitary, then, obseving that P defined by $P \begin{pmatrix} x \\ y \end{pmatrix} = \begin{pmatrix} x \\ 0 \end{pmatrix}$ is an orthogonal projection, it follows that $A = PV|_{U \oplus \{0\}}$, and we have

$$\|A\| = \|PV|_{U \oplus \{0\}}\| \leq \|V\| = 1,$$

that is, A is necessarily a contraction. The following theorem focuses on the converse.

Theorem 7.60. *A linear transformation A can be embedded in a 2×2 block unitary matrix if and only if it is a contraction.*

Proof. That embeddability in a unitary matrix implies that A is contractive has been shown before.

Thus, assume A is contractive, i.e., $\|Ax\| \leq \|x\|$ for all $x \in U$. Since

$$\|x\|^2 - \|Ax\|^2 = (x, x) - (Ax, Ax) = (x, x) - (A^*Ax, x)$$
$$= ((I - A^*A)x, x) = \|(I - A^*A)^{\frac{1}{2}}x\|^2.$$

Here we used the fact that the nonnegative operator $I - A^*A$ has a unique nonnegative square root. This implies that, setting $C = (I - A^*A)^{\frac{1}{2}}$, the transformation given by the block matrix $\begin{pmatrix} A \\ C \end{pmatrix}$ is isometric, i.e., satisfies $\begin{pmatrix} A^* & C^* \end{pmatrix} \begin{pmatrix} A \\ C \end{pmatrix} = I$.

Similarly, using the requirement that $\begin{pmatrix} A & B \end{pmatrix} \begin{pmatrix} A^* \\ B^* \end{pmatrix} = I$, we set $B = (I - AA^*)^{\frac{1}{2}}$. Computing now

$$
\begin{pmatrix} I & 0 \\ 0 & I \end{pmatrix} = \begin{pmatrix} A & (I-AA^*)^{\frac{1}{2}} \\ (I-A^*A)^{\frac{1}{2}} & D \end{pmatrix} \begin{pmatrix} A^* & (I-A^*A)^{\frac{1}{2}} \\ (I-AA^*)^{\frac{1}{2}} & D^* \end{pmatrix},
$$

we have $(I-A^*A)^{\frac{1}{2}}A^* + D(I-AA^*)^{\frac{1}{2}} = 0$. Using the equality $(I-A^*A)^{\frac{1}{2}}A^* = A^*(I-AA^*)^{\frac{1}{2}}$, we get $(A^*+D)(I-AA^*)^{\frac{1}{2}} = 0$. This indicates that we should choose $D = -A^*$. It remains to check that

$$
V = \begin{pmatrix} A & (I-AA^*)^{\frac{1}{2}} \\ (I-A^*A)^{\frac{1}{2}} & -A^* \end{pmatrix} \tag{7.23}
$$

is a required embedding.   ∎

The embedding (7.23) is not unique. In fact, if K and L are unitary operators in U, then $\begin{pmatrix} A & (I-AA^*)^{\frac{1}{2}}L \\ K(I-A^*A)^{\frac{1}{2}} & -KA^*L \end{pmatrix}$ parametrize all such embeddings.

The embedding presented in the previous theorem is possibly redundant, since it requires a space of twice the dimension of the original space U. For example, if A is unitary to begin with, then no extension is necessary. Thus the dimension of the minimal space where a unitary extension is possible seems to be related to the distance of the contraction A from unitary operators. This is generally measured by the ranks of the operators $(I-A^*A)^{\frac{1}{2}}$ and $(I-AA^*)^{\frac{1}{2}}$. The next theorem handles this situation. However, before embarking on that route, we give an important dilation result. This is an extension of a result originally obtained in Halmos (1950). That construction played a central role, via the Schäffer matrix, see Sz.-Nagy and Foias (1970), in the theory of unitary dilations.

Theorem 7.61. *Given a contraction A in $\mathbb{C}^n$, there exists a unitary matrix $\begin{pmatrix} A & B \\ C & D \end{pmatrix}$, with B injective and C surjective, if and only if this matrix has the form*

$$
\begin{pmatrix} A & (I-AA^*)^{\frac{1}{2}}V \\ U(I-A^*A)^{\frac{1}{2}} & -UA^*V \end{pmatrix}, \tag{7.24}
$$

where $U : \mathbb{C}^n \longrightarrow \mathbb{C}^p$ is a partial isometry with initial space $\mathrm{Im}\,(I-A^*A)^{\frac{1}{2}}$ and where $V : \mathbb{C}^p \longrightarrow \mathbb{C}^n$ is an isometry with final space $\mathrm{Im}\,(I-AA^*)^{\frac{1}{2}}$.

Proof. Let $P_{\mathrm{Im}\,(I-AA^*)^{\frac{1}{2}}}$ and $P_{\mathrm{Im}\,(I-A^*A)^{\frac{1}{2}}}$ be the orthogonal projections on the appropriate spaces. Assume U, V are partial isometries as described in the theorem. Then we have

$$
UU^* = I, \qquad U^*U = P_{\mathrm{Im}\,(I-A^*A)^{\frac{1}{2}}},
$$

$$
V^*V = I, \qquad VV^* = P_{\mathrm{Im}\,(I-AA^*)^{\frac{1}{2}}}. \tag{7.25}
$$

These identities imply the following:

$$VV^*(I-AA^*)^{\frac{1}{2}} = (I-AA^*)^{\frac{1}{2}},$$
$$(I-AA^*)^{\frac{1}{2}}VV^* = (I-AA^*)^{\frac{1}{2}}. \tag{7.26}$$

We compute now the product

$$\begin{pmatrix} A & (I-AA^*)^{\frac{1}{2}}V \\ U(I-A^*A)^{\frac{1}{2}} & -UA^*V \end{pmatrix} \begin{pmatrix} A^* & (I-A^*A)^{\frac{1}{2}}U^* \\ V^*(I-AA^*)^{\frac{1}{2}} & -V^*AU^* \end{pmatrix}.$$

We use the identities in (7.26). So

$$AA^* + (I-AA^*)^{\frac{1}{2}}VV^*(I-AA^*)^{\frac{1}{2}} = AA^* + (I-AA^*)^{\frac{1}{2}}(I-AA^*)^{\frac{1}{2}} = I.$$

Next,

$$A(I-A^*A)^{\frac{1}{2}}U^* - (I-AA^*)^{\frac{1}{2}}VV^*AU^* = A(I-A^*A)^{\frac{1}{2}}U^* - (I-AA^*)^{\frac{1}{2}}AU^* = 0.$$

Similarly,

$$U(I-A^*A)^{\frac{1}{2}}A^* - UA^*VV^*(I-AA^*)^{\frac{1}{2}} = U[(I-A^*A)^{\frac{1}{2}}A^* - A^*(I-AA^*)^{\frac{1}{2}}] = 0.$$

Finally,

$$U(I-A^*A)^{\frac{1}{2}}(I-A^*A)^{\frac{1}{2}}U^* + UA^*VV^*AU^* = U(I-A^*A)U^* + UA^*AU^* = UU^* = I.$$

This shows that the matrix in (7.24) is indeed unitary.

Conversely, suppose we are given a contraction A. If C satisfies $A^*A + C^*C = I$, then $C^*C = I - A^*A$. This in turn implies that for every vector x, we have

$$\|Cx\|^2 = \|(I-A^*A)^{\frac{1}{2}}x\|^2.$$

Therefore, there exists a partial isometry U, defined on $\mathrm{Im}\,(I-A^*A)^{\frac{1}{2}}$, for which

$$C = U(I-A^*A)^{\frac{1}{2}}. \tag{7.27}$$

Obviously, $\mathrm{Im}\,C \subset \mathrm{Im}\,U$. Since C is surjective this forces U to be surjective. Thus $UU^* = I$ and $U^*U = P_{\mathrm{Im}(I-A^*A)^{\frac{1}{2}}}$. In a similar fashion, we start from the equality $AA^* + BB^* = I$ and see that for every x,

$$\|B^*x\|^2 = \|(I-AA^*)^{\frac{1}{2}}x\|^2.$$

So we conclude that there exists a partial isometry V^* with initial space $\text{Im}\,(I - AA^*)^{\frac{1}{2}}$ for which

$$B^* = V^*(I - AA^*)^{\frac{1}{2}},$$

or

$$B = (I - AA^*)^{\frac{1}{2}}V. \tag{7.28}$$

Since B is injective, so is V. So we get $V^*V = I$, and $VV^* = P_{\text{Im}\,(I-AA^*)^{\frac{1}{2}}}$.

Next we use the equality $DB^* + CA^* = 0$, to get

$$DV^*(I - AA^*)^{\frac{1}{2}} + U(I - A^*A)^{\frac{1}{2}}A^* = (DV^* + UA^*)(I - AA^*)^{\frac{1}{2}} = 0,$$

so $DV^* + UA^*|_{\text{Im}\,(I-AA^*)^{\frac{1}{2}}} = 0$. Now we have $\mathbb{C}^n = \text{Im}\,(I - AA^*)^{\frac{1}{2}} \oplus \text{Ker}\,(I - AA^*)^{\frac{1}{2}}$. Moreover, the identity $A(I - A^*A)^{\frac{1}{2}} = (I - AA^*)^{\frac{1}{2}}A$ implies $A\text{Im}\,(I - A^*A)^{\frac{1}{2}} \subset \text{Im}\,(I - AA^*)^{\frac{1}{2}}$ as well as $A\text{Ker}\,(I - A^*A)^{\frac{1}{2}} \subset \text{Ker}\,(I - AA^*)^{\frac{1}{2}}$. Similarly, $A^*\text{Ker}\,(I - AA^*)^{\frac{1}{2}} \subset \text{Ker}\,(I - A^*A)^{\frac{1}{2}}$. Letting now $x \in \text{Ker}\,(I - AA^*)^{\frac{1}{2}}$, then $A^*x \in \text{Ker}\,(I - A^*A)^{\frac{1}{2}}$. Since U is a partial isometry with initial space $\text{Im}\,(I - A^*A)^{\frac{1}{2}}$, necessarily $UA^*x = 0$. On the other hand V^* is a partial isometry with initial space $(I - AA^*)^{\frac{1}{2}}$ and hence $V^*|_{\text{Ker}\,(I-AA^*)^{\frac{1}{2}}} = 0$. So $DV^* + UA^*|_{\text{Ker}\,(I-AA^*)^{\frac{1}{2}}} = 0$, and hence we can conclude that $DV^* + UA^* = 0$. Using the fact that $V^*V = I$, this implies

$$D = -UA^*V. \tag{7.29}$$

It is easy to check now, as in the sufficiency part, that with B, C, D satisfying (7.27), (7.28), and (7.29) respectively, all other equations are satisfied. ∎

7.6 Exercises

1. Show that the rational function $q(z) = \sum_{k=-n}^{n} q_k z^k$ is nonnegative on the unit circle if and only if $q(z) = p(z)p^{\#}(z)$ for some polynomial $p(z)$. Here $p^{\#}(z) = \overline{p(\overline{z}^{-1})}$. This theorem, the simplest example of **spectral factorization**, is due to Fejér.

2. Show that a polynomial $q(z)$ is nonnegative on the imaginary axis if and only if for some polynomial $p(z)$, we have $q(z) = p(z)\overline{p(-\overline{z})}$.

3. Given vectors $x_1, \ldots, x_k$ in an inner product space $\mathscr{U}$, the Gram determinant is defined by $G(x_1, \ldots, x_k) = \det((x_i, x_j))$. Show that $x_1, \ldots, x_k$ are linearly independent if and only if $\det G(x_1, \ldots, x_k) \neq 0$.

4. Let $\mathscr{X}$ be an inner product space of functions. Let $f_0, f_1, \ldots$ be a sequence of linearly independent functions in $\mathscr{X}$ and let $\phi_0, \phi_1, \ldots$ be the sequence of

functions obtained from it by way of the Gram–Schmidt orthonormalization procedure.

a. Show that

$$\phi_n(x) = \begin{vmatrix} (f_0, f_0) & \cdots & (f_n, f_0) \\ \cdot & \cdots & \cdot \\ \cdot & \cdots & \cdot \\ \cdot & \cdots & \cdot \\ (f_0, f_{n-1}) & \cdots & (f_n, f_{n-1}) \\ f_0(x) & \cdots & f_n(x) \end{vmatrix}.$$

b. Let

$$\phi_n(x) = \sum_{i=0}^{n} c_{n,i} f_i(x),$$

and define

$$D_{-1} = 1,$$

$$D_n = \begin{vmatrix} (f_0, f_0) & \cdots & (f_n, f_0) \\ \cdot & \cdots & \cdot \\ \cdot & \cdots & \cdot \\ \cdot & \cdots & \cdot \\ (f_0, f_n) & \cdots & (f_n, f_n) \end{vmatrix}.$$

Show that

$$c_{n,n} = \left(\frac{D_{n-1}}{D_n} \right)^{\frac{1}{2}}.$$

5. In the space of real polynomials $\mathbb{R}[x]$, an inner product is defined by

$$(p,q) = \int_{-1}^{1} p(x)q(x)dx.$$

Apply the Gram-Schmidt orthogonalization procedure to the sequence of polynomials $1, x, x^2, \ldots$. Show that up to multiplicative constants, the resulting polynomials coincide with the Legendre polynomials $P_n(x) = \frac{1}{2^n n!} \frac{d^n(x^2-1)^n}{dx^n}$.

6. Show that the rank of a skew-symmetric matrix in a real inner product space is even.

7. Let K be a skew-Hermitian operator, i.e., $K^* = -K$. Show that $U = (I + K)^{-1}(I - K)$ is unitary. Show that every unitary operator is of the form $U = \lambda(I+K)^{-1}(I-K)$, where λ is a complex number of absolute value 1.

8. Show that if A and B are normal and $\operatorname{Im} A \perp \operatorname{Im} B$, then $A + B$ is normal.

9. Show that A is normal if and only if for some unitary U, we have $A^* = AU$.

10. Let N be a normal operator and let $AN = NA$. Show that $AN^* = N^*A$.
11. Given a normal operator N and any positive integer k, show that there exists an operator A satisfying $A^k = N$.
12. Show that a circulant matrix $A \in \mathbb{C}^{n \times n}$ is normal.
13. Let T be a real orthogonal matrix, i.e., it satisfies $\tilde{T}T = I$. Show that T is similar to a block diagonal matrix $\operatorname{diag}(I, R_1, \ldots, R_k)$, where

$$ R_i = \begin{pmatrix} \cos \theta_i & \sin \theta_i \\ -\sin \theta_i & \cos \theta_i \end{pmatrix} . $$

14. Show that a linear operator A in an inner product space having eigenvalues $\lambda_1, \ldots, \lambda_n$ is normal if and only if $\operatorname{Trace}(A^*A) = \sum_{i=1}^{n} |\lambda_i|^2$.
15. Given contractive operators A and B in finite-dimensional inner product spaces $\mathcal{U}_1$ and $\mathcal{U}_2$ respectively, show that there exists an operator $X : \mathcal{U}_2 \longrightarrow \mathcal{U}_1$ for which $\begin{pmatrix} A & X \\ 0 & B \end{pmatrix}$ is contractive if and only if $X = (I - AA^*)^{\frac{1}{2}} C (I - B^*B)^{\frac{1}{2}}$ and $C : \mathcal{U}_2 \longrightarrow \mathcal{U}_1$ is a contraction.
16. Define the Jacobi tridiagonal matrices by

$$ J_k = \begin{pmatrix} a_1 & -b_1 & . & . & & 0 \\ -c_1 & a_2 & . & & & . \\ . & . & . & . & & . \\ . & & . & . & -b_{k-1} \\ 0 & & . & . & -c_{k-1} & a_k \end{pmatrix}, $$

assuming $b_i, c_i > 0$. Let $D_k(\lambda) = \det(\lambda I - J_k)$.

a. Show that the following recurrence relation is satisfied:

$$ D_k(\lambda) = (\lambda - a_k)D_{k-1}(\lambda) - b_{k-1}c_{k-1}D_{k-2}(\lambda). $$

b. Show that J_k is similar to a self-adjoint transformation.

17. **Singular value decomposition**. Let $A \in \mathbb{C}^{m \times n}$ and let $\sigma_1 \geq \cdots \geq \sigma_r$ be its nonzero singular values. Show that there exist unitary matrices $U \in \mathbb{C}^{m \times m}$ and $V \in \mathbb{C}^{n \times n}$ such that

$$ A = \begin{pmatrix} D & 0 \\ 0 & 0 \end{pmatrix}, $$

where $D = \operatorname{diag}(\sigma_1, \ldots, \sigma_r)$. Compare with Theorem 7.58.
18. Show that the eigenvalues of $\begin{pmatrix} 0 & A \\ A^* & 0 \end{pmatrix}$ are $\sigma_1, \ldots, \sigma_n, -\sigma_1, \ldots, -\sigma_n$.

7.7 Notes and Remarks

Most of the results in this chapter go over in one way or another to the context of Hilbert spaces, where, in fact, most of them were proved initially. Young (1988) is a good, very readable modern source for the basics of Hilbert space theory.

The spectral theorem in Hilbert space was one of the major early achievements of functional analysis and operator theory. For an extensive source on these topics and the history of the subject, see Dunford and Schwartz (1958, 1963).

Singular values were introduced by E. Schmidt in 1907. Because of the importance of singular values as approximation numbers, equation (7.22) is sometimes taken as the definition of singular values. In this connection, see Young (1988).

Chapter 8
Tensor Products and Forms

8.1 Introduction

In this chapter, we discuss the general theory of quadratic forms and the notion of congruence and its invariants, as well as the applications of the theory to the analysis of special forms. We will focus primarily on quadratic forms induced by rational functions, most notably the Hankel and Bezout forms, because of their connection to system-theoretic problems such as stability and signature symmetric realizations. These forms use as their data different representations of rational functions, namely power series and coprime factorizations respectively. But we will discuss also the partial fraction representation in relation to the computation of the Cauchy index of a rational function and the proof of the Hermite–Hurwitz theorem and the continued fraction representation as a tool in the computation of signatures of Hankel matrices as well as in the problem of Hankel matrix inversion. Thus, different representations of rational functions, i.e., different encodings of the information carried by a rational function, provide efficient starting points for different methods. The results obtained for rational functions will be applied to root-location problems for polynomials in the next chapter.

In the latter part of the chapter, we approach the study of bilinear forms from the point of view of tensor products and their representations as spaces of homomorphisms. This we specialize to the tensor product of polynomial and rational models, emphasizing concrete, functional identifications for the tensor products. Since polynomial models carry two natural structures, namely that of vector spaces over the underlying field $\mathbb{F}$ as well as that of modules over the polynomial ring $\mathbb{F}[z]$, it follows that there are two related tensor products. The relation between the two tensor products illuminates the role of the Bezoutians and the Bezout map. In turn, this leads to the characterization of the maps that intertwine two polynomial models, yielding a generalization of Theorem 5.17.

P.A. Fuhrmann, *A Polynomial Approach to Linear Algebra*, Universitext,
DOI 10.1007/978-1-4614-0338-8_8, © Springer Science+Business Media, LLC 2012

8.2   Basics

8.2.1   Forms in Inner Product Spaces

Given a linear transformation T in a finite-dimensional complex inner product space $\mathcal{U}$, we define a field-valued function ϕ on $\mathcal{U} \times \mathcal{U}$ by

$$\phi(x,y) = (Tx,y). \tag{8.1}$$

Clearly, ϕ is linear in the variable x and antilinear in the variable y. Such a function will be called a **sesquilinear form** or just a **form**. If the field is the field $\mathbb{R}$ of real numbers, then a form is actually linear in both variables, i.e., a **bilinear form**.

It might seem that the forms defined by (8.1) are rather special. This is not the case, as is seen from the following.

Theorem 8.1. *Let $\mathcal{U}$ be a finite-dimensional complex inner product space. Then ϕ is a form on $\mathcal{U}$ if and only if there exists a linear transformation T in $\mathcal{U}$ such that (8.1) holds. The operator T is uniquely determined by ϕ.*

Proof. We utilize Theorem 7.14 on the representation of linear functionals on inner product spaces. Thus, if we fix a vector $y \in \mathcal{U}$, the function $\phi(x,y)$ is a linear functional, hence given by an inner product with a vector η_y, that is, $\phi(x,y) = (x,\eta_y)$. It is clear that η_y depends linearly on y. Therefore, there exists a linear transformation S in $\mathcal{U}$ for which $\eta_y = Sy$. We complete the proof by defining $T = S^*$.

To see uniqueness, assume T_1, T_2 represent the same form, i.e., $(T_1x,y) = (T_2x,y)$. This implies $((T_1 - T_2)x,y) = 0$, for all $x,y \in \mathcal{U}$. Choosing $y = (T_1 - T_2)x$, we get $\|(T_1 - T_2)x\| = 0$ for all $x \in \mathcal{U}$. This shows that $T_2 = T_1$. $\blacksquare$

Suppose we choose a basis $\mathcal{B} = \{f_1, \ldots, f_n\}$ in $\mathcal{U}$. Then arbitrary vectors in $\mathcal{U}$ can be written as $x = \sum_{j=1}^{n} \xi_j f_j$ and $y = \sum_{i=1}^{n} \eta_i f_i$. In terms of the coordinates, we can compute the form by

$$\phi(x,y) = (Tx,y) = \left(T\sum_{j=1}^{n}\xi_j f_j, \sum_{i=1}^{n}\eta_i f_i\right) = \sum_{j=1}^{n}\sum_{i=1}^{n}\xi_j\overline{\eta_i}(Tf_j,f_i)$$
$$= \sum_{j=1}^{n}\sum_{i=1}^{n}\phi_{ij}\xi_j\overline{\eta_i},$$

where we define $\phi_{ij} = (Tf_j, f_i)$. We call the matrix (ϕ_{ij}) the **matrix representation of the form** ϕ in the basis $\mathcal{B}$, and denote this matrix by $[\phi]_{\mathcal{B}}^{\mathcal{B}}$. Using the standard inner product in $\mathbb{C}^n$, we can write

$$\phi(x,y) = ([\phi]_{\mathcal{B}}^{\mathcal{B}}[x]^{\mathcal{B}}, [x]^{\mathcal{B}}).$$

Next, we consider how the matrix of a quadratic form changes with a change of basis in $\mathcal{U}$. Let $\mathcal{B}' = \{g_1, \ldots, g_n\}$ be another basis in $\mathcal{U}$. Then $[x]^{\mathcal{B}} = [I]_{\mathcal{B}}^{\mathcal{B}'}[x]^{\mathcal{B}'}$ and therefore

$$\phi(x,y) = ([\phi]_{\mathscr{B}}^{\mathscr{B}}[x]^{\mathscr{B}}, [y]^{\mathscr{B}}) = ([\phi]_{\mathscr{B}}^{\mathscr{B}}[I]_{\mathscr{B}}^{\mathscr{B}'}[x]^{\mathscr{B}'}, [I]_{\mathscr{B}}^{\mathscr{B}'}[y]^{\mathscr{B}'})$$

$$= (([I]_{\mathscr{B}}^{\mathscr{B}'})^* [\phi]_{\mathscr{B}}^{\mathscr{B}}[I]_{\mathscr{B}}^{\mathscr{B}'}[x]^{\mathscr{B}'}, [y]^{\mathscr{B}'}) = ([\phi]_{\mathscr{B}'}^{\mathscr{B}'}[x]^{\mathscr{B}'}, [y]^{\mathscr{B}'}),$$

which shows that

$$[\phi]_{\mathscr{B}'}^{\mathscr{B}'} = ([I]_{\mathscr{B}}^{\mathscr{B}'})^* [\phi]_{\mathscr{B}}^{\mathscr{B}}[I]_{\mathscr{B}}^{\mathscr{B}'}, \tag{8.2}$$

i.e., in a change of basis, the matrix of a quadratic form changes by congruence.

The next result clarifies the connection between the matrix representation of a form and the matrix representation of the corresponding linear transformation. We say that a complex square matrix A_1 is **congruent** to a matrix A if there exists a nonsingular matrix R such that $A_1 = R^* A R$. In case we deal with the real field, the Hermitian adjoint R^* is replaced by the transpose $\tilde{R}$.

The following proposition is standard, and we omit the proof.

Proposition 8.2. *In the ring of square real or complex matrices, congruence is an equivalence relation.*

A **quadratic form** in a complex inner product space $\mathscr{U}$ is a function of the form

$$\hat{\phi}(x) = \phi(x,x), \tag{8.3}$$

where $\phi(x,y)$ is a symmetric bilinear form on $\mathscr{U}$. We say that a form on $\mathscr{U}$ is a **symmetric form** if we have $\phi(x,y) = \phi(y,x)$ for all $x,y \in \mathscr{U}$ and a **Hermitian form** if we have the additional property

$$\phi(x,y) = \overline{\phi(y,x)}.$$

Proposition 8.3. *A form ϕ on a complex inner product space $\mathscr{U}$ is Hermitian if and only if there exists a Hermitian operator T in $\mathscr{U}$ for which $\phi(x,y) = (Tx,y)$.*

Proof. Assume T is Hermitian, i.e., $T^* = T$. Then

$$\phi(x,y) = (Tx,y) = (x,Ty) = \overline{(Ty,x)} = \overline{\phi(y,x)}.$$

Conversely, assume $\phi(x,y) = \overline{\phi(y,x)}$. Let T be the linear transformation for which $\phi(x,y) = (Tx,y)$. We compute

$$(Tx,y) = \phi(x,y) = \overline{\phi(y,x)} = \overline{(Ty,x)} = (x,Ty),$$

and this for all $x,y \in \mathscr{U}$. This shows that T is Hermitian. ∎

Proposition 8.4. *Let $\mathscr{U}$ be a complex inner product space and ϕ a Hermitian form on $\mathscr{U}$. Let T be the uniquely defined linear transformation in $\mathscr{U}$ for which $\phi(x,y) = (Tx,y)$, and let $\mathscr{B} = \{f_1,\ldots,f_n\}$ be a basis for $\mathscr{U}$. Then $[\phi]_{\mathscr{B}}^{\mathscr{B}} = [T]_{\mathscr{B}}^{\mathscr{B}*}$.*

Proof. Let $\mathscr{B}^* = \{g_1, \ldots, g_n\}$ be the basis dual to $\mathscr{B}$. Then $[T]_{\mathscr{B}}^{\mathscr{B}^*} = (t_{ij})$ where the t_{ij} are defined through $T f_j = \sum_{k=1}^n t_{kj} g_k$. Computing

$$(T f_j, f_i) = \left(\sum_{k=1}^n t_{kj} g_k, f_i \right) = \sum_{k=1}^n t_{kj} \delta_{ik} = t_{ij},$$

the result follows.   ∎

We are interested in the possibility of reducing a Hermitian form, using congruence transformations, to a diagonal form.

Theorem 8.5. *Let ϕ be a Hermitian form on a finite-dimensional complex inner product space $\mathscr{U}$. Then there exists an orthonormal basis $\mathscr{B} = \{e_1, \ldots, e_n\}$ in $\mathscr{U}$ for which $[\phi]_{\mathscr{B}}^{\mathscr{B}}$ is diagonal with real entries.*

Proof. Let $T : \mathscr{U} \longrightarrow \mathscr{U}$ be the Hermitian operator representing the form ϕ. By Theorem 7.33, there exists an orthonormal basis $\mathscr{B} = \{e_1, \ldots, e_n\}$ in $\mathscr{U}$ consisting of eigenvectors of T corresponding to the real eigenvalues $\lambda_1, \ldots, \lambda_n$. With respect to this basis, the matrix of the form is given by

$$\phi_{ij} = (T e_j, e_i) = \lambda_j (e_j, e_i) = \lambda_j \delta_{ij}.$$ ∎

Just as positive operators are a subset of self-adjoint operators, so positive forms are a subset of Hermitian forms. We say that a form ϕ on a complex inner product space $\mathscr{U}$ is **nonnegative** if it is Hermitian and $\phi(x,x) \geq 0$ for all $x \in \mathscr{U}$. We say that a form ϕ on an inner product space $\mathscr{U}$ is **positive** if it is Hermitian and $\phi(x,x) > 0$ for all $0 \neq x \in \mathscr{U}$.

Corollary 8.6. *Let ϕ be a nonnegative Hermitian form on the complex inner product space $\mathscr{U}$. Then there exists an orthogonal basis $\mathscr{B} = \{f_1, \ldots, f_n\}$ for which the matrix $[\phi]_{\mathscr{B}}^{\mathscr{B}}$ is diagonal with*

$$\phi_{ii} = \begin{cases} 1, & 0 \leq i \leq r, \\ 0, & r < i \leq n. \end{cases}$$

Proposition 8.7. *A form ϕ on a complex inner product space $\mathscr{U}$ is positive if and only if there exists a positive operator P for which $\phi(x,y) = (Px, y)$.*

Of course, to check positivity of a form, it suffices to compute all the eigenvalues and check them for positivity. However, computing eigenvalues is not, in general, an algebraic process. So our next goal is to search for a computable criterion for positivity.

8.2.2 Sylvester's Law of Inertia

If we deal with real forms, then due to the different definition of congruence, we cannot apply the complex result directly.

The proof of Theorem 8.5 is particularly simple. However, it uses the spectral theorem, and that theorem is constructive only if we have access to the eigenvalues of the appropriate self-adjoint operator. Generally, this is not the case. Thus we would like to re-prove the diagonalization by congruence result in a different way. It is satisfying that this can be done algebraically, and the next theorem, known as the **Lagrange reduction method**, accomplishes this.

Theorem 8.8. *Let ϕ be a symmetric bilinear form on a real inner product space $\mathscr{U}$. Then there exists a basis $\mathscr{B} = \{f_1, \ldots, f_n\}$ of $\mathscr{U}$ for which $[\phi]_{\mathscr{B}}^{\mathscr{B}}$ is diagonal.*

Proof. Let $\phi(x,y)$ be a symmetric bilinear form on $\mathscr{U}$. We prove the theorem by induction on the dimension n of $\mathscr{U}$. If $n = 1$, then obviously $[\phi]_{\mathscr{B}}^{\mathscr{B}}$, being a 1×1 matrix, is diagonal.

Assume we have proved the theorem for spaces of dimension $\leq n - 1$. Let now ϕ be a symmetric bilinear form in an n-dimensional space $\mathscr{U}$. Assume $\mathscr{B} = \{f_1, \ldots, f_n\}$ is an arbitrary basis of $\mathscr{U}$.

We consider two cases. In the first, we assume that not all the diagonal elements of $[\phi]_{\mathscr{B}}^{\mathscr{B}}$ are zero, i.e., there exists an index i for which $\phi(f_i, f_i) \neq 0$. The second case is the complementary one, in which all the $\phi(f_i, f_i)$ are zero.

Case 1: Without loss of generality, using symmetric transpositions of rows and columns, we will assume that $\phi(f_1, f_1) \neq 0$. We define a new basis $\mathscr{B}' = \{f_1', \ldots, f_n'\}$, where

$$f_i' = \begin{cases} f_1, & i = 1, \\[2ex] f_i - \dfrac{\phi(f_1, f_i)}{\phi(f_1, f_1)} f_1, & 1 < i \leq n. \end{cases}$$

Clearly, $\mathscr{B}'$ is indeed a basis for $\mathscr{U}$ and

$$\phi(f_1', f_i') = \begin{cases} \phi(f_1, f_1), & i = 1, \\[2ex] 0, & 1 < i \leq n. \end{cases}$$

Thus, with the subspace $M = L(f_2', \ldots, f_n')$, we have

$$[\phi]_{\mathscr{B}'}^{\mathscr{B}'} = \begin{pmatrix} \phi(f_1, f_1) & 0 \\ 0 & [\phi|M]_{\mathscr{B}_0'}^{\mathscr{B}_0'} \end{pmatrix}.$$

Here $\phi|M$ is the restriction of ϕ to M and $\mathscr{B}_0' = \{f_2', \ldots, f_n'\}$ is a basis for M. The proof, in this case, is completed using the induction hypothesis.

Case 2: Assume now that $\phi(f_i, f_i) = 0$ for all i. If $\phi(f_i, f_j) = 0$ for all i, j, then ϕ is the zero form, and hence $[\phi]_{\mathscr{B}}^{\mathscr{B}}$ is the zero matrix, which is certainly diagonal. So we assume there exists a pair of distinct indices i, j for which $\phi(f_i, f_j) \neq 0$. Without loss of generality, we will assume that $i = 1, j = 2$. To simplify notation we will write $a = \phi(f_1, f_2)$. First, we note that defining $f_1' = \frac{f_1 + f_2}{2}$ and $f_2' = \frac{f_1 - f_2}{2}$, we have

$$\phi(f_1', f_1') = \frac{1}{4}\phi(f_1 + f_2, f_1 + f_2) = \frac{a}{2},$$

$$\phi(f_1', f_2') = \frac{1}{4}\phi(f_1 + f_2, f_1 - f_2) = 0,$$

$$\phi(f_2', f_2') = \frac{1}{4}\phi(f_1 - f_2, f_1 - f_2) = -\frac{a}{2}.$$

We choose the other vectors in the new basis to be of the form $f_i' = f_i - \alpha_i f_1 - \beta_i f_2$, with the requirement that for $i = 3, \ldots, n$, we have $\phi(f_1', f_i') = \phi(f_2', f_i') = 0$. Therefore we get the pair of equations

$$0 = \phi(f_1 + f_2, f_i - \alpha_i f_1 - \beta_i f_2) = \phi(f_1, f_i) + \phi(f_2, f_i) - \alpha_i \phi(f_2, f_1) - \beta_i \phi(f_2, f_1)$$
$$0 = \phi(f_1 - f_2, f_i - \alpha_i f_1 - \beta_i f_2) = \phi(f_1, f_i) - \phi(f_2, f_i) + \alpha_i \phi(f_2, f_1) - \beta_i \phi(f_2, f_1).$$

This system has the solution $\alpha_i = \frac{\phi(f_2, f_i)}{a}$ and $\beta_i = \frac{\phi(f_1, f_i)}{a}$. We define now $\mathscr{B}' = \{f_1', \ldots, f_n'\}$, where

$$
\begin{aligned}
f_1' &= \frac{f_1 + f_2}{2}, \\
f_2' &= \frac{f_1 - f_2}{2}, \\
f_i' &= f_i - \frac{\phi(f_2, f_i)}{a} f_1 - \frac{\phi(f_1, f_i)}{a} f_2.
\end{aligned}
\tag{8.4}
$$

It is easy to check that $\mathscr{B}'$ is a basis for $\mathscr{U}$. With respect to this basis we have the block diagonal matrix

$$[\phi]_{\mathscr{B}'}^{\mathscr{B}'} = \begin{pmatrix} \frac{a}{2} & 0 & \\ 0 & -\frac{a}{2} & \\ & & [\phi|M]_{\mathscr{B}_0'}^{\mathscr{B}_0'} \end{pmatrix}.$$

Here $\mathscr{B}_0' = \{f_3', \ldots, f_n'\}$ and $M = L(f_3', \ldots, f_n')$. We complete the proof by applying the induction hypothesis to $\phi|M$. ∎

Corollary 8.9. *Let ϕ be a quadratic form defined on a real inner product space $\mathscr{U}$. Let $\mathscr{B} = \{f_1, \ldots, f_n\}$ be an arbitrary basis of $\mathscr{U}$ and let $\mathscr{B}' = \{g_1, \ldots, g_n\}$ be a* *basis of $\mathscr{U}$ that diagonalizes ϕ. Let $\phi_i = \phi(g_i, g_i)$. If*

$$[x]^{\mathscr{B}} = \begin{pmatrix} \xi_1 \\ \vdots \\ \xi_n \end{pmatrix}, \qquad [x]^{\mathscr{B}'} = \begin{pmatrix} \eta_1 \\ \vdots \\ \eta_n \end{pmatrix},$$

and $[I]_{\mathscr{B}}^{\mathscr{B}'} = (a_{ij})$, then

$$\phi(x,x) = \sum_{i=1}^{n} \phi_i \left(\sum_{j=1}^{n} a_{ij} \xi_j \right)^2. \tag{8.5}$$

Proof. We have

$$\phi(x,x) = ([\phi]_{\mathscr{B}}^{\mathscr{B}}[x]^{\mathscr{B}}, [x]^{\mathscr{B}}) = ([\phi]_{\mathscr{B}'}^{\mathscr{B}'}[x]^{\mathscr{B}'}, [x]^{\mathscr{B}'}) = ([\phi]_{\mathscr{B}'}^{\mathscr{B}'}[I]_{\mathscr{B}}^{\mathscr{B}'}[x]^{\mathscr{B}'}, [[I]_{\mathscr{B}}^{\mathscr{B}'} x]^{\mathscr{B}'}).$$

Also $[x]^{\mathscr{B}'} = [I]_{\mathscr{B}}^{\mathscr{B}'}[x]^{\mathscr{B}}$ implies $\eta_i = \sum_{j=1}^{n} a_{ij} \xi_j$. So

$$\phi(x,x) = \left([\phi]_{\mathscr{B}'}^{\mathscr{B}'}[x]^{\mathscr{B}'}, [x]^{\mathscr{B}'} \right) = \sum_{i=1}^{n} \phi_i \eta_i^2 = \sum_{i=1}^{n} \phi_i \left(\sum_{j=1}^{n} a_{ij} \xi_j \right)^2.$$

■

We say that (8.5) is a representation of the quadratic form as a **sum of squares**. Such a representation is far from unique. In different representations, the diagonal elements as well as the linear forms may differ. We have, however, the following result, known as Sylvester's law of inertia, which characterizes the congruence invariant.

Theorem 8.10. *Let $\mathscr{U}$ be a real inner product space and $\phi(x,y)$ a symmetric bilinear form on $\mathscr{U}$. Then there exists a basis $\mathscr{B} = \{f_1, \ldots, f_n\}$ for $\mathscr{U}$ such that the matrix $[\phi]_{\mathscr{B}}^{\mathscr{B}}$ is diagonal, with $\phi(f_i, f_j) = \varepsilon_i \delta_{ij}$, with*

$$\varepsilon_i = \begin{cases} 1, & 0 \leq i \leq k, \\ -1, & k < i \leq r, \\ 0, & r < i \leq n. \end{cases}$$

Moreover, the numbers k and r are uniquely determined by ϕ.

Proof. By Theorem 8.8, there exists a basis $\mathscr{B}$ such that $[\phi]_{\mathscr{B}}^{\mathscr{B}}$ is real and diagonal, with $\lambda_1, \ldots, \lambda_n$ the diagonal elements. Without loss of generality, we can assume $\lambda_i > 0$ for $1 \leq i \leq k$, $\lambda_i < 0$ for $k < i \leq r$ and $\lambda_i = 0$ for $r < i \leq n$. We set $\mu_i = |\lambda_i|^{\frac{1}{2}}$ and $\varepsilon_i = \text{sign} \lambda_i$. Hence, we have, for all i, $\lambda_i = \varepsilon_i \mu_i^2$. We consider now the basis $\mathscr{B}_1 = \{\mu_1^{-1} e_1, \ldots, \mu_n^{-1} e_n\}$. Clearly, with respect to this basis, the matrix $[\phi]_{\mathscr{B}_1}^{\mathscr{B}_1}$ has the required form.

To prove the uniqueness of k and r, we note first that r is equal to the rank of $[\phi]_{\mathscr{B}}^{\mathscr{B}}$. So we have to prove uniqueness for k only. Suppose we have another basis $\mathscr{B}' = \{g_1, \dots, g_n\}$ such that $\phi(g_j, g_i) = \varepsilon_i' \delta_{ij}$, where

$$\varepsilon_i' = \begin{cases} 1, & 0 \le i \le k', \\ -1, & k' < i \le r, \\ 0, & r < i \le n. \end{cases}$$

Without loss of generality, we assume $k < k'$. Given $x \in \mathscr{U}$, we have $x = \sum_{i=1}^n \xi_i f_i = \sum_{i=1}^n \eta_i g_i$, that is, the ξ_i and η_i are the coordinates of x with respect to $\mathscr{B}$ and $\mathscr{B}'$ respectively. Computing the quadratic form, we get

$$\phi(x,x) = \sum_{i=1}^n \varepsilon_i \xi_i^2 = \sum_{i=1}^n \varepsilon_i' \eta_i^2,$$

or

$$\xi_1^2 + \cdots + \xi_k^2 - \xi_{k+1}^2 - \cdots - \xi_r^2 = \eta_1^2 + \cdots + \eta_{k'}^2 - \eta_{k'+1}^2 - \cdots - \eta_r^2.$$

This can be rewritten as

$$\xi_1^2 + \cdots + \xi_k^2 + \eta_{k'+1}^2 + \cdots + \eta_r^2 = \eta_1^2 + \cdots + \eta_{k'}^2 + \xi_{k+1}^2 + \cdots + \xi_r^2. \qquad (8.6)$$

Now we consider the set of equations $\xi_1 = \cdots = \xi_k = \eta_{k'+1} = \cdots = \eta_r = 0$. We have here $r - k' + k = r - (k' - k) < r$ equations. Thus there exists a nontrivial vector x for which not all the coordinates $\xi_{k+1} = \cdots = \xi_r$ are zero. However, by equation (8.6), we get $\xi_{k+1} = \cdots = \xi_r = 0$, which is a contradiction. Thus we must have $k' \ge k$, and by symmetry we get $k' = k$. ∎

Definition 8.11. The numbers r and $s = 2k - r$ determined by Sylvester's law of inertia are called the **rank** and **signature** respectively of the quadratic form ϕ. We will denote the signature of a quadratic form by $\sigma(\phi)$. Similarly, the signature of a symmetric (Hermitian) matrix A will be denoted by $\sigma(A)$. The triple (π, ν, δ), with $\pi = \frac{r+s}{2}$, $\nu = \frac{r-s}{2}$, $\delta = n - r$, is called the **inertia** of the form, and denoted by $\mathrm{In}(A)$.

Corollary 8.12. *1. Two quadratic forms on a real inner product space $\mathscr{U}$ are congruent if and only if they have the same rank and signature.*
2. Two $n \times n$ symmetric matrices are congruent if and only if they have the same rank and signature.

Proposition 8.13. *Let A be an $m \times m$ real symmetric matrix, X an $m \times n$ matrix of full row rank, and B the $n \times n$ symmetric matrix defined by*

$$B = \tilde{X} A X. \qquad (8.7)$$

Then the rank and signature of A and B coincide.

Proof. The statement concerning ranks is trivial. From (8.7) we obtain the two inclusions $\operatorname{Ker}B \supset \operatorname{Ker}X$ and $\operatorname{Im}B \subset \operatorname{Im}\tilde{X}$. Let us choose a basis for $\mathbb{R}^n$ that is compatible with the direct sum decomposition

$$\mathbb{R}^n = \operatorname{Ker}X \oplus \operatorname{Im}\tilde{X}.$$

In that basis, $X = \begin{pmatrix} 0 & X_1 \end{pmatrix}$ with X_1 invertible and

$$B = \begin{pmatrix} 0 & 0 \\ 0 & B_1 \end{pmatrix} = \begin{pmatrix} 0 \\ \tilde{X}_1 \end{pmatrix} A \begin{pmatrix} 0 & X_1 \end{pmatrix}$$

or $B_1 = \tilde{X}_1 A X_1$. Thus $\sigma(B) = \sigma(B_1) = \sigma(A)$, which proves the theorem. ∎

The positive definiteness of a quadratic form can be determined from any of its matrix representations. The following theorem gives a determinantal characterization for positivity.

Theorem 8.14. *Let ϕ be a symmetric bilinear form on an n-dimensional vector space $\mathcal{U}$ and let $[\phi]_{\mathcal{B}}^{\mathcal{B}} = (\phi_{ij})$ be its matrix representation with respect to the basis $\mathcal{B}$. Then the following statements are equivalent:*

1. *ϕ is positive definite.*
2. *$[\phi]_{\mathcal{B}}^{\mathcal{B}} = (\phi_{ij}) = X\tilde{X}$, with X invertible and lower triangular.*
3. *All principal minors are positive, i.e.,*

$$\begin{vmatrix} \phi_{11} & \cdots & \phi_{1k} \\ \cdot & \cdots & \cdot \\ \cdot & \cdots & \cdot \\ \cdot & \cdots & \cdot \\ \phi_{k1} & \cdots & \phi_{kk} \end{vmatrix} > 0,$$

   for $k = 1,\dots,n$.

Proof. $(2) \Rightarrow (1)$
This is obvious.
$(3) \Rightarrow (2)$
We prove this by induction on k. For $k = 1$, this is immediate. Assume it holds for all $i = 1,\dots,k-1$ and that $A = \begin{pmatrix} A_{11} & b \\ \tilde{b} & d \end{pmatrix}$ is $k \times k$ with $A_{11} \in \mathbb{R}^{(k-1)\times(k-1)}$ positive. By the induction hypothesis, there exists a matrix $X \in \mathbb{R}^{(k-1)\times(k-1)}$ for which $A_{11} = X_1\tilde{X}_1$, with X_1 nonsingular. We write

$$\begin{pmatrix} A_{11} & b \\ \tilde{b} & d \end{pmatrix} = \begin{pmatrix} X_1 & 0 \\ 0 & 1 \end{pmatrix} \begin{pmatrix} I & X_1^{-1}b \\ \widetilde{bX_1}^{-1} & d \end{pmatrix} \begin{pmatrix} \widetilde{X_1} & 0 \\ 0 & 1 \end{pmatrix},$$

with $\begin{pmatrix} I & v \\ \tilde{v} & d \end{pmatrix} > 0$, where $v = X_1^{-1}b$. Using the computation

$$\begin{pmatrix} I & v \\ \tilde{v} & d \end{pmatrix} = \begin{pmatrix} I & 0 \\ \tilde{v} & d \end{pmatrix} \begin{pmatrix} I & 0 \\ 0 & e \end{pmatrix} \begin{pmatrix} I & v \\ 0 & 1 \end{pmatrix} = \begin{pmatrix} I & v \\ \tilde{v} & e + \tilde{v}v \end{pmatrix},$$

i.e., $e + \tilde{v}v = d$ and necessarily $e = d - \tilde{v}v > 0$, and setting $e = \varepsilon^2$, it follows that $A = X\tilde{X}$, where $X = \begin{pmatrix} X_1 & 0 \\ 0 & 1 \end{pmatrix} \begin{pmatrix} I & 0 \\ \tilde{v} & \varepsilon \end{pmatrix}$.

$(1) \Rightarrow (3)$

Assume all principal minors have positive determinants. Let (ϕ_{ij}) be the matrix of the qudratic form. Let R be the matrix that reduces (ϕ_{ij}) to upper triangular form. If we do not normalize the diagonal elements, then R is lower triangular with all elements on the diagonal equal to 1. Thus $R\phi\tilde{R}$ is diagonal with the determinants of the principal minors on the digonal. Since $R\phi\tilde{R}$ is positive by assumption, so is ϕ.   ∎

The same result holds for Hermitian forms, where the transpose is replaced by the Hermitian adjoint.

Theorem 8.15. *1. Let $v_1, \ldots, v_k \in \mathbb{C}^n$, $\lambda_1, \ldots, \lambda_k \in \mathbb{R}$, and let A be the Hermitian matrix $\sum_{i=1}^{k} \lambda_i v_i v_i^*$. Then A is Hermitian congruent to $\Lambda = \mathrm{diag}(\lambda_1, \ldots, \lambda_k, 0, \ldots, 0)$.*

2. If A_1 and A_2 are Hermitian and

$$\mathrm{rank}(A_1 + A_2) = \mathrm{rank}(A_1) + \mathrm{rank}(A_2),$$

then

$$\sigma(A_1 + A_2) = \sigma(A_1) + \sigma(A_2).$$

Proof. 1. Extend $\{v_1, \ldots, v_k\}$ to a basis $\{v_1, \ldots, v_n\}$ of $\mathbb{C}^n$. Let V be the matrix whose columns are $v_1, \ldots, v_n$. Then $V\Lambda V^* = A$.

2. Let $\lambda_1, \ldots, \lambda_k$ be the nonzero eigenvalues of A_1, with corresponding eigenvectors $v_1, \ldots, v_k$ and let $\lambda_{k+1}, \ldots, \lambda_{k+l}$ be the nonzero eigenvalues of A_2, with corresponding eigenvectors $v_{k+1}, \ldots, v_{k+l}$. Thus $A_1 = \sum_{i=1}^{k} \lambda_i v_i v_i^*$ and $A_2 = \sum_{i=k+1}^{k+l} \lambda_i v_i v_i^*$. By our assumption, $\mathrm{Im}\,A_1 \cap \mathrm{Im}\,A_2 = \{0\}$. Therefore, the vectors $v_1, \ldots, v_{k+l}$ are linearly independent. Hence, by part (1), $A_1 + A_2$ is Hermitian congruent to $\mathrm{diag}(\lambda_1, \ldots, \lambda_{k+l}, 0, \ldots, 0)$.   ∎

In the sequel, we will focus our interest on some important quadratic forms induced by rational functions.

8.3   Some Classes of Forms

We pass now from the general theory of forms to some very important examples. We will focus mainly on the Hankel and Bezoutian forms and their interrelations. These forms arise out of different representations of rational functions. Since positivity is not essential to the principal results, we assume that the field $\mathbb{F}$ is arbitrary. Assume $\mathcal{V}$ is a finite-dimensional $\mathbb{F}$-vector space. A billinear form $\phi(x,y)$ on $\mathcal{V} \times \mathcal{V}$ is called **symmetric** if, for all $x, y \in \mathcal{V}$, we have

$$\phi(x,y) = \phi(y,x). \tag{8.8}$$

8.3.1 Hankel Forms

We begin our study of special forms by focusing on Hankel operators and forms. Though our interest is mainly in Hankel operators with rational symbols, we need to enlarge the setting in order to give Kronecker's characterization of rationality in terms of infinite Hankel matrices.

Let $g(z) = \sum_{j=-\infty}^{n} g_j z^j \in \mathbb{F}((z^{-1}))$ and let π_+ and π_- be the projections, defined in (1.23), that correspond to the direct sum representation $\mathbb{F}((z^{-1})) = \mathbb{F}[z] \oplus z^{-1}\mathbb{F}[[z^{-1}]]$. We define the **Hankel operator** $H_g : \mathbb{F}[z] \longrightarrow z^{-1}\mathbb{F}[[z^{-1}]]$ by

$$H_g f = \pi_- g f, \qquad f(z) \in \mathbb{F}[z]. \tag{8.9}$$

Here $g(z)$ is called the **symbol** of the Hankel operator. Clearly, H_g is a well-defined linear transformation.

We recall, from Chapter 1, that $\mathbb{F}[z], z^{-1}\mathbb{F}[[z^{-1}]]$ and $\mathbb{F}((z^{-1}))$ carry also natural $\mathbb{F}[z]$-module structures. In $\mathbb{F}[z]$ this is simply given in terms of polynomial multiplication. However, $z^{-1}\mathbb{F}[[z^{-1}]]$ is not an $\mathbb{F}[z]$-submodule of $\mathbb{F}((z^{-1}))$. Therefore, we define for $h(z) \in z^{-1}\mathbb{F}[[z^{-1}]]$,

$$p \cdot h = \pi_- p h.$$

For the polynomial z we denote the corresponding operators by $S_+ : \mathbb{F}[z] \longrightarrow \mathbb{F}[z]$ and $S_- : z^{-1}\mathbb{F}[[z^{-1}]] \longrightarrow z^{-1}\mathbb{F}[[z^{-1}]]$, defined by

$$\begin{cases} (S_+ f)(z), = z f(z) & f(z) \in \mathbb{F}[z], \\ (S_- h)(z), = \pi_- z h(z) & h(z) \in z^{-1}\mathbb{F}[[z^{-1}]]. \end{cases} \tag{8.10}$$

Proposition 8.16. *Let $g(z) \in \mathbb{F}((z^{-1}))$ and let $H_g : \mathbb{F}[z] \longrightarrow z^{-1}\mathbb{F}[[z^{-1}]]$ be the Hankel operator defined in (8.9). Then*

1. H_g is an $\mathbb{F}[z]$-module homomorphism.
2. Any $\mathbb{F}[z]$-module homomorphism from $\mathbb{F}[z]$ to $z^{-1}\mathbb{F}[[z^{-1}]]$ is a Hankel operator.
3. Let $g_1(z), g_2(z) \in \mathbb{F}((z^{-1}))$. Then $H_{g_1} = H_{g_2}$ if and only if $g_1(z) - g_2(z) \in \mathbb{F}[z]$.

Proof. 1. It suffices to show that

$$H_g S_+ = S_- H_g. \tag{8.11}$$

This we do by computing

$$H_g S_+ f = \pi_- g z f = \pi_- z(gf) = \pi_- z \pi_-(gf) = S_- H_g f.$$

Here we used the fact that $S_+ \operatorname{Ker} \pi_- \subset \operatorname{Ker} \pi_-$.

2. Let $H : \mathbb{F}[z] \longrightarrow z^{-1} \mathbb{F}[[z^{-1}]]$ be an $\mathbb{F}[z]$-module homomorphism. We set $g = H1 \in z^{-1} \mathbb{F}[[z^{-1}]]$. Then, since H is a homomorphism, for any polynomial $f(z)$ we have

$$Hf = H(f \cdot 1) = f \cdot H1 = \pi_- fg = \pi_g f = H_g f.$$

Thus we have identified H with a Hankel operator.

3. Since $H_{g_1} - H_{g_2} = H_{g_1 - g_2}$, it suffices to show that $H_g = 0$ if and only if $g(z) \in \mathbb{F}[z]$. Assume therefore $H_g = 0$. Then $\pi_- g1 = 0$, which means that $g(z) \in \mathbb{F}[z]$. The converse is obvious. ∎

Equation (8.11) expresses the fact that $H_g : \mathbb{F}[z] \longrightarrow z^{-1} \mathbb{F}[[z^{-1}]]$ is an $\mathbb{F}[z]$-module homomorphism, where both spaces are equipped with the $\mathbb{F}[z]$-module structure. We call (8.11) the **functional equation of Hankel operators** .

Corollary 8.17. *Let* $g(z) \in \mathbb{F}((z^{-1}))$. *Then* H_g *as a map from* $\mathbb{F}[z]$ *to* $z^{-1} \mathbb{F}[[z^{-1}]]$ *is not invertible.*

Proof. Were H_g invertible, it would follow from (8.11) that S_+ and S_- are similar transformations. This is impossible, since for S_- each $\alpha \in \mathbb{F}$ is an eigenvalue, whereas S_+ has no eigenvalues at all. ∎

Corollary 8.18. *Let* H_g *be a Hankel operator. Then* $\operatorname{Ker} H_g$ *is a submodule of* $\mathbb{F}[z]$ *and* $\operatorname{Im} H_g$ *a submodule of* $z^{-1} \mathbb{F}[[z^{-1}]]$.

Proof. Follows from (8.11). ∎

Submodules of $\mathbb{F}[z]$ are ideals, hence of the form $q\mathbb{F}[z]$. Also, we have a characterization of finite-dimensional submodules of $\mathbb{F}_-(z) \subset z^{-1} \mathbb{F}[[z^{-1}]]$, the subspace of rational, strictly proper functions. This leads to Kronecker's theorem.

Theorem 8.19 (Kronecker). *Let* $g(z) \in z^{-1} \mathbb{F}[[z^{-1}]]$.

1. *A Hankel operator* H_g *has finite rank if and only if* $g(z)$ *is a rational function.*
2. *If* $g(z) = \frac{p(z)}{q(z)}$ *with* $p(z), q(z)$ *coprime, then*

$$\operatorname{Ker} H_{\frac{p}{q}} = q\mathbb{F}[z] \tag{8.12}$$

and

$$\operatorname{Im} H_{\frac{p}{q}} = X^q. \tag{8.13}$$

Proof. Assume $g(z)$ is rational. So $g(z) = p(z)/q(z)$, and we may assume that $p(z)$ and $q(z)$ are coprime. Clearly $q\mathbb{F}[z] \subset \operatorname{Ker} H_g$, and because of coprimeness, we actually have the equality $q\mathbb{F}[z] = \operatorname{Ker} H_g$. This proves (8.12). Now we have $\mathbb{F}[z] = X_q \oplus q\mathbb{F}[z]$ as $\mathbb{F}$-vector spaces and hence $\operatorname{Im} H_g = \{\pi_- q^{-1} pf \mid f \in X_q\}$. We compute

$$\pi_- q^{-1} pf = q^{-1}(q\pi_- q^{-1} pf) = q^{-1}\pi_- qpf \in X^q.$$

Using coprimeness, and Theorem 5.18, we have $\{\pi_q pf \mid f \in X_q\} = X_q$ and therefore $\operatorname{Im} H_g = X^q$, which shows that H_g has finite rank.

Conversely, assume H_g has finite rank. So $M = \operatorname{Im} H_g$ is a finite-dimensional subspace of $z^{-1}\mathbb{F}[[z^{-1}]]$. Let $A = S_-|M$ and let $q(z)$ be the minimal polynomial of A. Therefore, for every $f(z) \in \mathbb{F}[z]$ we have $0 = \pi_- q\pi_- gf = \pi_- qgf$. Choosing $f(z) = 1$ we conclude that $q(z)g(z) = p(z)$ is a polynomial and hence $g(z) = p(z)/q(z)$ is rational. ∎

In terms of expansions in powers of z, since $g(z) = \sum_{j=1}^{\infty} g_j z^{-j}$, the Hankel operator has the infinite matrix representation

$$H = \begin{pmatrix} g_1 & g_2 & g_3 & \cdot & \cdot \\ g_2 & g_3 & g_4 & \cdot & \cdot \\ g_3 & g_4 & g_5 & \cdot & \cdot \\ \cdot & \cdot & \cdot & \cdot & \cdot \\ \cdot & \cdot & \cdot & \cdot & \cdot \end{pmatrix}. \tag{8.14}$$

Any infinite matrix (g_{ij}), satisfying $g_{ij} = g_{i+j-1}$ is called a **Hankel matrix**. Clearly, it is the matrix representation of the Hankel operator H_g with respect to the basis $\{1, z, z^2, \ldots\}$ in $\mathbb{F}[z]$ and $\{z^{-j} \mid j \geq 1\}$ in $z^{-1}\mathbb{F}[[z^{-1}]]$. Thus H has finite rank if and only if the associated Hankel operator has finite rank. In terms of Hankel matrices, Kronecker's theorem can be restated as follows.

Theorem 8.20 (Kronecker). *An infinite Hankel matrix H has finite rank if and only if its symbol $g(z) = \sum_{j=1}^{\infty} g_j z^{-j}$ is a rational function.*

Definition 8.21. Given a proper rational function $g(z)$, we define its **McMillan degree**, $\delta(g)$, by

$$\delta(g) = \operatorname{rank} H_g. \tag{8.15}$$

Proposition 8.22. *Let $g(z) = p(z)/q(z)$, with $p(z), q(z)$ coprime, be a proper rational function. Then*

$$\delta(g) = \deg q. \tag{8.16}$$

Proof. By Theorem 8.19 we have $\operatorname{Im} H_{p/q} = X^q$, and so

$$\delta(g) = \operatorname{rank} H_g = \dim X^q = \deg q. \tag{8.17}$$

∎

Proposition 8.23. *Let $g_i(z) = p_i(z)/q_i(z)$ be rational functions, with $p_i(z), q_i(z)$ coprime. Then*

1. We have

$$\operatorname{Im} H_{g_1+g_2} \subset \operatorname{Im} H_{g_1} + \operatorname{Im} H_{g_2} \tag{8.18}$$

and

$$\delta(g_1 + g_2) \leq \delta(g_1) + \delta(g_2). \tag{8.19}$$

2. We have

$$\delta(g_1 + g_2) = \delta(g_1) + \delta(g_2) \tag{8.20}$$

if and only if $q_1(z), q_2(z)$ are coprime.

Proof. 1. Obvious.
2. We have

$$g_1(z) + g_2(z) = \frac{q_1(z)p_2(z) + q_2(z)p_1(z)}{q_1(z)q_2(z)}.$$

Assume $q_1(z), q_2(z)$ are coprime. Then by Proposition 5.25, we have $X^{q_1} + X^{q_2} = X^{q_1 q_2}$. On the other hand, the coprimeness of $q_1(z), q_2(z)$ implies that of $q_1(z)q_2(z), q_1(z)p_2(z) + q_2(z)p_1(z)$. Therefore

$$\operatorname{Im} H_{g_1+g_2} = X^{q_1 q_2} = X^{q_1} + X^{q_2} = \operatorname{Im} H_{g_1} + \operatorname{Im} H_{g_2}.$$

Conversely, assume $\delta(g_1 + g_2) = \delta(g_1) + \delta(g_2)$. This implies

$$\operatorname{Im} H_{g_1} \cap \operatorname{Im} H_{g_2} = X^{q_1} \cap X^{q_2} = \{0\}.$$

But invoking once again Proposition 5.25, this implies the coprimeness of $q_1(z), q_2(z)$. ∎

For the special case of the real field, the Hankel operator defines a quadratic form on $\mathbb{R}[z]$. Since $\mathbb{R}[z]^* = z^{-1}\mathbb{R}[[z^{-1}]]$, it follows from our definition of self-dual maps that H_g is self-dual. Thus we have also an induced quadratic form on $\mathbb{R}[z]$ given by

$$[H_g f, f] = \sum_{i=0}^{\infty} \sum_{j=0}^{\infty} g_{i+j+1} f_i f_j. \tag{8.21}$$

Since $f(z)$ is a polynomial, the sum in (8.21) is well defined. Thus, with the Hankel map is associated the infinite Hankel matrix (8.14). With $g(z)$ we also associate the finite Hankel forms H_k, $k = 1, 2, \ldots$, defined by the matrices

$$H_k = \begin{pmatrix} g_1 & \cdots & g_k \\ \cdot & \cdots & \cdot \\ \cdot & \cdots & \cdot \\ \cdot & \cdots & \cdot \\ g_k & \cdots & g_{2k-1} \end{pmatrix}.$$

In particular, assuming as we did that $p(z)$ and $q(z)$ are coprime and that $\deg(q) = n$, then $\operatorname{rank}(H_k) = \deg(q) = n = \delta(g)$ for $k \geq n$.

Naturally, not only the rank information for H_n is derivable from $g(z)$ through its polynomial representation but also the signature information. This is most conveniently done by the use of Bezoutians or the Cauchy index. We turn next to these topics.

8.3.2 Bezoutians

A quadratic form is associated with each symmetric matrix. We will focus now on quadratic forms induced by polynomials and rational functions. In this section we present the basic theory of Bezout forms, or Bezoutians. Bezoutians are a special class of quadratic forms that were introduced to facilitate the study of root location of polynomials. Except for the study of signature information, we do not have to restrict the field. The study of Bezoutians is intimately connected to that of Hankel forms. The study of this connection will be undertaken in Section 8.3.5.

Let $p(z) = \sum_{i=0}^n p_i z^i$ and $q(z) = \sum_{i=0}^n q_i z^i$ be two polynomials and let z and w be two, generally noncommuting, variables. Then

$$p(z)q(w) - q(z)p(w) = \sum_{i=0}^n \sum_{j=0}^n p_i q_j (z^i w^j - z^j w^i)$$

$$= \sum_{0 \leq i < j \leq n} (p_i q_j - q_i p_j)(z^i w^j - z^j w^i). \tag{8.22}$$

Observe now that

$$z^i w^j - z^j w^i = \sum_{\nu=0}^{j-i-1} z^{i+\nu}(w - z)w^{j-\nu-1}.$$

Thus, equation (8.22) can be rewritten as

$$p(z)q(w) - q(z)p(w) = \sum_{0 \leq i < j \leq n} (p_i q_j - p_j q_i) \sum_{\nu=0}^{j-i-1} z^{i+\nu}(w - z)w^{j-\nu-1},$$

or

$$q(z)p(w) - p(z)q(w) = \sum_{i=1}^{n} \sum_{j=1}^{n} b_{ij} z^{i-1} (z - w) w^{j-1}. \tag{8.23}$$

The last equality is obtained by changing the order of summation and properly defining the coefficients b_{ij}. Equation (8.23) can also be written as

$$\frac{q(z)p(w) - p(z)q(w)}{z - w} = \sum_{i=1}^{n} \sum_{j=1}^{n} b_{ij} z^{i-1} w^{j-1}. \tag{8.24}$$

Definition 8.24. 1. The matrix (b_{ij}) defined by equation (8.23) or (8.24) is called the **Bezout form** associated with the polynomials $q(z)$ and $p(z)$ or just the **Bezoutian** of $q(z)$ and $p(z)$ and is denoted by $B(q,p)$. The function $\frac{q(z)p(w)-p(z)q(w)}{z-w}$ is called the **generating function** of the Bezoutian form.
2. If $g(z) = p(z)/q(z)$ is the unique coprime representation of a rational function $g(z)$, with $q(z)$ monic, then we will write $B(g)$ for $B(q,p)$. We shall call $B(g)$ the Bezoutian of $g(z)$.

It should be noted that equation (8.23) holds even when the variables z and w are noncommuting. This observation is extremely useful in the derivation of matrix representations of Bezoutians.

Note that $B(q,p)$ defines a bilinear form on $\mathbb{F}^n$ by

$$B(q,p)(\xi,\eta) = \sum_{i=1}^{n} \sum_{j=1}^{n} b_{ij} \xi_i \eta_j, \qquad \xi, \eta \in \mathbb{F}^n. \tag{8.25}$$

No distinction will be made in the notation between the matrix and the bilinear form. The following theorem summarizes the elementary properties of the Bezoutian.

Theorem 8.25. *Let $q(z), p(z) \in \mathbb{F}[z]$ with $\max(\deg q, \deg p) \leq n$. Then*

1. *The Bezoutian matrix $B(q,p)$ is a symmetric matrix.*
2. *The Bezoutian $B(q,p)$ is linear in $q(z)$ and $p(z)$, separately.*
3. *$B(p,q) = -B(q,p)$.*

It is these basic properties of the Bezoutian, in particular the linearity in both variables, that turn the Bezoutian into such a powerful tool. In conjunction with the Euclidean algorithm, it is the starting point for many fast matrix inversion algorithms.

Lemma 8.26. *Let $q(z)$ and $r(z)$ be two coprime polynomials and $p(z)$ an arbitrary nonzero polynomial. Then*

$$\operatorname{rank}(B(qp, rp)) = \operatorname{rank}(B(q,r)) \tag{8.26}$$

and

$$\sigma(B(qp, rp)) = \sigma(B(q,r)). \tag{8.27}$$

Proof. The Bezoutian $B(qp, rp)$ is determined by the polynomial expansion of

$$p(z)\frac{q(z)r(w) - r(z)q(w)}{z - w}p(w),$$

which implies $B(qp, pr) = \tilde{X}B(q, r)X$ with

$$X = \begin{pmatrix} p_0 & \cdots & p_m & & \\ & \ddots & & \ddots & \\ & & \ddots & & \ddots \\ & & & \ddots & \\ & & p_0 & \cdots & p_m \end{pmatrix}.$$

The result follows now by an application of Theorem 8.13 . ∎

The following theorem is central in the study of Bezout and Hankel forms. It gives a characterization of the Bezoutian as a matrix representation of an intertwining map. Thus, the study of Bezoutians is reduced to that of intertwining maps of the form $p(S_q)$, studied in Chapter 5. These are easier to handle and yield information on Bezoutians. This also leads to a conceptual theory of Bezoutians, in contrast to the purely computational approach that has been prevalent in their study for a long time.

Theorem 8.27. *Let $p(z), q(z) \in \mathbb{F}[z]$, with $\deg p \leq \deg q$. Then the Bezoutian $B(q, p)$ of $q(z)$ and $p(z)$ satisfies*
$$B(q, p) = [p(S_q)]_{co}^{st}. \tag{8.28}$$

Here $\mathscr{B}_{st}$ and $\mathscr{B}_{co}$ are the standard and control bases of X_q respectively defined in Proposition 5.3.

Proof. Note that

$$\pi_+ w^{-k}(z - w)w^{j-1} = \begin{cases} 0 & j < k \\ -1 & j = k \\ (z - w)w^{j-k-1} & j > k. \end{cases}$$

So, from equation (8.23), it follows that

$$q(z)\pi_+ w^{-k}p(w) - p(z)\pi_+ w^{-k}q(w)|_{w=z} = \sum_{i=1}^{n} b_{ik}z^{i-1},$$

or, with $\{e_1(z), \ldots, e_n(z)\}$ the control basis of X_q, and $p_k(z)$ defined by

$$p_k(z) = \pi_+ z^{-k}p(z),$$

we have

$$q(z)p_k(z) - p(z)e_k(z) = \sum_{i=1}^{n} b_{ik}z^{i-1}.$$

Applying the projection π_q, we obtain

$$\pi_q p e_k = p(S_q)e_k = \sum_{i=1}^{n} b_{ik}z^{i-1}, \tag{8.29}$$

which, from the definition of a matrix representation, completes the proof. ∎

Corollary 8.28. *Let* $p(z), q(z) \in \mathbb{F}[z]$, *with* $\deg p < \deg q$. *Then the last row and column of the Bezoutian* $B(q, p)$ *consist of the coefficients of* $p(z)$.

Proof. By equation (8.29) and the fact that the last element of the control basis satisfies $e_n(z) = 1$, we have

$$\sum_{i=1}^{n} b_{in}z^{i-1} = (\pi_q p e_n)(z) = (\pi_q p)(z) = p(z) = \sum_{i=0}^{n-1} p_i z^i, \tag{8.30}$$

i.e.,

$$b_{in} = p_{i-1}, \qquad i = 1, \dots, n. \tag{8.31}$$

The statement for rows follows from the symmetry of the Bezoutian. ∎

As an immediate consequence, we derive some well-known results concerning Bezoutians.

Corollary 8.29. *Let* $p(z), q(z) \in \mathbb{F}[z]$, *with* $\deg p \leq \deg q$. *Assume* $p(z)$ *has a factorization* $p(z) = p_1(z)p_2(z)$. *Let* $q(z) = \sum_{i=0}^{n} q_i z^i$ *be monic, i.e.,* $q_n = 1$. *Let* $C_q^\sharp$ *and* $C_q^\flat$ *be the companion matrices of* $q(z)$, *given by (5.9) and (5.10) respectively, and let* K *be the Hankel matrix*

$$K = \begin{pmatrix} q_1 & \cdots & q_{n-1} & 1 \\ \cdot & \cdots & 1 & 0 \\ \cdot & \cdots & & \cdot \\ \cdot & \cdots & & \cdot \\ q_{n-1} & 1 & & \\ 1 & 0 \cdots & & 0 \end{pmatrix}. \tag{8.32}$$

Then

1.

$$B(q, p_1 p_2) = B(q, p_1)p_2(C_q^\flat) = p_1(C_q^\sharp)B(q, p_2). \tag{8.33}$$

2.

$$B(q, p) = Kp(C_q^\flat) = p(C_q^\sharp)K. \tag{8.34}$$

3.

$$B(q,p)C_q^\flat = C_q^\sharp B(p,q). \tag{8.35}$$

Proof. Note that for the standard and control bases of X_q, we have $C_q^\sharp = [S_q]_{st}^{st}$ and $C_q^\flat = \tilde{C}_q^\sharp = [S_q]_{co}^{co}$. Note also that the matrix K in equation (8.32) satisfies $K = B(q, 1)$ and is a Bezoutian as well as a Hankel matrix.

1. This follows from the equality $p(S_q) = p_1(S_q)p_2(S_q)$ and the fact that

$$B(q, p_1 p_2) = [(p_1 p_2)(S_q)]_{co}^{st} = [p_1(S_q)p_2(S_q)]_{co}^{st}$$

$$= [p_1(S_q)]_{co}^{st}[p_2(S_q)]_{co}^{co} = [p_1(S_q)]_{st}^{st}[p_2(S_q)]_{co}^{st}.$$

2. This follows from Part (1) for the trivial factorization $p(z) = p(z) \cdot 1$.
3. From the commutativity of S_q and $p(S_q)$, it follows that

$$[p(S_q)]_{co}^{st}[S_q]_{co}^{co} = [S_q]_{st}^{st}[p(S_q)]_{co}^{st}.$$

We note that $[S_q]_{co}^{co} = \widetilde{([S_q]_{st}^{st})}$. ∎

Representation (8.34) for the Bezoutian is sometimes referred to as the **Barnett factorization**.

Already in Section 3.4, we derived a determinantal test, using the resultant, for the coprimeness of two polynomials. Now, with the introduction of the Bezoutian, we have another such test. We can classify the resultant approach as geometric, for basically it gives a condition for a polynomial model to be a direct sum of two shift-invariant subspaces. The Bezoutian approach, on the other hand, is arithmetic, inasmuch as it is based on the invertibility of intertwining maps. We shall see in the sequel yet another such test, given in terms of Hankel matrices, and furthermore, we shall clarify the connection between the various tests.

Theorem 8.30. *Given two polynomials $p(z), q(z) \in \mathbb{F}[z]$, then*

1. $\dim(\operatorname{Ker} B(q, p))$ is equal to the degree of the g.c.d. of $q(z)$ and $p(z)$.
2. $B(q, p)$ is invertible if and only if $q(z)$ and $p(z)$ are coprime.

Proof. Part (1) follows from Theorem 5.18. Part (2) follows from Theorem 5.18 and Theorem 8.27. ∎

8.3.3 Representation of Bezoutians

Relation (8.23) leads easily to some interesting representation formulas for the Bezoutian. For a polynomial $a(z)$ of degree n, we define the **reverse polynomial** $a^\sharp(z)$ by

$$a^\sharp(z) = a(z^{-1})z^n. \tag{8.36}$$

Theorem 8.31. *Let $p(z) = \sum_{i=0}^{n} p_i z^i$ and $q(z) = \sum_{i=0}^{n} q_i z^i$ be polynomials of degree n. Then the Bezoutian has the following representations:*

$$
\begin{aligned}
B(q,p) &= [p(\tilde{S})q^{\#}(S) - q(\tilde{S})p^{\#}(S)]J \\
&= J[p(S)q^{\#}(\tilde{S}) - q(S)p^{\#}(\tilde{S})] \\
&= -J[p^{\#}(\tilde{S})q(S) - q^{\#}(\tilde{S})p(S)] \\
&= -[p^{\#}(S)q(\tilde{S}) - q^{\#}(S)p(\tilde{S})]J.
\end{aligned}
\tag{8.37}
$$

Proof. We define the $n \times n$ shift matrix S by

$$
S = \begin{pmatrix}
0 & 1 & \ldots & 0 \\
0 & . & 1 & . & . & 0 \\
0 & . & & . & . & 0 \\
0 & & . & . & . & 1 & 0 \\
0 & . & & . & . & 1 \\
0 & . & & . & . & 0
\end{pmatrix}
\tag{8.38}
$$

and the $n \times n$ transposition matrix J by

$$
J = \begin{pmatrix}
0 & \ldots & 1 \\
0 & . & . & 1 & 0 \\
0 & \ldots & 0 \\
0 & 1 & . & . & 0 \\
1 & \ldots & 0
\end{pmatrix}.
\tag{8.39}
$$

Note also that for an arbitrary polynomial $a(z)$, we have $Ja(S)J = a(\tilde{S})$. From equation (8.23), we easily obtain

$$
q(z)p^{\#}(w) - p(z)q^{\#}(w) = [q(z)p(w^{-1}) - p(z)q(w^{-1})]w^n
$$

$$
= \sum_{i=1}^{n} \sum_{j=1}^{n} b_{ij} z^{i-1}(z - w^{-1})w^{-j+1}w^n
$$

$$
= \sum_{i=1}^{n} \sum_{j=1}^{n} b_{ij} z^{i-1}(zw - 1)w^{n-j}.
$$

In this identity, we substitute now $z = \tilde{S}$ and $w = S$. Thus, we get for the central term

$$
(\tilde{S}S - I) = - \begin{pmatrix}
1 & 0 & \ldots & 0 \\
0 & 0 & \ldots & 0 \\
0 & 0 & \ldots & 0 \\
0 & 0 & \ldots & 0 \\
0 & 0 & \ldots & 0
\end{pmatrix},
$$

and $\tilde{S}^{i-1}(\tilde{S}S - I)S^{n-j}$ is the matrix whose only nonzero term is -1 in the $i, n - j$ position. This implies the identity $B(q, p) = \{p(\tilde{S})q^{\sharp}(S) - q(\tilde{S})p^{\sharp}(S)\}J$. The other identities are similarly derived. ∎

From the representations (8.37) of the Bezoutian we obtain, by expanding the matrices, the **Gohberg–Semencul formulas**, of which we present one only:

$$B(q, p) = (p(\tilde{S})q^{\sharp}(S) - q(\tilde{S})p^{\sharp}(S))J$$

$$
= \left[
\begin{pmatrix}
p_0 & & & \\
\cdot & p_0 & & \\
\cdot & \cdot & p_0 & \\
\cdot & \cdot & \cdot & p_0 \\
p_{n-1} & \cdot & \cdot & p_0
\end{pmatrix}
\begin{pmatrix}
q_n & q_{n-1} & \cdots & q_1 \\
& \cdots & & \\
& & \cdots & \\
& & & \\
& & & q_n
\end{pmatrix}
-
\begin{pmatrix}
q_0 & & & \\
\cdot & & & \\
\cdot & \cdot & & \\
\cdot & \cdot & \cdot & \\
q_{n-1} & \cdots & q_0
\end{pmatrix}
\begin{pmatrix}
p_n & p_{n-1} & \cdots & p_1 \\
& \cdots & & \\
& & \cdots & \\
& & & \\
& & & p_n
\end{pmatrix}
\right] J
$$

$$
= \left[
\begin{pmatrix}
p_0 & & & \\
\cdot & p_0 & & \\
\cdot & \cdot & p_0 & \\
\cdot & \cdot & \cdot & p_0 \\
p_{n-1} & \cdot & \cdot & p_0
\end{pmatrix}
\begin{pmatrix}
q_1 & \cdot & \cdots & q_{n-1} & q_n \\
& \cdot & & \cdot & q_n \\
& & \cdot & & \\
q_{n-1} & q_n & & & \\
q_n & & & &
\end{pmatrix}
-
\begin{pmatrix}
q_0 & & & \\
\cdot & & & \\
\cdot & \cdot & & \\
\cdot & \cdot & \cdot & \\
q_{n-1} & \cdots & q_0
\end{pmatrix}
\begin{pmatrix}
p_1 & \cdot & \cdots & p_{n-1} & p_n \\
& \cdot & & \cdot & p_n \\
& & \cdot & & \\
p_{n-1} & p_n & & & \\
p_n & & & &
\end{pmatrix}
\right].
$$

$$(8.40)$$

Given two polynomials $p(z)$ and $q(z)$ of degree n, we let their **Sylvester resultant**, $\mathrm{Res}\,(p, q)$, be defined by

$$
\mathrm{Res}\,(p, q) =
\begin{vmatrix}
p_0 & \cdots & p_{n-1} & p_n & & & \\
& \cdots & & & \cdot & \cdot & \\
& & \cdots & & & \cdot & \cdots \\
& & & \cdot & \cdot & & \\
& & & p_0 & p_1 & \cdots & p_n \\
& q_0 & \cdots & q_{n-1} & q_n & & \\
& \cdots & & & & \cdot & \\
& \cdots & & & & \cdot & \cdots \\
& & \cdot & \cdot & & & \\
& & q_0 & q_1 & \cdots & q_n &
\end{vmatrix}.
$$

$$(8.41)$$

It has been proved, in Theorem 3.18, that the resultant $\mathrm{Res}\,(p, q)$ is nonsingular if and only if $p(z)$ and $q(z)$ are coprime. Equation (8.41) can be rewritten as the 2×2 block matrix

$$\mathrm{Res}\,(p, q) = \begin{pmatrix} p(S) & p^{\sharp}(\tilde{S}) \\ q(S) & q^{\sharp}(\tilde{S}) \end{pmatrix},$$

where the reverse polynomials $p^{\sharp}(z)$ and $q^{\sharp}(z)$ are defined in equation (8.36).

Based on the preceeding, we can state the following.

Theorem 8.32. *Let* $p(z) = \sum_{i=0}^n p_i z^i$ *and* $q(z) = \sum_{i=0}^n q_i z^i$ *be polynomials of degree n. Then*

$$\widetilde{\operatorname{Res}(p,q)} \begin{pmatrix} 0 & J \\ -J & 0 \end{pmatrix} \operatorname{Res}(p,q) = \begin{pmatrix} 0 & B(p,q) \\ -B(p,q) & 0 \end{pmatrix}. \qquad (8.42)$$

Proof. By expanding the left side of (8.42) we have

$$\begin{pmatrix} p(\tilde{S}) & q(\tilde{S}) \\ p^\sharp(S) & q^\sharp(S) \end{pmatrix} \begin{pmatrix} 0 & J \\ -J & 0 \end{pmatrix} \begin{pmatrix} p(S) & p^\sharp(\tilde{S}) \\ q(S) & q^\sharp(\tilde{S}) \end{pmatrix}$$

$$= \begin{pmatrix} (p(\tilde{S})Jq(S) - q(\tilde{S})Jp(S)) & (p(\tilde{S})Jq^\sharp(\tilde{S}) - q(\tilde{S})Jp^\sharp(\tilde{S})) \\ (p^\sharp(S)Jq(S) - q^\sharp(S)Jp(S)) & (p^\sharp(S)Jq^\sharp(\tilde{S}) - q^\sharp(S)Jp^\sharp(\tilde{S})) \end{pmatrix}$$

Now

$$p(\tilde{S})Jq(S) - q(\tilde{S})Jp(S) = J[p(S)q(S) - q(S)p(S)] = 0.$$

Similarly,

$$p^\sharp(S)Jq^\sharp(\tilde{S}) - q^\sharp(S)Jp^\sharp(\tilde{S}) = [p^\sharp(S)q^\sharp(S) - q^\sharp(S)p^\sharp(S)]J = 0.$$

In the same way,

$$p(\tilde{S})Jq^\sharp(\tilde{S}) - q(\tilde{S})Jp^\sharp(\tilde{S}) = J[p(S)q^\sharp(\tilde{S}) - q(S)p^\sharp(\tilde{S})] = B(q,p).$$

Finally, we compute

$$p^\sharp(S)Jq(S) - q^\sharp(S)Jp(S) = J[p(\tilde{S})q^\sharp(S) - q^\sharp(\tilde{S})p(S)] = -B(q,p).$$

This proves the theorem. ∎

The nonsingularity of both the Bezoutian and resultant matrices is equivalent to the coprimeness of the two generating polynomials. As an immediate corollary to the previous theorem, we obtain the equality, up to a sign, of their determinants. Yet another determinantal test for coprimeness, given in terms of Hankel matrices, will be given in Corollary 8.41.

Corollary 8.33. *Let* $p(z)$ *and* $q(z)$ *be polynomials of degree n. Then*

$$|\det \operatorname{Res}(p,q)| = |\det B(p,q)|. \qquad (8.43)$$

In particular, the nonsingularity of either matrix implies that of the other. As we already know, these conditions are equivalent to the coprimeness of the polynomials $p(z)$ and $q(z)$.

Actually, we can be more precise about the relationship between the determinants of the resultant and Bezoutian matrices.

Corollary 8.34. *Let $p(z)$ and $q(z)$ be polynomials of degree n. Then*

1.

$$\det B(q,p) = (-1)^{\frac{n(n+1)}{2}} \det \mathrm{Res}\,(q,p). \qquad (8.44)$$

2.

$$\det p(S_q) = \det p(C_q^\sharp) = \det \mathrm{Res}\,(p,q). \qquad (8.45)$$

Proof. 1. From the representation $B(q,p) = \{p(\tilde{S})q^\sharp(S) - q(\tilde{S})p^\sharp(S)\}J$ of the Bezoutian we get, by applying Lemma 3.17, $\det B(q,p) = \det \mathrm{Res}\,(q,p)\det J$. Since $B(q,p) = -B(p,q)$, it follows that $\det B(q,p) = (-1)^n \det B(p,q)$. On the other hand, for J the $n \times n$ matrix defined by (8.39), we have $\det J = (-1)^{\frac{n(n-1)}{2}}$. Therefore (8.44) follows.

2. We can compute the determinant of a linear transformation in any basis. Thus

$$\det p(S_q) = \det p(C_q^\sharp) = \det p(C_q^\flat).$$

Also, by Theorem 8.27, $B(q,p) = [p(S_q)]_{co}^{st} = [I]_{co}^{st}[p(S_q)]_{co}^{co}$, and since $\det[I]_{co}^{st} = (-1)^{\frac{n(n-1)}{2}}$, we get

$$\begin{aligned}
\det p(S_q) &= (-1)^{\frac{n(n-1)}{2}} \det B(q,p) \\
&= (-1)^{\frac{n(n-1)}{2}} (-1)^n \det B(p,q) \\
&= (-1)^{\frac{n(n-1)}{2}} (-1)^n (-1)^{\frac{n(n+1)}{2}} \det \mathrm{Res}\,(p,q) \\
&= \det \mathrm{Res}\,(p,q).
\end{aligned}$$

■

8.3.4 Diagonalization of Bezoutians

Since the Bezoutian of a pair of polynomials is a symmetric matrix, it can be diagonalized by a congruence transformation. The diagonalization procedure described here is based on polynomial factorizations and related direct sum decompositions. We begin by analyzing a special case, in which the congruence transformation is given in terms of a Vandermonde matrix. This is related to Proposition 5.15. In the proof of the following proposition, we shall make use of four distinct bases of the polynomial model X_q. These are the standard basis $\mathscr{B}_{st}$, the control basis $\mathscr{B}_{co}$, the interpolation basis $\mathscr{B}_{in}$ consisting of the Lagrange interpolation polynomials, and the spectral basis $\mathscr{B}_{sp} = \{s_1(z),\ldots,s_n(z)\}$, where $s_i(z) = \Pi_{j\neq i}(z-\lambda_j)$.

Proposition 8.35. *Let $q(z)$ be a monic polynomial of degree n having distinct zeros $\lambda_1,\ldots,\lambda_n$, and let $p(z)$ be a polynomial of degree $\leq n$. Let $V(\lambda_1,\ldots,\lambda_n)$ be the Vandermonde matrix*

$$V(\lambda_1,\ldots,\lambda_n) = [I]_{st}^{in} = \begin{pmatrix} 1 & \lambda_1 & .. & \lambda_1^{n-1} \\ . & . & ... & . \\ . & . & ... & . \\ . & . & ... & . \\ 1 & \lambda_n & .. & \lambda_n^{n-1} \end{pmatrix}.$$

Then the Bezoutian $B(q,p)$ satisfies the following identity:

$$VB(q,p)\tilde{V} = R,$$

where R is the diagonal matrix $\mathrm{diag}\,(r_1,\ldots,r_n)$ and

$$r_i = p(\lambda_i)s_i(\lambda_i) = p(\lambda_i)q'(\lambda_i).$$

Proof. The trivial operator identity, $Ip(S_q)I = p(S_q)$, implies the matrix equality

$$[I]_{st}^{in}[p(S_q)]_{co}^{st}[I]_{sp}^{co} = [p(S_q)]_{sp}^{in}. \tag{8.46}$$

Since $S_q s_i = \lambda_i s_i$, it follows that $p(S_q)p_i = p(\lambda_i)p_i = p(\lambda_i)p_i(\lambda_i l_i)$. Now $s_i(\lambda_i) = \prod_{j\neq i}(\lambda_i - \lambda_j)$, but $q'(z) = \sum_{i=1}^n s_i(z)$, and hence $q'(\lambda_i) = \prod_{j\neq i}(\lambda_i - \lambda_j) = s_i(\lambda_i)$. Next, we use the duality relations $\mathscr{B}_{st}^* = \mathscr{B}_{co}$ and $\mathscr{B}_{sp}^* = \mathscr{B}_{in}$. Thus $R = [p(S_q)]_{sp}^{in}$, and the result follows. ∎

The previous proposition uses a factorization of $q(z)$ into distinct linear factors. However, such a factorization does not always exist. In case it does exist, then the polynomials $p_i(z) = \prod_{j\neq i}(z - \lambda_j)$, which, up to a constant factor, are equal to the Lagrange interpolation polynomials, form a basis for X_q. Moreover, we have $X_q = p_1 X_{z-\lambda_1} \oplus \cdots \oplus d_n X_{z-\lambda_n}$. The fact that the sum is direct is implied by the assumption that the zeros λ_i are distinct, or that the polynomials $z - \lambda_i$ are pairwise coprime. In this case the internal direct sum decomposition is isomorphic to the external direct sum $X_{z-\lambda_1} \oplus \cdots \oplus X_{z-\lambda_n}$. This can be siginifically extended and generalized.

To this end, let $q_1(z),\ldots,q_s(z)$ be monic polynomials in $\mathbb{F}[z]$. Let $X_{q_1} \oplus \cdots \oplus X_{q_s}$ be the external direct sum of the corresponding polynomial models. We would like to compare this space with $X_{q_1\cdots q_s}$. We define polynomials by $d_i(z) = \prod_{j\neq i}q_j(z)$ and the map $Z : X_{q_1} \oplus \cdots \oplus X_{q_s} \longrightarrow X_{q_1\cdots q_s}$ by

$$Z\begin{pmatrix} f_1 \\ \vdots \\ f_s \end{pmatrix} = \sum_{i=1}^s d_i f_i. \tag{8.47}$$

Then we have the following result, related to the Chinese remainder theorem.

Proposition 8.36. *Let* $Z : X_{q_1} \oplus \cdots \oplus X_{q_s} \longrightarrow X_{q_1 \cdots q_s}$ *be defined by (8.47). Let* $\mathscr{B}_{ST}$ *be the standard basis in* $X_{q_1 \cdots q_s}$ *and let* $\mathscr{B}_{st}$ *be the basis of* $X_{q_1} \oplus \cdots \oplus X_{q_s}$ *constructed from the union of the standard bases in the* X_{q_i}. *Then*

1. *The adjoint map to* Z, *that is,* $Z^* : X_{q_1 \cdots q_s} \longrightarrow X_{q_1} \oplus \cdots \oplus X_{q_s}$, *is given by*

$$Z^* f = \begin{pmatrix} \pi_{q_1} f \\ \vdots \\ \pi_{q_s} f \end{pmatrix}. \tag{8.48}$$

2. *The map* Z *defined by (8.47) is invertible if and only if the* $q_i(z)$ *are mutually coprime.*
3. *For the case* $s = 2$ *we have*

$$[Z]_{st}^{ST} = \operatorname{Res}(q_2, q_1), \tag{8.49}$$

where $\operatorname{Res}(q_2, q_1)$ *is the resultant matrix defined in (3.16) and* $\mathscr{B}_{ST}$ *is the standard basis of* $X_{q_1} \oplus X_{q_2}$ *constructed from the union of the standard bases of* X_{q_1} *and* X_{q_2}.
4. *Assume* $q_i(z) = z - \lambda_i$. *Then*

$$[Z^*]_{ST}^{st} = \begin{pmatrix} 1 & \lambda_1 & .. & \lambda_1^{s-1} \\ . & . & .. & . \\ . & . & .. & . \\ . & . & .. & . \\ 1 & \lambda_n & .. & \lambda_n^{s-1} \end{pmatrix}. \tag{8.50}$$

5. *We have*

$$[Z^*]_{ST}^{st} = [Z^*]_{in}^{st} [I]_{ST}^{in}$$

and $[Z^*]_{in}^{st} = I$.

Proof. 1. We compute

$$\begin{aligned} \langle Zf, g \rangle &= \langle \textstyle\sum_{i=1}^{s} d_i f_i, g \rangle = [q^{-1} \textstyle\sum_{i=1}^{s} d_i f_i, g] \\ &= \textstyle\sum_{i=1}^{s} [q^{-1} d_i f_i, g] \\ &= \textstyle\sum_{i=1}^{s} [q_i^{-1} f_i, g] = \textstyle\sum_{i=1}^{s} \langle f_i, \pi_{q_i} g \rangle \\ &= \langle f, Z^* g \rangle. \end{aligned}$$

2. Note that Z^* is exactly the map given by the Chinese remainder theorem, namely, it maps an element of X_q into the vector of its remainders modulo the polynomials $q_i(z)$. It follows that Z^* is an isomorphism if and only if the $q_i(z)$ are mutually coprime. Finally, the invertibility of Z^* is equivalent to the invertibility of Z.

3. In this case, $d_1(z) = q_2(z)$ and $d_2(z) = q_1(z)$.

4. This follows from the fact that for an arbitrary polynomial $p(z)$, we have
$\pi_{z-\lambda_i} p = p(\lambda_i)$.
5. That $[I]_{ST}^{in}$ is the Vandermonde matrix has been proved in Corollary 2.36. That
$[Z^*]_{in}^{st} = I$ follows from the fact that for the Lagrange interpolation polynomials,
we have $\pi_i(\lambda_j) = \delta_{ij}$.   ∎

Next, we proceed to study a factorization of the map Z defined in (8.47). This
leads to, and generalizes, from a completely different perspective, the classical formula, given by (3.17), for the computation of the determinant of the Vandermonde
matrix.

Proposition 8.37. *Let $q_1(z), \ldots, q_s(z)$ be polynomials in $\mathbb{F}[z]$. Then,*

1. *The map $Z : X_{q_1} \oplus \cdots \oplus X_{q_s} \longrightarrow X_{q_1 \cdots q_s}$, defined in (8.47), has the factorization*

$$Z = Z_{s-1} \cdots Z_1, \tag{8.51}$$

where $Z_i : X_{q_1 \cdots q_i} \oplus X_{q_{i+1}} \oplus \cdots \oplus X_{q_s} \longrightarrow X_{q_1 \cdots q_{i+1}} \oplus X_{q_{i+2}} \oplus \cdots \oplus X_{q_s}$ is defined by

$$Z \begin{pmatrix} g \\ f_{i+1} \\ \vdots \\ f_s \end{pmatrix} = \begin{pmatrix} (q_1 \cdots q_i) f_{i+1} + q_{i+1} g \\ f_{i+2} \\ \vdots \\ f_s \end{pmatrix}.$$

2. *Choosing the standard basis in all spaces, we get*

$$[Z_i]_{st}^{st} = \begin{pmatrix} \mathrm{Res}\,(q_{i+1}, q_i \cdots q_1) & 0 \\ 0 & I \end{pmatrix}.$$

3. *For the determinant of $[Z_i]_{st}^{st}$ we have*

$$\det[Z_i]_{st}^{st} = \prod_{j=1}^{i} \det \mathrm{Res}\,(q_{i+1}, q_j).$$

4. *For the determinant of $[Z]_{st}^{st}$ we have*

$$\det[Z]_{st}^{st} = \prod_{1 \le i < j \le s} \det \mathrm{Res}\,(q_j, q_i). \tag{8.52}$$

5. *For the special case $q_i(z) = z - \lambda_i$, we have*

$$\det V(\lambda_1, \ldots, \lambda_s) = \prod_{1 \le i < j \le s} (\lambda_j - \lambda_i). \tag{8.53}$$

Proof. 1. We prove this by induction. For $s = 2$ this is trivial. Assume we have proved it for all integers $\leq s$. Then we define the map $Z' : X_{q_1} \oplus \cdots \oplus X_{q_{s+1}} \longrightarrow X_{q_1 \cdots q_s} \oplus X_{q_{s+1}}$ by

$$Z' \begin{pmatrix} f_1 \\ \vdots \\ f_s \\ f_{s+1} \end{pmatrix} = \begin{pmatrix} \sum_{j=1}^{s} d'_j f_j \\ f_{s+1} \end{pmatrix}.$$

Here $d'_i(z) = \prod_{j \neq i, s+1} q_j(z)$. Now $d_{s+1}(z) = \prod_{j \neq s+1} q_j(z)$ and

$$q_{s+1}(z) d'_i(z) = q_{s+1}(z) \prod_{j \neq i, s+1} q_j(z) = \prod_{j \neq i} q_j(z) = d_i(z).$$

So

$$Z_s Z' \begin{pmatrix} f_1 \\ \vdots \\ f_s \\ f_{s+1} \end{pmatrix} = (q_1 \cdots q_s) f_{s+1} + q_{s+1} \sum_{j=1}^{s} d'_j f_j = \sum_{j=1}^{s+1} d_j f_j = Z \begin{pmatrix} f_1 \\ \vdots \\ f_s \\ f_{s+1} \end{pmatrix}.$$

2. We use Proposition 8.36.
3. From equation (8.49) it follows that

$$\det[Z_i]_{st}^{st} = \det \operatorname{Res}(q_{i+1}, q_i \cdots q_1).$$

Now we use (8.45), i.e., the equality $\det \operatorname{Res}(q, p) = \det p(S_q)$, to get

$$\begin{aligned}
\det \operatorname{Res}(q_{i+1}, q_i \cdots q_1) &= (-1)^{(\sum_{j=1}^{i} m_j) m_{i+1}} \det \operatorname{Res}(q_i \cdots q_1, q_{i+1}) \\
&= (-1)^{(\sum_{j=1}^{i} m_j) m_{i+1}} \det(q_i \cdots q_1)(S_{q_{i+1}}) \\
&= \prod_{j=1}^{i} (-1)^{m_j m_{i+1}} \det(q_j)(S_{q_{i+1}}) \\
&= \prod_{j=1}^{i} (-1)^{m_j m_{i+1}} \det \operatorname{Res}(q_j, q_{i+1}) \\
&= \prod_{j=1}^{i} \det \operatorname{Res}(q_{i+1}, q_j).
\end{aligned}$$

4. Clearly, by the factorization (8.51) and the product rule of determinants, we have

$$\begin{aligned}
\det[Z]_{st}^{st} &= \prod_{j=1}^{s-1} \det[Z_i]_{st}^{st} = \prod_{j=1}^{s-1} \prod_{j=1}^{i} \det \operatorname{Res}(q_{i+1}, q_j) \\
&= \prod_{1 \leq i < j \leq s} \det \operatorname{Res}(q_j, q_i).
\end{aligned}$$

5. By Proposition 8.36, we have in this case $V(\lambda_1, \ldots, \lambda_s) = [Z^*]_{ST}^{st}$. Now $\det Z^* = \det Z$ implies that

$$\det V(\lambda_1, \ldots, \lambda_s) = \det[Z^*]_{ST}^{st} = \det[Z^*]_{co}^{CO} = \det \widetilde{[Z]_{st}^{ST}} = \det[Z]_{st}^{ST}.$$

Also, in this case,

$$\det \mathrm{Res}\,(z - \lambda_i, z - \lambda_j) = \begin{vmatrix} -\lambda_i & -\lambda_j \\ 1 & 1 \end{vmatrix}.$$

Thus (8.53) follows from (8.50).

This agrees with the computation of the Vandermonde determinant given previously in Chapter 3.   ∎

We can proceed now with the block diagonalization of the Bezoutian. Here we assume an arbitrary factorization of $q(z)$.

Proposition 8.38. *Let* $q(z) = q_1(z) \cdots q_s(z)$ *and let* $Z : X_{q_1} \oplus \cdots \oplus X_{q_s} \longrightarrow X_{q_1 \cdots q_s}$ *be the map defined by (8.47). Then*

1. The following is a commutative diagram:

$$
\begin{array}{ccc}
X_{q_1} \oplus \cdots \oplus X_{q_s} & \xrightarrow{\quad Z \quad} & X_{q_1 \cdots q_s} \\[2mm]
\Big\downarrow {\scriptstyle (pd_1)(S_{q_1}) \oplus \cdots \oplus (pd_s)(S_{q_k})} & & \Big\downarrow {\scriptstyle p(S_q)} \\[2mm]
X_{q_1} \oplus \cdots \oplus X_{q_s} & \xleftarrow{\quad Z^* \quad} & X_{q_1 \cdots q_s}
\end{array}
$$

i.e.,

$$Z^* p(S_q) Z = (pd_1)(S_{q_1}) \oplus \cdots \oplus (pd_s)(S_{q_k}). \tag{8.54}$$

2. Let $\mathscr{B}_{CO}$ *be the control basis in* $X_{q_1 \cdots q_s}$ *and let* $\mathscr{B}_{co}$ *be the basis of* $X_{q_1} \oplus \cdots \oplus X_{q_s}$ *constructed from the ordered union of the control bases in the* X_{q_i}, $i = 1, \ldots, s$. *We have*

$$\tilde{V} B(q, p) V = \mathrm{diag}\,(B(q_1, pd_1), \ldots, B(q_s, pd_s)),$$

where $d_i(z) = \prod_{j \neq i} q_j(z)$ *and* $V = [Z]_{co}^{CO}$.

Proof. 1. We compute

$$
Z^* p(S_q) Z \begin{pmatrix} f_1 \\ \vdots \\ f_s \end{pmatrix} = Z^* p(S_q) \sum_{j=1}^{s} d_j f_j = Z^* \pi_q p \sum_{j=1}^{s} d_j f_j
$$

$$
= \begin{pmatrix} \pi_{q_1} \sum_{j=1}^{s} \pi_q p d_j f_j \\ \vdots \\ \pi_{q_s} \sum_{j=1}^{s} \pi_q p d_j f_j \end{pmatrix} = \begin{pmatrix} \pi_{q_1} \pi_q p d_1 f_1 \\ \vdots \\ \pi_{q_s} \pi_q p d_s f_s \end{pmatrix} = \begin{pmatrix} (p d_1)(S_{q_1}) f_1 \\ \vdots \\ (p d_s)(S_{q_s}) f_s \end{pmatrix}
$$

$$
= ((p d_1)(S_{q_1}) \oplus \cdots \oplus (p d_s)(S_{q_k})) \begin{pmatrix} f_1 \\ \vdots \\ f_s \end{pmatrix}.
$$

2. We start from (8.54) and take the following matrix representations:

$$
[Z^*]_{ST}^{st} [p(S_q)]_{CO}^{ST} [Z]_{co}^{CO} = [(p d_1)(S_{q_1}) \oplus \cdots \oplus (p d_s)(S_{q_k})]_{co}^{st}.
$$

The control basis in $X_{q_1} \oplus \cdots \oplus X_{q_s}$ refers to the ordered union of the control bases in the X_{q_i}. Of course, we use Theorem 4.38 to infer the equality $[Z^*]_{ST}^{st} = \widetilde{[Z]_{co}^{CO}}$. ∎

8.3.5 Bezout and Hankel Matrices

Given a strictly proper rational function $g(z)$, we have considered two different representations of it. The first is the coprime factorization $g(z) = \frac{p(z)}{q(z)}$, where we assume $\deg q = n$, whereas the second is the expansion at infinity, i.e., the representation $g(z) = \sum_{i=1}^{\infty} \frac{g_i}{z^i}$. We saw that these two representations give rise to two different matrices, namely the Bezoutian matrix on the one hand and the Hankel matrix on the other. Our next goal is to study the relationship between these two objects. We already saw that the analysis of Bezoutians is facilitated by considering them as matrix representations of intertwining maps; see Theorem 5.18. The next theorem interprets the finite Hankel matrix in the same spirit.

Theorem 8.39. *Let $p(z), q(z)$ be polynomials, with $q(z)$ monic and $\deg p < \deg q = n$. Let $g(z) = \frac{p(z)}{q(z)} = \sum_{i=1}^{\infty} \frac{g_i}{z^i}$, and let $\overline{H} : X_q \longrightarrow X^q$ be defined by*

$$
\overline{H} f = H_g f, \qquad f \in X_q.
$$

Let $\rho_q : X^q \longrightarrow X_q$ be the map defined by (5.38), i.e., $\rho_q h = qh$. Let $\mathscr{B}_{st}, \mathscr{B}_{co}$ be the standard and control bases of X_q and let $\mathscr{B}_{rc} = \{\frac{e_1(z)}{q(z)}, \ldots, \frac{e_n(z)}{q(z)}\}$ be the **rational control basis** of X^q defined as the image of the control basis in X_q under the map ρ_q^{-1}. Then

1.

$$H_n = [\overline{H}]_{st}^{rc} = \begin{pmatrix} g_1 & \cdots & g_n \\ \cdot & \cdots & \cdot \\ \cdot & \cdots & \cdot \\ \cdot & \cdots & \cdot \\ g_n & \cdots & g_{2n-1} \end{pmatrix}. \tag{8.55}$$

2. H_n is invertible if and only if $p(z)$ and $q(z)$ are coprime.
3. If $p(z)$ and $q(z)$ are coprime, then $H_n^{-1} = B(q,a)$, where $a(z)$ arises out of any solution to the Bezout equation $a(z)p(z) + b(z)q(z) = 1$.

Proof. 1. We compute, using the representation of π_q given in (5.35),

$$\rho_q^{-1} p(S_q) f = q^{-1} \pi_q(pf) = q^{-1} q\pi_- q^{-1} pf = \pi_- gf = H_g f = \overline{H} f,$$

i.e., we have

$$\overline{H} = \rho_q^{-1} p(S_q). \tag{8.56}$$

This implies that the following diagram is commutative:

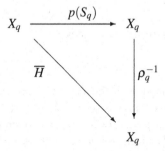

To compute the matrix representation of $\overline{H}$, let h_{ij} be the ij element of $[\overline{H}]_{st}^{rc}$. Then we have, by the definition of a matrix representation of a linear transformation, that

$$\overline{H} z^{j-1} = \pi_- p q^{-1} z^{j-1} = \sum_{i=1}^{n} h_{ij} e_i / q.$$

So

$$\sum_{i=1}^{n} h_{ij} e_i = q\pi_- q^{-1} p z^{j-1} = \pi_q p z^{j-1}.$$

Using the fact that B_{co} is the dual basis of B_{st} under the pairing (5.64), we have

$$h_{kj} = \sum_{i=1}^n h_{ij}\langle e_i, z^{k-1}\rangle = \langle \pi_q p z^{j-1}, z^{k-1}\rangle$$

$$= [q^{-1} q \pi_- q^{-1} p z^{j-1}, z^{k-1}] = [\pi_- q^{-1} p z^{j-1}, z^{k-1}]$$

$$= [g z^{j-1}, z^{k-1}] = [g, z^{j+k-2}] = g_{j+k-1}.$$

2. The invertibility of $\overline{H}$ and (8.55) implies the invertibility of H_n.
3. From the equality $\overline{H} = \rho_q^{-1} p(S_q)$ it is clear that $\overline{H}$ is invertible if and only if $p(S_q)$ is, and this, by Theorem 5.18, is the case if and only if $p(z)$ and $q(z)$ are coprime. ∎

The Bezout and Hankel matrices corresponding to a rational function $g(z)$ are both symmetric matrices arising from the same data. It is therefore natural to conjecture that not only are their ranks equal but also the signatures of the corresponding quadratic forms. The next theorem shows that indeed they are congruent, and moreover, it provides the relevant congruence transformations.

Theorem 8.40. *Let* $g(z) = p(z)/q(z)$ *with* $p(z)$ *and* $q(z)$ *coprime polynomials satisfying* $\deg p < \deg q = n$. *Assume* $g(z) = \sum_{i=1}^{\infty} g_i z^{-i}$ *and* $q(z) = z^n + q_{n-1} z^{n-1} + \cdots + q_0$. *Then the Hankel matrix* H_n *and the Bezoutian* $B(q, p) = (b_{ij})$ *are congruent. Specifically, we have*

1.

$$
\begin{pmatrix} g_1 \cdots g_n \\ \cdot \ \cdot \cdot \cdot \cdot \\ \cdot \ \cdot \cdot \cdot \cdot \\ \cdot \ \cdot \cdot \cdot \cdot \\ g_n \cdots g_{2n-1} \end{pmatrix}
$$

$$
= \begin{pmatrix} & & 1 \\ & \cdot & \psi_1 \\ & \cdot \cdot \cdot & \\ & \cdot \cdot \cdot & \\ 1 \ \psi_1 \ \cdot \cdot \ \psi_{n-1} \end{pmatrix} \begin{pmatrix} b_{11} \cdots b_{1n} \\ \cdot \ \cdot \cdot \cdot \cdot \\ \cdot \ \cdot \cdot \cdot \cdot \\ \cdot \ \cdot \cdot \cdot \cdot \\ b_{n1} \cdots b_{nn} \end{pmatrix} \begin{pmatrix} & & 1 \\ & \cdot & \psi_1 \\ & \cdot \cdot \cdot & \\ & \cdot \cdot \cdot & \\ 1 \ \psi_1 \ \cdot \cdot \ \psi_{n-1} \end{pmatrix}. \quad (8.57)
$$

Here, the ψ_i *are given by the polynomial characterization of Proposition 5.41.*

2. In the same way, we have

$$
\begin{pmatrix} b_{11} & \cdots & b_{1n} \\ \cdot & \cdots & \cdot \\ \cdot & \cdots & \cdot \\ \cdot & \cdots & \cdot \\ b_{n1} & \cdots & b_{nn} \end{pmatrix}
$$

$$
= \begin{pmatrix} q_1 & \cdot\cdot & q_{n-1} & 1 \\ \cdot & & \cdot\cdot & 1 \\ \cdot & & \cdot\cdot & \\ q_{n-1} & 1 & & \\ 1 & & & \end{pmatrix} \begin{pmatrix} g_1 & \cdots & g_n \\ \cdot & \cdots & \\ \cdot & \cdots & \\ \cdot & \cdots & \\ g_n & \cdots & g_{2n-1} \end{pmatrix} \begin{pmatrix} q_1 & \cdot\cdot & q_{n-1} & 1 \\ \cdot & & \cdot\cdot & 1 \\ \cdot & & \cdot\cdot & \\ q_{n-1} & 1 & & \\ 1 & & & \end{pmatrix}. \qquad (8.58)
$$

Note that equation (8.58) can be written more compactly as

$$
B(q,p) = B(q,1)H_{p/q}B(q,1). \qquad (8.59)
$$

3. We have

$$
\operatorname{rank}(H_n) = \operatorname{rank}(B(q,p))
$$

and

$$
\sigma(H_n) = \sigma(B(q,p)).
$$

Proof. 1. To this end, recall that (8.56) and the definition of the bases B_{st}, B_{co}, and B_{rc} enables us to compute a variety of matrix representations, depending on the choice of bases. In particular, it implies

$$
[\overline{H}]_{st}^{rc} = [\rho_q^{-1}]_{st}^{rc}[p(S_q)]_{co}^{st}[I]_{st}^{co} = [\rho_q^{-1}]_{co}^{rc}[I]_{st}^{co}[p(S_q)]_{co}^{st}[I]_{st}^{co}.
$$

It was proved in Theorem 8.39 that

$$
[\overline{H}]_{st}^{rc} = H_n = \begin{pmatrix} g_1 & \cdots & g_n \\ \cdot & \cdots & \\ \cdot & \cdots & \\ \cdot & \cdots & \\ g_n & \cdots & g_{2n-1} \end{pmatrix}.
$$

Obviously, $[\rho_q^{-1}]_{co}^{rc} = I$ and, using matrix representations for dual maps, $[I]_{co}^{st} = \widetilde{[I^*]}_{co}^{st} = \widetilde{[I]}_{co}^{st}$. Here, we used the fact that B_{st} and B_{co} are dual bases. So we see that $[I]_{co}^{st}$ is symmetric. In fact, it is a Hankel matrix, easily computed to be

$$[I]_{co}^{st} = \begin{pmatrix} q_1 & \cdots & q_{n-1} & 1 \\ \cdot & & \cdots & \\ \cdot & & \cdots & \\ \cdot & & \cdot \cdot & \\ q_{n-1} & \cdot & & \\ 1 & & & \end{pmatrix}.$$

Finally, from the identification of the Bezoutian as the matrix representation of an intertwining map for S_q, i.e., from $B(q,p) = [p(S_q)]_{co}^{st}$, and defining $R = [I]_{st}^{co} = \widetilde{[I]_{st}^{co}}$, we obtain the equality

$$H_n = \tilde{R}B(q,p)R. \tag{8.60}$$

2. Follows from part 1 by inverting the congruence matrix. See Proposition 5.41.
3. Follows from the congruence relations. ∎

The following corollary gives another determinantal test for the coprimeness of two polynomials, this time in terms of a Hankel matrix.

Corollary 8.41. *Let* $g(z) = p(z)/q(z)$ *with* $p(z)$ *and* $q(z)$ *polynomials satisfying* $\deg p < \deg q = n$. *If* $g(z) = \sum_{i=1}^{\infty} g_i z^{-i}$ *and*

$$H_n = \begin{pmatrix} g_1 & \cdots & g_n \\ \cdot & \cdots & \\ \cdot & \cdots & \\ \cdot & \cdots & \\ g_n & \cdots & g_{2n-1} \end{pmatrix},$$

then

$$\det H_n = \det B(q,p). \tag{8.61}$$

As a consequence, $p(z)$ *and* $q(z)$ *are coprime if and only if* $\det H_n \neq 0$.

Proof. Clearly, $\det R = (-1)^{n(n-1)/2}$, and so $(\det R)^2 = 1$, and hence (8.61) follows from (8.60). ∎

The matrix identities appearing in Theorem 8.40 are interesting also for other reasons. They include in them some other identities for the principal minors of H_n, which we state in the following way.

Corollary 8.42. *Under the previous assumption we have, for $k \leq n$,*

$$
\begin{pmatrix} g_1 & \cdots & g_k \\ \cdot & \cdots & \cdot \\ \cdot & \cdots & \cdot \\ \cdot & \cdots & \cdot \\ g_k & \cdots & g_{2k-1} \end{pmatrix}
$$

$$
= \begin{pmatrix} & & 1 \\ & & \cdot & \psi_1 \\ & & \cdots \\ & & \cdot & \cdots \\ 1 & \psi_1 & \cdots & \psi_{k-1} \end{pmatrix} \begin{pmatrix} b_{(n-k+1)(n-k+1)} & \cdots & b_{(n-k+1)n} \\ \cdot & \cdots & \cdot \\ \cdot & \cdots & \cdot \\ \cdot & \cdots & \cdot \\ b_{n(n-k+1)} & \cdots & b_{nn} \end{pmatrix} \begin{pmatrix} & & 1 \\ & & \cdot & \psi_1 \\ & & \cdots \\ & & \cdot & \cdots \\ 1 & \psi_1 & \cdots & \psi_{k-1} \end{pmatrix}.
$$

$$(8.62)$$

∎

In particular, we obtain the following.

Corollary 8.43. *The rank of the Bezoutian of the polynomials $q(z)$ and $p(z)$ is equal to the rank of the Hankel matrix $H_{p/q}$.*

Proof. Equation (8.62) implies the following equality:

$$\det(g_{i+j-1})_{i,j=1}^{k} = \det(b_{ij})_{i,j=n-k+1}^{n}. \tag{8.63}$$

∎

At this point, we recall the connection, given in equation (8.61), between the determinant of the Bezoutian and that of the resultant of the polynomials $p(z)$ and $q(z)$. In fact, not only can this be made more precise, we can find relations between the minors of the resultant and those of the corresponding Hankel and Bezout matrices. We state this as follows.

Theorem 8.44. *Let $q(z)$ be of degree n and $p(z)$ of degree $\leq n$. Then the lower $k \times k$ right-hand corner of the Bezoutian $B(q, p)$ is given by*

$$
\begin{pmatrix} b_{(n-k+1)(n-k+1)} & \cdots & b_{(n-k+1)n} \\ \cdot & \cdots & \cdot \\ \cdot & \cdots & \cdot \\ \cdot & \cdots & \cdot \\ b_{n(n-k+1)} & \cdots & b_{nn} \end{pmatrix}
$$

$$
= \begin{pmatrix} p_{n-k} & \cdot\ \cdot & p_{n-2k+1} \\ \cdot & \cdot\ \cdot\ \cdot & \cdot \\ \cdot & \cdot\ \cdot\ \cdot & \cdot \\ \cdot & \cdot\ \cdot\ \cdot & \cdot \\ p_{n-1} & \cdot\ \cdot\ \cdot & p_{n-k} \end{pmatrix} \begin{pmatrix} q_{n-k+1} & \cdot & & q_{n-1}\ q_n \\ \cdot & & \cdot\ \cdot & \cdot \\ & & \cdot\ \cdot & \\ & & q_{n-1} & q_n \\ & & & q_n \end{pmatrix}
$$

$$
- \begin{pmatrix} q_{n-k} & \cdot\ \cdot & q_{n-2k+1} \\ \cdot & \cdot\ \cdot\ \cdot & \\ \cdot & \cdot\ \cdot\ \cdot & \\ \cdot & \cdot\ \cdot\ \cdot & \\ q_{n-1} & \cdot\ \cdot\ \cdot & q_{n-k} \end{pmatrix} \begin{pmatrix} p_{n-k+1} & \cdot & & p_{n-1}\ p_n \\ \cdot & & \cdot\ \cdot & \cdot \\ & & \cdot\ \cdot & \\ & & p_{n-1} & p_n \\ & & p_n & \end{pmatrix}
$$

$$
= \left[\begin{pmatrix} p_{n-k} & \cdot\ \cdot\ \cdot & p_{n-2k+1} \\ \cdot & \cdot\ \cdot\ \cdot & \cdot \\ \cdot & \cdot\ \cdot\ \cdot & \cdot \\ \cdot & \cdot\ \cdot\ \cdot & \cdot \\ p_{n-1} & \cdot\ \cdot\ \cdot & p_{n-k} \end{pmatrix} \begin{pmatrix} q_n & q_{n-1} & \cdot\ \cdot & q_{n-k+1} \\ & \cdot & \cdot\ \cdot\ \cdot & \\ & & \cdot\ \cdot\ \cdot & \\ & & & \cdot\ \cdot & \\ & & & & q_n \end{pmatrix} \right.
$$

$$
\left. - \begin{pmatrix} q_{n-k} & \cdot\ \cdot\ \cdot & q_{n-2k+1} \\ \cdot & \cdot\ \cdot\ \cdot & \\ \cdot & \cdot\ \cdot\ \cdot & \\ \cdot & \cdot\ \cdot\ \cdot & \\ q_{n-1} & \cdot\ \cdot\ \cdot & q_{n-k} \end{pmatrix} \begin{pmatrix} p_n & p_{n-1} & \cdot\ \cdot & p_{n-k+1} \\ & \cdot & \cdot\ \cdot\ \cdot & \\ & & \cdot\ \cdot\ \cdot & \\ & & & \cdot\ \cdot & \\ & & & & p_n \end{pmatrix} \right] J_k,
$$

(8.64)

where we take $p_j = q_j = 0$ for $j < 0$. Moreover, we have

$$
\det(g_{i+j})_{i,j=1}^{k} = \det(b_{ij})_{i,j=n-k+1}^{n}
$$

$$
= \begin{vmatrix} p_{n-k} & q_{n-k} & p_{n-k-1} & q_{n-k-1} & & & q_{n-2k-1} \\ \cdot & \cdot & p_{n-k} & q_{n-k} & & & \\ \cdot & \cdot & \cdot & \cdot & & & \\ \cdot & \cdot & \cdot & \cdot & & \cdot\ \cdot\ p_{n-k} & q_{n-k} \\ p_{n-1} & q_{n-1} & \cdot & \cdot & & \cdot\ \cdot\ \cdot & \cdot \\ p_n & q_n & p_{n-1} & \cdot & & \cdot\ \cdot\ \cdot & \cdot \\ & & p_n & q_n & & \cdot\ \cdot\ \cdot & \cdot \\ & & & & & \cdot\ \cdot\ \cdot & \cdot \\ & & & & & p_{n-1} & q_{n-1} \\ & & & & & p_n & q_n \end{vmatrix}.
$$

(8.65)

Proof. We use equation (8.40) to infer (8.64). Using Lemma 3.17, we have

$$
\det(b_{ij})_{i,j=n-k+1}^{n} = \det J_k = \begin{vmatrix}
p_{n-k} \cdot & \cdots & p_{n-2k+1} & q_{n-k} & \cdots & q_{n-2k+1} \\
\cdot & \cdot & \cdots & \cdot & \cdots \\
\cdot & \cdot & \cdots & \cdot & \cdots \\
\cdot & \cdot & \cdots & \cdot & \cdots \\
p_{n-1} \cdot & \cdots & p_{n-k} & q_{n-1} & \cdots & q_{n-k} \\
p_n & p_{n-1} & \cdots p_{n-k+1} & q_n & \cdots & q_{n-k+1} \\
& & \cdots & & \cdots \\
& & \cdots & & \cdots \\
& & \cdots & & \cdots \\
& & \cdot\, p_{n-1} & & \cdot\, q_{n-1} \\
& & p_n & & q_n
\end{vmatrix} \cdot
$$

Since $\det J_k = (-1)^{\frac{k(k-1)}{2}}$, we can use this factor to rearrange the columns of the determinant on the right to get (8.65). An application of equation (8.63) yields equality (8.65). ∎

Equation (8.65) will be useful in the derivation of the Hurwitz stability test.

8.3.6   Inversion of Hankel Matrices

In this section we study the problem of inverting a finite nonsingular Hankel matrix

$$
H_n = \begin{pmatrix}
g_1 & \cdots & g_n \\
\cdot & \cdots & \\
\cdot & \cdots & \\
\cdot & \cdots & \\
g_n & \cdots & g_{2n-1}
\end{pmatrix}. \tag{8.66}
$$

We already saw, in Theorem 8.39, that if the g_i arise out of a rational function $g(z) = p(z)/q(z)$, with $p(z), q(z)$ coprime, via the expansion $g(z) = \sum_{i=1}^{\infty} g_i z^{-i}$, the corresponding Hankel matrix is invertible. Our aim here is to utilize the special structure of Hankel matrices in order to facilitate the computation of the inverse. What is more, we would like to provide a recursive algorithm, one that uses the information in the nth step for the computation in the next one. This is a natural entry point to the general area of fast algorithms for computing with specially structured matrices, of which the Hankel matrices are just an example.

The method we shall employ is to use the coprime factorization of $g(z)$ as the data of the problem. This leads quickly to concrete representations of the inverse. Going back to the data, given in the form of the coefficients $g_1, \ldots, g_{2n-1}$, we construct all possible rational functions with rank equal to n and whose first $2n - 1$ coefficients are prescribed by the g_i. Such a rational function is called a **minimal rational extension** of the sequence $g_1, \ldots, g_{2n-1}$. We use these rational functions to obtain inversion formulas. It turns out, as is to be expected, that the nature of a specific minimal extension is irrelevant as far as the inversion formulas are concerned. Finally, we will study the recursive structure of the solution, that is, how to use the solution to the inversion problem for a given n when the amount of data is increased.

Suppose now that rather than having a rational function $g(z)$ as our data, we are given only numbers $g_1, \ldots, g_{2n-1}$ for which the matrix H_n is invertible. We consider this to correspond to the rational function $\sum_{i=1}^{2n-1} g_i/z^i$. We look now for all other possible minimal rational extensions. It turns out that, preserving the rank condition, there is relatively little freedom in choosing the coefficients g_i, $i > 2n - 1$. In fact, we have the following.

Proposition 8.45. *Given $g_1, \ldots, g_{2n-1}$ for which the Hankel matrix H_n is invertible, any minimal rational extension is completely determined by the choice of $\xi = g_{2n}$.*

Proof. Since H_n is invertible, there exists a unique solution to the equation

$$
\begin{pmatrix} g_1 & \cdots & g_n \\ \cdot & \cdot \cdot \cdot \cdot \\ \cdot & \cdot \cdot \cdot \cdot \\ \cdot & \cdot \cdot \cdot \cdot \\ g_n & \cdots & g_{2n-1} \end{pmatrix} \begin{pmatrix} x_0 \\ \cdot \\ \cdot \\ \cdot \\ x_{n-1} \end{pmatrix} = \begin{pmatrix} g_{n+1} \\ \cdot \\ \cdot \\ \cdot \\ g_{2n} \end{pmatrix}.
$$

Since the rank of the infinite Hankel matrix does not increase, we have

$$
\begin{pmatrix} g_1 & \cdots & g_n \\ \cdot & \cdot \cdot \cdot \cdot \\ \cdot & \cdot \cdot \cdot \cdot \\ \cdot & \cdot \cdot \cdot \cdot \\ g_n & \cdots & g_{2n-1} \\ g_{n+1} & \cdots & g_{2n} \\ \cdot & \cdot \cdot \cdot \cdot \\ \cdot & \cdot \cdot \cdot \cdot \\ \cdot & \cdot \cdot \cdot \cdot \end{pmatrix} \begin{pmatrix} x_0 \\ \cdot \\ \cdot \\ \cdot \\ x_{n-1} \end{pmatrix} = \begin{pmatrix} g_{n+1} \\ \cdot \\ \cdot \\ \cdot \\ g_{2n} \\ g_{2n+1} \\ \cdot \\ \cdot \\ \cdot \end{pmatrix},
$$

which shows that the g_i, with $i > 2n$, are completely determined by the choice of g_{2n}, via the recurrence relation $g_k = \sum_{i=0}^{n-1} x_i g_{k-n+i}$. ∎

Incidentally, it is quite easy to obtain the monic-denominator polynomial of a minimal rational extension.

Theorem 8.46. *Given $g_1, \ldots, g_{2n-1}$ for which the Hankel matrix H_n is invertible. Let $g_\xi(z) = p_\xi(z)/q_\xi(z)$ be any minimal rational extension of the sequence $g_1, \ldots, g_{2n-1}$ determined by $\xi = g_{2n}$, $p_\xi(z)$ and $q_\xi(z)$ coprime, and $q_\xi(z)$ monic. Let $g(z) = p(z)/q(z)$ be the extension corresponding to $\xi = 0$. Also, let $a(z)$ and*

$b(z)$ be the unique polynomials of degree $< n$ and $< \deg p$ respectively that solve the Bezout equation

$$a(z)p(z) + b(z)q(z) = 1. \tag{8.67}$$

Then,

1. The polynomial $a(z)$ is independent of ξ, and its coefficients are given as a solution of the system of linear equations

$$
\begin{pmatrix}
g_1 & \cdots & g_n \\
\cdot & \cdots & \cdot \\
\cdot & \cdots & \cdot \\
\cdot & \cdots & \cdot \\
g_n & \cdots & g_{2n-1}
\end{pmatrix}
\begin{pmatrix}
a_0 \\
\cdot \\
\cdot \\
\cdot \\
a_{n-1}
\end{pmatrix}
= -
\begin{pmatrix}
0 \\
\cdot \\
\cdot \\
0 \\
1
\end{pmatrix}. \tag{8.68}
$$

2. The polynomial $b(z)$ is independent of ξ and has the representation

$$b(z) = -\sum_{i=1}^{n} g_i a_i(z),$$

where $a_i(z)$, $i = 1, \ldots, n-2$, are the polynomials defined by

$$a_i(z) = \pi_+ z^{-i} a.$$

3. The coefficients of the polynomial $q_\xi(z) = z^n + q_{n-1}(\xi)z^{n-1} + \cdots + q_0(\xi)$ are solutions of the system of linear equations

$$
\begin{pmatrix}
g_1 & \cdots & g_n \\
\cdot & \cdots & \cdot \\
\cdot & \cdots & \cdot \\
\cdot & \cdots & \cdot \\
g_n & \cdots & g_{2n-1}
\end{pmatrix}
\begin{pmatrix}
q_0(\xi) \\
\cdot \\
\cdot \\
\cdot \\
q_{n-1}(\xi)
\end{pmatrix}
= -
\begin{pmatrix}
g_{n+1} \\
\cdot \\
\cdot \\
g_{2n-1} \\
\xi
\end{pmatrix}. \tag{8.69}
$$

The polynomial q_ξ can also be given in determinantal form as

$$
q_\xi(z) = (\det H_n)^{-1}
\begin{vmatrix}
g_1 & \cdots & g_n & 1 \\
\cdot & \cdots & \cdot & z \\
\cdot & \cdots & \cdot & \cdot \\
\cdot & \cdots & \cdot & \cdot \\
g_n & \cdots & g_{2n-1} & z^{n-1} \\
g_{n+1} & \cdots & \xi & z^n
\end{vmatrix}. \tag{8.70}
$$

4. We have $q_\xi(z) = q(z) - \xi a(z)$.
5. We have $p_\xi(z) = p(z) + \xi b(z)$.

6. *With respect to the control basis polynomials corresponding to $q_\xi(z)$, i.e.,*

$$e_i(\xi,z) = \pi_+ z^{-i} q_\xi = q_i(\xi) + q_{i+1}(\xi)z + \cdots + z^{n-i},$$

$i = 1,\ldots,n$, *the polynomial $p_\xi(z)$ has the representation*

$$p_\xi(z) = \sum_{i=1}^{n} g_i e_i(\xi,z). \tag{8.71}$$

Proof. 1. Let $p_\xi(z), q_\xi(z)$ be the polynomials in the coprime representation $g_\xi(z) = p_\xi(z)/q_\xi(z)$ corresponding to the choice $g_{2n} = \xi$. Let $a(z)$ and $b(z)$ be the unique polynomials, assuming $\deg a < \deg q_\xi$, that solve the Bezout equation $a(z)p_\xi(z) + b(z)q_\xi(z) = 1$. We rewrite this equation as

$$a(z)\frac{p_\xi(z)}{q_\xi(z)} + b(z) = \frac{1}{q_\xi(z)}, \tag{8.72}$$

and note that the right-hand side has an expansion of the form

$$\frac{1}{q_\xi(z)} = \frac{1}{z^n} + \frac{\sigma_{n+1}}{z^{n+1}} + \cdots .$$

Equating coefficients of z^{-i}, $i = 1,\ldots,n$, in (8.72), we have

$$a_0 g_1 + \cdots + a_{n-1} g_n = 0,$$
$$a_0 g_2 + \cdots + a_{n-1} g_{n+1} = 0,$$
$$.$$
$$.$$
$$.$$
$$a_0 g_{n-1} + \cdots + a_{n-1} g_{2n-2} = 0,$$
$$a_0 g_n + \cdots + a_{n-1} g_{2n-1} = -1,$$

which is equivalent to (8.68). Since $g_1,\ldots,g_{2n-1}$ do not depend on ξ, this shows that the polynomial $a(z)$ is also independent of ξ.

2. Equating nonnegative indexed coefficients in (8.72), we obtain

$$a_{n-1} g_1 = -b_{n-2},$$
$$a_{n-1} g_2 + a_{n-2} g_1 = -b_{n-3},$$
$$.$$
$$.$$
$$.$$
$$a_{n-1} g_{n-1} + \cdots + a_1 g_1 = -b_0,$$

or in matrix form

$$
\begin{pmatrix} a_{n-1} & & \\ & \cdot & \cdot \\ & \cdot & \cdot \\ & & \cdot \\ a_1 & \cdots & a_{n-1} \end{pmatrix} \begin{pmatrix} g_0 \\ \cdot \\ \cdot \\ \cdot \\ g_{n-1} \end{pmatrix} = - \begin{pmatrix} b_{n-2} \\ \cdot \\ \cdot \\ \cdot \\ b_0 \end{pmatrix}. \tag{8.73}
$$

This means that the polynomial $b(z)$ is independent of ξ and has the representation $b(z) = -\sum_{i=1}^{n} g_i a_i(z)$.

3. From the representation $g_\xi(z) = p_\xi(z)/q_\xi(z)$, we get $p_\xi(z) = q_\xi(z)g_\xi(z)$. In the last equality we equate coefficients of $z^{-1}, \ldots, z^{-n}$ to get the system of equations (8.69). Since H_n is assumed to be invertible, this system has a unique solution, and the solution obviously depends linearly on the parameter ξ. This system can be solved by inverting H_n to get

$$
\begin{pmatrix} q_0(\xi) \\ \cdot \\ \cdot \\ \cdot \\ q_{n-1}(\xi) \end{pmatrix} = -H_n^{-1} \begin{pmatrix} g_{n+1} \\ \cdot \\ \cdot \\ g_{2n-1} \\ \xi \end{pmatrix} = -\frac{\mathrm{adj}\, H_n}{\det H_n} \begin{pmatrix} g_{n+1} \\ \cdot \\ \cdot \\ g_{2n-1} \\ \xi \end{pmatrix}.
$$

Multiplying on the left by the polynomial row vector $\begin{pmatrix} 1 & z & \cdots & z^{n-1} \end{pmatrix}$, and using equation (3.14), we get

$$
q_\xi(z) = z^n + q_{n-1}(\xi)z^{n-1} + \cdots + q_0(\xi)
$$

$$
= z^n - \begin{pmatrix} 1 & z & \cdots & z^{n-1} \end{pmatrix} \frac{\mathrm{adj}\, H_n}{\det H_n} \begin{pmatrix} g_{n+1} \\ \cdot \\ \cdot \\ g_{2n-1} \\ \xi \end{pmatrix}
$$

$$
= (\det H_n)^{-1} \left[\det H_n z^n - \begin{pmatrix} 1 & z & \cdots & z^{n-1} \end{pmatrix} \mathrm{adj}\, H_n \begin{pmatrix} g_{n+1} \\ \cdot \\ \cdot \\ g_{2n-1} \\ \xi \end{pmatrix} \right]
$$

$$
= (\det H_n)^{-1} \begin{vmatrix} g_1 & \cdots & g_n & g_{n+1} \\ \cdot & \cdots & \cdot & \cdot \\ \cdot & \cdots & \cdot & \cdot \\ \cdot & \cdots & \cdot & \cdot \\ g_n & \cdots & g_{2n-1} & \xi \\ 1 & z & \cdots z^{n-1} & z^n \end{vmatrix},
$$

which is equivalent to (8.70).

4. By writing the last row of the determinant in (8.70) as

$$\left(g_{n+1} \cdots \xi \ z^n \right) = \left(g_{n+1} \cdots 0 \ z^n \right) + \left(0 \ .. \ 0 \ \xi \ 0 \right),$$

and using the linearity of the determinant as a function of each of its columns, we get

$$q_\xi(z) = (\det H_n)^{-1} \begin{vmatrix} g_1 & \cdots & g_n & 1 \\ \cdot & \cdots & \cdot & z \\ \cdot & \cdots & \cdot & \cdot \\ \cdot & \cdots & \cdot & \cdot \\ g_n & \cdots & g_{2n-1} & z^{n-1} \\ g_{n+1} & \cdots & \xi & z^n \end{vmatrix}$$

$$= (\det H_n)^{-1} \begin{vmatrix} g_1 & \cdots & g_n & 1 \\ \cdot & \cdots & \cdot & z \\ \cdot & \cdots & \cdot & \cdot \\ \cdot & \cdots & \cdot & \cdot \\ g_n & \cdots & g_{2n-1} & z^{n-1} \\ g_{n+1} & \cdots & 0 & z^n \end{vmatrix}$$

$$+ (\det H_n)^{-1} \begin{vmatrix} g_1 & \cdots & g_n & 1 \\ \cdot & \cdots & \cdot & z \\ \cdot & \cdots & \cdot & \cdot \\ \cdot & \cdots & \cdot & \cdot \\ g_n & \cdots & g_{2n-1} & z^{n-1} \\ 0 & .. \ 0 & \xi & 0 \end{vmatrix}$$

$$= q(z) - \xi r(z),$$

where

$$r(z) = \begin{vmatrix} g_1 & \cdots & g_{n-1} & 1 \\ \cdot & \cdots & \cdot & z \\ \cdot & \cdots & \cdot & \cdot \\ \cdot & \cdots & \cdot & \cdot \\ g_n & \cdots & g_{2n-2} & z^{n-1} \end{vmatrix}.$$

We will show that $r(z) = a(z)$. From the Bezout equations $a(z)p(z) + b(z)q(z) = 1$ and $a(z)p_\xi(z) + b(z)q_\xi(z) = 1$ it follows by subtraction that $a(z)(p_\xi(z) - p(z)) + b(z)(q_\xi(z) - q(z)) = 0$. Since $q_\xi(z) = q(z) - \xi r(z)$, we must have $a(z)(p_\xi(z) - p(z)) = \xi b(z) r(z)$. The coprimeness of $a(z)$ and $b(z)$ implies that $a(z)$ is a divisor of $r(z)$. Setting $r(z) = a(z)d(z)$, it follows that $p_\xi(z) = p(z) + \xi b(z)d(z)$ and $q_\xi(z) = q(z) - \xi a(z)d(z)$. Now

$$g_\xi(z) - g(z) = \frac{p(z) + \xi b(z)d(z)}{q(z) - \xi a(z)d(z)} - \frac{p(z)}{q(z)} = \frac{\xi(a(z)p(z) + b(z)q(z))d(z)}{q(z)(q(z) - \xi a(z)d(z))}$$
$$= \frac{\xi d(z)}{q(z)(q(z) - \xi a(z)d(z))}.$$

Since $g_\xi(z)$ and $g(z)$ have expansions in powers of z^{-1} agreeing for the first $2n - 1$ terms, and the next term has to be ξz^{-2n}, this implies $d = 1$.

5. Follows from $p_\xi(z) = p(z) + \xi b(z)d(z)$, by substituting $d(z) = 1$.

6. Equating positively indexed coefficients in the equality $p_\xi(z) = q_\xi(z)g_\xi(z)$, we have

$$p_{n-1}(\xi) = g_1,$$
$$p_{n-2}(\xi) = g_2 + q_{n-1}(\xi)g_1,$$
$$\vdots$$
$$p_0(\xi) = g_n + q_{n-1}(\xi)g_{n-1} + \cdots + q_1(\xi)g_1,$$

or

$$\begin{pmatrix} p_0(\xi) \\ \cdot \\ \cdot \\ \cdot \\ p_{n-1}(\xi) \end{pmatrix} = \begin{pmatrix} q_1(\xi) & \cdots & q_{n-1}(\xi) & 1 \\ \cdot & & \cdots & \cdot \\ \cdot & & \cdots & \\ \cdot & & \cdot & \cdot \\ q_{n-1}(\xi) & 1 & & \\ 1 & & & \end{pmatrix} \begin{pmatrix} g_1 \\ \cdot \\ \cdot \\ \cdot \\ g_n \end{pmatrix}.$$

This is equivalent to

$$p_\xi(z) = g_1 e_1(\xi, z) + \cdots + g_n e_n(\xi, z),$$

which proves (8.71).   ■

Incidentally, parts (4) and (5) of the theorem provide a nice parametrization of all solutions to the minimal partial realization problem arising out of a nonsingular Hankel matrix.

Corollary 8.47. *Given the nonsingular Hankel matrix H_n, if $g(z) = p(z)/q(z)$ is the minimal rational extension corresponding to the choice $g_{2n} = 0$, then the minimal rational extension corresponding to $g_{2n} = \xi$ is given by*

$$g_\xi(z) = \frac{p(z) + \xi b(z)}{q(z) - \xi a(z)},$$

where $a(z), b(z)$ are the polynomials arising from the solution of the Bezout equation $a(z)p(z) + b(z)q(z) = 1$.

This theorem explains the mechanism of how the particular minimal rational extension chosen does not influence the computation of the inverse, as of course it should not. Indeed, we see that

$$B(q_\xi, a) = B(q - \xi a, a) = B(q, a) - \xi B(a, a) = B(q, a),$$

and so $H_n^{-1} = B(q_\xi, a) = B(q, a)$.

8.3.7 Continued Fractions and Orthogonal Polynomials

The Euclidean algorithm provides a tool not only for the computation of a g.c.d. of a pair of polynomials but also, if coprimeness is assumed, for the solution of the corresponding Bezout equation. The application of the Euclidean algorithm yields a representation of a rational function in the form of a continued fraction. It should be noted that a slight modification of the process makes the method applicable to an arbitrary (formal) power series. This is the theme of the present section.

Given an irreducible representation $p(z)/q(z)$ of a strictly proper rational function $g(z)$, we define, using the division rule for polynomials, a sequence of polynomials $q_i(z)$, nonzero constants β_i, and monic polynomials $a_i(z)$, by

$$\begin{cases} q_{-1}(z) = q(z), \quad q_0(z) = p(z), \\ q_{i+1}(z) = a_{i+1}(z)q_i(z) - \beta_i q_{i-1}(z), \end{cases} \tag{8.74}$$

with $\deg q_{i+1} < \deg q_i$. The procedure ends when $q_n = 0$, and then q_{n-1} is the g.c.d. of $p(z)$ and $q(z)$. Since $p(z)$ and $q(z)$ are assumed coprime, $q_{n-1}(z)$ is a nonzero constant. The pairs $\{\beta_{i-1}, a_i(z)\}$ are called the **atoms** of $g(z)$. We define $\alpha_i = \deg a_i$.

For $i = 0$, we rewrite equation (8.74) as

$$\frac{p}{q} = \frac{q_0}{q_{-1}} = \frac{\beta_0}{a_1 - \dfrac{q_1}{q_0}}.$$

Thus $g_1 = q_1/q_0$ is also a strictly proper rational function, and furthermore, this is an irreducible representation. By an inductive argument, this leads to the **continued fraction representation** of $g(z)$ in terms of the atoms β_{i-1} and $a_i(z)$:

$$g(z) = \cfrac{\beta_0}{a_1(z) - \cfrac{\beta_1}{a_2(z) - \cfrac{\beta_2}{a_3(z) - \cfrac{\cdots \beta_{n-2}}{a_{n-1}(z) - \cfrac{\beta_{n-1}}{a_n(z)}}}}}. \tag{8.75}$$

There are two ways to truncate the continued fraction. The first one is to consider the top part, i.e., let

$$g_i(z) = \cfrac{\beta_0}{a_1(z) - \cfrac{\beta_1}{a_2(z) - \cfrac{\beta_2}{a_3(z) - \cdots \cfrac{\beta_{i-2}}{a_{i-1}(z) - \cfrac{\beta_{i-1}}{a_i(z)}}}}}. \tag{8.76}$$

The other option is to consider the rational functions γ_i defined by the bottom part, namely

$$\gamma_i(z) = \cfrac{\beta_i}{a_{i+1}(z) - \cfrac{\beta_{i+1}}{a_{i+2}(z) - \cdots \cfrac{\beta_{n-2}}{a_{n-1}(z) - \cfrac{\beta_{n-1}}{a_n(z)}}}}. \tag{8.77}$$

Clearly, we have $g_n(z) = \gamma_0(z) = g(z) = \frac{p(z)}{q(z)}$. We also set $\gamma_n(z) = 0$.

Just as a rational function determines a unique continued fraction, it is also completely determined by it.

Proposition 8.48. *Let $g(z)$ be a rational function having the continued fraction representation (8.75). Then $\{\beta_i, a_{i+1}\}_{i=0}^{n-1}$ are the atoms of $g(z)$.*

Proof. Assume $g(z) = p(z)/q(z)$. Then

$$g(z) = \frac{p(z)}{q(z)} = \frac{\beta_0}{a_1 - \gamma_1},$$

with γ_i strictly proper. This implies $p(z)\gamma_1 = a_1(z)p(z) - \beta_0 q(z)$. This shows that $r(z) = p(z)\gamma_1(z)$ is a polynomial, and since $\gamma_1(z)$ is strictly proper, we have $\deg r < \deg p$. It follows that $\{\beta_0, a_1(z)\}$ is the first atom of $g(z)$. The rest follows by induction. ∎

Fractional linear transformations are crucial to the analysis of continued fractions and play a central role in a wide variety of parametrization problems.

Definition 8.49. A **fractional linear transformation**, or a **Möbius transformation**, is a function of the form

$$f(w) = \frac{aw + b}{cw + d},$$

under the extra condition $ad - bc \neq 0$. With this fractional linear transformation we associate the matrix $\begin{pmatrix} a & b \\ c & d \end{pmatrix}$.

The following lemma relates composition of fractional linear transformations to matrix multiplication.

Lemma 8.50. *Let $f_i(w) = \frac{a_i w + b_i}{c_i w + d_i}$, $i = 1, 2$, be two fractional linear transformations. Then their composition is given by $(f_1 \circ f_2)(w) = \frac{aw + b}{cw + d}$, where*

$$\begin{pmatrix} a & b \\ c & d \end{pmatrix} = \begin{pmatrix} a_1 & b_1 \\ c_1 & d_1 \end{pmatrix} \begin{pmatrix} a_2 & b_2 \\ c_2 & d_2 \end{pmatrix}.$$

Proof. By computation. ∎

We define now two sequences of polynomials $\{Q_k(z)\}$ and $\{P_k(z)\}$ by the three term recurrence formulas

$$\begin{cases} Q_{-1}(z) = 0, \quad Q_0(z) = 1, \\ Q_{k+1}(z) = a_{k+1}(z) Q_k(z) - \beta_k Q_{k-1}(z), \end{cases} \tag{8.78}$$

and

$$\begin{cases} P_{-1}(z) = -1, \quad P_0(z) = 0, \\ P_{k+1}(z) = a_{k+1}(z) P_k(z) - \beta_k P_{k-1}(z). \end{cases} \tag{8.79}$$

The polynomials $Q_i(z)$ and $P_i(z)$ so defined are called the **Lanczos polynomials** of the first and second kind respectively.

Theorem 8.51. *Let $g(z)$ be a strictly proper rational function having the continued fraction representation (8.75). Let the Lanczos polynomials be defined by (8.78) and (8.79), and define the polynomials $A_k(z), B_k(z)$ by*

$$A_k(z) = \frac{Q_{k-1}(z)}{\beta_0 \cdots \beta_{k-1}},$$

$$\tag{8.80}$$

$$B_k(z) = \frac{-P_{k-1}(z)}{\beta_0 \cdots \beta_{k-1}}.$$

Then

1. For the rational functions $\gamma_i(z)$, $i = 1, \ldots, n$, we have

$$a_{i+1}(z) \gamma_i(z) - \beta_i = \gamma_i(z) \gamma_{i+1}(z). \tag{8.81}$$

2. Defining the rational functions $E_i(z)$ by

$$E_i(z) = Q_i(z) P_n(z) / Q_n(z) - P_i(z) = (Q_i(z) P_n(z) - P_i(z) Q_n(z)) / Q_n(z), \tag{8.82}$$

they satisfy the following recurrence:

$$\begin{cases} E_{-1} = 1, \quad E_0 = g_0 = g, \\ E_{k+1}(z) = a_{k+1}(z)E_k(z) - \beta_k E_{k-1}(z). \end{cases} \tag{8.83}$$

3. *We have*

$$E_i(z) = \gamma_0(z) \cdots \gamma_i(z). \tag{8.84}$$

4. *In the expansion of $E_i(z)$ in powers of z^{-1}, the leading terms is $\frac{\beta_0 \cdots \beta_i}{z^{\alpha_1 + \cdots + \alpha_{i+1}}}$.*
5. *With $\alpha_i = \deg a_i$, we have $\deg Q_k = \sum_{i=1}^k \alpha_i$ and $\deg P_k = \sum_{i=2}^k \alpha_i$.*
6. *The polynomials $A_k(z), B_k(z)$ are coprime and solve the Bezout equation*

$$P_k(z)A_k(z) + Q_k(z)B_k(z) = 1. \tag{8.85}$$

7. *The rational function $P_k(z)/Q_k(z)$ is strictly proper, and the expansions of $p(z)/q(z)$ and $P_k(z)/Q_k(z)$ in powers of z^{-1} agree up to order $2\sum_{i=1}^k \deg a_i + \deg a_{k+1} = 2\sum_{i=1}^k \alpha_i + \alpha_{i+1}$.*
8. *For the unimodular matrices $\begin{pmatrix} 0 & \beta_i \\ -1 & a_{i+1} \end{pmatrix}$, $i = 0, \ldots, n-1$, we have*

$$\begin{pmatrix} 0 & \beta_0 \\ -1 & a_1 \end{pmatrix} \cdots \begin{pmatrix} 0 & \beta_{i-1} \\ -1 & a_i \end{pmatrix} = \begin{pmatrix} -P_{i-1} & P_i \\ -Q_{i-1} & Q_i \end{pmatrix}.$$

9. *Let T_i be the fractional linear transformation corresponding to the matrix $\begin{pmatrix} 0 & \beta_i \\ -1 & a_{i+1} \end{pmatrix}$. Then*

$$g(z) = (T_0 \circ \cdots \circ T_{i-1})(\gamma_i) = \frac{P_i - P_{i-1}\gamma_i}{Q_i - Q_{i-1}\gamma_i}. \tag{8.86}$$

In particular,

$$g(z) = (T_0 \circ \cdots \circ T_{n-1})(\gamma_n) = P_n(z)/Q_n(z),$$

i.e., we have

$$P_n(z) = p(z), \qquad Q_n(z) = q(z). \tag{8.87}$$

10. *The atoms of $P_k(z)/Q_k(z)$ are $\{\beta_{i-1}, a_i(z)\}_{i=1}^k$.*
11. *The generating function of the Bezoutian of $Q_k(z)$ and $A_k(z)$ satisfies*

$$\frac{Q_k(z)A_k(w) - A_k(z)Q_k(w)}{z - w}$$

$$= \sum_{j=1}^k (\beta_0 \cdots \beta_{j-1})^{-1} Q_j(z) \frac{a_{j+1}(z) - a_{j+1}(w)}{z - w} Q_j(w). \tag{8.88}$$

The formula for the expansion of the generating function is called the gen-
eralized Christoffel–Darboux formula . *The regular Christoffel–Darboux
formula is the generic case, in which all the atoms a_j have degree one.*

12. *The Bezoutian matrix $B(Q_k, A_k)$ can be written as*

$$B(Q_k, A_k) = \sum_{j=1}^{k-1} (\beta_0 \cdots \beta_{j-1})^{-1} \tilde{R}_j, B(a_{j+1}, 1) R_j,$$

where R_j is the $n_j \times n$ Toeplitz matrix

$$R_j = \begin{pmatrix} \sigma_0^{(j)} & \sigma_1^{(j)} & \cdots & \sigma_{n_j}^{(j)} & & & \\ & \cdot & \cdot\cdot\cdot & \cdot & \cdot & & \\ & & \cdot\cdot\cdot & \cdot & & \cdot\cdot & \\ & & \cdot\cdot & \cdot & & \cdot\cdot\cdot & \\ & & \cdot & \cdot & \cdot & \cdot\cdot\cdot & \\ & & & \sigma_0^{(j)} & \sigma_1^{(j)} & \cdots & \sigma_{n_j}^{(j)} \end{pmatrix} \tag{8.89}$$

and $Q_j(z) = \sum \sigma_\nu^{(j)} z^\nu$.

13. *Assume the Hankel matrix H_n of (8.66) to be nonsingular. Let $g = p/q$ be any
minimal rational extension of the sequence $g_1, \ldots, g_{2n-1}$. Then*

$$H_n^{-1} = \sum_{j=1}^{n} \frac{1}{\beta_0 \cdots \beta_{j-1}} \tilde{R}_j B(1, a_{j+1}) R_j.$$

Proof.   1. Clearly, by the definition of the $\gamma_i(z)$, we have, for $i = 1, \ldots, n-1$,

$$\gamma_i(z) = \frac{\beta_i}{a_{i+1}(z) - \gamma_{i+1}(z)}, \tag{8.90}$$

which can be rewritten as (8.81). Since $\gamma_{n-1}(z) = \beta_{n-1}/a_n(z)$ and $\gamma_n(z) = 0$,
equation (8.81) holds also for $i = n$.

2. We compute first the initial conditions, using the initial conditions of the
Lanczos polynomials. Thus

$$E_{-1}(z) = Q_{-1}(z)P_n(z)/Q_n(z) - P_{-1}(z) = 1,$$
$$E_0(z) = Q_0(z)P_n(z)/Q_n(z) - P_0(z) = P_n(z)/Q_n(z) = \gamma_0(z) = g(z).$$

Next, we compute

$$E_{i+1}(z) = Q_{i+1}(z)P_n(z)/Q_n(z) - P_{i+1}(z)$$
$$= (Q_{i+1}(z)P_n(z) - P_{i+1}(z)Q_n(z))/Q_n(z)$$
$$= [(a_{i+1}(z)Q_i(z) - \beta_i Q_{i-1}(z))P_n(z) - (a_{i+1}(z)P_i(z) - \beta_i P_{i-1}(z))Q_n(z)]/Q_n(z)$$
$$= a_{i+1}(z)(Q_i(z)P_n(z) - P_i(z)Q_n(z))/Q_n(z)$$
$$\quad - \beta_i (Q_{i-1}(z)P_n(z) - P_{i-1}(z)Q_n(z))/Q_n(z)$$
$$= a_{i+1}(z)E_i(z) - \beta_i E_{i-1}(z).$$

3. We use the recurrence formula (8.83) and equation (8.81). The proof goes by induction. For $i = 1$ we have

$$E_1 = a_1 E_0 - \beta_0 = a_1 \gamma_0 - \beta_0 = \gamma_0 \gamma_1.$$

Assuming that $E_i(z) = \gamma_0(z) \cdots \gamma_i(z)$, we compute

$$E_{i+1}(z) = a_{i+1}E_i(z) - \beta_i E_{i-1}(z)$$
$$= a_{i+1}(z)\gamma_0(z) \cdots \gamma_i(z) - \beta_i \gamma_0(z) \cdots \gamma_{i-1}(z)$$
$$= \gamma_0(z) \cdots \gamma_{i-1}(z)(a_{i+1}(z)\gamma_i(z) - \beta_i)$$
$$= \gamma_0(z) \cdots \gamma_{i-1}(z)(\gamma_i(z)\gamma_{i+1}(z)).$$

4. From (8.90) we have $\beta_i/z^{\alpha_{i+1}}$ for the leading term of the expansion of γ_i. Since $E_i(z) = \gamma_0(z) \cdots \gamma_i(z)$, we have $\beta_0 \cdots \beta_i/z^{\alpha_1 + \cdots + \alpha_{i+1}} = \beta_0 \cdots \beta_i/z^{v_{i+1}}$ for the leading term of the expansion of $E_i(z)$.
5. By induction, using the recurrence formulas as well as the initial conditions.
6. We compute the expression $Q_{k+1}(z)P_k(z) - Q_k(z)P_{k+1}(z)$ using the defining recurrence formulas. So

$$Q_{k+1}(z)P_k(z) - Q_k(z)P_{k+1}(z) = (a_{k+1}(z)Q_k(z) - \beta_k Q_{k-1}(z))P_k(z)$$
$$\quad - (a_{k+1}(z)P_k(z) - \beta_k P_{k-1}(z))Q_k(z)$$
$$= \beta_k (Q_k(z)P_{k-1}(z) - Q_{k-1}(z)P_k(z)),$$

and by induction, this implies

$$Q_{k+1}(z)P_k(z) - Q_k(z)P_{k+1}(z) = \beta_k \cdots \beta_0 (Q_0(z)P_{-1}(z) - Q_{-1}(z)P_0(z))$$
$$= -\beta_k \cdots \beta_0.$$

Therefore, for each k, $Q_k(z)$ and $P_k(z)$ are coprime. Moreover, dividing by $-\beta_k \cdots \beta_0$, we see that the following Bezout identities hold:

$$P_k(z)A_k(z) + Q_k(z)B_k(z) = 1.$$

7. With the $E_i(z)$ defined by (8.82), we compute

$$P_n(z)/Q_n(z) - P_i(z)/Q_i(z) = (P_n(z)Q_i(z) - Q_n(z)P_i(z))/Q_n(z)Q_i(z)$$
$$= E_i(z)/Q_i(z).$$

Since $\deg Q_i = \sum_{j=1}^{i} \alpha_j$ and using part (4), it follows that the leading term of $E_i(z)/Q_i(z)$ is $\beta_0 \cdots \beta_i / z^{2\sum_{j=1}^{i} \alpha_j + \alpha_{i+1}}$. This is in agreement with the statement of the theorem, since $\deg a_i = \sum_{j=1}^{i} \alpha_j$.

8. We will prove it by induction. Indeed, for $i = 0$, we have

$$\begin{pmatrix} 0 & \beta_0 \\ -1 & a_1 \end{pmatrix} = \begin{pmatrix} -P_0 & P_1 \\ -Q_0 & Q_1 \end{pmatrix}.$$

Next, using the induction hypothesis, we compute

$$\begin{pmatrix} -P_{i-1} & P_i \\ -Q_{i-1} & Q_i \end{pmatrix} \begin{pmatrix} 0 & \beta_i \\ -1 & a_{i+1} \end{pmatrix} = \begin{pmatrix} -P_i & a_{i+1}P_i - \beta_i P_{i-1} \\ -Q_i & a_{i+1}Q_i - \beta_i Q_{i-1} \end{pmatrix} = \begin{pmatrix} -P_i & P_{i+1} \\ -Q_i & Q_{i+1} \end{pmatrix}.$$

9. We clearly have

$$g = \gamma_0 = T_0(\gamma_1) = T_0(T_1(\gamma_2)) = (T_0 \circ T_1)(\gamma_2) = \cdots = (T_0 \circ \cdots \circ T_{i-1})(\gamma_i).$$

To prove (8.87), we use the fact that $\gamma_n = 0$.

10. From (8.76), and using (8.86), we get $g_k = (T_0 \circ \cdots \circ T_{k-1})(0) = P_k/Q_k$.

11. Obviously, the Bezout identity (8.85) implies also the coprimeness of the polynomials $A_k(z)$ and $Q_k(z)$. Since the Bezoutian is linear in each of its arguments we can get a recursive formula for $B(A_k, Q_k)$:

$$B(Q_k, A_k) = B(Q_k, (\beta_0 \cdots \beta_{k-1})^{-1} Q_{k-1})$$

$$= B(a_k Q_{k-1} - \beta_{k-1}Q_{k-2}, (\beta_0 \cdots \beta_{k-1})^{-1} Q_{k-1})$$

$$= (\beta_0 \cdots \beta_{k-1})^{-1} Q_{k-1}(z)B(a_k, 1)Q_{k-1}(w) \\ + B(Q_{k-1}, (\beta_0 \cdots \beta_{k-2})^{-1} Q_{k-2})$$

$$= \cdots = \Sigma(\beta_0 \cdots \beta_{k-1})^{-1} Q_j(z)B(a_{j+1}, 1)Q_j(w).$$

12. This is just the matrix representation of (8.88).

13. We use Theorem 8.39, the equalities $Q_n(z) = q(z)$, $P_n(z) = p(z)$, and the Bezout equation (8.85) for $k = n$. ∎

A continued fraction representation is an extremely convenient tool for the computation of the signature of the Hankel or Bezout quadratic forms.

Theorem 8.52 (Frobenius). *Let $g(z)$ be rational having the continued fraction representation (8.75). Then*

$$\sigma(H_g) = \sum_{i=1}^{r} \left(\text{sign} \prod_{j=0}^{i-1} \beta_j \right) \frac{1 + (-1)^{\deg a_i - 1}}{2}. \tag{8.91}$$

Proof. It suffices to compute the signature of the Bezoutian $B(q,p)$. Using the equations of the Euclidean algorithm, we have

$$B(q,p) = B(q_{-1},q_0) = B\left(\frac{a_1 q_0 - q_1}{\beta_0}, q_0\right)$$
$$= \frac{1}{\beta_0} q_0(z) B(a_1, 1) q_0(w) + \frac{1}{\beta_0} B(q_0, q_1)$$

Here we used the fact that the Bezoutian is alternating. Since $p(z)$ and $q(z)$ are coprime, we have

$$\text{rank}\, B(a_1, 1) + \frac{1}{\beta_0} \text{rank}\, B(q_0, q_1) = \deg a_1 + \deg q_0 = \deg q = \text{rank}\, B(q, p).$$

The additivity of the ranks implies, by Theorem 8.15, the additivity of the signatures, and therefore

$$\sigma(B(q,p)) = \sigma\left(\frac{1}{\beta_0} B(a_1, 1)\right) + \sigma\left(\frac{1}{\beta_0} B(q_0, q_1)\right).$$

We proceed by induction and readily derive

$$\sigma(B(p,q)) = \sum_{i=1}^{r} \text{sign}\left(\frac{1}{\beta_0 \cdots \beta_i}\right) \sigma(B(a_{i+1}, 1)). \tag{8.92}$$

Now, given a polynomial $c(z) = z^k + c_{k-1} z^{k-1} + \cdots + c_0$, we have

$$B(c,1) = \frac{c(z) - c(w)}{z - w} = \sum_{i=1}^{k} c_i \frac{z^i - w^i}{z - w},$$

with $c_k = 1$. Using the equality

$$\frac{z^i - w^i}{z - w} = \sum_{j=1}^{i} z^j w^{i-j-1},$$

we have

$$B(c,1) = \begin{pmatrix} c_1 & c_2 & \cdots & c_{i-1} & 1 \\ c_2 & & \cdots & & \\ \vdots & & \cdots & & \\ \vdots & & \cdots & & \\ c_{i-1} & 1 & & & \\ 1 & & & & \end{pmatrix}.$$

Hence

$$
\begin{cases}
1 & \text{if } \deg c \text{ is odd}, \\
0 & \text{if } \deg c \text{ is even}.
\end{cases}
$$

Equality (8.92) can now be rewritten as (8.91).   ∎

The Lanczos polynomials $Q_i(z)$ generated by the recurrence relation (8.78) satisfy certain orthogonality conditions. These orthogonality conditions depend of course on the rational function $g(z)$ from which the polynomials were derived. These orthogonality relations are related to an inner product induced by the Hankel operator H_g. We will explain now this relation. Starting from the strictly proper rational function $g(z)$, we have the bilinear form defined on the space of polynomials by

$$
\{x, y\} = [H_g x, y] = \sum_i \sum_j g_{i+j} \xi_i \eta_j, \tag{8.93}
$$

where $x(z) = \sum_i \xi_i z^i$ and $y(z) = \sum_j \eta_j z^j$. If $g(z)$ has the coprime representation $g(z) = p(z)/q(z)$, then $\mathrm{Ker}\, H_g = q\mathbb{F}[z]$, and hence we may as well restrict ourselves to the study of H_g as a map from X_q to $X^q = X_q^*$. However, we can view all the action as taking place in X_q, using the pairing $\langle \cdot, \cdot \rangle$ defined in (5.64). Thus, for $x(z), y(z) \in X_q$, we have

$$
\begin{aligned}
[H_g x, y] &= [\pi_- q^{-1} p x, y] \\
&= [q^{-1} q \pi_- q^{-1} p x, y] = \langle \pi_q p x, y \rangle \\
&= \langle p(S_q) x, y \rangle.
\end{aligned}
$$

In fact, $p(S_q)$ is, as a result of Theorem 5.39, a self-adjoint operator in X_q. The pairing

$$
\{x, y\} = \langle p(S_q) x, y \rangle
$$

can of course be evaluated by choosing an appropriate basis. In fact, we have $\langle p(S_q) x, y \rangle = ([p(S_q)]_{st}^{co}[x]^{st}, [y]^{st})$. Since $[p(S_q)]_{st}^{co} = H_n$, we recover (8.93).

Proposition 8.53. *Let $P_i(z), Q_i(z)$ be the Lanczos polynomials associated with the rational function $g(z)$. Relative to the pairing*

$$
\{x, y\} = \langle p(S_q) x, y \rangle = [H_g x, y] = \sum_i \sum_j g_{i+j} \xi_i \eta_j, \tag{8.94}
$$

the polynomial $Q_i(z)$ is orthogonal to all polynomials of lower degree. Specifically,

$$
\{Q_i, z^j\} = [H_g Q_i, z^j] = \begin{cases}
0, & j < \alpha_1 + \cdots + \alpha_i, \\
\beta_0 \cdots \beta_i, & j = \alpha_1 + \cdots + \alpha_i.
\end{cases} \tag{8.95}
$$

Proof. Recalling that the rational functions $E_i(z)$ are defined in (8.82) by $E_i(z) = g(z) Q_i(z) - P_i(z)$, this means that

$$\pi_- E_i = \pi_- g Q_i = H_g Q_i.$$

Since the leading term of $E_i(z)$ is $\frac{\beta_0 \cdots \beta_i}{z^{\alpha_1 + \cdots + \alpha_{i+1}}}$, this implies (8.95).   ∎

The set $\{z^{j_i}(\Pi_{k=0}^{i-1} Q_k(z)) \mid i = 1, \ldots, r-1; \ j_i = 0, \ldots, n_i - 1\}$, lexicographically ordered, is clearly a basis for X_q since it contains one polynomial for each degree between 0 and $n-1$. We denote it by B_{or} and refer to it as the **orthogonal basis** because of its relation to orthogonal polynomials. The properties of the Lanczos polynomials can be used to compute a matrix representation of the shift S_q relative to this basis.

Theorem 8.54. *Let the strictly proper transfer function* $g(z) = \frac{p(z)}{q(z)}$ *have the sequence of atoms* $\{a_i(z), \beta_{i-1}\}_{i=1}^r$, *and assume that* $\alpha_1, \ldots, \alpha_r$ *are the degrees of the atoms and*

$$a_k(z) = z^{\alpha_k} + \sum_{i=0}^{\alpha_k - 1} a_i^{(k)} z^i.$$

Then the shift operator S_q *has the following, block tridiagonal, matrix representation with respect to the orthogonal basis:*

$$[S_q]_{or}^{or} = \begin{pmatrix} A_{11} & A_{12} & & & \\ A_{21} & A_{22} & \cdot & & \\ & \cdot & \cdot & \cdot & \\ & & \cdot & \cdot & A_{r-1\, r} \\ & & & A_{r\, r-1} & A_{rr} \end{pmatrix},$$

where

$$A_{ii} = \begin{pmatrix} 0 & \ldots & & -a_0^{(i)} \\ 1 & & & \cdot \\ & \cdot & & \cdot \\ & & \cdot & \cdot \\ & & 1 & -a_{\alpha_i - 1}^{(i)} \end{pmatrix}, \ i = 1, \ldots, r,$$

$$A_{i+1\, i} = \begin{pmatrix} 0 & \ldots & 0 & 1 \\ & & & 0 \\ \cdot & & & \cdot \\ & & & \cdot \\ 0 & \ldots & & 0 \end{pmatrix}, \ i = 1, \ldots r-1,$$

and $A_{i\, i+1} = \beta_{i-1} A_{i+1\, i}.$

Proof. On the basis elements $z^i Q_{j-1}(z)$, $i = 0, \ldots, n_j - 2, j = 1, \ldots, r-1$, the map S_q indeed acts as the shift to the next basis element. For $i = n_j - 1$ we compute

$$S_q z^{n_j-1} Q_{j-1} = \pi_q z^{n_j} Q_{j-1}$$
$$= \pi_q \left(z^{n_j} + \sum_{k=0}^{n_j-1} a_k^{(j)} z^k - \sum_{k=0}^{n_j-1} a_k^{(j)} z^k \right) Q_{j-1}$$
$$= a_j Q_{j-1} - \sum_{k=0}^{n_j-1} a_k^{(j)} z^k \cdot Q_{j-1}$$
$$= Q_j + \beta_{j-1} Q_{j-2} - \sum_{k=0}^{n_j-1} a_k^{(j)} z^k \cdot Q_{j-1}. \qquad \blacksquare$$

In the generic case, in which all Hankel matrices H_i, $i = 1, \ldots, n$, are nonsingular, all polynomials $a_i(z)$ are linear, and we write $a_i(z) = z - \theta_i$. The recurrence equation for the orthogonal polynomials becomes $Q_{i+1}(z) = (z - \theta_{i+1}) Q_i(z) - \beta_i Q_{i-1}(z)$, and we obtain the following corollary.

Corollary 8.55. *Assuming all Hankel matrices H_i, $i = 1, \ldots, n$, are nonsingular, then with respect to the orthogonal basis $\{Q_0(z), \ldots, Q_{n-1}(z)\}$ of $X_q = X_{Q_n}$, the shift operator S_q has the tridiagonal matrix representation*

$$[S_q]_{or}^{or} = \begin{pmatrix} \theta_1 & \beta_1 & & & \\ 1 & \theta_2 & . & & \\ & . & . & . & \\ & & . & . & \\ & & & . & . & \beta_{n-1} \\ & & & & 1 & \theta_n \end{pmatrix}.$$

The Hankel matrix H_n is symmetric, hence can be diagonalized by a congruence transformation. In fact, we can exhibit explicitly a diagonalizing congruence. Using the same method, we can clarify the connection between $C_q^\sharp$, the companion matrix of $q(z)$, and the tridiagonal representation.

Proposition 8.56. *Let $g(z) = p(z)/q(z)$ be strictly proper, rational of McMillan degree $n = \deg q$. Assume all Hankel matrices H_i, $i = 1, \ldots, n$, are nonsingular and let $\{Q_0(z), \ldots, Q_{n-1}(z)\}$ be the orthogonal basis of X_q. Then*

1.

$$\begin{pmatrix} q_{0,0} & & & \\ q_{1,0} & q_{1,1} & & \\ . & & . & \\ . & & & . \\ q_{n-1,0} & . & . & \cdot q_{n-1,n-1} \end{pmatrix} \begin{pmatrix} g_1 & \cdots & g_n \\ . & \cdots & . \\ . & \cdots & . \\ . & \cdots & . \\ g_n & \cdots & g_{2n-1} \end{pmatrix} \begin{pmatrix} q_{0,0} & q_{1,0} & \cdots & q_{n-1,0} \\ & q_{1,1} & & . \\ & & . & . \\ & & & . \\ & & & q_{n-1,n-1} \end{pmatrix}$$

$$= \begin{pmatrix} \beta_0 & & & \\ & \beta_0 \beta_1 & & \\ & & . & \\ & & & \beta_0 \cdots \beta_{n-1} \end{pmatrix}.$$

$$(8.96)$$

2.

$$
\begin{pmatrix}
0 & & & -q_0 \\
1 & & \cdot & \\
& \cdot & \cdot & \\
& & \cdot & \cdot \\
& & 1 & -q_{n-1}
\end{pmatrix}
\begin{pmatrix}
q_{0,0} & q_{1,0} & \cdots & q_{n-1,0} \\
& q_{1,1} & & \cdot \\
& & \cdot & \cdot \\
& & & \cdot \\
& & & q_{n-1,n-1}
\end{pmatrix}
$$

$$
=
\begin{pmatrix}
q_{0,0} & q_{1,0} & \cdots & q_{n-1,0} \\
& q_{1,1} & & \cdot \\
& & \cdot & \cdot \\
& & & \cdot \\
& & & q_{n-1,n-1}
\end{pmatrix}
\begin{pmatrix}
\theta_1 & \beta_1 & & \\
1 & \theta_2 & \cdot & \\
& & \cdot & \cdot \\
& & \cdot & \beta_{n-1} \\
& & 1 & \theta_n
\end{pmatrix}
\quad (8.97)
$$

Proof. 1. This is the matrix representation of the orthogonality relations given in (8.95).

2. We use the trivial operator identity $S_q I = I S_q$ and take the matrix representation $[S_q]_{st}^{st}[I]_{or}^{st} = [I]_{or}^{st}[S_q]_{or}^{or}$. This is equivalent to (8.97). ∎

It is easy to read off from the diagonal representation (8.96) of the Hankel matrix the relevant signature information. A special case is the following.

Corollary 8.57. *Let* $g(z)$ *be a rational function with the continued fraction representation (8.75). Then* H_n *is positive definite if and only if all* $a_i(z)$ *are linear and all the* β_i *are positive.*

8.3.8   The Cauchy Index

Let $g(z)$ be a rational transfer function with real coefficients having the coprime representation $g(z) = p(z)/q(z)$. The **Cauchy index** of $g(z)$, denoted by I_g, is defined as the number of jumps, on the real axis, of $g(z)$ from $-\infty$ to $+\infty$ minus the number of jumps from $+\infty$ to $-\infty$. Thus, the Cauchy index is related to discontinuities of $g(z)$ on the real axis. These discontinuities arise from the real zeros of $q(z)$. However, the contributions to the Cauchy index come only from zeros of $q(z)$ of odd order.

In this section we will establish some connections between the Cauchy index, the signature of the Hankel map induced by $g(z)$, and the existence of signature-symmetric realizations of $g(z)$. The central result is the classical Hermite–Hurwitz theorem. However, before proving it, we state and prove an auxiliary result that is of independent interest.

Proposition 8.58. *Let $g(z)$ be a real rational function. Then the following scaling operations*

$$(1)\ g(z) \longrightarrow mg(z), \qquad m > 0,$$
$$(2)\ g(z) \longrightarrow g(z-a), \qquad a \in \mathbb{R},$$
$$(3)\ g(z) \longrightarrow g(rz), \qquad r > 0,$$

leave the rank and signature of the Hankel map as well as the Cauchy index invariant.

Proof. (1) is obvious. To prove the rank invariance, let $g(z) = p(z)/q(z)$ with $p(z)$ and $q(z)$ coprime. By the Euclidean algorithm, there exist polynomials $a(z)$ and $b(z)$ such that $a(z)p(z) + b(z)q(z) = 1$. This implies

$$a(z-a)p(z-a) + b(z-a)q(z-a) = 1$$

as well as

$$a(rz)p(rz) + b(rz)q(rz) = 1,$$

i.e., $p(z-a), q(z-a)$ are coprime and so are $p(rz), q(rz)$. Now $g(z-a) = p(z-a)/q(z-a)$ and $g(rz) = p(rz)/q(rz)$, which proves the invariance of McMillan degree, which is equal to the rank of the Hankel map. Now it is easy to check that given any polynomial $u(z)$, we have

$$[H_g u, u] = [H_{g_a} u_a, u_a],$$

where $g_a(z) = g(z-a)$. If we define a map $R_a : \mathbb{R}[z] \longrightarrow \mathbb{R}[z]$ by

$$(R_a u)(z) = u(z-a) = u_a(z),$$

then R_a is invertible, $R_a^{-1} = R_{-a}$, and

$$[H_g u, u] = [H_{g_a} u_a, u_a] = [H_{g_a} R_a u, R_a u] = [R_a * H_{g_a} R_a u, u],$$

which shows that $H_g = R_a^* H_{g_a} R_a$ and hence that $\sigma(H_g) = \sigma(H_{g_a})$. This proves (2). To prove (3), define, for $r > 0$, a map $P_r : \mathbb{R}[z] \longrightarrow \mathbb{R}[z]$ by

$$(P_r u)(z) = u(rz).$$

Clearly, P_r is invertible and $P_r^{-1} = P_{1/r}$. Letting $u_r = P_r u$, we have

$$[H_{g_r} u, u] = [\pi_- g(rz)u, u] = [\pi_- \Sigma(g_k/r^{k+1}z^{k+1})u, u]$$

$$= \Sigma\Sigma(g_{i+j}r^{-i-j}u_i)u_j = \Sigma\Sigma g_{i+j}(u_i r^{-i})(u_j r^{-j})$$

$$= [H_g P_r u, P_r u] = [P_r^* H_g P_r u, u],$$

hence $H_{g_r} = P_r^* H_g P_r$, which implies $\sigma(H_{g_r}) = \sigma(H_g)$. The invariance of the Cauchy index under these scaling operations is obvious.   ∎

Theorem 8.59 (Hermite–Hurwitz). *Let $g(z) = p(z)/q(z)$ be a strictly proper, real rational function with $p(z)$ and $q(z)$ coprime. Then*

$$I_g = \sigma(H_g) = \sigma(B(q,p)). \tag{8.98}$$

Proof. That $\sigma(H_g) = \sigma(B(q,p))$ has been proved in Theorem 8.40. So it suffices to prove the equality $I_g = \sigma(H_g)$.

Let us analyze first the case that $q(z)$ is a polynomial with simple real zeros, i.e., $q(z) = \prod_{j=1}^{n}(z - a_j)$ and $a_i \neq a_j$ for $i \neq j$. Let $d_i(z) = q(z)/(z - a_i)$. Given any polynomial $u(z) \in X_q$, it has a unique expansion $u(z) = \sum_{i=1}^{n} u_i d_i(z)$. We compute

$$[H_g u, u] = [\pi_- g u, u] = [\pi_- q^{-1} p u, u] = [q^{-1} q \pi_- q^{-1} p u, u]$$

$$= \langle \pi_q p u, u \rangle = \langle p(S_q) u, u \rangle = \sum_{i=1}^{n} \sum_{j=1}^{n} \langle p(S_q) d_i, d_j \rangle u_i u_j$$

$$= \sum_{i=1}^{n} \sum_{j=1}^{n} \langle p(a_i) d_i, d_j \rangle u_i u_j = \sum_{i=1}^{n} p(a_i) d_i(a_i) u_i^2,$$

since $d_i(z)$ are eigenfunctions of S_q corresponding to the eigenvalues a_i and since by Proposition 5.40, $\langle d_i, d_j \rangle = d_i(a_i) \delta_{ij}$. From this computation it follows, since $[p(S_q)]_{co}^{st} = B(q,p)$, that

$$\sigma(H_g) = \sigma(B(q,p)) = \sum_{i=1}^{n} \operatorname{sign}[p(a_i) d_i(a_i)].$$

On the other hand, we have the partial fraction decomposition

$$g(z) = \frac{p(z)}{q(z)} = \sum_{i=1}^{n} \frac{c_i}{z - a_i},$$

or

$$p(z) = \sum_{i=1}^{n} c_i \frac{q(z)}{z - a_i} = \sum_{i=1}^{n} c_i d_i(z),$$

which implies $p(a_i) = c_i d_i(a_i)$, or equivalently, that $c_i = \frac{p(a_i)}{d_i(a_i)}$. Now obviously

$$I_g = \sum_{i=1}^{n} \operatorname{sign}(c_i) = \sum_{i=1}^{n} \operatorname{sign}\left(\frac{p(a_i)}{d_i(a_i)}\right),$$

and, since $\operatorname{sign}(p(a_i) d_i(a_i)) = \operatorname{sign}(p(a_i)/d_i(a_i))$, the equality (8.98) is proved in this case.

We pass now to the general case. Let $q(z) = q_1(z) \cdots q_s(z)$ be the unique factorization of $q(z)$ into powers of relatively prime irreducible monic polynomials. As before, we define polynomials $d_i(z)$ by

$$d_i(z) = \frac{q(z)}{q_i(z)}.$$

Since we have the direct sum decomposition $X_q = d_1 X_{q_1} \oplus \cdots \oplus d_s X_{q_s}$, it follows that each $f(z) \in X_q$ has a unique representation of the form $f(z) = \sum_{i=1}^{s} d_i(z) u_i(z)$ with $u_i(z) \in X_{q_i}$. Relative to the indefinite metric of X_q, introduced in (5.64), this is an orthogonal direct sum decomposition, i.e.,

$$\langle d_i X_{q_i}, d_j X_{q_j} \rangle = 0 \qquad \text{for } i \neq j.$$

Indeed, if $f_i(z) \in X_{q_i}$ and $g_j(z) \in X_{q_j}$, then

$$\langle d_i f_i, d_j f_j \rangle = [q^{-1} d_i f_i, d_j f_j] = [d_j q^{-1} d_i f_i, f_j] = 0,$$

since $d_i(z) d_j(z)$ is divisible by $q(z)$ and $\mathbb{F}[z]^{\perp} = \mathbb{F}[z]$.

Let $g(z) = \sum_{i=1}^{s} p_i(z)/q_i(z)$ be the partial fraction decomposition of $g(z)$. Since the zeros of the $q_i(z)$ are distinct, it is clear that

$$I_g = I_{\frac{p}{q}} = \sum_{i=1}^{s} I_{p_i/q_i}.$$

Also, as a consequence of Proposition 8.23, it is clear that for the McMillan degree $\delta(g)$ of $g(z)$, we have $\delta(g) = \sum_{i=1}^{s} \delta(p_i/q_i)$, and hence, by Theorem 8.15, the signatures of the Hankel forms are additive, namely $\sigma(H_g) = \sum_{i=1}^{s} \sigma(H_{p_i/q_i})$. Therefore, to prove the Hermite–Hurwitz theorem it suffices to prove it in the case of $q(z)$ being the power of a monic prime. Since we are discussing the real case, the primes are of the form $z - a$ and $((z-a)^2 + b^2)$, with $a, b \in \mathbb{R}$. By applying the previous scaling result, the proof of the Hermite–Hurwitz theorem reduces to the two cases $q(z) = z^m$ or $q(z) = (z^2 + 1)^m$.

Case 1: $q(z) = z^m$.
Assuming $p(z) = p_0 + p_1 z + \cdots + p_{m-1} z^{m-1}$, then the coprimeness of $p(z)$ and $q(z)$ is equivalent to $p_0 \neq 0$. Therefore we have $g(z) = p_{m-1} z^{-1} + \cdots + p_0 z^{-m}$, which shows that

$$I_g = I_{(p/z^m)} = \begin{cases} 0 & \text{if } m \text{ is even,} \\ \text{sign}(p_0) & \text{if } m \text{ is odd.} \end{cases}$$

On the other hand, $\text{Ker}\, H_g = z^{m+1} \mathbb{R}[z]$, and so

$$\sigma(H_g) = \sigma(H_g | X_{z^m}).$$

Relative to the standard basis, the truncated Hankel map has the matrix representation

$$\begin{pmatrix} p_{m-1} & \cdots & p_1 & p_0 \\ \cdot & & \cdot & \cdot \cdot \cdot \\ \cdot & & & \cdot \cdot \cdot \\ \cdot & & & \cdot \cdot \\ p_1 & & & \cdot \\ p_0 & & & \end{pmatrix}.$$

Now, clearly, the previous matrix has the same signature as

$$\begin{pmatrix} 0 & \cdots & 0 & p_0 \\ \cdot & & \cdot & \cdot \cdot \cdot \\ \cdot & & & \cdot \cdot \cdot \\ \cdot & & & \cdot \cdot \\ 0 & & & \cdot \\ p_0 & & & \end{pmatrix},$$

and hence

$$\sigma(H_g) = \begin{cases} \text{sign}\,(p_0) & \text{if } m \text{ is odd,} \\ 0 & \text{if } m \text{ is even.} \end{cases}$$

Case 2: $q(z) = (z^2 + 1)^m$.

Since $q(z)$ has no real zeros, it follows that in this case $I_g = I_{\frac{p}{q}} = 0$. So it suffices to prove that also $\sigma(H_g) = 0$. Let $g(z) = p(z)/(z^2 + 1)^m$ with $\deg p < 2m$. Let us expand $p(z)$ in the form

$$p(z) = \sum_{k=0}^{m-1} (p_k + q_k z)(z^2 + 1)^k,$$

with the p_k and q_k uniquely determined. The coprimeness condition is equivalent to p_0 and q_0 not being zero together. The transfer function $g(z)$ has therefore the following representation:

$$g(z) = \sum_{k=0}^{m-1} \frac{p_k + q_k z}{(z^2 + 1)^{m-k}}.$$

In much the same way, every polynomial $u(z)$ can be written, in a unique way, as

$$u(z) = \sum_{i=0}^{m-1} (u_i + v_i z)(z^2 + 1)^i.$$

We compute now the matrix representation of the Hankel form with respect to the basis

$$\mathcal{B} = \{1, z, (z^2+1), z(z^2+1), \ldots, (z^2+1)^{m-1}, z(z^2+1)^{m-1}\}$$

of $X_{(z^2+1)^m}$. Thus, we need to compute

$$[H_g z^\alpha (z^2+1)^\lambda, z^\beta (z^2+1)^\mu] = [H_g z^{\alpha+\beta}(z^2+1)^{\lambda+\mu}, 1].$$

Now, with $0 \le \gamma \le 2$ and $0 \le \nu \le 2m-2$, we compute

$$[H_g z^\gamma(z^2+1)^\nu, 1] = \left[\frac{\sum_{i=0}^{m-1}(p_i+q_i z)(z^2+1)^i}{(z^2+1)^m} z^\gamma(z^2+1)^\nu, 1 \right]$$

$$= \left[\sum_{i=0}^{m-1} \frac{(p_i+q_i z)z^\gamma(z^2+1)^{\nu+i}}{(z^2+1)^m}, 1 \right].$$

The only nonzero contributions come from the terms in which $\nu+i = m-2, m-1$, or equivalently, when $i = m - \lambda - \mu - 2, m - \lambda - \mu - 1$. Now

$$\left[\frac{(p_i+q_i z)z^\gamma(z^2+1)^{\nu+i}}{(z^2+1)^m}, 1 \right] = \begin{cases} \begin{cases} q_i, & \gamma = 0, \\ p_i, & \gamma = 1, \quad i = m - \nu - 1, \\ -q_i, & \gamma = 2, \end{cases} \\[2mm] \begin{cases} 0, & \gamma = 0, \\ 0, & \gamma = 1, \quad i = m - \nu - 2. \\ q_i, & \gamma = 2, \end{cases} \end{cases}$$

Thus the matrix of the Hankel form in this basis has the following block triangular form:

$$M = \begin{pmatrix} q_{m-1} & p_{m-1} & & q_1 & p_1 & q_0 & p_0 \\ p_{m-1} & q_{m-2} - q_{m-1} & & p_1 & q_0 - q_1 & p_0 & -q_0 \\ & & & q_0 & p_0 & & \\ & & & p_0 & -q_0 & & \\ & & & & & & \\ & & & & & & \\ q_0 & p_0 & & & & & \\ p_0 & -q_0 & & & & & \end{pmatrix}.$$

By our coprimeness assumption, the matrix $\begin{pmatrix} q_0 & p_0 \\ p_0 & -q_0 \end{pmatrix}$ is nonsingular and its determinant is negative. Hence it has signature equal to zero. This implies that also the signature of M is zero, and with this, the proof of the theorem is complete. ∎

8.4   Tensor Products of Models

The underlying idea in this book is that the study of linear algebra is simplified and clarified when functional, or module-theoretic, methods are used. This motivated us in the introduction of polynomial and rational models. Naturally, we expect that the study of tensor products and forms by the use of these models will lead to similar simplifications, and this we proceed to do. In our approach, we emphasize the concreteness of specific representations of tensor products of polynomial models. Thus, we expect that the tensor products of polynomial models, whether over the underlying field $\mathbb{F}$ or over the polynomial ring $\mathbb{F}[z]$, will also turn out to be (isomorphic to) polynomial models in one or two variables as the case may be.

There are several different models we can use. Two of them arise from Kronecker products of multiplication maps, the difference being whether the product is taken over the field $\mathbb{F}$ or the ring $\mathbb{F}[z]$.

8.4.1   Bilinear Forms

Let $\mathscr{X}, \mathscr{Y}$ be $\mathbb{F}$-vector spaces. We shall denote by $L(\mathscr{X}, \mathscr{Y})$, as well as by $\mathrm{Hom}_{\mathbb{F}}(\mathscr{X}, \mathscr{Y})$, the space of $\mathbb{F}$-linear maps from $\mathscr{X}$ to $\mathscr{Y}$. We take now a closer look at the case of tensor products of two finite-dimensional $\mathbb{F}$-vector spaces $\mathscr{X}, \mathscr{Y}$. In this situation, and in order to clearly distinguish it from the algebraic dual M' of a module M, we will write $\mathscr{X}^* = \mathrm{Hom}_{\mathbb{F}}(\mathscr{X}, \mathbb{F})$ for the **vector space dual**. Let $\mathscr{B}_{\mathscr{X}} = \{f_i\}_{i=1}^n$, $\mathscr{B}_{\mathscr{Y}} = \{g_i\}_{i=1}^m$ be bases of $\mathscr{X}$ and $\mathscr{Y}$ respectively. Let $\mathscr{B}_{\mathscr{X}}^* = \{\phi_i\}_{i=1}^n$ be the basis of $\mathscr{X}^*$ that is dual to $\mathscr{B}_{\mathscr{X}}$, i.e., it satisfies $\phi_i(f_j) = \delta_{ij}$.

Let $\mathbb{F}$ be a field and let $\mathscr{X}, \mathscr{Y}, \mathscr{Z}$ be $\mathbb{F}$-vector spaces. A function $\phi : \mathscr{X} \times \mathscr{Y} \longrightarrow \mathscr{Z}$ is called a $\mathscr{Z}$-valued **bilinear form** on $\mathscr{X} \times \mathscr{Y}$ if ϕ is linear in both variables. We will denote by $\mathrm{Bilin}(\mathscr{X}, \mathscr{Y}; \mathscr{Z})$ the space of all $\mathscr{Z}$-valued bilinear functions. We will denote by $\mathrm{Bilin}(\mathscr{X}, \mathscr{Y}; \mathbb{F})$ the set of all $\mathbb{F}$ valued bilinear functions on $\mathscr{X} \times \mathscr{Y}$ and refer to its elements as **bilinear forms**.

There is a close connection between bilinear forms and linear transformations which we summarize in the following proposition.

Proposition 8.60. *Given $\mathbb{F}$-linear vector spaces $\mathscr{X}, \mathscr{Y}$, a necessary and sufficient condition for a function $\phi : \mathscr{X} \times \mathscr{Y} \longrightarrow \mathbb{F}$ to be a bilinear form is the existence of a, uniquely defined, linear map $T : \mathscr{X} \longrightarrow \mathscr{Y}^*$ for which*

$$\phi(x, y) = \langle Tx, y \rangle. \tag{8.99}$$

Proof. Assume $T \in \mathrm{Hom}_{\mathbb{F}}(\mathscr{X}, \mathscr{Y}^*)$. Defining $\phi : \mathscr{X} \times \mathscr{Y} \longrightarrow \mathbb{F}$ by (8.99), it clearly follows that $\phi(x, y)$ is an $\mathbb{F}$-valued bilinear form.

Conversely, if $\phi(x, y)$ is an $\mathbb{F}$-valued bilinear form on $\mathscr{X} \times \mathscr{Y}$, then for each $x \in \mathscr{X}$, $\phi(x, y)$ is a linear functional on $\mathscr{Y}$. Thus there exists $y_x^* \in \mathscr{Y}^*$ for which

$$\phi(x,y) = \langle y_x^*, y \rangle.$$

Define a map $T : \mathscr{X} \longrightarrow \mathscr{Y}^*$ by $Tx = y_x^*$. Using the linearity of $\phi(x,y)$ in the x variable, we compute

$$\langle y_{\alpha_1 x_1 + \alpha_2 x_2}^*, y \rangle = \phi(\alpha_1 x_1 + \alpha_2 x_2, y) = \alpha_1 \phi(x_1, y) + \alpha_2 \phi(x_2, y)$$
$$= \alpha_1 \langle y_{x_1}^*, y \rangle + \alpha_2 \langle y_{x_2}^*, y \rangle = \langle \alpha_1 y_{x_1}^* + \alpha_2 y_{x_2}^*, y \rangle.$$

This implies $y_{\alpha_1 x_1 + \alpha_2 x_2}^* = \alpha_1 y_{x_1}^* + \alpha_2 y_{x_2}^*$, which can be rewritten as $T(\alpha_1 x_1 + \alpha_2 x_2) = \alpha_1 T x_1 + \alpha_2 T x_2$, i.e., $T \in \mathrm{Hom}_{\mathbb{F}}(\mathscr{X}, \mathscr{Y}^*)$. Uniqueness of T follows by a standard argument. ∎

8.4.2 Tensor Products of Vector Spaces

The availability of duality pairings allows us to take another look at matrix representations. Given $\mathbb{F}$-vector spaces $\mathscr{X}, \mathscr{Y}$ and elements $f \in \mathscr{X}^*$ and $w \in \mathscr{Y}$, we use the notation $\langle v, f \rangle = f(v)$. Next, we define a map $w \otimes f : \mathscr{X} \longrightarrow \mathscr{Y}$ by

$$(w \otimes f)v = \langle v, f \rangle w, \quad v \in \mathscr{X}. \tag{8.100}$$

It is easily checked that $w \otimes f$ is the zero map if and only if $f = 0$ or $w = 0$; otherwise, it has rank one.

Proposition 8.61. *We have*

$$\mathrm{Hom}_{\mathbb{F}}(\mathscr{X}, \mathscr{Y}) = \left\{ \sum_{i=1}^{s} w_i \otimes v_i^* \mid w_i \in \mathscr{Y}, v_i^* \in \mathscr{X}^*, s \in \mathbb{N} \right\}. \tag{8.101}$$

Proof. Clearly, from (8.100) and the bilinearity of the pairing $\langle v, f \rangle$, it follows that $w \otimes f$ is a linear transformation. Since $\mathrm{Hom}_{\mathbb{F}}(\mathscr{X}, \mathscr{Y})$ is closed under linear combinations, we have the inclusion $\mathrm{Hom}_{\mathbb{F}}(\mathscr{X}, \mathscr{Y}) \supset \{\sum_i w_i \otimes f_i \mid w_i \in \mathscr{Y}, f_i \in \mathscr{X}^*\}$.

Conversely, let $T \in \mathrm{Hom}_{\mathbb{F}}(\mathscr{X}, \mathscr{Y})$. We consider the bases $\mathscr{B}_V = \{e_1, \ldots, e_n\}$ of $\mathscr{X}$ and $B_W = \{f_1, \ldots, f_m\}$ of $\mathscr{Y}$. A vector $v \in \mathscr{X}$ has a unique representation of the form $v = \sum_{j=1}^{n} \alpha_j e_j$. Similarly, $w = Tv \in \mathscr{Y}$ has a unique representation of the form $Tv = \sum_{i=1}^{m} \beta_i f_i$. With $\mathscr{B}_{V^*} = \{e_1^*, \ldots, e_n^*\}$ the dual basis to $\mathscr{B}_V$, we have $\langle e_j, e_i^* \rangle = \delta_{ij}$ and $\mathscr{B}_{V^*} = \mathscr{B}_V^*$. Let $E_{ij} \in \mathrm{Hom}_{\mathbb{F}}(\mathscr{X}, \mathscr{Y})$ be the map defined by (4.12), i.e., by $E_{ij} e_k = \delta_{jk} f_i$. Clearly, we have

$$E_{ij} v = \langle v, e_j^* \rangle f_i = (f_i \otimes e_j^*)v. \tag{8.102}$$

The computation (4.14) shows that

$$T = \sum_{i=1}^{m} \sum_{j=1}^{n} t_{ij} f_i \otimes e_j^*, \tag{8.103}$$

and hence the inclusion $\mathrm{Hom}_{\mathbb{F}}(\mathscr{X},\mathscr{Y}) \subset \{\sum_i w_i \otimes f_i | w_i \in \mathscr{Y}, f_i \in \mathscr{X}^*\}$. The two inclusions are equivalent to (8.101). ∎

We proceed to define the tensor product of two vector spaces. Our preference is to do it in a concrete way, so that the tensor product is identified with a particular representation. Given $\mathbb{F}$-vector spaces $\mathscr{X}, \mathscr{Y}$, let $\mathscr{X}^*$ be the vector space dual of $\mathscr{X}$. For $x^* \in \mathscr{X}^*$ and $y \in \mathscr{Y}$, we define a map $y \otimes x^* \in \mathrm{Hom}_{\mathbb{F}}(\mathscr{X},\mathscr{Y})$ by

$$(y \otimes x^*)\xi = (x^*\xi)y, \qquad \xi \in \mathscr{X}. \tag{8.104}$$

Although we can give an abstract definition of a **tensor product**, we prefer to give a concrete one.

Definition 8.62. Given $\mathbb{F}$-vector spaces $\mathscr{X}, \mathscr{Y}$, we define

$$\mathscr{Y} \otimes_{\mathbb{F}} \mathscr{X}^* = \mathrm{Hom}_{\mathbb{F}}(\mathscr{X},\mathscr{Y}). \tag{8.105}$$

We call $\mathscr{Y} \otimes_{\mathbb{F}} \mathscr{X}^* = \mathrm{Hom}_{\mathbb{F}}(\mathscr{X},\mathscr{Y})$ the **tensor product** of $\mathscr{Y}$ and $\mathscr{X}^*$. The map $\phi : \mathscr{Y} \times \mathscr{X}^* \longrightarrow \mathscr{Y} \otimes_{\mathbb{F}} \mathscr{X}^*$ defined by

$$\phi(y,x^*) = y \otimes x^* \tag{8.106}$$

is called the **canonical map**.

Clearly, in this case, the canonical map ϕ is bilinear and bijective.

Note that for finite-dimensional $\mathbb{F}$-vector spaces $\mathscr{X}, \mathscr{Y}$, and using reflexivity, that is, $(\mathscr{X}^*)^* \simeq \mathscr{X}$, we can also define $\mathscr{Y} \otimes_{\mathbb{F}} \mathscr{X} = \mathrm{Hom}_{\mathbb{F}}(\mathscr{X}^*,\mathscr{Y})$.

Proposition 8.63. *Let $\mathscr{B}_{\mathscr{X}} = \{e_1,\ldots,e_n\}$ and $B_{\mathscr{Y}} = \{f_1,\ldots,f_m\}$ be bases of $\mathscr{X}$ and $\mathscr{Y}$ respectively and let $\mathscr{B}_{\mathscr{X}}^* = \{e_1^*,\ldots,e_n^*\}$ be the dual basis to $\mathscr{B}_{\mathscr{X}}$. Then*

1. The set

$$\mathscr{B}_{\mathscr{Y}} \otimes \mathscr{B}_{\mathscr{X}}^* = \{f_i \otimes e_j^* \mid i = 1,\ldots,m, j = 1,\ldots,n\} \tag{8.107}$$

*is a basis for $\mathscr{Y} \otimes_{\mathbb{F}} \mathscr{X}^*$. We will refer to the basis $\mathscr{B}_{\mathscr{Y}} \otimes \mathscr{B}_{\mathscr{X}}^*$ as the **tensor product of the bases** $\mathscr{B}_{\mathscr{Y}}$ and $\mathscr{B}_{\mathscr{X}}^*$.*

2. We have

$$\dim \mathscr{Y} \otimes \mathscr{X}^* = \dim \mathscr{Y} \cdot \dim \mathscr{X}. \tag{8.108}$$

3. For $T \in \mathrm{Hom}_{\mathbb{F}}(\mathscr{X},\mathscr{Y})$, with the matrix representation $[T]_{\mathscr{B}_{\mathscr{X}}}^{\mathscr{B}_{\mathscr{Y}}}$ defined by (4.16), we have

$$[T]^{\mathscr{B}_{\mathscr{Y}} \otimes \mathscr{B}_{\mathscr{X}}^*} = [T]_{\mathscr{B}_{\mathscr{X}}}^{\mathscr{B}_{\mathscr{Y}}}. \tag{8.109}$$

Proof. 1. Note that we have the equality

$$f_i \otimes e_j^* = E_{ij}, \tag{8.110}$$

where the maps E_{ij} are defined by (4.7). Thus, the statement follows from Theorem 4.6.

2. The number of elements in the basis $\mathscr{B}_{\mathscr{Y}} \otimes \mathscr{B}_{\mathscr{X}}^*$ is mn; hence (8.108) follows.
3. Based on the identification (8.105), we have $T = \sum_{i=1}^m \sum_{j=1}^n t_{ij} f_i \otimes e_j^*$. Comparing this with (4.15), (8.109) follows. ∎

The representation $T = \sum_{i=1}^m \sum_{j=1}^n t_{ij} e_i \otimes f_j^*$ of $T \in \mathrm{Hom}_{\mathbb{F}}(\mathscr{X}, \mathscr{Y})$ can generally be simplified. If $\mathrm{rank}\, T = \dim \mathrm{Im}\, T = k$, then there exists a minimal-length representation

$$T = \sum_{i=1}^k \psi_i \otimes \phi_i, \tag{8.111}$$

where $\{\phi_i\}_{i=1}^k$ is a basis of $(\mathrm{Ker}\, T)^\perp \subset \mathscr{X}^*$ and $\{\psi_i\}_{i=1}^k$ is a basis for $\mathrm{Im}\, T \subset \mathscr{Y}$.

The following proposition describes the underlying property of tensor products.

Proposition 8.64. *Let* $\gamma : \mathscr{Y} \times \mathscr{X}^* \longrightarrow \mathscr{Z}$ *be a bilinear map. Then there exists a unique linear map* $\gamma_* : \mathscr{Y} \otimes_{\mathbb{F}} \mathscr{X}^* \longrightarrow \mathscr{Z}$ *for which the following diagram commutes:*

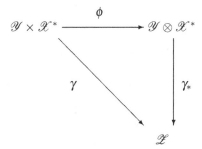

Proof. Given the bilinear map $\gamma \in \mathrm{Hom}_{\mathbb{F}}(\mathscr{Y} \times \mathscr{X}^*, \mathscr{Z})$, we define $\gamma_* \in \mathrm{Hom}_{\mathbb{F}}(\mathscr{Y} \otimes \mathscr{X}^*, \mathscr{Z})$ by

$$\gamma_*(y \otimes x^*) = \gamma(y, x^*). \tag{8.112}$$

It is easily checked that γ_* is well defined and linear and that $\gamma = \gamma_* \circ \phi$ holds. Since any element of $\mathscr{Y} \otimes \mathscr{X}^*$ is a finite sum $\sum_{i=1}^r y_i \otimes x_i^*$, the result follows. ∎

One of the fundamental properties of tensor products is the connection between maps defined on tensor products, bilinear forms (hence also quadratic forms), and module homomorphisms. In the special case of $\mathbb{F}$-vector spaces $\mathscr{X}$ and $\mathscr{Y}$, we have the following result.

Proposition 8.65. *1. Let $\mathscr{X}, \mathscr{Y}$ be finite-dimensional $\mathbb{F}$-vector spaces. Then we have the following isomorphisms:*

$$(\mathscr{X} \otimes_{\mathbb{F}} \mathscr{Y})^* \simeq \mathrm{Bilin}\,(\mathscr{X}, \mathscr{Y}; \mathbb{F}) \simeq \mathrm{Hom}_{\mathbb{F}}(\mathscr{X}, \mathscr{Y}^*). \qquad (8.113)$$

2. Let $\mathscr{X}, \mathscr{Y}$ be finite-dimensional $\mathbb{F}$-vector spaces. Then we have the following isomorphism

$$(\mathscr{X} \otimes_{\mathbb{F}} \mathscr{Y})^* \simeq \mathscr{X}^* \otimes_{\mathbb{F}} \mathscr{Y}^* \simeq \mathscr{Y}^* \otimes_{\mathbb{F}} \mathscr{X}^*. \qquad (8.114)$$

Proof. 1. We give only a sketch of the proof. Let $\phi(x,y)$ be a bilinear form on $\mathscr{X} \times \mathscr{Y}$. For fixed $x \in \mathscr{X}$, this defines a linear functional $\eta_x \in \mathscr{Y}^*$ by

$$\eta_x(y) = \phi(x,y). \qquad (8.115)$$

Clearly, η_x depends linearly on x; hence there exists a map $Z : \mathscr{X} \longrightarrow \mathscr{Y}^*$ for which $Zx = \eta_x$. So $\phi(x,y) = (Zx)(y)$.

Conversely, assume $Z \in \mathrm{Hom}_{\mathbb{F}}(\mathscr{X}, \mathscr{Y}^*)$. Then for every $x \in \mathscr{X}$, we have $Zx \in \mathscr{Y}^*$. Defining $\phi(x,y) = (Zx)(y)$, we get a bilinear form defined on $\mathscr{X} \times \mathscr{Y}$. The homomorphisms $\phi \mapsto Z$ and $Z \mapsto \phi$ are inverses to each other; hence the isomorphism $\mathrm{Bilin}\,(\mathscr{X}, \mathscr{Y}; \mathbb{F}) \simeq \mathrm{Hom}_{\mathbb{F}}(\mathscr{X}, \mathscr{Y}^*)$ follows.

Given a bilinear form $\gamma : \mathscr{X} \times \mathscr{Y} \longrightarrow \mathbb{F}$, by the definition of tensor products and with $\gamma : \mathscr{X} \times \mathscr{Y} \longrightarrow \mathscr{X} \otimes_{\mathbb{F}} \mathscr{Y}$ the canonical map, there exists a unique $\gamma_* \in (\mathscr{X} \otimes_{\mathbb{F}} \mathscr{Y})^*$ such that $\gamma = \gamma_* \phi$. The uniqueness implies that the map $\gamma \mapsto \gamma_*$ is injective. It is also surjective, since any $\gamma_* \in (\mathscr{X} \otimes_{\mathbb{F}} \mathscr{Y})^*$ induces a bilinear form γ on $\mathscr{X} \times \mathscr{Y}$ by letting $\gamma = \gamma_* \phi$. Thus we have also the isomorphism $(\mathscr{X} \otimes_{\mathbb{F}} \mathscr{Y})^* \simeq \mathrm{Bilin}\,(\mathscr{X}, \mathscr{Y}; \mathbb{F})$.

2. Assume $\xi_i \in \mathscr{X}^*, \eta_i \in \mathscr{Y}^*$. Then for every $x \in \mathscr{X}$, we define

$$\sum_{i=1}^{s} (\eta_i \otimes \xi_i) x = \sum_{i=1}^{s} \xi_i(x) \eta_i, \qquad (8.116)$$

which defines a map $\mathscr{Y}^* \otimes \mathscr{X}^* \longrightarrow \mathrm{Hom}_{\mathbb{F}}(\mathscr{X}, \mathscr{Y}^*)$.

Conversely, if $\mathscr{X}$ is finite-dimensional, the image of any $Z \in \mathrm{Hom}(\mathscr{X}, \mathscr{Y}^*)$ is finite-dimensional, hence has a basis $\{\eta_i\}_{i=1}^{q}$, with $\eta_i \in \mathscr{Y}^*$. Therefore we have, for suitable $\xi_i \in \mathscr{X}^*$, the representation

$$Zx = \sum_{i=1}^{q} \xi_i(x) \eta_i = \sum_{i=1}^{q} (\eta_i \otimes \xi_i) x. \qquad (8.117)$$

Thus $Z \mapsto \sum_{i=1}^{q} \xi_i \otimes \eta_i$ defines an isomorphism from $\mathrm{Hom}(\mathscr{X}, \mathscr{Y}^*)$ to $\mathscr{Y}^* \otimes \mathscr{X}^*$, and thus, by the isomorphism (8.113), also an isomorphism with $\mathscr{X}^* \otimes_{\mathbb{F}} \mathscr{Y}^*$. ∎

8.4.3 Tensor Products of Modules

Our definition of tensor products of vector spaces was an ad hoc one. To get into line with standard algebraic terminology, we use the characterization of tensor products, given in Proposition 8.64, as the basis for the following definition.

Definition 8.66. Let R be a commutative ring and $\mathcal{X}, \mathcal{Y}$ R-modules. An R-module $\mathcal{X} \otimes_R \mathcal{W}$ is called a **tensor product** over R if there exists a bilinear map $\phi : \mathcal{X} \times \mathcal{Y} \longrightarrow \mathcal{X} \otimes_R \mathcal{W}$ such that for any bilinear map $\gamma : \mathcal{X} \times \mathcal{Y} \longrightarrow \mathcal{Z}$, there exists a unique R-linear map $\gamma_* : \mathcal{X} \otimes_R \mathcal{W} \longrightarrow \mathcal{Z}$ for which

$$\gamma = \gamma_* \phi. \tag{8.118}$$

The following theorem summarizes the main properties of tensor products. We will not give a proof and refer the reader to Hungerford (1974) and Lang (1965) for the details.

Theorem 8.67. *Let R be a commutative ring and $\mathcal{V}, \mathcal{W}$ R-modules. Then*

1. *For any R-modules $\mathcal{V}, \mathcal{W}$, a tensor product $\mathcal{V} \otimes_R \mathcal{W}$ exists and is unique up to isomorphism.*
2. *We have the following isomorphisms:*

$$(\oplus_{i=1}^k M_i) \otimes_R N \simeq \oplus_{i=1}^k (M_i \otimes_R N),$$

$$M \otimes_R (\oplus_{j=1}^l N_j) \simeq \oplus_{j=1}^l (M \otimes_R N_j),$$

$$M \otimes_R (N \otimes_R P) \simeq (M \otimes_R N) \otimes_R P,$$

$$M \otimes_R N \simeq N \otimes_R M. \tag{8.119}$$

3. *If $S \subset R$ is any subring and M, N are R-modules, then we have a well-defined surjective S-linear map $M \otimes_S N \longrightarrow M \otimes_R N$ defined by $m \otimes_S n \mapsto m \otimes_R n$.*
4. *Let M_1, M_2 be R-modules, with R a commutative ring. Let $N_i \subset M_i$ be submodules. The quotient spaces M_i/N_i have a natural R-module structure. Let N be the submodule generated in $M_1 \otimes_R M_2$ by $N_1 \otimes_R M_2$ and $M_1 \otimes_R N_2$. Then we have the isomorphism*

$$M_1/N_1 \otimes_R M_2/N_2 \simeq (M_1 \otimes_R M_2)/N. \tag{8.120}$$

5. *Given R-modules M, N, L, we denote by $\mathrm{Bilin}_R(M, N; L)$ the set of all R-bilinear maps $\phi : M \times N \longrightarrow L$. We have the following isomorphisms:*

$$\mathrm{Hom}_R(M \otimes_R N, L) \simeq \mathrm{Bilin}_R(M, N; L) \simeq \mathrm{Hom}_R(M, \mathrm{Hom}_R(N, L)). \tag{8.121}$$

8.4.4   Kronecker Product Models

We define the $\mathbb{F}$-**Kronecker product** of two scalar polynomials $d_1(z), d_2(z) \in \mathbb{F}[z]$ as the map $d_1 \otimes_{\mathbb{F}} d_2 : \mathbb{F}[z, w] \longrightarrow \mathbb{F}[z, w]$ given by

$$(d_1 \otimes_{\mathbb{F}} d_2)q(z, w) = d_1(z)q(z, w)d_2(w).$$

This map induces a projection $\pi_{d_1 \otimes_{\mathbb{F}} d_2}$ in $\mathbb{F}[z, w]$ defined by

$$\pi_{d_1 \otimes_{\mathbb{F}} d_2} q(z, w) = (d_1 \otimes_{\mathbb{F}} d_2)(\pi_-^z \otimes \pi_-^w)(d_1 \otimes_{\mathbb{F}} d_2)^{-1} q(z, w). \tag{8.122}$$

Thus, if $q(z, w)$ is given as $q(z, w) = \sum_{i=1}^{r} a_i(z)b_i(w)$, then

$$\begin{aligned} \pi_{d_1 \otimes_{\mathbb{F}} d_2} q(z, w) &= \sum_{i=1}^{r} (\pi_{d_1} a_i)(z)(\pi_{d_2} b_i)(w) \\ &= \sum_{i=1}^{r} d_1(z)\pi_-(d_1^{-1} a_i)(z)\pi_-(d_2^{-1} b_i)(w)d_2(w). \end{aligned} \tag{8.123}$$

We obtain the $\mathbb{F}$-**Kronecker product model** as $X_{d_1 \otimes_{\mathbb{F}} d_2} = X_{d_1(z)d_2(w)} = \operatorname{Im} \pi_{d_1 \otimes_{\mathbb{F}} d_2}$. From now on, we will mainly use the more suggestive notation $X_{d_1(z)d_2(w)}$ for the $\mathbb{F}$-Kronecker product model. By inspection, one verifies that

$$X_{d_1(z)d_2(w)} = \left\{ \sum_{i=0}^{\deg d_1 - 1} \sum_{j=0}^{\deg d_2 - 1} f_{ij} z^i w^j \right\}. \tag{8.124}$$

From this description, it follows that we have the dimension formula

$$\dim_{\mathbb{F}}(X_{d_1(z)d_2(w)}) = \deg d_1 \cdot \deg d_2. \tag{8.125}$$

Similarly, we define the $\mathbb{F}[z]$-**Kronecker product** $d_1 \otimes_{\mathbb{F}[z]} d_2$ of two scalar polynomials $d_1(z), d_2(z)$ as the map $d_1 \otimes_{\mathbb{F}[z]} d_2 : \mathbb{F}[z] \longrightarrow \mathbb{F}[z]$ given by $d_1 \otimes_{\mathbb{F}[z]} d_2 q(z) = d_1(z)q(z)d_2(z)$. This defines the projection map $\pi_{d_1 \otimes d_2} : \mathbb{F}[z] \longrightarrow \mathbb{F}[z]$ given by

$$\pi_{d_1 \otimes_{\mathbb{F}[z]} d_2} f = \pi_{(d_1 d_2)} f. \tag{8.126}$$

In turn, this allows us to introduce the $\mathbb{F}[z]$-**Kronecker product model** as

$$X_{d_1 \otimes_{\mathbb{F}[z]} d_2} \simeq \mathbb{F}[z]/(d_1 d_2)\mathbb{F}[z] \simeq X_{d_1 d_2}. \tag{8.127}$$

Consequently, for the $\mathbb{F}[z]$-Kronecker product model, we have the dimension formula

$$\dim_{\mathbb{F}} X_{d_1 \otimes_{\mathbb{F}[z]} d_2} = \deg(d_1 d_2) = \deg d_1 + \deg d_2 = \dim_{\mathbb{F}} X_{d_1} + \dim_{\mathbb{F}} X_{d_2}. \tag{8.128}$$

Finally, from duality of polynomial and rational models, we achieve a representation of the dual to the Kronecker product model as

$$(X_{d_1 \otimes_{\mathbb{F}[z]} d_2})^* = (X_{d_1 d_2})^* = X^{d_1 d_2}. \tag{8.129}$$

8.4.5 Tensor Products over a Field

The tensor products were defined, in Definition 8.66, in a formal way. Our conviction is that a concrete representation for the tensor product is always preferable. Thus, we have the identifications, i.e., up to isomorphism, for tensor products of polynomial spaces, i.e.,

$$\mathbb{F}[z] \otimes_{\mathbb{F}} \mathbb{F}[z] \simeq \mathbb{F}[z, w], \tag{8.130}$$

as well as

$$\mathbb{F}[z] \otimes_{\mathbb{F}[z]} \mathbb{F}[z] = \mathbb{F}[z]. \tag{8.131}$$

For our purposes, this is less interesting than the identification of the tensor product of two polynomial models. This will be achieved by the use of Theorem 8.67, and in particular, equation (8.120). Central to this is the isomorphism (5.6), namely $X_q \simeq \mathbb{F}[z]/q\mathbb{F}[z]$. The availability of Kronecker product polynomial models allows us to give concrete representations for the tensor products. We show that the tensor product $X_{d_1} \otimes_{\mathbb{F}} X_{d_2}$ of two polynomial models over the field $\mathbb{F}$ can be interpreted as a polynomial model for the two-variable polynomial $d_1(z)d_2(w)$. Using (8.120) and the isomorphism (5.6), we have

$$X_{d_1} \otimes_{\mathbb{F}} X_{d_2} \simeq (\mathbb{F}[z]/d_1\mathbb{F}[z]) \otimes_{\mathbb{F}} (\mathbb{F}[z]/d_2\mathbb{F}[z])$$
$$\simeq \mathbb{F}[z, w]/(d_1(z)\mathbb{F}[z, w] + d_2(w)\mathbb{F}[z, w]).$$

The submodule $d_1(z)\mathbb{F}[z, w] + d_2(w)\mathbb{F}[z, w]$ can be seen as the kernel of the projection operator $\pi_{d_1(z)} \otimes \pi_{d_2(w)}$ acting in $\mathbb{F}[z, w]$ by

$$(\pi_{d_1(z)} \otimes \pi_{d_2(w)})(f(z)g(w)) = (\pi_{d_1(z)}f(z))(\pi_{d_2(w)}g(w)).$$

From this, the above representation of $X_{d_1} \otimes_{\mathbb{F}} X_{d_2}$ easily follows. Defining $X_{d_1(z)d_2(w)} = \mathrm{Im}\,(\pi_{d_1(z)} \otimes \pi_{d_2(w)})$ implies the isomorphism

$$X_{d_1} \otimes_{\mathbb{F}} X_{d_2} \simeq X_{d_1(z)d_2(w)}, \tag{8.132}$$

i.e., the $\mathbb{F}$-tensor product of polynomial models is a polynomial model in two variables.

Proposition 8.68. *Given nontrivial polynomials $d_1(z), d_2(z) \in \mathbb{F}[z]$, we have the isomorphism*

$$X_{d_1} \otimes_{\mathbb{F}} X_{d_2} \simeq X_{d_1 \otimes_{\mathbb{F}} d_2} = X_{d_1(z)d_2(w)}. \tag{8.133}$$

In particular, the $\mathbb{F}$*-tensor product of two polynomial models in one variable is the two-variable Kronecker product polynomial model, and the map* $\gamma_* : X_{d_1} \otimes_{\mathbb{F}} X_{d_2} \longrightarrow X_{d_1(z)d_2(w)}$ *induced by*

$$\gamma_*(f_1 \otimes f_2) = f_1(z)f_2(w) \tag{8.134}$$

is an $\mathbb{F}$*-isomorphism.*

Proof. Let $\phi : X_{d_1} \times X_{d_2} \longrightarrow X_{d_1} \otimes_{\mathbb{F}} X_{d_2}$ be the canonical map. Consider the $\mathbb{F}$-bilinear map $\gamma : X_{d_1} \times X_{d_2} \longrightarrow X_{d_1(z)d_2(w)}$ defined by $\gamma(f_1, f_2) = f_1(z)f_2(w)$. Thus γ_* exists and is $\mathbb{F}$-linear, leading to the following commutative diagram:

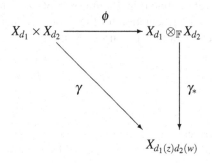

The description (8.124) shows that γ_* is surjective. From the dimension formula (8.125) we conclude that $X_{d_1} \otimes_{\mathbb{F}} X_{d_2}$ and $X_{d_1(z)d_2(w)}$ have the same $\mathbb{F}$-dimension, and the result follows. ∎

Being the tensor product of two vector spaces over the field $\mathbb{F}$, the tensor product model $X_{d_1} \otimes_{\mathbb{F}} X_{d_2}$ is clearly an $\mathbb{F}$-vector space. We will see later that it carries also a natural $\mathbb{F}[z, w]$-module structure.

We now connect the space of $\mathbb{F}$-linear maps between polynomial models X_{d_1}, X_{d_2}, namely $\mathrm{Hom}_{\mathbb{F}}(X_{d_1}, X_{d_2})$, with the tensor product $X_{d_2} \otimes_{\mathbb{F}} X_{d_1}$. Using the residue form on X_{d_1}, i.e., $\langle f, g \rangle = (d_1^{-1}fg)_{-1}$, where h_{-1} denotes the coefficient of z^{-1} in the Laurent series expansion of $h(z)$, we can get the concrete form of the isomorphism as follows.

Theorem 8.69. *We define a map* $\Psi : X_{d_2(z)d_1(w)} \longrightarrow \mathrm{Hom}_{\mathbb{F}}(X_{d_1}, X_{d_2})$ *as follows. For* $q(z, w) \in X_{d_2} \otimes_{\mathbb{F}} X_{d_1}$*, the map is* $q(z, w) \mapsto \Psi(q)$*, where* $\Psi(q) : X_{d_1} \longrightarrow X_{d_2}$ *is defined, for* $f \in X_{d_1}$*, by*

$$\Psi(q)f = \langle f, q(z, \cdot) \rangle = (q(z, \cdot)d_1(\cdot)^{-1}f(\cdot))_{-1}. \tag{8.135}$$

The map $q(z, w) \mapsto \Psi(q)$ *defines a vector space isomorphism, i.e., we have the following isomorphism:*

$$X_{d_2} \otimes_{\mathbb{F}} X_{d_1} \simeq \mathrm{Hom}_{\mathbb{F}}(X_{d_1}, X_{d_2}). \tag{8.136}$$

Proof. Let $q_1(z), \ldots, q_s(z)$ denote a basis of $X_{d_1(z)d_2(w)}$. Then it is easily seen that $\Psi(q_1), \ldots, \Psi(q_n)$ are linearly independent in $\mathrm{Hom}_{\mathbb{F}}(X_{d_1}, X_{d_2})$. The result follows because both spaces have the same dimension. ∎

Using the isomorphism

$$X_{d_2} \otimes_{\mathbb{F}} X_{d_1} \simeq X_{d_2(z)d_1(w)}, \tag{8.137}$$

we obtain the explicit isomorphism $\Psi : X_{d_2} \otimes_{\mathbb{F}} X_{d_1} \longrightarrow \mathrm{Hom}_{\mathbb{F}}(X_{d_1}, X_{d_2})$ given, for $g \in X_{d_1}$, by

$$\Psi(f_2 \otimes f_1)g = [d_1^{-1}g, f_1]f_2 = \langle g, f_1 \rangle f_2, \tag{8.138}$$

where the pairing $\langle g, f \rangle$ is defined by (5.64).

We consider next the $\mathbb{F}$-vector space dual of the tensor product.

Proposition 8.70. *Given nontrivial polynomials $d_1(z), d_2(z) \in \mathbb{F}[z]$, we have the following identification of the vector space dual of $X_{d_1} \otimes_{\mathbb{F}} X_{d_2}$:*

$$(X_{d_1} \otimes_{\mathbb{F}} X_{d_2})^* \simeq X^{d_1} \otimes_{\mathbb{F}} X^{d_2} \simeq X^{d_1(z)} \cap X^{d_2(w)}. \tag{8.139}$$

Proof. We compute, using vector space annihilators,

$$
\begin{aligned}
(X_{d_1} \otimes_{\mathbb{F}} X_{d_2})^* &\simeq (\mathbb{F}[z]/d_1\mathbb{F}[z] \otimes_{\mathbb{F}} \mathbb{F}[z]/d_2\mathbb{F}[z])^* \\
&\simeq (d_1(z)\mathbb{F}[z,w] + d_2(w)\mathbb{F}[z,w])^{\perp} \\
&\simeq (d_1(z)\mathbb{F}[z,w])^{\perp} \cap (d_2(w)\mathbb{F}[z,w])^{\perp} \\
&= X^{d_1 \otimes 1} \cap X^{1 \otimes d_2} \\
&= \left\{ h(z,w) \in z^{-1}\mathbb{F}[[z^{-1}, w^{-1}]]w^{-1} \,\middle|\, h(z,w) = \frac{p(z,w)}{d_1(z)d_2(w)} \right\}.
\end{aligned}
$$

∎

8.4.6 Tensor Products over the Ring of Polynomials

From Proposition 8.68, we have seen that the $\mathbb{F}$-tensor product of polynomial models, namely $X_{d_1} \otimes_{\mathbb{F}} X_{d_2}$, takes us out of the realm of single-variable polynomial spaces. The situation changes when the underlying ring is $\mathbb{F}[z]$. Since polynomial models X_{d_1}, X_{d_2} are $\mathbb{F}[z]$-modules, their tensor product, namely $X_{d_1} \otimes_{\mathbb{F}[z]} X_{d_2}$, is an $\mathbb{F}[z]$-module too. The tensor product is, by Definition 8.66, an abstract object. Again, for clarity, as well as computational purposes, we would like to have a concrete representation for it, and this is done next. In analogy with (8.133), we have the $\mathbb{F}[z]$-module isomorphism

$$X_{d_1} \otimes_{\mathbb{F}[z]} X_{d_2} \simeq \mathbb{F}[z]/d_1\mathbb{F}[z] \otimes_{\mathbb{F}[z]} \mathbb{F}[z]/d_2\mathbb{F}[z]$$
$$\simeq \mathbb{F}[z]/(d_1(z)\mathbb{F}[z] + d_2(z)\mathbb{F}[z]) \simeq \mathbb{F}[z]/((d_1 \wedge d_2)\mathbb{F}[z]) \qquad (8.140)$$
$$\simeq X_{d_1 \wedge d_2}.$$

Proposition 8.71. *Given nontrivial polynomials $d_1(z), d_2(z) \in \mathbb{F}[z]$, let $\phi : X_{d_1} \times X_{d_2} \longrightarrow X_{d_1} \otimes_{\mathbb{F}[z]} X_{d_2}$ be the canonical map and let $\gamma : X_{d_1} \times X_{d_2} \longrightarrow X_{d_1 \wedge d_2}$ be defined by*

$$\gamma(f_1, f_2) = \pi_{d_1 \wedge d_2}(f_1 f_2). \qquad (8.141)$$

Then the map $\gamma_* : X_{d_1} \otimes_{\mathbb{F}[z]} X_{d_2} \longrightarrow X_{d_1 \wedge d_2}$, defined by

$$\gamma_*(f_1 \otimes f_2) = \pi_{d_1 \wedge d_2}(f_1 f_2), \qquad (8.142)$$

is an $\mathbb{F}[z]$-module isomorphism that implies the isomorphism

$$X_{d_1} \otimes_{\mathbb{F}[z]} X_{d_2} \simeq X_{d_1 \wedge d_2}. \qquad (8.143)$$

The commutativity of the following diagram holds:

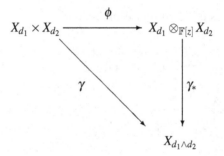

Proof. We check first that γ_* is well defined on the tensor product space. This follows immediately from the identity

$$\gamma_*(p \cdot f_1 \otimes f_2) = \pi_{d_1 \wedge d_2}(p f_1 f_2) = \pi_{d_1 \wedge d_2} p \pi_{d_1 \wedge d_2}(f_1 f_2) = p \cdot \gamma_*(f_1 \otimes f_2),$$

which shows that γ_* is a well defined $\mathbb{F}[z]$-homomorphism.

Clearly, the map γ is surjective, since for any $f(z) \in X_{d_1 \wedge d_2}$, we have $\gamma(f, 1) = f$. This implies also the surjectivity of γ_*. Next, we note that

$$\dim_{\mathbb{F}}(X_{d_1} \otimes_{\mathbb{F}[z]} X_{d_2}) = \deg(d_1 \wedge d_2) = \dim_{\mathbb{F}} X_{d_1 \wedge d_2}, \qquad (8.144)$$

which shows that γ_* is bijective. This proves the isomorphism (8.143). ∎

Clearly, we have the dimension formula

$$\dim_{\mathbb{F}} X_{d_1} \otimes_{\mathbb{F}[z]} X_{d_2} = \dim_{\mathbb{F}} X_{(d_1 \wedge d_2)} = \deg(d_1 \wedge d_2) \leq \max(\deg d_1, \deg d_2)$$
$$\leq \deg d_1 \cdot \deg d_2 = \dim_{\mathbb{F}}(X_{d_1} \otimes_{\mathbb{F}} X_{d_2}). \qquad (8.145)$$

This computation shows that in the case of tensor products over rings, a lot of dimension collapsing can occur. In fact, in the special case that $d_1(z), d_2(z)$ are coprime, then $X_{d_1} \otimes_{\mathbb{F}[z]} X_{d_2} = 0$.

The following theorem is the analogue of the isomorphism (8.136) for the case of $\mathbb{F}[z]$-tensor products. We give two versions of it, using both $X_{d_1} \otimes_{\mathbb{F}[z]} X_{d_2}$ and $X_{d_1 \wedge d_2}$.

Theorem 8.72. *Given nontrivial polynomials $d_1(z), d_2(z) \in \mathbb{F}[z]$, then*

1. We have the $\mathbb{F}[z]$-isomorphism of $\mathbb{F}[z]$-modules

$$X_{d_1} \otimes_{\mathbb{F}[z]} X_{d_2} \simeq \mathrm{Hom}_{\mathbb{F}[z]}(X_{d_1}, X_{d_2}). \qquad (8.146)$$

The isomorphism is given by the map $\psi' : X_{d_1} \otimes_{\mathbb{F}[z]} X_{d_2} \longrightarrow \mathrm{Hom}_{\mathbb{F}[z]}(X_{d_1}, X_{d_2})$ defined by $f_1 \otimes f_2 \mapsto \psi'_{f_1 \otimes f_2}$:

$$\psi'_{f_1 \otimes f_2}(g) = d_2 \pi_-((d_1 \wedge d_2)^{-1} f_1 f_2 g) = \left(\frac{d_2}{d_1 \wedge d_2} \right) \pi_{d_1 \wedge d_2}(f_1 f_2 g), \quad g \in X_{d_1}.$$
$$(8.147)$$

2. We have the $\mathbb{F}[z]$-isomorphism of $\mathbb{F}[z]$-modules

$$X_{d_1 \wedge d_2} \simeq \mathrm{Hom}_{\mathbb{F}[z]}(X_{d_1}, X_{d_2}). \qquad (8.148)$$

The isomorphism is given by the map $\psi : X_{d_1 \wedge d_2} \longrightarrow \mathrm{Hom}_{\mathbb{F}[z]}(X_{d_1}, X_{d_2})$ defined by $n \mapsto \psi_n$ defined, for $n \in X_{d_1 \wedge d_2}$, by

$$\psi_n(g) = \pi_{d_2}(n_2 g), \quad g \in X_{d_1}. \qquad (8.149)$$

Proof. 1. We use the isomorphism (8.143) to identify $X_{d_1} \otimes_{\mathbb{F}[z]} X_{d_2}$ with $X_{d_1 \wedge d_2}$. Choosing any element $f(z) \in X_{d_1 \wedge d_2}$, by construction ψ'_f is $\mathbb{F}[z]$-linear. To show injectivity of ψ', assume $\psi'_f = 0$, i.e., for all $g(z) \in X_{d_1}$ we have $0 = \psi'_f g = \left(\frac{d_2}{d_1 \wedge d_2} \right) \pi_{d_1 \wedge d_2}(fg)$. This holds in particular for $g(z) = 1$. Thus $\pi_{d_1 \wedge d_2} f = 0$ and necessarily $f(z) \in (d_1 \wedge d_2)\mathbb{F}[z]$. However, we assumed $f(z) \in X_{d_1 \wedge d_2}$. Thus the direct sum decomposition $\mathbb{F}[z] = X_{d_1 \wedge d_2} \oplus (d_1 \wedge d_2)\mathbb{F}[z]$ implies $f(z) = 0$, and the injectivity of ψ' follows.

Using the $\mathbb{F}$-linear isomorphism $(X_{d_1} \otimes_{\mathbb{F}[z]} X^{d_2})^* \simeq \mathrm{Hom}_{\mathbb{F}[z]}(X_{d_1}, X_{d_2})$, it follows that $\dim_{\mathbb{F}} \mathrm{Hom}_{\mathbb{F}[z]}(X_{d_1}, X_{d_2}) = \dim_{\mathbb{F}} X_{d_1} \otimes_{\mathbb{F}[z]} X_{d_2} = \dim_{\mathbb{F}} X_{d_1} \otimes_{\mathbb{F}[z]} X_{d_2}$. Hence injectivity implies surjectivity, and we are done.

2. Note that $n(z) \in X_{d_1 \wedge d_2}$ if and only if $h(z) = \frac{n(z)}{d_1(z) \wedge d_2(z)} \in X^{d_1 \wedge d_2} = X^{d_1} \cap X^{d_2}$.

In turn, $h(z)$ is rational and strictly proper and has a unique representation of the form $h(z) = \frac{n(z)}{d_1(z) \wedge d_2(z)}$. Writing $d_i(z) = (d_1(z) \wedge d_2(z))d_i'(z)$, we have, with $n_i(z) = n(z)d_i'(z)$,

$$h(z) = \frac{n(z)}{d_1(z) \wedge d_2(z)} = \frac{n_1(z)}{d_1(z)} = \frac{n_2(z)}{d_2(z)}, \tag{8.150}$$

which also implies the intertwining relation

$$n_2(z)d_1(z) = d_2(z)n_1(z). \tag{8.151}$$

We compute now, with $n(z) = \pi_{d_1(z) \wedge d_2(z)}(f_1(z)f_2(z))$,

$$\begin{aligned}
\psi'_{f_1 \otimes f_2} g &= d_2 \pi_-(d_1 \wedge d_2)^{-1} f_1 f_2 g = d_2 \pi_-(d_1 \wedge d_2)^{-1}(\pi_{d_1 \wedge d_2}(f_1 f_2))g \\
&= d_2 \pi_-(d_1 \wedge d_2)^{-1} n g = d_2 \pi_- d_2^{-1} n_2 g = \pi_{d_2} n_2 g \\
&= \psi_n(g).
\end{aligned}$$

∎

The most important aspect of Theorem 8.72 is the establishment of a concrete connection between the space of maps Z intertwining the shifts S_{d_1} and S_{d_2}, i.e., satisfying $ZS_{d_1} = S_{d_2}Z$ on the one hand, and the $\mathbb{F}[z]$-tensor product of the polynomial models X_{d_1} and X_{d_2} on the other. Note that (8.149) is the more familiar representation of an intertwining map; see Theorem 5.17.

Next, we turn to the identification of the dual to the tensor product.

Proposition 8.73. *Given nontrivial polynomials* $d_1(z), d_2(z) \in \mathbb{F}[z]$, *then we have the series of* $\mathbb{F}[z]$-*isomorphisms*

$$\begin{aligned}
(X_{d_1} \otimes_{\mathbb{F}[z]} X_{d_2})^* &\simeq (X_{d_1 \wedge d_2})^* \simeq X^{d_1 \wedge d_2} \simeq X^{d_1} \cap X^{d_2} \\
&\simeq X_{d_1} \otimes_{\mathbb{F}[z]} X_{d_2} \simeq \mathrm{Hom}_{\mathbb{F}[z]}(X_{d_1}, X_{d_2}).
\end{aligned} \tag{8.152}$$

Proof. We use the duality results on polynomial models, as well as the isomorphisms (8.143) and (8.148). ∎

8.4.7   The Polynomial Sylvester Equation

Note that the polynomial models X_{d_1} and X_{d_2} not only have a vector space structure but are actually, by (5.8), $\mathbb{F}[z]$-modules. This implies that $X_{d_2(z)d_1(w)}$, and hence also, using the isomorphism (8.132), $X_{d_2} \otimes_{\mathbb{F}} X_{d_1}$, have natural $\mathbb{F}[z, w]$-module structures. This is defined by

$$p(z,w) \cdot q(z,w) = \pi_{d_2(z) \otimes d_1(w)}(p(z,w)q(z,w)), \qquad q(z,w) \in X_{d_2(z) \otimes d_1(w)}, \tag{8.153}$$

where $p(z,w) \in \mathbb{F}[z,w]$.

Similarly, we define an $\mathbb{F}[z,w]$-module structure on the tensored rational model $X^{d_2(z)\otimes d_1(w)}$ by letting, for $p(z,w) = \sum_{i=1}^{k} \sum_{j=1}^{l} p_{ij} z^{i-1} w^{j-1} \in \mathbb{F}[z,w]$ and $h(z,w) \in X^{d_2(z)\otimes d_1(w)}$,

$$p(z,w) \cdot h(z,w) = \pi^{d_2(z)\otimes d_1(w)} [\sum_{i=1}^{k} \sum_{j=1}^{l} p_{ij} z^{i-1} h(z,w) w^{j-1}]. \qquad (8.154)$$

Proposition 8.74. *Let $d_2(z) \in \mathbb{F}[z]$ and $d_1(w) \in \mathbb{F}[w]$ be nonsingular polynomials. Then*

1. *The $\mathbb{F}[z,w]$-module structure on $X^{d_2(z)\otimes d_1(w)}$ defined by (8.154) can be rewritten as*

$$p(z,w) \cdot h(z,w) = (\pi^z_- \otimes \pi^w_-) \sum_{i=1}^{k} \sum_{j=1}^{l} p_{ij} z^{i-1} h(z,w) w^{j-1}. \qquad (8.155)$$

2. *With the $\mathbb{F}[z,w]$-module structure on $X_{d_2(z)\otimes d_1(w)}$ and $X^{d_2(z)\otimes d_1(w)}$, given by (8.155) and (8.154) respectively, the multiplication map*

$$d_2(z) \otimes d_1(w) : X^{d_2(z)\otimes d_1(w)} \longrightarrow X_{d_2(z)\otimes_{\mathbb{F}} d_1(w)}$$

is an $\mathbb{F}[z,w]$-module isomorphism, i.e., we have

$$X_{d_2(z)\otimes d_1(w)} \simeq X^{d_2(z)\otimes d_1(w)}. \qquad (8.156)$$

Proof. 1. Follows from (8.154).
2. Follows, using Proposition 8.74, from the fact that $h(z,w) \in X^{d_2(z)\otimes d_1(w)}$ if and only if $h(z,w) = \pi^{d_2(z)\otimes d_1(w)} h(z,w)$, equivalently, if and only if

$$\pi_{d_2(z)\otimes d_1(w)} d_2(z) h(z,w) d_1(w) = d_2(z) h(z,w) d_1(w),$$

i.e., $d_2(z) h(z,w) d_1(w) \in X_{d_2(z)\otimes d_1(w)}$. ∎

Special cases that are of interest are the single-variable shift operators $S_z, S_w : X_{d_2(z)\otimes d_1(w)} \longrightarrow X_{d_2(z)\otimes d_1(w)}$, defined by

$$S_z q(z,w) = \pi_{d_2(z)\otimes d_1(w)} z q(z,w) = \pi_{d_2(z)} z q(z,w),$$
$$S_w q(z,w) = \pi_{d_2(z)\otimes d_1(w)} q(z,w) w = \pi_{I\otimes d_1(w)} q(z,w) w. \qquad (8.157)$$

If we specialize definition (8.153) to the polynomial $p(z,w) = z - w$, we get, with $q(z,w) \in X_{d_1(z)\otimes_{\mathbb{F}} d_2(w)}$,

$$\mathscr{S} q(z,w) = (z - w) \cdot q(z,w) = \pi_{d_2(z)\otimes_{\mathbb{F}} d_1(w)} (z q(z,w) - q(z,w) w). \qquad (8.158)$$

In fact, with $A_2 \in \mathbb{F}^{p\times p}$ and $A_1 \in \mathbb{F}^{m\times m}$, if $D_2(z) = zI - A_2$ and $D_1(w) = wI - A_1$, then $Q(z,w) \in X_{D_2(z)\otimes \tilde{D}_1(w)}$ if and only if $Q(z,w) \in \mathbb{F}^{p\times m}$, i.e., $Q(z,w)$

is a constant matrix. In that case we have $X_{D_2(z) \otimes_{\mathbb{F}} \tilde{D}_1(w)} = \mathbb{F}^{p \times m}$ and

$$(z - w) \cdot Q = \pi_{(zI - A_2) \otimes (wI - \tilde{A}_1)} (z - w) Q = A_2 Q - Q A_1, \tag{8.159}$$

which is the standard **Sylvester operator**. The equation $(z - w) \cdot Q = R$ reduces in this case to the **Sylvester equation**

$$A_2 Q - Q A_1 = R. \tag{8.160}$$

We proceed now to a more detailed study of the Sylvester equation in the tensored polynomial model framework. We saw above that the classical Sylvester equation

$$AX - XB = C \tag{8.161}$$

corresponds to the equation

$$\mathscr{S} Q(z, w) = \pi_{(zI - A) \otimes (wI - B)} (z - w) Q = C, \tag{8.162}$$

with $Q, C \in X_{(zI - A) \otimes_{\mathbb{F}} (wI - B)}$ necessarily constant matrices.

Note that every $t(z, w) \in X_{d_2(z) \otimes d_1(w)}$ has a full row rank factorization of the form

$$t(z, w) = R_2(z) \widetilde{R_1}(w), \tag{8.163}$$

with $R_2(z) \in X_{D_2}$ and $R_1(w) \in X_{D_1}$ both of full row rank. The following theorem reduces the analysis of the general Sylvester equation to a polynomial equation of Bezout type. A special case is of course the homogeneous Sylvester equation which has a direct connection to the theory of Bezoutians.

Theorem 8.75. *Let $d_2(z) \in \mathbb{F}[z]$ and $d_1(w) \in \mathbb{F}[w]$ be nonsingular. Defining the Sylvester operator $\mathscr{S} : X_{d_2(z) \otimes d_1(w)} \longrightarrow X_{d_2(z) \otimes d_1(w)}$ by*

$$\mathscr{S} q(z, w) = \pi_{d_2(z) \otimes d_1(w)} (z - w) q(z, w) = R_2(z) \tilde{R}_1(w) \tag{8.164}$$

(8.164), then for $R_2(z) \in X_{d_2 \otimes I}$ and $\tilde{R}_1(w) \in X_{I \otimes d_1}$ we have

1. The Sylvester equation

$$S_{d_2} q - q S_{d_1} = t(z, w) = R_2(z) \tilde{R}_1(w) \tag{8.165}$$

or equivalently

$$\mathscr{S} q = R_2(z) \tilde{R}_1(w), \tag{8.166}$$

is solvable if and only if there exist polynomials $n_2(z) \in X_{d_2(z)}$ and $n_1(z) \in X_{d_1(z)}$ for which

$$d_2(z) n_1(z) - n_2(z) d_1(z) + R_2(z) \tilde{R}_1(z) = 0. \tag{8.167}$$

We will refer to (8.167) as the **polynomial Sylvester equation,** *or* **PSE.** *In that case, the solution is given by*

$$q(z,w) = \frac{d_2(z)n_1(w) - n_2(z)d_1(w) + R_2(z)\tilde{R}_1(w)}{z-w}. \tag{8.168}$$

2. $Q(z,w) \in X_{D_2(z) \otimes \tilde{D}_1(w)}$ *solves the* **homogeneous polynomial Sylvester equation,** *or* **HPSE,** *if and only if there exist polynomials* $n_2(z) \in X_{d_2}$ *and* $n_1(z) \in X_{d_1}$ *that satisfy*

$$d_2(z)n_1(z) - n_2(z)d_1(z) = 0, \tag{8.169}$$

in terms of which

$$q(z,w) = \frac{d_2(z)n_1(w) - n_2(z)d_1(w)}{z-w}, \tag{8.170}$$

i.e., $q(z,w)$ is a Bezoutian.

Proof. 1. Assume there exist polynomials $n_2(z) \in X_{d_2 \otimes I}$ and $n_1(z) \in X_{g_1}$ solving equation (8.167), and for which $q(z,w)$ is defined by (8.168). We note first that under our assumptions on $R_2(z), R_1(z)$,

$$d_2(z)^{-1}q(z,w)d_1(w)^{-1}$$
$$= \frac{n_1(w)d_1(w)^{-1} - d_2(z)^{-1}n_2(z) + d_2(z)^{-1}R_1(z)R_2(w)d_1(w)^{-1}}{z-w}$$

is strictly proper in both variables, i.e., $q(z,w)$ is in $X_{d_2(z) \otimes d_1(w)}$. We compute

$$\begin{aligned} \mathscr{S}q(z,w) &= \pi_{d_2(z) \otimes d_1(w)}(z-w)q(z,w) \\ &= \pi_{d_2(z) \otimes d_1(w)}(d_2(z)n_1(w) - n_2(z)d_1(w) + R_2(z)R_1(w)) \\ &= R_2(z)R_1(w), \end{aligned}$$

i.e., $q(z,w)$ is indeed a solution.

To prove the converse, we note that for the single-variable case, given a nonzero polynomial $d_2(z) \in \mathbb{F}[z]$, we have, for $f(z) \in X_{d_1}$, $S_{d_1}f = zf(z) - d_1(z)\xi_f$, where $\xi_f = (d_1^{-1}f)_{-1}$. This implies that for $q(z,w) \in X_{d_2(z) \otimes d_1(w)}$, we have $S_{z \otimes 1}q(z,w) = zq(z,w) - d_2(z)n_1(w)$ and $S_{1 \otimes w}q(z,w) = q(z,w)w - n_2(z)d_1(w)$ with $n_1 d_1^{-1}, d_2^{-1}n_2$ strictly proper. Thus, assuming $q(z,w)$ is a solution of the PSE, we have

$$\begin{aligned} S_{z-w}q(z,w) &= [zq(z,w) - d_2(z)n_1(w)] - [q(z,w)w - n_2(z)d_1(w)] \\ &= [zq(z,w) - q(z,w)w] - [d_2(z)n_1(w) - n_2(z)d_1(w)] \\ &= R_2(z)R_1(w). \end{aligned}$$

Equation (8.162) reduces to

$$[zQ(z,w) - Q(z,w)w] - [d_2(z)n_1(w) - n_2(z)d_1(w)] = R_2(z)R_1(w), \quad (8.171)$$

or

$$q(z,w) = \frac{d_2(z)n_1(w) - n_2(z)d_1(w) + R_2(z)R_1(w)}{z - w}. \quad (8.172)$$

However, since $q(z,w) \in X_{d_2(z) \otimes d_1(w)}$ is a polynomial, we must have that (8.167) holds.

2. Follows from the previous part.   ∎

8.4.8 Reproducing Kernels

In Theorem 8.69, we proved the isomorphism (8.136), which shows that any linear transformation $K : X_{d_1} \longrightarrow X_{d_2}$ can be represented by a unique polynomial $k(z,w) \in X_{d_2(z)d_1(w)}$, with the representation given, for $f(z) \in X_{d_1}$, by

$$(Kf)(z) = \langle f(\cdot), k(z, \cdot) \rangle. \quad (8.173)$$

With the identification $X_q^* = X_q$, given in terms of the pairing (5.64), we can introduce reproducing kernels for polynomial models. To this end we consider the ring $\mathbb{F}[z,w]$ of polynomials in the two variables z,w. Let now $q(z)$ be a monic polynomial of degree n. Given a polynomial $k(z,w) \in \mathbb{F}[z,w]$ of degree less than n in each variable, it can be written as $k(z,w) = \sum_{i=1}^{n} \sum_{j=1}^{n} k_{ij} z^{i-1} w^{j-1}$. Such a polynomial $k(z,w)$ induces a linear transformation $K : X_d \longrightarrow X_d$, defined by

$$(Kf)(z) = \langle f(\cdot), k(z, \cdot) \rangle = \sum_{i=1}^{n} \sum_{j=1}^{n} k_{ij} z^{i-1} \langle w^{j-1}, f \rangle. \quad (8.174)$$

We will say that the polynomial $k(z,w)$ is the **kernel** of the map K defined in (8.174).

Now, any $f(w) \in X_d$ has an expansion $f(w) = \sum_{k=1}^{n} f_k e_k(w)$, with $\{e_1(z),\ldots, e_n(z)\}$ the control basis in X_q. Thus, we compute

$$
\begin{aligned}
(Kf)(z) &= \sum_{i=1}^{n} \sum_{j=1}^{n} k_{ij} z^{i-1} \langle w^{j-1}, f \rangle \\
&= \sum_{i=1}^{n} \sum_{j=1}^{n} \sum_{k=1}^{n} k_{ij} z^{i-1} f_k \langle w^{j-1}, e_k \rangle \\
&= \sum_{i=1}^{n} \sum_{j=1}^{n} \sum_{k=1}^{n} k_{ij} f_k z^{i-1} \delta_{jk} \\
&= \sum_{i=1}^{n} \sum_{j=1}^{n} k_{ij} f_j z^{i-1}.
\end{aligned}
$$

This implies that $[K]_{co}^{st} = (k_{ij})$.

Given a polynomial $q(z)$ of degree n, we consider now the special kernel, defined by

$$k(z,w) = \frac{q(z) - q(w)}{z - w}. \quad (8.175)$$

Proposition 8.76. *Given a monic polynomial* $q(z) = z^n + q_{n-1}z^{n-1} + \cdots + q_0$ *of degree n, then, with* $k(z,w)$ *defined by (8.175):*

1. *We have*

$$k(z,w) = \sum_{j=1}^{n} w^{j-1}e_j(z) = \sum_{j=1}^{n} z^{j-1}e_j(w). \qquad (8.176)$$

2. *For any* $f(z) \in X_d$, *we have*

$$\langle f(\cdot), k(z,\cdot) \rangle = f(z). \qquad (8.177)$$

3. *We have the following matrix representations:*

$$[K]_{co}^{co} = [K]_{st}^{st} = I. \qquad (8.178)$$

Proof. 1. We use the fact that $z^j - w^j = (z-w)\sum_{k=1}^{j-1} z^k w^{j-k-1}$ to get

$$\begin{aligned}
q(z) - q(w) &= z^n - w^n + \sum_{j=1}^{n-1} q_j(z^j - w^j) \\
&= (z-w)\sum_{k=1} z^k w^{n-k-1} + (z-w)\sum_{j=1} q_j \sum_{k=1} z^k w^{j-k-1} \\
&= (z-w)\sum_{j=1} w^{j-1}e_j(z)
\end{aligned}$$

and hence $\frac{q(z)-q(w)}{z-w} = \sum_{j=1}^{n} w^{j-1}e_j(z)$. The equality $\frac{q(z)-q(w)}{z-w} = \sum_{j=1}^{n} z^{j-1}e_j(w)$ follows by symmetry.

2. Let $f(z) \in X_q$ and $f(z) = \sum_{j=1}^{n} f_j e_j(z)$. Then

$$\begin{aligned}
\langle k(z,\cdot), f \rangle &= \langle \sum_{j=1}^{n} w^{j-1}e_j(z), \sum_{k=1}^{n} f_k e_k(w) \rangle \\
&= \sum_{j=1}^{n} \sum_{k=1}^{n} f_k e_j(z)\langle w^{j-1}, e_k(w) \rangle \\
&= \sum_{j=1}^{n} \sum_{k=1}^{n} f_k e_j(z)\delta_{jk} = \sum_{k=1}^{n} f_k e_k(z) = f(z).
\end{aligned}$$

3. Obviously, by the previous part, K acts as the identity; hence its matrix representation in any basis is the identity matrix. The matrix representations in (8.178) can also be verified by direct computations. ∎

A kernel with the properties described in Proposition 8.76 is called a **reproducing kernel** for the space X_q.

We note that the map $K : X_q \longrightarrow X_q$, defined by Proposition 8.76, is the identity map. As such, it clearly satisfies $IS_q = S_q I$, i.e., it is an intertwining map. Of course, not every map in $\text{Hom}_{\mathbb{F}}(X_q, X_q)$ is intertwining. What makes K an intertwining map is the special form of the kernel given in (8.175), namely the fact that the kernel is one of the simplest Bezoutians, a topic we will pick up in the sequel.

Given a polynomial $M(z,w) = \sum_{i=1}^{n} \sum_{j=1}^{n} M_{ij}z^{i-1}w^{j-1}$ in the variables z,w, we have an induced bilinear form in $X_q \times X_q$ defined by

$$\phi(f,g) = \langle Mf, g \rangle = \langle \langle M(z,\cdot), f \rangle_w g \rangle_z. \qquad (8.179)$$

In turn, this induces a quadratic form given by $\phi(f,f) = \langle Mf,f \rangle = \langle\langle M(z,\cdot),f \rangle_w f \rangle_z$.

Proposition 8.77. *Let $q(z) \in \mathbb{F}[z]$ and let $\mathscr{B}_{st}$ and $\mathscr{B}_{co}$ be the standard and control bases respectively of X_q, defined in Chapter 5.*

1. *The matrix representation of the bilinear form $\phi(f,g)$, given by (8.179), with respect to the control and standard bases of X_q is (M_{ij}).*
2. *ϕ is symmetric if and only if (M_{ij}) is symmetric.*
3. *The form ϕ is positive definite if and only if (M_{ij}) is a positive definite matrix.*

Proof. 1. M_{ij}, the ij entry of $[M]_{co}^{st}$, is given by

$$
\begin{aligned}
\langle Me_j, e_i \rangle &= \langle\langle \textstyle\sum_{k=1}^n \sum_{l=1}^n M_{ij} z^{k-1} w^{l-1}, e_j \rangle_w, e_i \rangle_z \\
&= \langle \textstyle\sum_{k=1}^n \sum_{l=1}^n M_{ij} z^{k-1} \langle w^{l-1}, e_j \rangle_w, e_i \rangle_z \\
&= \langle \textstyle\sum_{k=1}^n M_{kj} z^{k-1}, e_i \rangle_z \\
&= \textstyle\sum_{k=1}^n M_{kj} \delta_{ik} = M_{ij}.
\end{aligned}
$$

2. Follows from (8.179) and the previous computation.
3. Follows from the equality $\phi(f,f) = ([M]_{co}^{st}[f]^{co}, [f]^{co})$. ∎

In the next subsection, we will return to the study of representations of general linear transformations in terms of kernels. This will be done using the polynomial representations of the tensor products of polynomial models, both over the underlying field $\mathbb{F}$ and over the polynomial ring $\mathbb{F}[z]$.

8.4.9 The Bezout Map

Since a map intertwining the polynomial models X_{q_1} and X_{q_2} is automatically linear, there is a natural embedding of $\mathrm{Hom}_{\mathbb{F}[z]}(X_{d_1}, X_{d_2})$ in $\mathrm{Hom}_{\mathbb{F}}(X_{d_1}, X_{d_2})$. This, taken together with the isomorphisms (8.137) and (8.146), shows that there exists a natural embedding of $X_{d_1} \otimes_{\mathbb{F}[z]} X_{d_2}$ in $X_{d_1} \otimes_{\mathbb{F}} X_{d_2}$. In order to make this embedding specific, we use concrete representations for the two tensor products, namely the isomorphisms (8.136) and (8.146). One result of using concrete representations for the tensor product is that they reveal, in the clearest possible conceptual way, the role of the Bezoutians in the analysis of maps intertwining two shifts.

Theorem 8.78. *Given $d_1(z), d_2(z) \in \mathbb{F}[z] - \{0\}$, let $\Psi : X_{d_1(z)d_2(w)} \longrightarrow \mathrm{Hom}_{\mathbb{F}}(X_{d_1}, X_{d_2})$ be given by (8.135) and let $\psi : X_{d_1} \otimes_{\mathbb{F}[z]} X_{d_2} \longrightarrow \mathrm{Hom}_{\mathbb{F}[z]}(X_{d_1}, X_{d_2})$ be given by (8.147). Let $i : \mathrm{Hom}_{\mathbb{F}[z]}(X_{d_1}, X_{d_2}) \longrightarrow \mathrm{Hom}_{\mathbb{F}}(X_{d_1}, X_{d_2})$ be the natural embedding. For $n(z) \in X_{d_1 \wedge d_2}$ having the representation (8.150), we define the **Bezout map** $\beta : X_{d_1 \wedge d_2} \longrightarrow X_{d_1(z)d_2(w)}$ by*

$$
\beta(n) = q(z,w) = \frac{d_2(z)n_1(w) - n_2(z)d_1(w)}{z-w}, \tag{8.180}
$$

i.e., $q(z,w)$, denotes the Bezoutian based on the intertwining relation (8.151). Then β is injective, and we have the equality

$$\Psi \circ \beta = i \circ \psi, \tag{8.181}$$

i.e., the Bezout map is the concretization of the embedding of $X_{d_1} \otimes_{\mathbb{F}[z]} X_{d_2}$ in $X_{d_1} \otimes_{\mathbb{F}} X_{d_2}$.

Proof. Note, as above, that any element $n(z) \in X_{d_1 \wedge d_2}$ yields a unique strictly proper function $h(z) = \frac{n(z)}{d_1(z) \wedge d_2(z)} = \frac{n_1(z)}{d_1(z)} = \frac{n_2(z)}{d_2(z)}$. Thus $n(z)$ determines unique polynomials $n_1(z), n_2(z)$ satisfying the intertwining relation (8.151). By inspection, $d_1(z)^{-1} q(z,w) d_2(w)^{-1} \in z^{-1} \mathbb{F}_{sep}[[z^{-1}, w^{-1}]] w^{-1}$ and thus $q(z,w) \in X_{d_1(z) d_2(w)}$. This shows that β is well defined and $\mathbb{F}$-linear. The map β is injective, since $q(z,w) = 0$ implies, by (8.150), that $\frac{n_1(z)}{d_1(z)} = \frac{n_2(w)}{d_2(w)}$ is constant. In turn, we conclude that $h(z) = \frac{n(z)}{d_1(z) \wedge d_2(z)}$ is a constant which, by strict properness, is necessarily zero.

To prove the equality (8.181), we compute, with $\beta(n) = q(z,w)$ given by (8.180),

$$
\begin{aligned}
\langle g, q(z,\cdot) \rangle &= \left\langle g, \frac{d_2(z) n_1(w) - n_2(z) d_1(w)}{z - w} \right\rangle \\
&= \left[d_1(w)^{-1} g(w), \frac{d_2(z) n_1(w) - n_2(z) d_1(w)}{z - w} \right] \\
&= \left(\frac{d_2(z) n_1(w) - n_2(z) d_1(w)}{z - w} d_1(w)^{-1} g(w) \right)^w_{-1} \\
&= \left(\frac{d_2(z) n_1(w) d_1(w)^{-1} g(w)}{z - w} - \frac{n_2(z) g(w)}{z - w} \right)^w_{-1} \\
&= -d_2 \pi_+ n_1 d_1^{-1} g + n_2 g = n_2 g - d_2 \pi_+ d_2^{-1} n_2 g \\
&= \pi_{d_2} n_2 g. \tag{8.182}
\end{aligned}
$$

Here $(F(z,w))^w_{-1}$ denotes the coefficient of w^{-1} in the Laurent expansion of $F(z,w)$ with respect to the variable w. ∎

The availability of the Bezout map and the representation (8.182) allow us to extend Theorem 5.17, giving a characterization of elements of $\text{Hom}_{\mathbb{F}[z]}(X_{d_1}, X_{d_2})$.

Theorem 8.79. *Given nonzero polynomials $d_1(z), d_2(z) \in \mathbb{F}[z]$, then $Z \in \text{Hom}(X_{d_1}, X_{d_2})$, i.e., it satisfies $Z S_{d_1} = S_{d_2} Z$, if and only if there exist polynomials $n_1(z), n_2(z)$ satisfying*

$$n_2(z) d_1(z) = d_2(z) n_1(z), \tag{8.183}$$

and for which Z has the representation

$$Z g = \pi_{d_2} n_2 g, \quad g(z) \in X_{d_1}. \tag{8.184}$$

Proof. Follows from Theorem 8.78. ∎

8.5   Exercises

1. **Jacobi's signature rule**. Let $A(x,x)$ be a Hermitian form and let Δ_i be the determinants of the principal minors. If the rank of A is r and $\Delta_1,\ldots,\Delta_r$ are nonzero then

$$\pi = P(1,\Delta_1,\ldots,\Delta_r), \qquad v = V(1,\Delta_1,\ldots,\Delta_r),$$

where P and V denote, respectively, the number of sign permanences and the number of sign changes in the sequence $1,\Delta_1,\ldots,\Delta_r$.

2. Let $p(z) = \sum_{i=0}^{m} p_i z^i = p_m \prod_{i=1}^{m}(z - \alpha_i)$ and $q(z) = \sum_{i=0}^{n} q_i z^i = q_n \prod_{i=1}^{m}(z - \beta_i)$. Show that $\det \mathrm{Res}\,(p,q) = p_m^n q_n^m \prod_{i,j}(\beta_i - \alpha_j)$.

3. Show that a real Hankel form

$$H_n(x,x) = \sum_{i=1}^{n} \sum_{j=1}^{n} s_{i+j-1} \xi_i \xi_j$$

is positive if and only if the parameters s_k allow a representation of the form

$$s_k = \sum_{j=1}^{n} \rho_j \theta_j^k,$$

with $\rho_j > 0$ and θ_j distinct real numbers.

4. Given $g_1,\ldots,g_{2n-1}$, consider the Hankel matrix H_n given by (8.66). Let $a(z) = \sum_{i=0}^{n-1} a_i z^i$ and $x(z) = \sum_{i=0}^{n-1} x_i z^i$ be the polynomials arising out of the solutions of the system of linear equations

$$\begin{pmatrix} g_1 & \cdots & g_n \\ \cdot & \cdot\,\cdot\,\cdot\,\cdot & \\ \cdot & \cdot\,\cdot\,\cdot\,\cdot & \\ \cdot & \cdot\,\cdot\,\cdot\,\cdot & \\ g_n & \cdots & g_{2n-1} \end{pmatrix} \begin{pmatrix} x_0 \\ \cdot \\ \cdot \\ \cdot \\ x_{n-1} \end{pmatrix} = \begin{pmatrix} r_1 \\ \cdot \\ \cdot \\ \cdot \\ r_n \end{pmatrix}$$

with the right-hand side being given respectively by

$$r_i = \begin{cases} 0, & i = 1,\ldots,n-1, \\ 1, & i = n, \end{cases}$$

and

$$r_i = \begin{cases} 1, & i = 1, \\ 0, & i = 2,\ldots,n. \end{cases}$$

Show that if $a_0 \neq 0$, we have $x_{n-1} = a_0$, the Hankel matrix H_n is invertible, and its inverse is given by $H_n^{-1} = B(y,a)$, where the polynomial y is defined by $y(z) = a(0)^{-1} z x(z) = q(z) - a(0)^{-1} q(0) a(z)$.

5. Show that a Hermitian **Toeplitz form**

$$T_n(x,x) = \sum_{i=1}^{n} \sum_{j=1}^{n} c_{i-j} \xi_i \overline{\xi}_j$$

with the matrix

$$T_n = \begin{pmatrix} c_0 & \cdot\cdot & c_{-n+1} & c_{-n} \\ & \cdot & \cdot\cdot & \cdot & \cdot \\ & \cdot & \cdot\cdot & \cdot & \cdot \\ & \cdot & \cdot\cdot & \cdot & \cdot \\ c_n & \cdot\cdot & & \cdot & c_0 \end{pmatrix}$$

is positive definite if and only if the parameters $c_k = \overline{c}_{-k}$ allow a representation of the form

$$c_k = \sum_{j=1}^{n} \rho_j \theta_j^k,$$

with $\rho_j > 0$, $|\theta_j| = 1$, and the θ_j distinct.

6. We say a sequence $c_0, c_1, \ldots$ of complex numbers is a **positive sequence** if for all $n \geq 0$, the corresponding Toeplitz form is positive definite. An inner product on $\mathbb{C}[z]$ is defined by

$$\langle a,b \rangle_C = \begin{pmatrix} a_0 & \ldots & a_n \end{pmatrix} \begin{pmatrix} c_0 & \ldots & c_n \\ & \cdot\cdot\cdot\cdot \\ & \cdot\cdot\cdot\cdot \\ & \cdot\cdot\cdot\cdot \\ c_n & \ldots & c_0 \end{pmatrix} \begin{pmatrix} \overline{b}_0 \\ \cdot \\ \cdot \\ \overline{b}_n \end{pmatrix} = \sum_{i=0}^{n} \sum_{j=0}^{n} c_{(i-j)} a_i \overline{b}_j.$$

Under our assumption of positivity, this is a definite inner product on $\mathbb{C}[z]$.

a. Show that the inner product defined above has the following properties:

$$\langle z^i, z^j \rangle_C = c_{(i-j)}$$

and

$$\langle za, zb \rangle_C = \langle a,b \rangle_C.$$

b. Let ϕ_n be the sequence of polynomials, obtained from the polynomials $1, z, z^2, \ldots$ by an application of the Gram-Schmidt orthogonalization procedure. Assume all the ϕ_n are monic. Show that

$$\phi_0(z) = 1,$$

$$\phi_n(z) = \frac{1}{\det T_{n-1}} \begin{vmatrix} c_0 & \cdots & \cdot & c_{-n} \\ \cdot & \cdot & \cdot & \cdot \\ \cdot & \cdot & \cdot & \cdot \\ c_{n-1} & \cdots & c_0 & c_{-1} \\ 1 & \cdots & z^{n-1} & z^n \end{vmatrix}.$$

c. Show that

$$\langle \phi_n^\sharp, z^{n-i} \rangle_C = 0, \qquad i = 0, \ldots, n-1.$$

d. Show that

$$\phi_n^\sharp(z) = 1 - z \sum_{i=0}^{n-1} \gamma_{i+1} \phi_i(z),$$

where the γ_i are defined by

$$\gamma_{i+1} = \frac{\langle 1, z\phi_i \rangle}{\|\phi_i\|^2}.$$

The γ_i are called the **Schur parameters** of the sequence $c_0, c_1, \ldots$.
e. Show that the orthogonal polynomials satisfy the following recurrences

$$\phi_n(z) = z\phi_{n-1}(z) - \gamma_n \phi_{n-1}^\sharp(z),$$

$$\phi_n^\sharp(z) = \phi_{n-1}^\sharp(z) - z\gamma_n \phi_{n-1}(z),$$

with the initial conditions for the recurrences given by $\phi_0(z) = \phi_0^\sharp(z) = 1$. Equivalently,

$$\begin{pmatrix} \phi_n(z) \\ \phi_n^\sharp(z) \end{pmatrix} = \begin{pmatrix} z & -\gamma_n \\ -z\gamma_n & 1 \end{pmatrix} \begin{pmatrix} \phi_{n-1}(z) \\ \phi_{n-1}^\sharp(z) \end{pmatrix}, \quad \begin{pmatrix} \phi_0(0) \\ \phi_0^\sharp(0) \end{pmatrix} = \begin{pmatrix} 1 \\ 1 \end{pmatrix}.$$

f. Show that

$$\gamma_n = -\phi_n(0).$$

g. Show that

$$\|\phi_n\|^2 = (1 - \gamma_n^2)\|\phi_{n-1}\|^2 /$$

h. Show that the Schur parameters γ_n satisfy

$$|\gamma_n| < 1.$$

i. The **Levinson algorithm.** Let $\{\phi_n(z)\}$ be the monic orthogonal polynomials associated with the positive sequence $\{c_n\}$. Assume $\phi_n(z) = z^n + \sum_{i=0}^{n-1} \phi_{n,i} z^i$.

Show that they can be computed recursively by

$$r_{n+1} = (1 - \gamma_n^2)r_n, \qquad\qquad r_0 = c_0,$$

$$\gamma_{n+1} = \frac{1}{r_n} \sum_{i=0}^{n} c_{i+1}\phi_{n,i},$$

$$\phi_{n+1}(z) = z\phi_n(z) - \gamma_{n+1}\phi_n^{\#}(z), \qquad \phi_0(z) = 1.$$

j. Show that the shift operator S_{ϕ_n} acting in X_{ϕ_n} satisfies

$$S_{\phi_n}\phi_i = \phi_{i+1} - \gamma_{i+1} \sum_{j=1}^{i} \gamma_j \Pi_{v=j+1}^{i}(1 - \gamma_v^2)\phi_j + \gamma_{i+1}\Pi_{v=1}^{i}(1 - \gamma_v^2)\phi_0.$$

k. Let $c_0, c_1, \ldots$ be a positive sequence. Define a new sequence $\hat{c}_0, \hat{c}_1, \ldots$ through the infinite system of equations given by

$$\begin{pmatrix} c_0/2 & 0 & \cdots \\ c_1 & c_0/2 & 0 & \cdots \\ c_2 & c_1 & \cdots \\ \cdot & \cdot & \cdots \end{pmatrix} \begin{pmatrix} \hat{c}_0/2 \\ \hat{c}_1 \\ \cdot \\ \cdot \end{pmatrix} = \begin{pmatrix} 1 \\ 0 \\ \cdot \\ \cdot \end{pmatrix}.$$

Show that $\hat{c}_0, \hat{c}_1, \ldots$ is also a positive sequence.

l. Show that the Schur parameters for the orthogonal polynomials ψ_n, corresponding to the new sequence, are given by $-\gamma_n$.

m. Show that ψ_n satisfy the recurrence relation

$$\psi_n = z\psi_{n-1} + \gamma_n \psi_{n-1}^{\#},$$

$$\psi_n^{\#} = \psi_{n-1}^{\#} + z\gamma_n \psi_{n-1}.$$

The initial conditions for the recurrences are $\psi_0(z) = \psi_0^{\#}(z) = 1$.

8.6 Notes and Remarks

The study of quadratic forms owes much to the work of Sylvester and Cayley. Independently, Hermite studied both quadratic and Hermitian forms and, more or less simultaneously, Sylvester's law of inertia. Bezout forms and matrices were studied already by Sylvester and Cayley. The connection of Hankel and Bezout forms was known to Jacobi. For some of the history of the subject, we recommend Krein and Naimark (1936), as well as Gantmacher (1959).

Continued fraction representations need not be restricted to rational functions. In fact, most applications of continued fractions are in the area of rational approximations of functions, or of numbers. Here is the **generalized Euclidean algorithm**; see Magnus (1962). Given $g(z) \in \mathbb{F}((z^{-1}))$, we set $g(z) = a_0(z) + g_0(z)$, with $a_0(z) = \pi_+ g(z)$. Define recursively

$$\frac{\beta_{n-1}}{g_{n-1}(z)} = a_n(z) - g_n(z),$$

where β_{n-1} is a normalizing constant, $a_n(z)$ a monic polynomial, and $g_n(z) \in z^{-1}\mathbb{F}[[z^{-1}]]$. This leads, in the nonrational case, to an infinite continued fraction, which can be applied to approximation problems.

In the proof of Proposition 8.65, we follow Lang (1965). Theorem 8.32 is due to Kravitsky (1980). The Gohberg–Semencul representation formula (8.40) for the Bezoutian is from Gohberg and Semencul (1972). The interpretation of the Bezoutian as a matrix representation of an intertwining map is due to Fuhrmann (1981b) and Helmke and Fuhrmann (1989). For the early history of the Bezoutian, see also Wimmer (1990).

The approach to tensor products of polynomial models and the Bezout map are based on Fuhrmann and Helmke (2010). The analysis of the polynomial Sylvester equation extends the method, introduced in Willems and Fuhrmann (1992), for the analysis of the Lyapunov equation. In turn, this was influenced by Kalman (1969). The characterization of the inverse of finite nonsingular Hankel matrices as Bezoutians, given in Theorem 8.39, is due to Lander (1974).

For a more detailed discussion of the vectorial versions of many of the results given in this chapter, in particular to the tensor products of vectorial functional models and the corresponding version of the Bezout map, we refer to Fuhrmann and Helmke (2010).

Chapter 9
Stability

9.1 Introduction

The formalization of mathematical stability theory is generally traced to the classic paper Maxwell (1868). The problem is this: given a system of linear time-invariant, differential (or difference) equations, find a characterization for the asymptotic stability of all solutions. This problem reduces to finding conditions on a given polynomial that guarantee that all its zeros, i.e., roots, lie in the open left half-plane (or open unit disk).

9.2 Root Location Using Quadratic Forms

The problem of root location of algebraic equations has served to motivate the introduction and study of quadratic and Hermitian forms. The first result in this direction seems to be that of Borchardt (1847). This in turn motivated Jacobi (1857), as well as Hermite (1856), who made no doubt the greatest contribution to this subject. We follow, as much as possible, the masterful exposition of Krein and Naimark (1936), which every reader is advised to consult.

   The following theorem summarizes one of the earliest applications of quadratic forms to the problem of zero location.

Theorem 9.1. *Let* $q(z) = \sum_{i=0}^{n} q_i z^i$ *be a real polynomial,* $g(z) = \frac{q'(z)}{q(z)}$, *and* H_g *the infinite Hankel matrix of* $g(z)$. *Then*

1. *The number of distinct, real or complex, roots of* $q(z)$ *is equal to* $\mathrm{rank}\,(H_g)$.
2. *The number of distinct real roots of* $q(z)$ *is equal to the signature* $\sigma(H_g)$, *or equivalently to the Cauchy index* I_g *of* $g(z)$.

Proof. 1. Let $\alpha_1, \ldots, \alpha_m$ be the distinct roots of $q(z) = \sum_{i=0}^{n} q_i z^i$, whether real or complex. So

P.A. Fuhrmann, *A Polynomial Approach to Linear Algebra*, Universitext,
DOI 10.1007/978-1-4614-0338-8_9, © Springer Science+Business Media, LLC 2012

$$q(z) = q_n \prod_{i=1}^{m} (z - \alpha_i)^{\nu_i},$$

with $\nu_1 + \cdots + \nu_m = n$. Hence the rational function $g(z)$ defined below satisfies

$$g(z) = \frac{q'(z)}{q(z)} = \sum_{i=1}^{m} \frac{\nu_i}{(z - \alpha_i)}.$$

From this it is clear that m is equal to $\delta(g)$, the McMillan degree of $g(z)$, and so in turn to rank (H_g).

2. On the other hand, the Cauchy index of $g(z)$ is obviously equal to the number of real zeros, since all residues ν_i are positive. But by the Hermite-Hurwitz theorem, we have $I_g = \sigma(H_g)$.   ∎

The analysis of Bezoutians, undertaken in Chapter 8, can be efficiently applied to the problem of root location of polynomials. We begin by applying it to the same problem as above.

Theorem 9.2. *Let $q(z)$ be as in the previous theorem. Then*

1. *The number of distinct, real or complex, roots of $q(z)$ is equal to* codim Ker $B(q, q')$.

2. *The number of distinct real roots of $q(z)$ is equal to $\sigma(B(q, q'))$.*

3. *All roots of $q(z)$ are real if and only if $B(q, q') \geq 0$.*

Proof. 1. From our study of Bezoutians and intertwining maps we know that $\dim \operatorname{Ker} B(q, q') = \deg r$, where $r(z)$ is the g.c.d. of $q(z)$ and $q'(z)$. It is easy to check that the g.c.d. of $q(z)$ and $q'(z)$ is equal to $r(z) = q_n \prod_{i=1}^{m} (z - \alpha_i)^{\nu_i - 1}$; hence its degree is $n - m$.

2. This follows as above.

3. Note that if the inertia, see Definition 8.11, is $\operatorname{In}(B(q, q')) = (\pi, \nu, \delta)$, then $\nu = (\operatorname{rank}(B(q, q')) - \sigma(B(q, q')))/2$. So $\nu = 0$ if and only if $\operatorname{rank}(B(q, q')) = \sigma(B(q, q'))$, i.e., if and only if all roots of $q(z)$ are real.   ∎

Naturally, in Theorem 9.1, the infinite quadratic form H_g can be replaced by H_n, the $n \times n$ truncated form. In fact, it is this form, in various guises, that appears in the early studies. We digress a little on this point. If we expand $1/(z - \alpha_i)$ in powers of z^{-1}, we obtain

$$g(z) = \frac{q'(z)}{q(z)} = \sum_{i=1}^{m} \frac{\nu_i}{z - \alpha_i} = \sum_{i=1}^{m} \nu_i \sum_{k=1}^{\infty} \frac{\alpha_i^k}{z^{k+1}} = \sum_{k=1}^{\infty} \frac{s_k}{z^{k+1}},$$

with $s_k = \sum_{i=1}^{m} \nu_i \alpha_i^k$. If all the zeros are simple, the numbers s_k are called the **Newton sums** of $q(z)$. The finite Hankel quadratic form $\sum_{i=0}^{n-1} s_{i+j} \xi_i \xi_j$ is easily seen to be a different representation of $\sum_{i=1}^{n} (\sum_{j=0}^{n-1} \xi_j \alpha_i^j)^2$.

The result stated in Theorem 9.1 far from exhausts the power of the method of quadratic forms. We address ourselves now to the problem of determining, for an arbitrary complex polynomial $q(z)$, the number of its zeros in an open half-plane. We begin with the problem of determining the number of zeros of $q(z) = \sum_{k=0}^{n} q_k z^k$ in the open upper half-plane. Once this is accomplished, we can apply the result to other half-planes, most importantly to the open left half-plane. To solve this problem, Hermite introduced the notion of Hermitian forms, which had far-reaching consequences in mathematics, Hilbert space theory being one offshoot.

Given a complex polynomial $q(z)$ of degree n, we define

$$\tilde{q}(z) = \overline{q(\bar{z})}. \tag{9.1}$$

Clearly, (9.1) implies that $\tilde{q}(\alpha) = 0$ if and only if $q(\bar{\alpha}) = 0$. Thus, α is a common zero of $q(z)$ and $\tilde{q}(z)$ if and only if α is a real zero of $q(z)$ or $\alpha, \bar{\alpha}$ is a pair of complex conjugate zeros of $q(z)$. Therefore the degree of $q(z) \wedge \tilde{q}(z)$ counts the number of real and complex conjugate zeros of $q(z)$. The next theorem gives a complete analysis.

Theorem 9.3. *Given a complex polynomial $q(z)$ of degree n, we define a polynomial $Q(z,w)$ in two variables by*

$$Q(z,w) = -iB(q,\tilde{q}) = -i\frac{q(z)\tilde{q}(w) - \tilde{q}(z)q(w)}{z - w}$$

$$= \sum_{j=1}^{n}\sum_{k=1}^{n} Q_{jk} z^{j-1} w^{k-1}. \tag{9.2}$$

*The polynomial $Q(z,w)$ is called the **generating function** of the quadratic form, defined on $\mathbb{C}^n$, by*

$$Q = \sum_{j=1}^{n}\sum_{k=1}^{n} Q_{jk} \xi_j \overline{\xi_k}. \tag{9.3}$$

Then

1. *The form Q defined in (9.3) is Hermitian, i.e., we have $Q_{jk} = \overline{Q}_{kj}$.*
2. *Let (π, ν, δ) be the inertia of the form Q. Then the number of real zeros of $q(z)$ together with the number of zeros of $q(z)$ arising from complex conjugate pairs is equal to δ. There are π more zeros of $q(z)$ in the open upper half-plane and ν more in the open lower half-plane. In particular, all zeros of $q(z)$ are in the open upper half-plane if and only if Q is positive definite.*

Proof. 1. We compute

$$\overline{Q(\overline{w},\overline{z})} = -i\overline{\left[\frac{q(\overline{w})\tilde{q}(\overline{z}) - \tilde{q}(\overline{w})q(\overline{z})}{\overline{w} - \overline{z}}\right]}$$

$$= i\left[\frac{\tilde{q}(w)q(z) - q(w)\tilde{q}(z)}{w - z}\right] = -i\left[\frac{\tilde{q}(z)q(w) - q(z)\tilde{q}(w)}{z - w}\right]$$

$$= Q(z,w).$$

In turn, this implies

$$\sum_{j=1}^{n}\sum_{k=1}^{n}Q_{jk}z^{j-1}w^{k-1} = \overline{\sum_{j=1}^{n}\sum_{k=1}^{n}\overline{Q}_{jk}\overline{w}^{j-1}\overline{z}^{k-1}} = \sum_{j=1}^{n}\sum_{k=1}^{n}\overline{Q}_{jk}z^{k-1}w^{j-1}$$

$$= \sum_{j=1}^{n}\sum_{k=1}^{n}\overline{Q}_{kj}z^{j-1}w^{k-1}.$$

Comparing coefficients, the equality $Q_{jk} = \overline{Q}_{kj}$ is obtained.

2. Let $d(z)$ be the g.c.d. of $q(z)$ and $\tilde{q}(z)$. We saw previously that the zeros of $d(z)$ are the real zeros of $q(z)$ and all pairs of complex conjugate zeros of $q(z)$. Thus, we may assume that $d(z)$ is a real monic polynomial. Write $q(z) = d(z)q'(z)$ and $\tilde{q}(z) = d(z)\tilde{q}'(z)$. Using Lemma 8.26, we have $B(q,\tilde{q}) = B(dq',d\tilde{q}')$ and hence $\operatorname{rank}B(q,\tilde{q}) = \operatorname{rank}B(q',\tilde{q}')$ and $\sigma(-iB(q,\tilde{q})) = \sigma(-iB(q',\tilde{q}'))$. This shows that $\delta = \deg d$ is the number of real zeros added to the number of zeros arising from complex conjugate zeros of $q(z)$. Therefore, for the next step, we may as well assume that $q(z)$ and $\tilde{q}(z)$ are coprime. Let $q(z) = q_1(z)q_2(z)$ be any factorization of $q(z)$ into real or complex factors. We do not assume that $q_1(z)$ and $q_2(z)$ are coprime. We compute

$$-i\frac{q(z)\tilde{q}(w) - \tilde{q}(z)q(w)}{z - w} = -iq_2(z)\frac{q_1(z)\tilde{q}_1(w) - \tilde{q}_1(z)q_1(w)}{z - w}\tilde{q}_2(w)$$

$$-i\tilde{q}_1(z)\frac{q_2(z)\tilde{q}_2(w) - \tilde{q}_2(z)q_2(w)}{z - w}q_1(w).$$

Since by construction, the polynomials $q_i(z), \tilde{q}_i(z)$, $i = 1,2$ are coprime, the ranks of the two Hermitian forms on the right are $\deg q_1$ and $\deg q_2$ respectively. By Theorem 8.15, the signature of Q is equal to the sum of the signatures of the forms

$$Q_1 = -i\frac{q_1(z)\tilde{q}_1(w) - \tilde{q}_1(z)q_1(w)}{z - w}$$

and

$$Q_2 = -i\frac{q_2(z)\tilde{q}_2(w) - \tilde{q}_2(z)q_2(w)}{z - w}.$$

Let $\alpha_1, \ldots, \alpha_n$ be the zeros of $q(z)$. Then by an induction argument, it follows that

$$\sigma(-iB(q,\tilde{q})) = \sum_{k=1}^{n} \sigma(-iB(z-\alpha_k, z-\overline{\alpha}_k))$$
$$= \sum_{k=1}^{n} \operatorname{sign}(2\operatorname{Im}(\alpha_k)) = \pi - v.$$

Thus π is the number of zeros of $q(z)$ in the upper half-plane and v the number in the lower half-plane. ∎

We can restate the previous theorem in terms of either the Cauchy index or the Hankel form.

Theorem 9.4. *Let $q(z)$ be a complex polynomial of degree n. Define its real and imaginary parts by*

$$q_r(z) = \frac{q(z) + \tilde{q}(z)}{2},$$

$$q_i(z) = \frac{q(z) - \tilde{q}(z)}{2i}. \tag{9.4}$$

We assume $\deg q_r \geq \deg q_i$. Define the proper rational function $g(z) = q_i(z)/q_r(z)$. Then

1. *The Cauchy index of $g(z)$, $I_g = \sigma(H_g)$, is the difference between the number of zeros of $q(z)$ in the lower half-plane and in the upper half-plane.*
2. *We have $I_g = \sigma(H_g) = -n$, i.e., H_g or equivalenly $B(q_r, q_i)$ are negative definite if and only if all the zeros of $q(z)$ lie in the open upper half-plane.*

Proof. 1. Note that the assumption that $\deg q_r \geq \deg q_i$ does not limit generality, since if this condition is not satisfied, we can consider instead the polynomial $iq(z)$. The Hermitian form $-iB(q, \tilde{q})$ can be also represented as the Bezoutian of two real polynomials. Indeed, if $q(z) = q_r(z) + iq_i(z)$ with q_r and q_i real polynomials, then $\tilde{q}(z) = q_r(z) - iq_i(z)$ and

$$-iB(q, \tilde{q}) = -iB(q_r + iq_i, q_r - iq_i)$$
$$= -i[B(q_r, q_r) + iB(q_i, q_r) - iB(q_r, q_i) + B(q_i, q_i)]$$
$$= 2B(q_i, q_r),$$

or

$$-iB(q, \tilde{q}) = 2B(q_i, q_r). \tag{9.5}$$

From equality (9.5) it follows that $\sigma(-iB(q, \tilde{q})) = \sigma(B(q_i, q_r))$ and, by the Hermite–Hurwitz theorem, that

$$\sigma(B(q_i, q_r)) = -\sigma(B(q_r, q_i)) = -\sigma(H_g) = -I_g.$$

The result follows from Theorem 9.3.
2. Follows from part 1. ∎

Definition 9.5. A complex polynomial $q(z)$ is called **stable**, or a **Hurwitz polynomial**, if all its zeros lie in the open left half-plane Π_-.

To obtain stability characterizations, we need to replace the open upper half-plane by the left open half-plane and this can easily be done by a change of variable.

Theorem 9.6. *Let $q(z)$ be a complex polynomial. Then a necessary and sufficient condition for $q(z)$ to be a Hurwitz polynomial is that the* **Hermite–Fujiwara quadratic form** *with generating function*

$$H = \frac{q(z)\tilde{q}(w) - \tilde{q}(-z)q(-w)}{z+w} = \sum_{j=1}^{n}\sum_{k=1}^{n} H_{jk}z^{j-1}w^{k-1} \tag{9.6}$$

be positive definite.

Proof. Obviously, a zero α of $q(z)$ is in the open left half-plane if and only if $-i\alpha$, which is in the upper half-plane, is a zero of $f(z) = q(iz)$. Thus, by Theorem 9.4, all the zeros of $q(z)$ are in the left half-plane if and only if the Hermitian form $-iB(f, \tilde{f})$ is positive definite.

Now, since $\tilde{f}(z) = \overline{f(\bar{z})} = \overline{q(i\bar{z})} = q(\overline{(-iz)}) = \tilde{q}(-iz)$, it follows that

$$-iB(f, \tilde{f}) = -i\frac{f(z)\tilde{f}(w) - \tilde{f}(z)f(w)}{z-w} = \frac{q(iz)\tilde{q}(-iw) - \tilde{q}(-iz)q(iw)}{i(z-w)}.$$

If we substitute $\zeta = iz$ and $\omega = -iw$, then we get the Hermite-Fujiwara generating function (9.6), and this form has to be positive for all the zeros of $q(z)$ to be in the upper half plane.

Definition 9.7. Let $q(z)$ be a real monic polynomial of degree m with simple real zeros

$$\alpha_1 < \alpha_2 < \cdots < \alpha_m$$

and let $p(z)$ be a real polynomial of degree $\leq m$ with positive leading coefficient and zeros

$$\begin{aligned}\beta_1 < \beta_2 < \cdots < \beta_{m-1} \quad &\text{if} \quad \deg p = m-1,\\ \beta_1 < \beta_2 < \cdots < \beta_m \quad &\text{if} \quad \deg p = m.\end{aligned}$$

Then

1. We say that $q(z)$ and $p(z)$ are a **real pair** if the zeros satisfy the **interlacing property**

$$\begin{aligned}\alpha_1 < \beta_1 < \alpha_2 < \beta_2 < \cdots < \beta_{m-1} < \alpha_m \quad &\text{if} \quad \deg p = m-1,\\ \beta_1 < \alpha_1 < \beta_2 < \cdots < \beta_m < \alpha_m \quad &\text{if} \quad \deg p = m.\end{aligned} \tag{9.7}$$

2. We say that $q(z)$ and $p(z)$ form a **positive pair** if they form a real pair and $\alpha_m < 0$ holds.

Theorem 9.8. *Let $q(z)$ and $p(z)$ be a pair of real polynomials as above. Then*

1. $q(z)$ and $p(z)$ form a real pair if and only if $B(q,p) > 0$.
2. $q(z)$ and $p(z)$ form a positive pair if and only if $B(q,p) > 0$ and $B(zp,q) > 0$.

Proof. 1. Assume $q(z)$ and $p(z)$ form a real pair. By our assumption, all zeros α_i of $q(z)$ are real and simple. Define a rational function by $g(z) = p(z)/q(z)$. Clearly $q(z) = \prod_{j=1}^{m}(z - \alpha_j)$. Hence $g(z) = \sum_{j=1}^{m} \frac{c_j}{z - \alpha_j}$, and it is easily checked that $c_j = p(\alpha_i)/q'(\alpha_i)$. Using the Hermite-Hurwitz theorem, to show that $B(q,p) > 0$ it suffices to show that $\sigma(B(q,p)) = I_g = m$, or equivalently that $p(\alpha_i)/q'(\alpha_i) > 0$, for all i. Since both $p(z)$ and $q(z)$ have positive leading coefficients, we have, except for the trivial case $m = 1$ and $\deg p = 0$, $\lim_{x \to \infty} p(x) = \lim_{x \to \infty} q(x) = +\infty$. This implies $q'(\alpha_m) > 0$ and $p(\alpha_m) > 0$. As a result, we have $p(\alpha_m)/q'(\alpha_m) > 0$. The simplicity of the zeros of $q(z)$ and $p(z)$ and the interlacing property forces the signs of the $q'(\alpha_i)$ and $p(\alpha_i)$ to alternate, and this forces all other residues $p(\alpha_i)/q'(\alpha_i)$ to be positive. The reader is advised to draw a picture in order to gain a better understanding of the argument.

Conversely, assume $B(q,p) > 0$. Then by the Hermite-Hurwitz theorem, $I_{\frac{p}{q}} = m$, which means that all the zeros of $q(z)$ have to be real and simple and the residues satisfy $p(\alpha_i)/q'(\alpha_i) > 0$. Since all the zeros of $q(z)$ are simple necessarily the signs of the $q'(\alpha_i)$ alternate. Thus $p(z)$ has values of different signs at neighboring zeros of $q(z)$, and hence its zeros are located between zeros of $q(z)$. If $\deg p = m - 1$, we are done. If $\deg p = m$, there is an extra zero of $p(z)$. Since $q'(\alpha_m) > 0$, necessarily $p(\alpha_m) > 0$. Now $\beta_m > \alpha_m$ would imply $\lim_{x \to \infty} p(x) = -\infty$, contrary to the assumption that p has positive leading coefficient. So necessarily we have $\beta_1 < \alpha_1 < \beta_2 < \cdots < \beta_m < \alpha_m$, i.e., $q(z)$ and $p(z)$ form a real pair.

2. Assume now $q(z)$ and $p(z)$ are a positive pair. By our assumption all α_i are negative. Moreover, since in particular $q(z)$ and $p(z)$ are a real pair, it follows by part (1) that $B(q,p) > 0$. This means that $p(\alpha_i)/q'(\alpha_i) > 0$ for all $i = 1, \ldots, m$ and hence that $\alpha_i p(\alpha_i)/q'(\alpha_i) < 0$. But these are the residues of $zp(z)/q(z)$, and so $I_{\frac{zp}{q}} = \sigma(B(q,zp)) = -m$. So $B(q,zp)$ is negative definite and $B(zp,q)$ positive definite.

Conversely, assume the two Bezoutians $B(q,p)$ and $B(zp,q)$ are positive definite. By part (1), the polynomials $q(z)$ and $p(z)$ form a real pair. Thus the residues satisfy $p(\alpha_i)/q'(\alpha_i) > 0$ and $\alpha_i p(\alpha_i)/q'(\alpha_i) < 0$. So in particular all zeros of $q(z)$ are negative. In particular, $\alpha_m < 0$, and so $q(z)$ and $p(z)$ form a positive pair. ∎

As a corollary we obtain the following characterization.

Theorem 9.9. *Let $q(z)$ be a complex polynomial and let $q_r(z), q_i(z)$ be its real and imaginary parts, defined in (9.4). Then all the zeros of $q(z) = q_r(z) + iq_i(z)$ are in the open upper half-plane if and only if all the zeros of $q_r(z)$ and $q_i(z)$ are simple and real and separate each other.*

Proof. Follows from Theorems 9.4 and 9.8.   ∎

In most applications what is needed are stability criteria for real polynomials. The assumption of realness of the polynomial $q(z)$ leads to further simplifications. These simplifications arise out of the decomposition of a real polynomial into its even and odd parts. We say that a polynomial $p(z)$ is even if $p(z) = p(-z)$ and odd if $p(z) = -p(-z)$. With $q(z) = \sum q_j z^j$, the even and odd parts of $q(z)$ are determined by

$$q_+(z^2) = \frac{q(z) + q(-z)}{2} = \sum_{j \geq 0} q_{2j} z^{2j},$$

$$q_-(z^2) = \frac{q(z) - q(-z)}{2} = \sum_{j \geq 0} q_{2j+1} z^{2j}. \tag{9.8}$$

This is equivalent to writing

$$q(z) = q_+(z^2) + z q_-(z^2). \tag{9.9}$$

In preparation for the next theorem, we state the following lemma, describing a condition for stability.

Lemma 9.10. *Let $q(z)$ be a monic real polynomial with all its roots real. Then $q(z) = z^n + q_{n-1} z^{n-1} + \cdots + q_0$ is stable if and only if $q_i > 0$ for $i = 0, \ldots, n-1$.*

Proof. Assume $q(z)$ is stable and let $-\beta_i < 0$ be its zeros, i.e., $q(z) = \Pi_{i=1}^n (z + \beta_i)$. This implies the positivity of the coefficients q_i.

Since, by assumption, all zeros of $q(z)$ are real, for stability it suffices to check that $q(z)$ has no nonnegative zeros. But clearly, for every $x \geq 0$ we have $q(x) = x^n + q_{n-1} x^{n-1} + \cdots + q_0 > 0$.   ∎

The following theorem sums up the central results on the characterization of stability of real polynomials.

Theorem 9.11. *Let $q(z)$ be a real monic polynomial of degree n, and let $q_+(z), q_-(z)$ be its even and odd parts respectively, as defined in (9.8). Then the following statements are equivalent:*

1. *$q(z)$ is a stable, or Hurwitz, polynomial.*
2. *The Hermite–Fujiwara form is positive definite.*
3. *The two Bezoutians $B(q_+, q_-)$ and $B(zq_-, q_+)$ are positive definite.*
4. *The polynomials $q_+(z)$ and $q_-(z)$ form a positive pair.*
5. *The Bezoutian $B(q_+, q_-)$ is positive definite and all q_i are positive.*

Proof. (1) ⟺ (2) By Theorem 9.6 the stability of $q(z)$ is equivalent to the positive definiteness of the Hermite Fujiwara form.

(2) ⟺ (3) Since the polynomial $q(z)$ is real, we have in this case $\tilde{q}(z) = q(z)$. From (9.9) it follows that $q(-z) = q_+(z^2) - z q_-(z^2)$. Therefore

$$\frac{q(z)q(w) - q(-z)q(-w)}{(z+w)}$$

$$= \frac{(q_+(z^2)+zq_-(z^2))(q_+(w^2)+wq_-(w^2))-(q_+(z^2)-zq_-(z^2))(q_+(w^2)-wq_-(w^2))}{z+w}$$

$$= 2\frac{zq_-(z^2)q_+(w^2) + q_+(z^2)wq_-(w^2)}{z+w}$$

$$= 2\frac{z^2q_-(z^2)q_+(w^2) - q_+(z^2)w^2q_-(w^2)}{z^2 - w^2} + 2zw\frac{q_+(z^2)q_-(w^2) - q_-(z^2)q_+(w^2)}{z^2 - w^2}.$$

One form contains only even-indexed terms, the other only odd ones. So the positive definiteness of the form (9.6) is equivalent to the positive definiteness of the two Bezoutians $B(q_+,q_-)$ and $B(zq_-,q_+)$.

(3) $\Leftrightarrow$ (4) This is the content of Theorem 9.8.

(4) $\Rightarrow$ (5) Since the first four conditions are equivalent, from (3) it follows that $B(q_+,q_-) > 0$. Also from (1), applying Lemma 9.10, it follows that all q_i are positive.

(5) $\Rightarrow$ (4) $B(q_+,q_-) > 0$ implies, by Theorem 9.8, that $q_+(z), q_-(z)$ are a real pair. In particular, all zeros of both polynomials are real and simple. The positivity of all q_i implies the positivity of all coefficients of $q_+(z)$. Applying Lemma 9.10, all zeros of $q_+(z)$ are negative, which from the interlacing properties (9.7) shows that the polynomials $q_+(z)$ and $q_-(z)$ form a positive pair. ∎

The next corollary gives the stability characterization in terms of the Cauchy index.

Corollary 9.12. *Let $q(z)$ be a real monic polynomial of degree n, having the expansion $q(z) = z^n + q_{n-1}z^{n-1} + \cdots + q_0$, and let*

$$q(z) = q_+(z^2) + zq_-(z^2)$$

be its decomposition into its even and odd parts. Then $q(z)$ is a Hurwitz polynomial if and only if the Cauchy index of the function $g(z)$ defined by

$$g(z) = \frac{q_{n-1}z^{n-1} - q_{n-3}z^{n-3} + \cdots}{z^n - q_{n-2}z^{n-2} + \cdots}$$

is equal to n.

Proof. We distinguish two cases.

Case I: n is even.
Let $n = 2m$. Clearly, in this case

$$q_+(-z^2) = (-1)^m(q_{2m}z^{2m} - q_{2m-2}z^{2m-2} + \cdots)$$
$$= (-1)^m(z^n - q_{n-2}z^{n-2} + \cdots),$$

$$-zq_-(-z^2) = (-1)^m(q_{2m-1}z^{2m-1} - q_{2m-3}z^{2m-3} + \cdots)$$
$$= (-1)^m(q_{n-1}z^{n-1} - q_{n-3}z^{n-3} + \cdots).$$

So we have, in this case, $g(z) = \frac{-zq_-(-z^2)}{q_+(-z^2)}$. By the Hermite-Hurwitz theorem, the Cauchy index of $g(z)$ is equal to the signature of the Bezoutian of the polynomials $q_+(-z^2), -zq_-(-z^2)$, and moreover, $I_g = n$ if and only if that Bezoutian is positive definite. We compute

$$\frac{-q_+(-z^2)wq_-(-w^2) + zq_-(-z^2)q_+(-w^2)}{z-w}$$

$$= \frac{[-q_+(-z^2)wq_-(-w^2) + zq_-(-z^2)q_+(-w^2)](z+w)}{z^2 - w^2}$$

$$= \frac{z^2q_-(-z^2)q_+(-w^2) - q_+(-z^2)w^2q_-(-w^2)}{z^2 - w^2}$$

$$+ zw\frac{q_-(-z^2)q_+(-w^2) - q_+(-z^2)q_-(-w^2)}{z^2 - w^2}$$

So the Bezoutian of $q_+(-z^2)$ and $-zq_-(-z^2)$ is isomorphic to $B(zq_-, q_+) \oplus B(q_+, q_-)$. In particular, the Cauchy index of $g(z)$ is equal to n if and only if both Bezoutians are positive definite. This, by Theorem 9.11, is equivalent to the stability of $q(z)$.

Case II: n is odd.
Let $n = 2m+1$. In this case

$$q_+(-z^2) = (-1)^m(q_{2m}z^{2m} - q_{2m-2}z^{2m-2} + \cdots)$$
$$= (-1)^m(q_{n-1}z^{n-1} - q_{n-3}z^{n-3} + \cdots),$$

$$-zq_-(-z^2) = (-1)^{m+1}(q_{2m+1}z^{2m+1} - q_{2m-1}z^{2m-1} + \cdots)$$
$$= (-1)^{m+1}(z^n - q_{n-2}z^{n-2} + \cdots).$$

So we have, in this case, $g(z) = \frac{q_+(-z^2)}{zq_-(-z^2)}$. We conclude the proof as before. ∎

Since Bezout and Hankel forms are related by congruence, we expect that stability criteria can also be given in terms of Hankel forms. We present next such a result.

As before, let $q(z) = q_+(z^2) + zq_-(z^2)$. Since for $n = 2m$, we have $\deg q_+ = m$ and $\deg q_- = m - 1$, whereas for $n = 2m + 1$, we have $\deg q_+ = \deg q_- = m$, the rational function $g(z)$ defined by

$$g(z) = \frac{q_-(z)}{q_+(z)}$$

is proper for odd n and strictly proper for even n. Thus $g(z)$ can be expanded in a power series in z^{-1},

$$g(z) = g_0 + g_1 z^{-1} + \cdots,$$

with $g_0 = 0$ in case n is even.

Theorem 9.13. *Let $q(z)$ be a real monic polynomial of degree n and let $q_+(z), q_-(z)$ be its even and odd parts respectively. Let $m = [n/2]$. Define the rational function $g(z)$ by*

$$g(z) = \frac{-q_-(-z)}{q_+(-z)} = -g_0 + \sum_{i=1}^{\infty} \frac{g_i}{z^i}.$$

Then $q(z)$ is stable if and only if the two Hankel forms

$$H_m = \begin{pmatrix} g_1 & \cdots & g_m \\ \cdot & \cdot & \cdot & \cdot \\ \cdot & \cdot & \cdot & \cdot \\ \cdot & \cdot & \cdot & \cdot \\ g_m & \cdots & g_{2m-1} \end{pmatrix}$$

and

$$(\sigma H)_m = \begin{pmatrix} g_2 & \cdots & g_{m+1} \\ \cdot & \cdot & \cdot & \cdot \\ \cdot & \cdot & \cdot & \cdot \\ \cdot & \cdot & \cdot & \cdot \\ g_{m+1} & \cdots & g_{2m} \end{pmatrix}$$

are positive definite.

Proof. For the purpose of the proof we let $\hat{p}(z) = p(-z)$. By Theorem 8.40, the Hankel forms H_m and $(\sigma H)_m$ are congruent to the Bezoutians $-B(\hat{q}_+, \hat{q}_-)$ and $B(\hat{q}_+, \widehat{zq_-})$ respectively. It is easily computed, by simple change of variables, that $-B(\hat{q}_+, \hat{q}_-) = B(q_+, q_-)$ and $B(\hat{q}_+, \widehat{zq_-}) = B(zq_-, q_+)$. By Theorem 9.11, the stability of $q(z)$ is equivalent to the positive definiteness of the Bezoutians $B(q_+, q_-)$ and $B(zq_-, q_+)$, and the result follows. ∎

We wish to remark that if we do not assume $q(z)$ to be monic or to have its highest coefficient positive, we need to impose the extra condition $g_0 > 0$ if n is odd.

We conclude our discussion with the Hurwitz determinantal stability criterion.

Theorem 9.14. *All the zeros of the real polynomial $q(z) = \sum_{i=0}^{n} q_i z^i$ lie in the open left half plane if and only if, under the assumption $q_n > 0$, the n determinantal conditions*

$$\mathbf{H}_1 = |q_{n-1}| > 0, \quad \mathbf{H}_2 = \begin{vmatrix} q_{n-2} & q_{n-3} \\ q_n & q_{n-1} \end{vmatrix} > 0, \quad \mathbf{H}_3 = \begin{vmatrix} q_{n-3} & q_{n-4} & q_{n-5} \\ q_{n-1} & q_{n-2} & q_{n-3} \\ 0 & q_n & q_{n-1} \end{vmatrix} > 0, \dots,$$

$$\mathbf{H}_n = \begin{vmatrix} q_0 & 0 & . & & . & 0 \\ . & . & . & . & & . \\ . & . & . & & . & . \\ . & . & q_{n-3} & q_{n-4} & q_{n-5} \\ . & . & q_{n-1} & q_{n-2} & q_{n-3} \\ 0 & . & . & 0 & q_n & q_{n-1} \end{vmatrix} > 0,$$

are satisfied. We interpret $q_{n-k} = 0$ for $k > n$. These determinants are called the **Hurwitz determinants**.

Proof. Clearly, the assumption $q_n > 0$ involves no loss of generality; otherwise, we consider the polynomial $-q(z)$. By Theorem 9.8, the polynomial $q(z)$ is stable if and only if the two Bezoutians $B(q_+, q_-)$ and $B(zq_-, q_+)$ are positive definite. This, by Theorem 8.14, is equivalent to the positive definiteness of all principal minors of both Bezoutians. We find it convenient to check for positivity all the lower-right-hand-corner minors rather than all upper-left-hand-corner minors. In the proof, we will use the Gohberg-Semencul formula (8.40) for the representation of the Bezoutian. The proof will be split in two parts according to $n = \deg q$ being even or odd. We will prove only the case that n is even. The case that n is odd proceeds along similar lines, and we omit it.

Assume n is even and $n = 2m$. The even and odd parts of $q(z)$ are given by

$$q_+(z) = q_0 + \cdots + q_{2m} z^m,$$

$$q_-(z) = q_1 + \cdots + q_{2m-1} z^{m-1}.$$

The Bezoutian $B(q_+, q_-)$ has therefore the representation

$$\left[\begin{pmatrix} q_1 & & \\ . & . & \\ . & . & . \\ . & . & . & . \\ q_{2m-1} & \cdots & q_1 \end{pmatrix} \begin{pmatrix} q_{2m} & q_{2m-2} & \cdots & q_2 \\ & . & . & . \\ & & . & . & . \\ & & & . & . \\ & & & & q_{2m} \end{pmatrix} - \begin{pmatrix} q_0 & & \\ . & . & \\ . & . & . \\ . & . & . & . \\ q_{2m-2} & \cdots & q_0 \end{pmatrix} \begin{pmatrix} 0 & q_{2m-1} & \cdots & q_3 \\ & . & . & . \\ & & . & . & . \\ & & & . & q_{2m-1} \\ & & & & 0 \end{pmatrix} \right] J_m.$$

We consider now the lower right-hand $k \times k$ submatrix. This is given by

$$\left[\begin{pmatrix} q_{2m-2k+1} & & \\ . & . & \\ . & . & . \\ . & . & . & . \\ q_{2m-1} & \cdots & q_{2m-2k+1} \end{pmatrix} \begin{pmatrix} q_{2m} & \cdots & q_{2m-2k+2} \\ & . & . & . \\ & & . & . & . \\ & & & . & . \\ & & & & q_{2m} \end{pmatrix} - \begin{pmatrix} q_{2m-2k} & & \\ . & . & \\ . & . & . \\ . & . & . & . \\ q_{2m-2} & \cdots & q_{2m-2k} \end{pmatrix} \begin{pmatrix} 0 & q_{2m-1} & \cdots & q_{2m-2k+3} \\ & . & . & . \\ & & . & . & . \\ & & & . & q_{2m-1} \\ & & & & 0 \end{pmatrix} \right] J_k.$$

Applying Lemma 3.17, we have

$$\det(B(q_+,q_-))_{i,j=m-k+1}^{m}$$

$$=
\begin{vmatrix}
q_{2m-2k+1} & & & & q_{2m-2k} & & \\
\cdot & \cdot & & & & \cdot & \cdot \\
\cdot & & \cdot & & & \cdot & & \cdot \\
\cdot & & & \cdot & & \cdot & & & \cdot \\
q_{2m-1} & \cdots & q_{2m-2k+1} & q_{2m-2} & \cdots & q_{2m-2k} & \\
0 & \cdots & q_{2m-2k+3} & q_{2m} & \cdots & q_{2m-2k+2} & \\
\cdot & \cdot & \cdot & & \cdot & \cdot & \cdot & \cdot \\
& \cdot & \cdot & \cdot & & & \cdot & \cdot & \cdot \\
& & \cdot & q_{2m-1} & & & & \cdot & \cdot \\
& & & 0 & & & & & q_{2m}
\end{vmatrix} \cdot \det J_k,
$$

Which, using the factor $\det J_k = (-1)^{\frac{k(k-1)}{2}}$, can be rearranged in the form

$$
\begin{vmatrix}
q_{2m-2k+1} & q_{2m-2k} & & 0 & & & & & \cdot \\
\cdot & q_{2m-2k+2} & q_{2m-2k+1} & & & & & & \cdot \\
\cdot & & \cdot & & & & & & \cdot \\
\cdot & & & \cdot & \cdot & & & & 0 \\
q_{2m-1} & q_{2m-2} & & \cdot & & \cdot & \cdots & \cdot & q_{2m-2k+1} & q_{2m-2k} & & \cdot \\
0 & q_{2m} & & & \cdot & & \cdot & \cdots & \cdot & q_{2m-2k+3} & q_{2m-2k+2} & & \cdot \\
\cdot & & & 0 & & \cdot & \cdot & & & & & \cdot \\
\cdot & & & & \cdot & & \cdot & \cdot & \cdot & \cdot & & \cdot \\
\cdot & & & & & & 0 & q_{2m} & q_{2m-1} & q_{2m-2} \\
\cdot & & & & & & & \cdot & 0 & q_{2m}
\end{vmatrix}
$$

Since $q_{2m} = q_n > 0$, it follows that the following determinantal conditions hold:

$$
\begin{vmatrix}
q_{2m-2k+1} & q_{2m-2k} & & 0 & & & \\
& q_{2m-2k+2} & q_{2m-2k+1} & & & & \\
\cdot & & \cdot & & & & \\
\cdot & & & \cdot & \cdot & & \\
q_{2m-1} & q_{2m-2} & & \cdot & & \cdots & \cdot & q_{2m-2k+1} \\
0 & q_{2m} & & \cdot & & \cdot & \cdot & q_{2m-2k+3} \\
\cdot & & & 0 & & \cdot & \cdot & \\
\cdot & & & & \cdot & \cdot & \cdot & \cdot \\
\cdot & & & & & 0 & q_{2m} & q_{2m-1}
\end{vmatrix} > 0.
$$

Using the fact that $n = 2m$, we can write the last determinantal condition as

$$
\begin{vmatrix}
q_{n-2k+1} & q_{n-2k} & 0 & & & & \\
\cdot & q_{n-2k+2} & q_{n-2k+1} & & & & \\
\cdot & & & & & & \\
\cdot & & \cdot & \cdot & & & \\
q_{n-1} & q_{n-2} & \cdot & & \cdot \cdot \cdot & q_{n-2k+1} \\
0 & q_n & \cdot & & \cdot \cdot & q_{n-2k+3} \\
& & 0 & & \cdot \cdot & \\
& & & \cdot \cdot & \cdot & \cdot \\
\cdot & & \cdot & & & 0 \; q_n \; q_{n-1}
\end{vmatrix} > 0.
$$

Note that the order of these determinants is $1, \ldots, 2m - 1$.

Next, we consider the Bezoutian $B(zq_-, q_+)$, which has the representation

$$
\left[
\begin{pmatrix}
q_0 & & \\
\cdot & \cdot & \\
\cdot & \cdot \cdot & \\
\cdot & & \cdot \cdot \cdot \\
q_{2m-2} & \cdots & q_0
\end{pmatrix}
\begin{pmatrix}
q_{2m-1} & q_{2m-3} & \cdot \cdot & q_1 \\
& \cdot & \cdot \cdot \cdot & \cdot \\
& & \cdot \cdot & \cdot \\
& & & \cdot \\
& & & q_{2m-1}
\end{pmatrix}
-
\begin{pmatrix}
0 & & \\
q_1 & \cdot & \\
\cdot & \cdot \cdot & \\
\cdot & & \cdot \cdot \\
q_{2m-3} & \cdot \cdot & q_1 \; 0
\end{pmatrix}
\begin{pmatrix}
q_{2m} & \cdot \cdot & q_2 \\
& \cdot \cdot \cdot & \cdot \\
& & \cdot \cdot & \cdot \\
& & & \cdot \\
& & & q_{2m}
\end{pmatrix}
\right] J_m.
$$

The lower right hand $k \times k$ submatrix is given by

$$
\left[
\begin{pmatrix}
q_{2m-2k} & & \\
\cdot & \cdot & \\
\cdot & \cdot \cdot & \\
\cdot & & \cdot \cdot \cdot \\
q_{2m-2} & \cdots & q_{2m-2k}
\end{pmatrix}
\begin{pmatrix}
q_{2m-1} & q_{2m-3} & \cdot \cdot & q_{2m-2k+1} \\
& \cdot & \cdot \cdot \cdot & \cdot \\
& & \cdot \cdot & \cdot \\
& & & \cdot \\
& & & q_{2m-1}
\end{pmatrix}
-
\begin{pmatrix}
q_{2m-2k-1} & & \\
\cdot & \cdot & \\
\cdot & \cdot \cdot & \\
\cdot & & \cdot \cdot \cdot \\
q_{2m-3} & \cdots & q_{2m-2k-1}
\end{pmatrix}
\begin{pmatrix}
q_{2m} & \cdot \cdot & q_{2m-2k+2} \\
& \cdot \cdot \cdot & \cdot \\
& & \cdot \cdot & \cdot \\
& & & \cdot \\
& & & q_{2m}
\end{pmatrix}
\right] J_k.
$$

Again, by the use of Lemma 3.17, we have

$$
\det\big(B(zq_-, q_+)\big)_{i,j=m-k+1}^m
$$

$$
= \begin{vmatrix}
q_{2m-2k} & & & q_{2m-2k-1} & & \\
\cdot & \cdot & & & \cdot & \cdot \\
\cdot & \cdot & & & & \cdot \\
\cdot & \cdot & & & \cdot & \\
q_{2m-2} & \cdots & q_{2m-2k} & q_{2m-3} & \cdots & q_{2m-2k-1} \\
q_{2m} & \cdots & q_{2m-2k+2} & q_{2m-1} & \cdots & q_{2m-2k+1} \\
& \cdot \cdot \cdot & \cdot & & \cdot \cdot \cdot & \cdot \\
& \cdot \cdot & & & \cdot \cdot & \\
& \cdot & q_{2m-1} & & & \cdot \\
& & q_{2m} & & & q_{2m-1}
\end{vmatrix} \cdot \det J_k
$$

$$
= \begin{vmatrix}
q_{2m-2k} & q_{2m-2k-1} & 0 & & & & & & \cdot \\
\cdot & q_{2m-2k+1} & q_{2m-2k} & & & & & & \cdot \\
\cdot & & \cdot & & & & & & \cdot \\
\cdot & & & \cdot & \cdot & & & & 0 \\
q_{2m-1} & q_{2m-2} & \cdot & \cdot & \cdot & \cdot & q_{2m-2k+1} & q_{2m-2k} \\
q_{2m} & q_{2m-1} & \cdot & \cdot & \cdot & \cdot & q_{2m-2k+3} & q_{2m-2k+2} & \cdot \\
\cdot & & & \cdot & \cdot & \cdot & & & \\
\cdot & & & \cdot & \cdot & \cdot & & & \\
\cdot & & & q_{2m} & q_{2m-1} & q_{2m-2} & q_{2m-3} & \\
\cdot & & & & \cdot & & q_{2m} & q_{2m-1}
\end{vmatrix}
$$

Using $n = 2m$, this determinant can be rewritten as

$$
\begin{vmatrix}
q_{n-2k} & q_{n-2k-1} & 0 & & & & & & \cdot \\
\cdot & q_{n-2k+1} & q_{n-2k} & & & & & & \cdot \\
\cdot & & \cdot & & & & & & \cdot \\
\cdot & & & \cdot & \cdot & & & & 0 \\
q_{n-1} & q_{n-2} & \cdot & \cdot & \cdot & \cdot & q_{n-2k+1} & q_{n-2k} \\
q_n & q_{n-1} & \cdot & \cdot & \cdot & \cdot & q_{n-2k+3} & q_{n-2k+2} & \cdot \\
\cdot & & & \cdot & \cdot & \cdot & & & \\
\cdot & & & \cdot & \cdot & \cdot & & & \\
\cdot & & & q_n & q_{n-1} & q_{n-2} & q_{n-3} & \\
\cdot & & & & \cdot & & q_n & q_{n-1}
\end{vmatrix}
$$

Case II: This is the case in which $n = 2m + 1$ is odd. The proof proceeds along similar lines. ∎

9.3 Exercises

1. Let the real polynomial $q(z) = \sum_{i=0}^{n} q_i z^i$ have zeros $\mu_1, \ldots, \mu_n$. Prove **Orlando's formula**, i.e., that for the Hurwitz determinants defined in Theorem 9.14, we have

$$
\det \mathbf{H}_{n-1} = (-1)^{\frac{n(n-1)}{2}} q_n^{n-1} \prod_{i<k} (\mu_i + \mu_k). \tag{9.10}
$$

2. a. Show that a real polynomial $p(z)$ is a Hurwitz polynomial if and only if the zeros and poles of

$$
f(z) = \frac{p(z) - p(-z)}{p(z) + p(-z)}
$$

are simple, located on the imaginary axis, and mutually separate each other.

b. Show that a real polynomial $p(z) = \sum_{i=0}^{n} p_i z^i$ has all its zeros in the open unit disk if and only if $|p_n| > |p_0|$ and the zeros and poles of

$$f(z) = \frac{p(z) - p^\sharp(z)}{p(z) + p^\sharp(z)}$$

are simple, located on the unit circle, and mutually separate each other. Here $p^\sharp(z) = z^n p(z^{-1})$ is the reciprocal polynomial to p.

3. Given a polynomial $f(z) \in \mathbb{R}[z]$ of degree n, we define

$$S(f) = f(z)f''(z) - f'(z)^2.$$

Show that $f(z)$ has n distinct real roots if and only if

$$S(f) < 0, \ \ S(f') < 0, \ \ \ldots, \ \ S(f^{(n-2)}) < 0.$$

Show that $f(z) = z^3 + 3uz + 2v$ has three distinct real roots if and only if $u < 0$ and $v^2 + u^3 < 0$.

9.4   Notes and Remarks

The approach to problems of root location via the use of quadratic forms goes back to Jacobi. The greatest impetus to this line of research was given by Hermite (1856). A very thorough exposition of the method of quadratic forms is given in the classic paper Krein and Naimark (1936). This paper also contains a comprehensive bibliography.

Theorem 9.11.4 is usually known as the Hermite–Biehler theorem.

There are other approaches to the stability analysis of polynomials. Some are based on complex analysis, and in particular on the argument principle. A very general approach to stability was developed by Liapunov (1893). For the case of a linear system $\dot{x} = Ax$, this leads to the celebrated Lyapunov matrix equation $AX + XA^* = -Q$. Surprisingly, it took half a century to clarify the connection between Lyapunov stability theory and the algebraic stability criteria. For such a derivation, in the spirit of this book, see Willems and Fuhrmann (1992).

Chapter 10
Elements of Linear System Theory

10.1 Introduction

This chapter is devoted to a short introduction to algebraic system theory. We shall focus on the main conceptual underpinnings of the theory, more specifically on the themes of external and internal representations of systems and the associated realization theory. We feel that these topics are to be considered an essential part of linear algebra. In fact, the notions of reachability and observability, introduced by Kalman, fill a gap that the notion of cyclicity left open. Also, they have such a strong intuitive appeal that it would be rather perverse not to use them and search instead for sterile, but "pure," substitute terms.

We saw the central role that polynomials play in the structure theory of linear transformations. The same role is played by rational functions in the context of algebraic system theory. In fact, realization theory is, for rational functions, what the shift operator and the companion matrices are for polynomials.

From the purely mathematical point of view, realization theory has, in the algebraic context, as its main theme a special type of representation for rational functions. Note that given a quadruple of matrices (A, b, c, d) of sizes $n \times n, n \times 1, 1 \times n$, and 1×1 respectively, the function $g(z)$ defined by

$$g(z) = d + c(zI - A)^{-1}b \tag{10.1}$$

is a scalar proper rational function. The realization problem is the corresponding inverse problem. Namely, given a scalar proper rational function $g(z)$, we want to find a quadruple of matrices (A, b, c, d) for which (10.1) holds.

In this sense, the realization problem is an extension of a simpler problem that we solved earlier. That problem was the construction of an operator, or a matrix, that had a given polynomial as its characteristic polynomial. In the solution to that problem, shift operators, and their matrix representations in terms of companion matrices, played a central role. Therefore, it is natural to expect that the same objects will play a similar role in the solution of the realization problem, and this in fact is the case.

P.A. Fuhrmann, *A Polynomial Approach to Linear Algebra*, Universitext,
DOI 10.1007/978-1-4614-0338-8_10, © Springer Science+Business Media, LLC 2012

In Sections 10.4 and 10.5 we use the Hardy space $\mathbf{RH}_+^\infty$. For the reader unfamiliar with the basic results on Hardy spaces and the operators acting on them, it is advisable to read first Chapter 11.

10.2 Systems and Their Representations

Generally, we associate the word system with dynamics, that is, with the way an object evolves with time. Time itself can be modeled in various ways, most commonly as continuous or discrete, as the case may be. In contrast to some other approaches to the study of systems, we will focus here on the proverbial black box approach, given by the schematic diagram

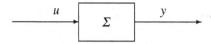

where Σ denotes the system, u the input or control signal, and y the output or observation signal. The way the output signal depends on the input signal is called the **input/output relation**. Such a description is termed an **external representation**. An **internal representation** of a system is a model, usually given in terms of difference or differential equations, that explains, or is compatible with, the external representation. Unless further assumptions are made on the properties of the input/output relations, i.e., linearity and continuity, there is not much of interest that can be said. Because of the context of linear algebra and the elementary nature of our approach, with the tendency to emphasize linear algebraic properties, it is natural for us to restrict ourselves to linear, time-invariant, finite-dimensional systems, which we define below. For discrete-time systems, with an eye to applications such as coding theory, we choose to work over an arbitrary field. On the other hand, for continuous-time systems, requiring derivatives, we work over the real or complex fields.

Definition 10.1. 1. A discrete time, finite-dimensional linear time-invariant system is a triple $(\mathcal{U}, \mathcal{X}, \mathcal{Y})$ of finite-dimensional vector spaces over a field $\mathbb{F}$ and a quadruple of linear transformations $A \in L(\mathcal{X}, \mathcal{X})$, $B \in L(\mathcal{U}, \mathcal{X})$, $C \in L(\mathcal{X}, \mathcal{Y})$, and $D \in L(\mathcal{U}, \mathcal{Y})$, with the system equations given by

$$x_{n+1} = Ax_n + Bu_n,$$
$$y_n = Cx_n + Du_n. \tag{10.2}$$

2. A continuous-time, finite-dimensional linear time-invariant system is a triple $(\mathcal{U}, \mathcal{X}, \mathcal{Y})$ of finite-dimensional vector spaces over the real or complex field and a quadruple of linear transformations $A \in L(\mathcal{X}, \mathcal{X})$, $B \in L(\mathcal{U}, \mathcal{X})$, $C \in L(\mathcal{X}, \mathcal{Y})$, and $D \in L(\mathcal{U}, \mathcal{Y})$, with the system equations given by

$$\dot{x} = Ax + Bu,$$

$$y = Cx + Du. \tag{10.3}$$

The spaces $\mathscr{U}, \mathscr{X}, \mathscr{Y}$ are called the **input space**, **state space**, and **output space** respectively. Usually, we identify these spaces with $\mathbb{F}^m, \mathbb{F}^n, \mathbb{F}^p$ respectively. In that case, the transformations A, B, C, D are given by matrices.

In both cases, we will use the notation $\left(\dfrac{A \,|\, B}{C \,|\, D} \right)$ to describe the system. Such representations are called **state space-realizations**.

Since the development of discrete-time linear systems does not depend on analysis, we will concentrate our attention on this class.

Let us start from the state equation $x_{n+1} = Ax_n + Bu_n$. Substituting in this $x_n = Ax_{n-1} + Bu_{n-1}$, we get $x_{n+1} = A^2 x_{n-1} + ABu_{n-1} + Bu_n$, and proceeding by induction, we get

$$x_{n+1} = A^k x_{n-k+1} + A^{k-1} Bu_{n-k+1} + \cdots + Bu_n. \tag{10.4}$$

Our standing assumption is that before a finite time, all signals were zero, that is, in the remote past the system was at rest. Thus, for some n_0, $u_n = 0, x_n = 0$ for $n < n_0$. With this, equation (10.4) reduces to

$$x_{n+1} = \sum_{j=0}^{\infty} A^j Bu_{n-j},$$

and in particular,

$$x_0 = \sum_{j=0}^{\infty} A^j Bu_{-j-1}.$$

Thus, for $n \geq 0$, equation (10.4) could also be written as

$$\begin{aligned}
x_{n+1} &= \sum_{j=0}^{n} A^j Bu_{n-j} + \sum_{j=n+1}^{\infty} A^j Bu_{n-j} \\
&= A^{n+1} \sum_{j=0}^{\infty} A^j Bu_{-j-1} + \sum_{j=0}^{n} A^j Bu_{n-j} \\
&= A^{n+1} x_0 + \sum_{j=0}^{n} A^j Bu_{n-j}.
\end{aligned}$$

This is the state evolution equation based on initial conditions at time zero.

From the preceding, the input/output relations become

$$y_{n+1} = \sum_{j=0}^{\infty} CA^j Bu_{n-j}. \tag{10.5}$$

We write $y = \tilde{f}(u)$ and call $\tilde{f}$ the input/output map of the system.

Suppose we look now at sequences of input and output signals shifted by one time unit. Let us write $\theta_n = u_{n-1}$ and denote the corresponding states and outputs by ξ_n and η_n respectively. Then

$$\xi_n = \sum_{j=0}^{\infty} A^j B\theta_{n-j-1} = \sum_{j=0}^{\infty} A^j Bu_{n-j-2} = x_{n-1}.$$

and

$$\eta_n = C\xi_n = Cx_{n-1} = y_{n-1}.$$

Let us introduce now the shift operator σ acting on time signals by $\sigma(u_j) = u_{j-1}$. Then the input/output map satisfies

$$\tilde{f}(\sigma(u)) = \sigma(y) = \sigma(\tilde{f}(u)). \tag{10.6}$$

If the signal, zero in the remote past, goes over to the truncated Laurent series $\sum_{j=-\infty}^{\infty} u_{-j} z^j$, we conclude that $\sigma(u)$ is mapped into

$$\sum_{j=-\infty}^{\infty} u_{-j-1} z^j = \sum_{j=-\infty}^{\infty} u_{-j-1} z^{j+1} = z \cdot \sum_{j=-\infty}^{\infty} u_{-j} z^j,$$

that is, in $\mathscr{U}((z^{-1}))$, the shift σ acts as multiplication by z. Since $\tilde{f}$ is a linear map, we get by induction that for an arbitrary polynomial p, we have

$$\tilde{f}(p \cdot u) = p \cdot \tilde{f}(u).$$

This means that with the polynomial module structure induced by σ, the input/output map $\tilde{f}$ is an $\mathbb{F}[z]$-module homomorphism.

We find it convenient to associate with the infinite sequence $\{u_j\}_{-\infty}^{n_0}$ the truncated Laurent series $\sum_{j \geq n_0} u_j z^{-j}$. Thus positive powers of z are associated with past signals, whereas negative powers are associated with future signals. With this convention applied also to the state and output sequences, the input/output relation (10.5) can be written now as

$$y(z) = G(z)u(z), \tag{10.7}$$

where $y(z) = \sum_{j \geq n_0} y_j z^{-j}$ and

$$G(z) = D + \sum_{i=1}^{\infty} \frac{CA^{i-1}B}{z^i} = D + C(zI - A)^{-1}B. \tag{10.8}$$

The function $G(z)$ defined in (10.8) is called the **transfer function** of the system. From this we can conclude the following result.

Proposition 10.2. *Let* $\Sigma = \left(\begin{array}{c|c} A & B \\ \hline C & D \end{array} \right)$ *be a finite-dimensional, linear time-invariant system. Then its* **transfer function** $G(z) = D + C(zI - A)^{-1}B$ *is proper rational.*

Proof. Follows from the equality $(zI - A)^{-1} = \frac{\mathrm{adj}\,(zI-A)}{\det(zI-A)}$. ∎

From the system equations (10.2) it is clear that given that the system is at rest in the remote past, there will be no output from the system prior to an input being sent into the system. In terms of signal spaces, the space of future inputs is mapped into the space of future output signals. This is the concept of causality, and it is built into the internal representation. We formalize this as follows.

Definition 10.3. Let $\mathcal{U}$ and $\mathcal{Y}$ be finite-dimensional linear spaces over the field $\mathbb{F}$.

1. An input/output map is an $\mathbb{F}[z]$-module homomorphism $\tilde{f} : \mathcal{U}((z^{-1})) \longrightarrow \mathcal{Y}((z^{-1}))$.
2. An input/output map $\tilde{f}$ is **causal** if $\tilde{f}(\mathcal{U}[[z^{-1}]]) \subset \mathcal{Y}[[z^{-1}]]$ and **strictly causal** if $\tilde{f}(\mathcal{U}[[z^{-1}]]) \subset z^{-1}\mathcal{Y}[[z^{-1}]]$.

We have the following characterization.

Proposition 10.4. *1. $\tilde{f}(\mathcal{U}((z^{-1}))) \subset \mathcal{Y}((z^{-1}))$ is an $\mathbb{F}[z]$-module homomorphism if and only if there exists $G \in L(\mathcal{U},\mathcal{Y})((z^{-1}))$ for which $\tilde{f}(u) = G \cdot u$.*
2. $\tilde{f}$ is causal if and only if $G \in L(\mathcal{U},\mathcal{Y})[[z^{-1}]]$ and strictly causal if and only if $G \in z^{-1}L(\mathcal{U},\mathcal{Y})[[z^{-1}]]$.

Proof. 1. Assume $\tilde{f} : \mathcal{U}((z^{-1})) \longrightarrow \mathcal{Y}((z^{-1}))$ is defined by $\tilde{f}(u) = G \cdot u$ for some $G \in L(\mathcal{U},\mathcal{Y})((z^{-1}))$. Then

$$z \cdot \tilde{f}(u) = z(Gu) = G(zu) = \tilde{f}(z \cdot u),$$

i.e., $\tilde{f}$ is an $\mathbb{F}[z]$-module homomorphism.

Conversely, assume $\tilde{f}$ is an $\mathbb{F}[z]$-module homomorphism. Let $e_1, \ldots, e_m$ be a basis in U. Set $g_i = \tilde{f}(e_i)$. Let G have columns g_i. Then, using the fact that $\tilde{f}$ is a homomorphism, we compute

$$\tilde{f}(u) = \tilde{f}\sum_{i=1}^{m} u_i z^i = \sum_{i=1}^{m} z^i \tilde{f}(u_i) = \sum_{i=1}^{m} z^i G u_i = Gu.$$

2. Assume $G(z)$ has the expansion $G(z) = \sum_{i=-n}^{\infty} G_i/z^i$ and $u(z) = \sum_{i=0}^{\infty} u_i/z^i$. The polynomial part, not including the constant term, of Gu is given by $\sum_{k=0}^{n}\sum_{i=0}^{k} G_{-n+k-i}u_i z^k$. This vanishes if and only if $\sum_{i=0}^{k} G_{-n+k-i}u_i = 0$ for all choices of $u_0, \ldots, u_{n-1}$. This in turn is equivalent to $G_{-n} = \cdots = G_{-1} = 0$. Strict causality is handled similarly. ∎

Let us digress a bit on the notion of state. Heuristically, the state of the system is the minimum amount of information we need to know about the system at present so that its future outputs can be computed, given that we have access to future inputs. In a sense, the present state of the system is the combined effect of all past inputs. But this type of information is overly redundant. In particular, many past inputs may lead to the same state. Since the input/output map is linear, we can easily eliminate the future inputs from the definition. Therefore, given an input/output map $\tilde{f}$, it is natural to introduce the following notion of equivalence of past inputs. We say that, with $u(z), v(z) \in \mathcal{U}[z]$, $u \simeq_f v$ if $\pi_- \tilde{f}u = \pi_- \tilde{f}v$. This means that the past input sequences $u(z)$ and $v(z)$ have the same future outputs; thus they are indistinguishable based on future observations. The moral of this is that in order to elucidate the notion of state, it is convenient to introduce an auxiliary input/output map.

Definition 10.5. Let $\tilde{f}(\mathscr{U}((z^{-1}))) \subset \mathscr{Y}((z^{-1}))$ be an input/output map. Then the **restricted input/output map** $f : \mathscr{U}[z] \longrightarrow z^{-1}\mathscr{Y}[[z^{-1}]]$ is defined by $f(u) = \pi_{-}\tilde{f}(u)$.

If $\hat{f}$ has the transfer function $G(z)$, then the restricted input/output map is given by $f(u) = \pi_{-}Gu$, i.e., it is the Hankel operator defined by $G(z)$. The relation between the input/output map and the restricted input/output map is best described by the following commutative diagram:

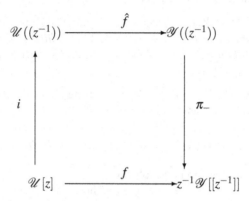

Here $i : \mathscr{U}[z] \longrightarrow \mathscr{U}((z^{-1}))$ is the natural embedding.

10.3   Realization Theory

We turn now to the realization problem. Given a causal input/output map $\hat{f}$ with proper transfer function $G(z)$, we want to find a state-space realization of it, i.e., we want to find linear transformations A, B, C, D such that

$$G(z) = \left(\begin{array}{c|c} A & B \\ \hline C & D \end{array}\right) = D + C(zI - A)^{-1}B.$$

Since $G(z) = \sum_{i=0}^{\infty} G_i/z^i$, we must have for the realization,

$$D = G_0,$$
$$CA^{i-1}B = G_i, \ i \geq 1.$$

The coefficients G_i are called the **Markov parameters** of the system.

To this end, we go back to a state space system $\left(\begin{array}{c|c} A & B \\ \hline C & D \end{array}\right)$. We define the **reachability map** $\mathscr{R} : \mathscr{U}[z] \longrightarrow \mathscr{X}$ by

$$\mathscr{R} \sum_{i=0}^{n} u_i z^i = \sum_{i=0}^{n} A^i B u_i, \tag{10.9}$$

and the **observability map** $\mathscr{O} : \mathscr{X} \longrightarrow z^{-1} \mathscr{Y}[[z^{-1}]]$ by

$$\mathscr{O}x = \sum_{i=1}^{\infty} \frac{CA^{i-1}x}{z^i}. \tag{10.10}$$

Proposition 10.6. *Given the state space system* $\left(\begin{array}{c|c} A & B \\ \hline C & D \end{array}\right)$ *with state space* $\mathscr{X}$, *let* $\mathscr{X}$ *carry the* $\mathbb{F}[z]$-*module structure induced by A as in (4.41). Then*

1. *The reachability and observability maps are* $\mathbb{F}[z]$-*module homomorphisms.*
2. *If f is the restricted input/output map of the system and $G(z) = D + C(zI - A)^{-1}B$ the transfer function, then*

$$f = H_G = \mathscr{O}\mathscr{R}. \tag{10.11}$$

Proof. 1. We compute

$$\mathscr{R}z\sum_{i=0}^{n} u_i z^i = \mathscr{R}\sum_{i=0}^{n} u_i z^{i+1} = \sum_{i=0}^{n} A^{i+1} B u_i$$

$$= A\sum_{i=0}^{n} A^i B u_i = A\mathscr{R}\sum_{i=0}^{n} u_i z^i. \tag{10.12}$$

Also

$$\mathscr{O}Ax = \sum_{i=1}^{\infty} \frac{CA^{i-1}A\xi}{z^i} = \sum_{i=1}^{\infty} \frac{CA^i \xi}{z^i}$$

$$= S_- \sum_{i=1}^{\infty} \frac{CA^{i-1}\xi}{z^i} = S_- \mathscr{O}\xi. \tag{10.13}$$

2. Let $\xi \in \mathscr{U}$ and consider it as a constant polynomial in $\mathscr{U}[z]$. Then we have

$$\mathscr{O}\mathscr{R}\xi = \mathscr{O}(B\xi) = \sum_{i=1}^{\infty} \frac{CA^{i-1}Bx}{z^i} = G(z)\xi.$$

Since both $\mathscr{O}$ and $\mathscr{R}$ are $\mathbb{F}[z]$-module homomorphisms, it follows that

$$\mathscr{O}\mathscr{R}(z^i \xi) = \mathscr{O}(A^i B\xi) = \sum_{j=1}^{\infty} \frac{CA^{j-1}(A^i B\xi)}{z^j}$$

$$= \sum_{j=1}^{\infty} \frac{CA^{i+j-1}B\xi}{z^j} = \pi_- z^i \sum_{j=1}^{\infty} \frac{CA^{j-1}B\xi}{z^j}$$

$$= \pi_- z^i G(z)\xi = \pi_- G(z) z^i \xi.$$

By linearity we get, for $u \in \mathscr{U}[z]$, that (10.11) holds. ∎

So far, nothing has been assumed that would imply further properties of the reachability and observability maps. In particular, recalling the canonical factorizations of maps discussed in Chapter 1, we are interested in the case that $\mathscr{R}$ is surjective and $\mathscr{O}$ is injective. This leads to the following definition.

Definition 10.7. Given the state space system Σ with transfer function $G = \begin{pmatrix} A & B \\ \hline C & D \end{pmatrix}$ with state space $\mathscr{X}$, then

1. The system Σ is called **reachable** if the reachability map $\mathscr{R}$ is surjective.
2. The system Σ is called **observable** if the observability map $\mathscr{O}$ is injective.
3. The system is called **canonical** if it is both reachable and observable.

Proposition 10.8. *Given the state space system Σ with transfer function $G = \begin{pmatrix} A & B \\ \hline C & D \end{pmatrix}$ with the n-dimensional state space $\mathscr{X}$, then*

1. The following statements are equivalent:

   a. The system Σ is reachable.
   b. The zero state can be steered to an arbitrary state $\xi \in \mathscr{X}$ in a finite number of steps.
   c. We have

$$\operatorname{rank}(B, AB, \ldots, A^{n-1}B) = n. \tag{10.14}$$

   d.

$$\cap_{i=0}^{n-1} \operatorname{Ker} B^*(A^*)^i = \{0\}. \tag{10.15}$$

   e.

$$\cap_{i=0}^{\infty} \operatorname{Ker} B^*(A^*)^i = \{0\}. \tag{10.16}$$

2. The following statements are equivalent:

   a. The system Σ is observable.
   b. The only state with all future outputs zero is the zero state.
   c.

$$\cap_{i=0}^{\infty} \operatorname{Ker} CA^i = \{0\}. \tag{10.17}$$

   d.

$$\cap_{i=0}^{n-1} \operatorname{Ker} CA^i = \{0\}. \tag{10.18}$$

   e. We have

$$\operatorname{rank}(C^*, A^*C^*, \ldots, (A^*)^{n-1}C^*) = n. \tag{10.19}$$

Proof. 1. $(a) \Rightarrow (b)$ Given a state $\xi \in \mathscr{X}$, there exists $v(z) = \sum_{i=0}^{k-1} v_i z^i$ such that $\xi = \mathscr{R}v = \sum_{i=0}^{k-1} A^i B v_i$. Setting $u_i = v_{k-1-i}$ and using this control sequence, the solution to the state equation, with zero initial condition, is given by $x_k = \sum_{i=0}^{k-1} A^i B u_{k-1-i} = \sum_{i=0}^{k-1} A^i B v_i = \xi$.

$(b) \Rightarrow (c)$ Since every state $\xi \in \mathscr{X}$ has a representation $\xi = \sum_{i=0}^{k-1} A^i B u_i$ for some k, it follows by an application of the Cayley–Hamilton theorem that we can assume without loss of generality that $k = n$. Thus the map $(B, AB, \ldots, A^{n-1}B) : \mathscr{U}^n \longrightarrow \mathscr{X}$ defined by $(u_0, \ldots, u_n) \mapsto \sum_{i=0}^{n-1} A^i B u_i$ is surjective. Hence $\operatorname{rank}(B, AB, \ldots, A^{n-1}B) = n$.

$(c) \Rightarrow (d)$ The adjoint to the map $(B, AB, \ldots, A^{n-1}B)$ is

$$
\begin{pmatrix} B^* \\ B^* A^* \\ \cdot \\ \cdot \\ B^* (A^*)^{n-1} \end{pmatrix} : \mathscr{X}^* \longrightarrow (\mathscr{U}^*)^n.
$$

Applying Theorem 4.36, we have

$$
\operatorname{Ker} \begin{pmatrix} B^* \\ B^* A^* \\ \cdot \\ \cdot \\ B^* (A^*)^{n-1} \end{pmatrix} = \bigcap_{i=0}^{n-1} \operatorname{Ker} B^* (A^*)^i = \{0\}.
$$

$(d) \Rightarrow (e)$ This follows from the inclusion $\cap_{i=0}^{\infty} \operatorname{Ker} B^* (A^*)^i \subset \cap_{i=0}^{n-1} \operatorname{Ker} B^* (A^*)^i$.

$(e) \Rightarrow (a)$ Let $\phi \in (\operatorname{Im} \mathscr{R})^{\perp} = \operatorname{Ker} \mathscr{R}^* = \cap_{i=0}^{\infty} \operatorname{Ker} B^* (A^*)^i = \{0\}$. Then $\phi = 0$, and the system is reachable.

2. This can be proved similarly. Alternatively, it follows from the first part by duality considerations.

∎

Proposition 10.6 implied that any realization of an input/output map leads to a factorization of the restricted input/output map. We proceed to prove the converse.

Theorem 10.9. *Let $\bar{f} : \mathscr{U}((z^{-1})) \longrightarrow \mathscr{Y}((z^{-1}))$ be an input/output map. Then to any realization of f corresponds a unique factorization $f = hg$, with $h : \mathscr{X} \longrightarrow \mathscr{Y}((z^{-1}))$ and $g : \mathscr{U}((z^{-1})) \longrightarrow \mathscr{X}$ $\mathbb{F}[z]$-module homomorphisms.*

Conversely, given any factorization $f = hg$ into a product of $\mathbb{F}[z]$-module homomorphisms, there exists a unique associated realization. The factorization is canonical, that is, g is surjective and h injective, if and only if the realization is canonical.

Proof. We saw that a realization leads to a factorization $f = \mathscr{O}\mathscr{R}$ with $\mathscr{R}, \mathscr{O}$ the reachability and observability maps respectively.

Conversely, assume we are given a factorization

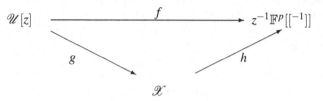

with $\mathscr{X}$ an $\mathbb{F}[z]$-module and g, h both $\mathbb{F}[z]$-module homomorphisms. We define a triple of maps $A : \mathscr{X} \longrightarrow \mathscr{X}$, $B : \mathscr{U} \longrightarrow \mathscr{X}$ and $C : \mathscr{X} \longrightarrow \mathscr{Y}$ by

$$
\begin{aligned}
Ax &= z \cdot x, \\
Bu &= g(i(u)), \\
Cx &= \pi_1 h(x).
\end{aligned}
\tag{10.20}
$$

Here $i : \mathscr{U} \longrightarrow \mathscr{U}[z]$ is the embedding of $\mathscr{U}$ in $\mathscr{U}[z]$ that identifies a vector with the corresponding constant polynomial vector, and $\pi_1 : z^{-1}\mathscr{Y}[[z^{-1}]]$ is the map that reads the coefficient of z^{-1} in the expansion of an element of $z^{-1}\mathscr{Y}[[z^{-1}]]$. It is immediate to check that these equations constitute a realization of f with g and h the reachability and observability maps respectively. In particular, the realization is canonical if and only if the factorization is canonical.   ∎

There are natural concepts of isomorphism for both factorizations as well as realizations.

Definition 10.10. 1. Let $f = h_i g_i$, $i = 1, 2$ be two factorizations of the restricted input/output map f through the $\mathbb{F}[z]$-modules $\mathscr{X}_i$ respectively. We say that the two factorizations are **isomorphic** if there exists an $\mathbb{F}[z]$-module isomorphism ϕ for which the following diagram is commutative:

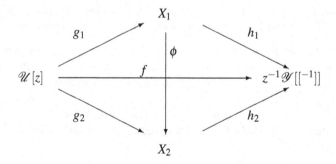

2. We say that two realizations $\left(\dfrac{A_1 \mid B_1}{C_1 \mid D_1} \right)$ and $\left(\dfrac{A_2 \mid B_2}{C_2 \mid D_2} \right)$ are **isomorphic** if there exists an isomorphism $Z : \mathscr{X}_1 \longrightarrow \mathscr{X}_2$ such that the following diagram is commutative.

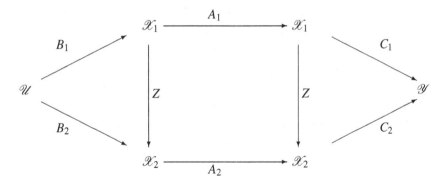

Theorem 10.11 (State space-isomorphism). *1. Given two canonical factorizations of a restricted input/output map, then they are isomorphic. Moreover, the isomorphism is unique.*

2. Assume $G = \left(\begin{array}{c|c} A_1 & B_1 \\ \hline C_1 & D_1 \end{array}\right) = \left(\begin{array}{c|c} A_2 & B_2 \\ \hline C_2 & D_2 \end{array}\right)$ and the two realizations are canonical. Then $D_1 = D_2$ and the two realizations are isomorphic, and moreover, the isomorphism is unique.

Proof. 1. Let $f = h_1 g_1 = h_2 g_2$ be two canonical factorizations through the $\mathbb{F}[z]$-modules $\mathscr{X}_1, \mathscr{X}_2$ respectively. We define $\phi : \mathscr{X}_1 \longrightarrow \mathscr{X}_2$ by $\phi(g_1(u)) = g_2(u)$. First we show that ϕ is well defined. We note that the injectivity of h_1, h_2 implies $\mathrm{Ker}\, g_i = \mathrm{Ker}\, f$. So $g_1(u_1) = g_1(u_2)$ implies $u_1 - u_2 \in \mathrm{Ker}\, g_1 = \mathrm{Ker}\, f = \mathrm{Ker}\, g_2$ and hence also $g_2(u_1) = g_2(u_2)$.

Next we compute

$$\phi(z \cdot g_1(u)) = \phi(g_1(z \cdot u)) = g_2(z \cdot u) = z \cdot g_2(u) = z \cdot \phi(g_1(u)).$$

This shows that ϕ is an $\mathbb{F}[z]$-homomorphism.

It remains to show that $h_2 \phi = h_1$. Now, given $x \in \mathscr{X}_1$, we have $x = g_1(u)$ and $h_1(x) = h_1 g_1(u) = f(u)$. On the other hand, $h_2 \phi(x) = h_2 \phi(g_1(u)) = h_2 g_2(u) = f(u)$, and the equality $h_2 \phi = h_1$ is proved.

To prove uniqueness, assume ϕ_1, ϕ_2 are two isomorphisms from $\mathscr{X}_1$ to $\mathscr{X}_2$ that make the isomorphism diagram commutative. Then $f = h_2 g_2 = h_2 \phi_1 g_1 = h_2 \phi_2 g_1$, which implies $h_2(\phi_2 - \phi_1)g_1 = 0$. Since g_1 is surjective and h_2 injective, the equality $\phi_2 = \phi_1$ follows.

2. Let $\mathscr{R}_1, \mathscr{R}_2$ be the reachability maps of the two realizations respectively and $\mathscr{O}_1, \mathscr{O}_2$ the respective observability maps. All of these four maps are $\mathbb{F}[z]$-homomorphisms and by our assumptions, the reachability maps are surjective while the observability maps are injective. Thus the factorizations $f = \mathscr{O}_i \mathscr{R}_i$ are canonical factorizations. By the first part, there exists a unique $\mathbb{F}[z]$-isomorphism $Z : X_1 \longrightarrow X_2$ such that $Z\mathscr{R}_1 = \mathscr{R}_2$ and $\mathscr{O}_2 Z = \mathscr{O}_1$. That Z is a module isomorphism implies that Z is an invertible map satisfying $ZA_1 = A_2 Z$. Restricting $Z\mathscr{R}_1 = \mathscr{R}_2$ to constant polynomials implies $ZB_1 = B_2$.

Finally, looking at coefficients of z^{-1} in the equality $\mathcal{O}_2 Z = \mathcal{O}_1$ implies $C_2 Z = C_1$. These equalities, taken together, imply the commutativity of the diagram. Uniqueness follows from part 1. ∎

We have used the concept of canonical factorizations, but so far have not demonstrated their existence. Actually, this is easy and we take it up next.

Proposition 10.12. *Let $f : \mathcal{U}[z] \longrightarrow z^{-1}\mathcal{Y}[[z^{-1}]]$ be an $\mathbb{F}[z]$-homomorphism, and $f = H_G$ the Hankel operator of the transfer function. Then canonical factorizations exist. Specifically, the following factorizations are canonical:*

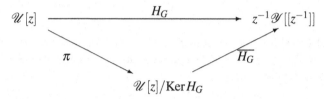

Here, the map π is the canonical projection and $\overline{H_G}$ the map induced by H_G on the quotient space $\mathcal{U}[z]/\mathrm{Ker}\,H_G$.

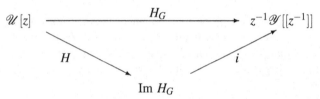

Here, the map i is the embedding of $\mathrm{Im}\,H_G$ in $z^{-1}\mathcal{Y}[[z^{-1}]]$ and H is the Hankel operator considered as a map from $\mathcal{U}[z]$ onto $\mathrm{Im}\,H_G$.

Proof. We saw in Chapter 8 that the Hankel operator is an $\mathbb{F}[z]$-homomorphism. Thus we can apply Exercise 6 in Chapter 1 to conclude that $H_G|_{\mathcal{U}[z]/\mathrm{Ker}\,H_G}$ is an injective $\mathbb{F}[z]$-homomorphism. ∎

From now on, we will treat only the case of single-input single-output systems. These are the systems with scalar-valued transfer functions. If we further assume that the systems have finite-dimensional realizations, then by Kronecker's theorem, this is equivalent to the transfer function being rational. In this case both $\mathrm{Ker}\,H_g$ and $\mathrm{Im}\,H_g$ are submodules of $\mathbb{F}[z]$ and $z^{-1}\mathbb{F}[[z^{-1}]]$ respectively. Moreover, as linear spaces, $\mathrm{Ker}\,H_g$ has finite codimension and $\mathrm{Im}\,H_g$ is finite-dimensional. We can take advantage of the results on representation of submodules contained in Theorems 1.28 and 5.24 to go from the abstract formulation of realization theory to very concrete state space representations. Since $\mathrm{Im}\,H_g$ is a finite-dimensional S_--invariant subspace, it is necessarily of the form X^q for some, uniquely determined, monic polynomial $q(z)$. This leads to the coprime factorization $g(z) = p(z)q(z)^{-1} = q(z)^{-1}p(z)$. A similar argument applies to $\mathrm{Ker}\,H_g$, which is necessarily of the form $q\mathbb{F}[z]$, and this leads to the same coprime factorization.

However, rather than start with a coprime factorization, our starting point will be any factorization

$$g(z) = p(z)q(z)^{-1} = q(z)^{-1}p(z), \qquad (10.21)$$

with no coprimeness assumptions. We note that $\pi_1(f) = [f, 1]$, with the form defined by (5.62), in particular $\pi_1(f) = 0$ for any polynomial $f(z)$. The reader may wonder, since we are in a commutative setting, why distinguish between the two factorizations in (10.21). The reason for this is that in our approach to realization theory, it makes a difference if we consider $q(z)$ as a right denominator or left denominator. In the first case, it determines the input pair (A, B) up to similarity, whereas in the second case it determines the output pair (C, A) up to similarity.

Theorem 10.13. *Let g be a strictly proper rational function and assume the factorization (10.21) with $q(z)$ monic. Then the following are realizations of $g(z)$.*

1. *Assume $g(z) = q(z)^{-1}p(z)$. In the state space X_q define*

$$\begin{aligned} Af &= S_q f, \\ B\xi &= p \cdot \xi, \\ Cf &= \langle f, 1 \rangle. \end{aligned} \qquad (10.22)$$

Then $g(z) = \left(\begin{array}{c|c} A & B \\ \hline C & 0 \end{array} \right)$. This realization is observable. The reachability map $\mathscr{R} : \mathbb{F}[z] \longrightarrow X_q$ is given by

$$\mathscr{R}u = \pi_q(pu). \qquad (10.23)$$

The reachable subspace is $\operatorname{Im} \mathscr{R} = q_1 X_{q_2}$, where $q(z) = q_1(z)q_2(z)$ and $q_1(z)$ is the greatest common divisor of $p(z)$ and $q(z)$. Hence the realization is reachable if and only if $p(z)$ and $q(z)$ are coprime.

2. *Assume $g(z) = p(z)q(z)^{-1}$. In the state space X_q define*

$$\begin{aligned} Af &= S_q f, \\ B\xi &= \xi, \\ Cf &= \langle f, p \rangle. \end{aligned} \qquad (10.24)$$

Then $g(z) = \left(\begin{array}{c|c} A & B \\ \hline C & 0 \end{array} \right)$. This realization is reachable, and the reachability map $\mathscr{R} : \mathbb{F}[z] \longrightarrow X_q$ is given by

$$\mathscr{R}u = \pi_q u, \qquad u \in \mathbb{F}[z]. \qquad (10.25)$$

The unobservable subspace is given by $\operatorname{Ker} \mathscr{O} = q_1 X_{q_2}$, where $q(z) = q_1(z)q_2(z)$ and $q_2(z)$ is the greatest common divisor of $p(z)$ and $q(z)$. The system is observable if and only if $p(z)$ and $q(z)$ are coprime.

3. *Assume $g(z) = q(z)^{-1}p(z)$. In the state space X^q define*

$$Ah = S^q h,$$
$$B\xi = g \cdot \xi,$$
$$Ch = [h, 1]. \qquad (10.26)$$

Then $g(z) = \left(\frac{A|B}{C|0} \right)$. *This realization is observable. It is reachable if and only if* $p(z)$ *and* $q(z)$ *are coprime.*

4. *Assume $g(z) = p(z)q(z)^{-1}$. In the state space X^q define*

$$Ah = S^q h$$
$$B\xi = q^{-1} \cdot \xi$$
$$Ch = [h, p] \qquad (10.27)$$

Then $g(z) = \left(\frac{A|B}{C|0} \right)$. *This realization is reachable. It is observable if and only if* $p(z)$ *and* $q(z)$ *are coprime.*

We refer to these realizations as **shift realizations**.

5. *If $p(z)$ and $q(z)$ are coprime, then all four realizations are isomorphic. In particular, the following commutative diagram shows the isomorphism of realizations (10.22) and (10.24):*

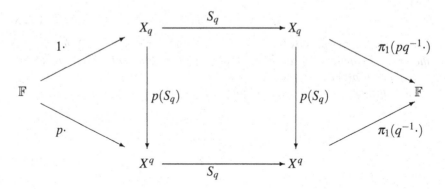

Proof. 1. Let $\xi \in \mathbb{F}$. We compute

$$CA^{i-1}B\xi = \langle S_q^{i-1}p\xi, 1 \rangle = [q^{-1}q\pi_- q^{-1}z^{i-1}p\xi, 1]$$
$$= [z^{i-1}g\xi, 1] = [g, z^{i-1}]\xi = g_i\xi.$$

Hence (10.22) is a realization of $g(z)$.

To show the observability of this realization, assume $f \in \cap_{i=0}^{\infty} \operatorname{Ker} CA^i$. Then, for $i \geq 0$,

$$0 = \langle S_q^i f, 1 \rangle = [q^{-1}q\pi_- q^{-1}z^i f, 1] = [q^{-1}f, z^i].$$

This shows that $q^{-1}(z)f(z)$ is necessarily a polynomial. However $f(z) \in X_q$ implies that $q(z)^{-1}f(z)$ is strictly proper. So $q(z)^{-1}f(z) = 0$ and hence also $f(z) = 0$.

We proceed to compute the reachability map $\mathscr{R}$. We have, for $u(z) = \sum_i u_i z^i$,

$$\mathscr{R}u = \mathscr{R}\sum_i u_i z^i = \sum_i S_q^i p u_i = \pi_q \sum_i p u_i z^i = \pi_q(pu).$$

Clearly, since $\mathscr{R}$ is an $\mathbb{F}[z]$-module homomorphism, $\operatorname{Im}\mathscr{R}$ is a submodule of X_q hence of the form $\operatorname{Im}\mathscr{R} = q_1 X_{q_2}$ for a factorization $q(z) = q_1(z)q_2(z)$ into monic factors. Since $p(z) \in \operatorname{Im}\mathscr{R}$, it follows that $p(z) = q_1(z)p_1(z)$, and therefore $q_1(z)$ is a common factor of $p(z)$ and $q(z)$.

Let now $q'(z)$ be any other common factor of $p(z)$ and $q(z)$. Then $q(z) = q'(z)q''(z)$ and $p(z) = q'(z)p'(z)$. For any polynomial $u(z)$ we have, by Lemma (1.24),

$$\mathscr{R}u = \pi_q(pu) = q'\pi_{q''}(p'u) \in q'X_{q''}.$$

So we have obtained the inclusion $q_1 X_{q_2} \subset q'X_{q''}$, which, by Proposition 5.7, implies $q'(z) \mid q_1(z)$, i.e., $q_1(z)$ is the greatest common divisor of $p(z)$ and $q(z)$.

Clearly $\operatorname{Im}\mathscr{R} = q_1 X_{q_2} = X_q$ if and only if $q_1(z) = 1$, i.e., $p(z)$ and $q(z)$ are coprime.

2. Let $\xi \in \mathbb{F}$. We compute

$$\begin{aligned} CA^{i-1}B\xi &= \langle S_q^{i-1}\xi, p \rangle = [q^{-1}q\pi_- q^{-1}z^{i-1}\xi, p] \\ &= [gz^{i-1}, 1]\xi = g_i\xi. \end{aligned}$$

Hence (10.24) is a realization of g.

We compute the reachability map of this realization:

$$\begin{aligned} \mathscr{R}u &= \mathscr{R}\sum_i u_i z^i = \sum_i S_q^i u_i = \sum_i \pi_q z^i u_i \\ &= \pi_q \sum_i z^i u_i = \pi_q u. \end{aligned}$$

This implies the surjectivity of $\mathscr{R}$ and hence the reachability of this realization.

Now the unobservable subspace is a submodule of X_q, hence of the form $q_1 X_{q_2}$ for a factorization $q(z) = q_1(z)q_2(z)$. So for every $f(z) \in q_1 X_{q_2}$ we have

$$\begin{aligned} 0 = CA^i f &= \langle S_q^i f, p \rangle = [q^{-1}q\pi_- q^{-1}z^i f, p] = [q^{-1}z^i f, p] \\ &= [q_2^{-1}q_1^{-1}z^i q_1 f_2, p] = [pq_2^{-1}f_2, z^i]. \end{aligned}$$

In particular, choosing $f_1(z) = 1$, we conclude that $p(z)q_2(z)^{-1}$ is a polynomial, say $p_2(z)$. So $p(z) = p_2(z)q_2(z)$ and $q_2(z)$ is a common factor of $p(z)$ and $q(z)$. Let now $q''(z)$ be any other common factor of $p(z)$ and $q(z)$ and set $q(z) = q'(z)q''(z)$, $p(z) = p'(z)q''(z)$. For $f = q'f' \in q'X_{q''}$ we have

$$CA^i f = \langle S^i_q q' f', p \rangle = [q^{-1} q \pi_- q^{-1} z^i q' f', p]$$
$$= [p(q")^{-1} z^i f', 1] = [p' f', z^i] = 0.$$

So we get the inclusion $q' X_{q''} \subset q_1 X_{q_2}$ and hence $q''(z) \mid q_2(z)$, which shows that $q_2(z)$ is indeed the greatest common factor of $p(z)$ and $q(z)$.

Clearly, this realization is observable if and only if $\mathrm{Ker}\,\mathcal{O} = q_1 X_{q_2} = \{0\}$, which is the case only if $q_2(z)$ is constant and hence $p(z)$ and $q(z)$ coprime.

3. Using the isomorphism (5.38) of the polynomial and rational models, we get the isomorphisms of the realizations (10.22) and (10.26).

4. By the same method, we get the isomorphisms of the realizations (10.24) and (10.27). These isomorphisms are independent of coprimeness assumptions.

5. By parts (3) and (4), it suffices to show that the realizations (10.22) and (10.24) are isomorphic. By the coprimeness of $p(z)$ and $q(z)$ and Theorem 5.18, the transformation $p(S_q)$ is invertible and certainly satisfies $p(S_q)S_q = S_q p(S_q)$, i.e., it is a module ismorphism. Since also $p(S_q)1 = \pi_q(p \cdot 1) = p$ and $\langle p(S_q)f, 1 \rangle = \langle f, p \rangle$, it follows that the diagram is commutative. ∎

Based on the shift realizations, we can, by a suitable choice of bases whether in X_q or X^q, get concrete realizations given in terms of matrices.

Theorem 10.14. *Let $g(z)$ be a strictly proper rational function and assume $g(z) = p(z)q(z)^{-1} = q(z)^{-1}p(z)$ with $q(z)$ monic. Assume $q(z) = z^n + q_{n-1}z^{n-1} + \cdots + q_0$, $p(z) = p_{n-1}z^{n-1} + \cdots + p_0$, and $g(z) = \sum_{i=1}^{\infty} g_i/z^i$. Then the following are realizations of $g(z)$:*

1. **Controller realization**

$$A = \begin{pmatrix} 0 & 1 & & \\ & & \cdot & \\ & & \cdot & \\ & & & 1 \\ -q_0 & \cdots & & -q_{n-1} \end{pmatrix}, \quad B = \begin{pmatrix} 0 \\ \cdot \\ \cdot \\ 0 \\ 1 \end{pmatrix}, \quad C = \begin{pmatrix} p_0 & \cdots & p_{n-1} \end{pmatrix}.$$

2. **Controllability realization**

$$A = \begin{pmatrix} 0 & \cdots & & -q_0 \\ 1 & & & \cdot \\ & \cdot & & \cdot \\ & & \cdot & \cdot \\ & & 1 & -q_{n-1} \end{pmatrix}, \quad B = \begin{pmatrix} 1 \\ 0 \\ \cdot \\ \cdot \\ 0 \end{pmatrix}, \quad C = \begin{pmatrix} g_1 & \cdots & g_n \end{pmatrix}.$$

3. Observability realization

$$A = \begin{pmatrix} 0 & 1 & & \\ & & \cdot & \\ & & & \cdot \\ & & & & 1 \\ -q_0 & \cdots & & -q_{n-1} \end{pmatrix}, \quad B = \begin{pmatrix} g_1 \\ \cdot \\ \cdot \\ \cdot \\ g_n \end{pmatrix}, \quad C = \begin{pmatrix} 1 & 0 & .. & 0 \end{pmatrix}.$$

4. Observer realization

$$A = \begin{pmatrix} 0 & \cdots & & -q_0 \\ 1 & & & \cdot \\ & \cdot & & \cdot \\ & & \cdot & \cdot \\ & & 1 & -q_{n-1} \end{pmatrix}, \quad B = \begin{pmatrix} p_0 \\ \cdot \\ \cdot \\ \cdot \\ p_{n-1} \end{pmatrix}, \quad C = \begin{pmatrix} 0 & .. & 0 & 1 \end{pmatrix}.$$

5. *If $p(z)$ and $q(z)$ are coprime, then the previous four realizations are all canonical, hence isomorphic. The isomorphisms are given by means of the following diagram:*

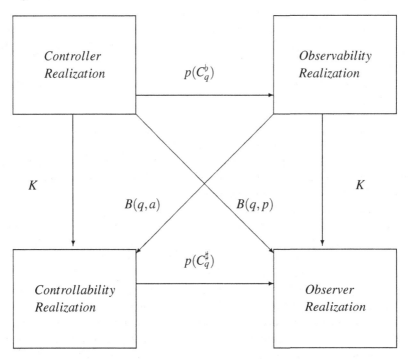

Here $B(q,p)$ and $B(q,a)$ are Bezoutians with the polynomial $a(z)$ arising from the solution to the Bezout equation $a(z)p(z) + b(z)q(z) = 1$, and K is the Hankel matrix

$$K = [I]_{co}^{st} = \begin{pmatrix} q_1 & \cdots & q_{n-1} & 1 \\ \cdot & & \cdots & 1 \\ \cdot & & \cdots & \\ \cdot & & \cdot & \\ q_{n-1} & 1 & & \\ 1 & & & \end{pmatrix}.$$

Proof. 1. We use the representation $g(z) = p(z)q(z)^{-1}$ and take the matrix representation of the shift realization (10.24) with respect to the control basis $\{e_1, \ldots, e_n\}$. In fact, we have $A = [S_q]_{co}^{co} = C_q^\flat$, and since $e_n(z) = 1$, also B has the required form. Finally, the matrix representation of C follows from $Cf = \langle p, f \rangle = \widetilde{[p]}^{st} [f]^{co}$ and the obvious fact that $\widetilde{[p]}^{st} = (p_0 \cdots p_{n-1})$.

2. We use the representation $g(z) = p(z)q(z)^{-1}$ and take the matrix representation of the shift realization (10.24) with respect to the standard basis $\{1, z, \ldots, z^{n-1}\}$. In this case, $A = [S_q]_{st}^{st} = C_q^\sharp$. The form of B is immediate. To conclude, we use equation (8.71), that is, $p(z) = \sum_{i=1}^{n} g_i e_i(z)$, to get $\widetilde{[p]}^{co} = (g_1 \cdots g_n)$.

3. We use the representation $g(z) = q(z)^{-1}p(z)$ and take the matrix representation of the shift realization (10.22) with respect to the control basis. Here again $A = [S_q]_{co}^{co} = C_q^\flat$ and B is again obtained by applying (8.71). The matrix C is obtained as $\widetilde{[1]}^{st}$.

4. We use the representation $g(z) = q(z)^{-1}p(z)$ and take the matrix representation of the shift realization (10.22) with respect to the standard basis. The derivation follows as before.

5. We note that the controller and controllability realizations are different matrix representations of realization (10.22) taken with respect to the control and standard bases. Thus the isomorphism is given by $K = [I]_{co}^{st}$, which has the required Hankel form. A similar observation holds for the isomorphism of the observability and observer realizations.

   On the other hand, the controller and observability realizations are matrix representations of realizations (10.22) and (10.24), both taken with respect to the control basis. By Theorem 10.13, under the assumption of coprimeness for p and q, these realizations are isomorphic, and the isomorphism is given by the map $p(S_q)$. Taking the matrix representation of this map with respect to the control basis, we get $[p(S_q)]_{co}^{co} = p([S_q]_{co}^{co}) = p(C_q^\flat)$. A similar derivation holds for the isomorphism of the controllability and observer realizations.

   The controller and observer realizations are matrix representations of (10.22) and (10.24), taken with respect to the control and standard bases respectively. Hence based on the diagram in Theorem 10.13 and using Theorem 8.27, the isomorphism is given by $[p(S_q)]_{co}^{st} = B(q,p)$.

Finally, since by Theorem 5.18 the inverse of $p(S_q)$ is $a(S_q)$, the polynomial $a(z)$ coming from a solution to the Bezout equation, the same reasoning as before shows that the isomorphism between the observability and controllability realizations is given by $[a(S_q)]^{st}_{co} = B(q,a)$. ∎

Note that the controller and observer realizations are dual to each other, and the same is true for the controllability and observability realizations.

The continued fraction representation (8.75) of a rational function can be translated immediately into a canonical form realization that incorporates the atoms.

Theorem 10.15. *Let the strictly proper transfer function* $g(z) = p(z)/q(z)$ *have the sequence of atoms* $\{\beta_i, a_{i+1}(z)\}$, *and assume that* $\alpha_1, \ldots, \alpha_r$ *are the degrees of the atoms and*

$$a_k(z) = z^{\alpha_k} + \sum_{i=0}^{\alpha_k - 1} a_i^{(k)} z^i. \tag{10.28}$$

Then $g(z)$ *has a realization* (A, b, c) *of the form*

$$A = \begin{pmatrix} A_{11} & A_{12} & & & \\ A_{21} & A_{22} & \cdot & & \\ & \cdot & \cdot & \cdot & \\ & & \cdot & \cdot & A_{r-1\,r} \\ & & & A_{r\,r-1} & A_{rr} \end{pmatrix}, \tag{10.29}$$

where

$$A_{ii} = \begin{pmatrix} 0 & \cdots & & -a_0^{(i)} \\ 1 & & & \cdot \\ & \cdot & & \cdot \\ & & \cdot & \cdot \\ & & 1 & -a_{\alpha_i - 1}^{(i)} \end{pmatrix}, \quad i = 1, \ldots r,$$

$$A_{i+1\,i} = \begin{pmatrix} 0 & \cdots & 0 & 1 \\ \cdot & & & 0 \\ \cdot & & & \cdot \\ \cdot & & & \cdot \\ 0 & \cdots & & 0 \end{pmatrix} \tag{10.30}$$

and $A_{i\,i+1} = \beta_{i-1} A_{i+1\,i}$,

$$b = \begin{pmatrix} 1 \\ 0 \\ \cdot \\ \cdot \\ \cdot \\ 0 \end{pmatrix}, \quad c = \begin{pmatrix} 0 & .. & \beta_0 & 0 & .. & 0 \end{pmatrix} \tag{10.31}$$

with β_0 being in the α_1 position.

Proof. We use the shift realization (10.24) of $g(z)$ which, by the coprimeness of p and q, is canonical. Taking its matrix representation with respect to the orthogonal basis $B_{or} = \{1, z, \ldots, z^{n_1 - 1}, Q_1, zQ_1, \ldots, z^{n_2 - 1}Q_1, \ldots, Q_{r-1}, \ldots, z^{n_r - 1 - 1}Q_{r-1}\}$, with Q_i the Lanczos polynomials of the first kind, proves the theorem. In the computation of the matrix representation we lean heavily on the recurrence formula (8.78) for the Lanczos polynomials. ∎

10.4   Stabilization

The purpose of modeling, and realization theory is part of that process, is to gain a better understanding of the system in order to improve its performance by controlling it.

The most fundamental control problem is that of stabilization. We say that a system is stable if given any initial condition, the system tends to its rest point with increasing time. If the system is given by a difference equation $x_{n+1} = Ax_n$, the stability condition is that all eigenvalues of A lie in the open unit disk. If we have a continuous-time system given by $\dot{x} = Ax$, the stability condition is that all eigenvalues of A lie in the open left half-plane.

Since all the information needed for the computation of the future behavior of the system is given in the system equations and the present state, that is the initial condition of the system, we expect that a control law should be a function of the state, that is, a map, linear in the context we work in, $F : \mathscr{X} \longrightarrow \mathscr{U}$. In the case of a system given by equations (10.2), we have

$$x_{n+1} = Ax_n + Bu_n,$$
$$u_n = Fx_n + v_n, \tag{10.32}$$

and the closed-loop equation becomes

$$x_{n+1} = (A + BF)x_n + Bv_n.$$

In terms of flow diagrams we have the feedback configuration

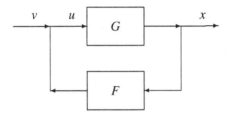

There are two extreme idealizations inherent in the assumptions of this scheme. The first is the assumption that we have a precise model of our system, which is hardly ever the case. The other idealization is the assumption that we have access to full state information. This difficulty is more easily overcome via the utilization of observers, that is auxiliary systems that reconstruct, asymptotically, the state, or more generally by the use of dynamic controllers. The first problem is more serious, and one way to cope with it is to develop the theory of robust control. This, however, is outside the scope of this book.

It is still instructive to study the original problem of stabilizing a given linear system with full state information. The following theorem on pole-shifting states that using state feedback, we have full freedom in assigning the characteristic polynomial of the closed-loop system, provided the pair (A,B) is reachable. Here we handle only the single-input case.

Theorem 10.16. *Let the pair (A,b) be reachable with A an $n \times n$ matrix and b an n vector. Then for each monic polynomial s of degree n, there exists a control law $k : \mathbb{F}^n \longrightarrow \mathbb{F}$ for which the characteristic polynomial of $A - bk$ is s.*

Proof. The characteristic polynomial is independent of the basis chosen. We choose to work in the control basis corresponding to $q(z) = \det(zI - A) = z^n + q_{n-1}z^{n-1} + \cdots + q_0$. With respect to this basis, the pair (A,b) has the matrix representation

$$
A = \begin{pmatrix} 0 & 1 & & \\ & & \ddots & \\ & & & 1 \\ -q_0 & \cdots & & -q_{n-1} \end{pmatrix}, \quad b = \begin{pmatrix} 0 \\ \vdots \\ \vdots \\ 1 \end{pmatrix}.
$$

Let $s(z) = z^n + s_{n-1}z^{n-1} + \cdots + s_0$. Then $k = \begin{pmatrix} q_0 - s_0 & \cdots & q_{n-1} - s_{n-1} \end{pmatrix}$ is the required control law. This theorem explains the reference to the control basis. ∎

We pass now to the analysis of stabilization when we have no direct access to full state information. We assume that the system is given by the equations

$$x_{n+1} = Ax_n + Bu_n,$$

$$y_n = Cx_n. \tag{10.33}$$

We construct an auxiliary system, called an observer, that utilizes all the available information, namely the control signal and the observation signal:

$$\xi_{n+1} = A\xi_n + Bu_n - H(y_n - C\xi_n),$$
$$\eta_n = \xi_n. \tag{10.34}$$

We define the error signal by

$$e_n = x_n - \xi_n.$$

The object of the construction of an observer is to choose the matrix H so that the error e_n tends to zero. This means that with increasing time, the output of the observer gives an approximation of the current state of the system. Subtracting the two system equations, we get

$$x_{n+1} - \xi_{n+1} = A(x_n - \xi_n) + H(Cx_n - C\xi_n),$$

or

$$e_{n+1} = (A + HC)e_n.$$

Thus, we see that the observer construction problem is exactly dual to that of stabilization by state feedback.

Theorem 10.17. *An observable linear single-output system has an observer, or equivalently, an asymptotic state estimator.*

Proof. We apply Theorem 10.16. This shows that the characteristic polynomial of $A + HC$ can be arbitrarily chosen. ∎

Of course, state estimation and the state feedback law can be combined into a single framework. Thus we use the observer to get an asymptotic approximation for the state, and we feed this back into the system as if it were the true state. The next theorem shows that as regards stabilization, this is sufficient.

Theorem 10.18. *Given a canonical single-input/single-output system $\left(\begin{array}{c|c} A & B \\ \hline C & 0 \end{array}\right)$, choose linear maps H, F so that $A + BF$ and $A + HC$ are both stable matrices. Then*

1. The closed-loop system given by

$$x_{n+1} = Ax_n + Bu_n,$$
$$y_n = Cx_n,$$

$$\xi_{n+1} = A\xi_n + Bu_n - H(y_n - C\xi_n),$$
$$u_n = -F\xi_n + v_n, \tag{10.35}$$

is stable.

2. *The transfer function of the closed-loop system, that is, the transfer function from*
   v to y, is $G_c(z) = C(zI - A + BF)^{-1}B$.

Proof. 1. We can rewrite the closed-loop system equations as

$$\begin{pmatrix} x_{n+1} \\ \xi_{n+1} \end{pmatrix} = \begin{pmatrix} A & -BF \\ HC & A - HC - BF \end{pmatrix} \begin{pmatrix} x_n \\ \xi_n \end{pmatrix},$$

$$y_n = \begin{pmatrix} C & 0 \end{pmatrix} \begin{pmatrix} x_n \\ \xi_n \end{pmatrix},$$

and it remains to compute the characteristic polynomial of the new state matrix. We use the similarity

$$\begin{pmatrix} I & 0 \\ -I & I \end{pmatrix} \begin{pmatrix} A & -BF \\ HC & A - HC - BF \end{pmatrix} \begin{pmatrix} I & 0 \\ I & I \end{pmatrix} = \begin{pmatrix} A - BF & -BF \\ 0 & A - HC \end{pmatrix}.$$

Therefore, the characteristic polynomial of $\begin{pmatrix} A & -BF \\ HC & A - HC - BF \end{pmatrix}$ is the product of the characteristic polynomials of $A - BF$ and $A - HC$. This shows also that the controller and observer for the original system can be designed independently.

2. We compute the transfer function of the closed-loop system. The closed-loop system transfer function is given by

$$G_c(z) = \left(\begin{array}{cc|c} A & -BF & B \\ HC & A - HC - BF & B \\ \hline C & 0 & 0 \end{array} \right).$$

Utilizing the previous similarity, we get

$$G_c(z) = \left(\begin{array}{cc|c} A - BF & -BF & B \\ 0 & A - HC & 0 \\ \hline C & 0 & 0 \end{array} \right) = C(zI - A + BF)^{-1}B.$$

Note that the closed-loop transfer function does not depend on the specifics of the observer. This is the case because the transfer function desribes the steady state of the system. The observer does affect the transients in the system.   ∎

Let us review the previous observer-controller construction from a transfer function point of view. Let $G(z) = C(zI - A)^{-1}B$ have the polynomial coprime factorization p/q. Going back to the closed-loop system equations, we can write $\xi_{n+1} = (A - HC)\xi_n + Hy_n + Bu_n$. Hence the transfer function from (u, y) to the control signal is given by $\left(F(zI - A + HC)^{-1}B \quad F(zI - A + HC)^{-1}H \right)$, and the overall transfer function is determined from

$$y = Gu,$$
$$u = v - H_u u - H_y y,$$

(10.36)

where

$$H_u = F(zI - A + HC)^{-1}B,$$
$$H_y = F(zI - A + HC)^{-1}H.$$

The flow chart of this configuration is given by

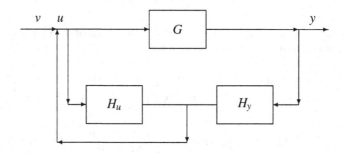

Note now that given that we chose H to stabilize $A - HC$, the two transfer functions H_u, H_y have the following representations:

$$H_u = \frac{p_u}{q_c}, \qquad H_y = \frac{p_y}{q_c},$$

where $q_c(z) = \det(zI - A + HC)$ is a stable polynomial with $\deg q_c = \deg q$. From (10.36) we get $(I + H_u)u = v - H_y y$, or $u = (I + H_u)^{-1}v - (I + H_u)^{-1}H_y y$. This implies $y = Gu = G(I + H_u)^{-1}v - G(I + H_u)^{-1}H_y y$. So the closed-loop transfer function is given by

$$G_c = (I + G(I + H_u)^{-1}H_y)^{-1}G(I + H_u)^{-1}.$$

Since we are dealing here with single-input single-output systems, all transfer functions are scalar-valed and have polynomial coprime factorizations. When these are substituted in the previous expression, we obtain

$$G_c = \frac{1}{1 + \dfrac{p}{q} \cdot \dfrac{1}{1 + \dfrac{p_u}{q_c}} \cdot \dfrac{p_y}{q_c}} \cdot \frac{p}{q} \cdot \frac{1}{1 + \dfrac{p_u}{q_c}}$$

$$= \frac{1}{1 + \dfrac{p}{q} \cdot \dfrac{p_y}{q_c + p_u}} \cdot \frac{p_y}{q_c} \cdot \frac{p}{q} \cdot \frac{q_c}{q_c + p_u} = \frac{pq_c}{q(q_c + p_u) + pp_y}.$$

The denominator polynomial has degree $2 \deg q$ and has to be chosen to be stable. So let $t(z)$ be an arbitrary stable polynomial of degree $2 \deg q$. Since $p(z)$ and $q(z)$ are coprime, the equation $a(z)p(z) + b(z)q(z) = t(z)$ is solvable, with an infinite number of solutions. We can specify a unique solution by imposing the extra requirement that $\deg a < \deg q$. Since $\deg a p < 2 \deg q - 1$, we must have $\deg(bq) = \deg t$. So $\deg b = \deg q$, and we can write $b(z) = q_c + p_u(z)$ and $a(z) = p_y(z)$. Moreover, $p_u(z), p_y(z)$ have degrees smaller than $\deg q$.

A very special case is obtained if we choose a stable polynomial $s(z)$ of degree equal to the degree of $q(z)$, and let $t(z) = s(z) \cdot q_c$. The closed-loop transfer function reduces in this case to $p(z)/s(z)$, which is the case in the observer-controller configuration.

We go back now to the equation

$$q(z)(q_c + p_u(z)) + p(z)p_y(z) = s(z)q_c$$

and divide by the right-hand side to obtain

$$\frac{q(z)}{s(z)} \cdot \frac{q_c + p_u}{q_c} + \frac{p(z)}{s(z)} \cdot \frac{p_y(z)}{q_c(z)} = 1. \tag{10.37}$$

Now all four rational functions appearing in this equation are proper and stable, that is, they belong to $\mathbf{RH}_+^\infty$. Moreover, equation (10.37) is just a Bezout identity in $\mathbf{RH}_+^\infty$, and $g(z) = \frac{p(z)/s(z)}{q(z)/s(z)}$ is a coprime factorization of $g(z)$ over the ring $\mathbf{RH}_+^\infty$.

This analysis is important, since we can now reverse our reasoning. We see that for stabilization it was sufficient to solve a Bezout equation over $\mathbf{RH}_+^\infty$ that is based on a coprime factorization of $g(z)$ over the same ring. We can take this as a starting point of a more general approach.

10.5   The Youla–Kucera Parametrization

We go back now to the stabilization problem. We would like to design a compensator, that is, an auxiliary system, that takes as its input the output of the original system, and its output is taken as an additional control signal, in such a way that the closed-loop system is stable. In terms of flow charts we are back to the configuration

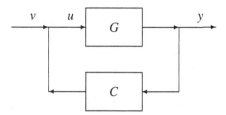

with the sole difference that we no longer assume the state to be accessible. It is a simple exercise to check that the closed-loop transfer function is given by $G_c = (I - GC)^{-1}G = G(I - CG)^{-1}$.

Now a natural condition for stabilization would be that G_c should be stable. Indeed, from an external point of view, this seems the right condition. However, this condition does not guarantee that internal signals stay bounded. To see what the issue is, we note that there are two natural entry points for disturbances. One is the observation errors and the other is that of errors in the control signals. Thus we are led to the standard control configuration

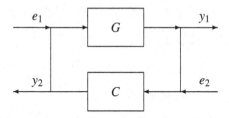

and the internal stability requirement is that the four transfer functions from the error variables u_1, u_2 to the internal variables e_1, e_2 should be stable. Since the system equations are

$$e_2 = u_2 + Ge_1,$$
$$e_1 = u_1 + Ce_2,$$

we get

$$\begin{pmatrix} e_1 \\ e_2 \end{pmatrix} = \begin{pmatrix} (I-CG)^{-1} & (I-CG)^{-1}C \\ (I-GC)^{-1}G & (I-GC)^{-1} \end{pmatrix}.$$

We denote the 2×2 block transfer function by $H(G,C)$. This leads to the following.

Definition 10.19. Given a linear system with transfer function G and a compensator C, we say that the pair (G,C) is **internally stable** if $H(G,C) \in \mathbf{RH}_+^\infty$.

Thus internal stability is a stronger requirement than just closed loop stability, which only requires $(I - G(z)C(z))^{-1}G(z)$ to be stable.

The next result gives a criterion for internal stability via a coprimeness condition over the ring $\mathbf{RH}_+^\infty$. We use the fact that any rational transfer function has a coprime factorization over the ring $\mathbf{RH}_+^\infty$.

Proposition 10.20. *Given the rational transfer function $G(z)$ and the compensator $C(z)$, let $G(z) = N(z)/M(z)$ and $C(z) = U(z)/V(z)$ be coprime factorizations over $\mathbf{RH}_+^\infty$. Then $(G(z),C(z))$ is internally stable if and only if $\Delta(z) = M(z)V(z) - N(z)U(z)$ is invertible in $\mathbf{RH}_+^\infty$.*

Proof. Assume $\Delta^{-1} \in \mathbf{RH}_+^\infty$. Then clearly $H(G,C) = \Delta^{-1}\begin{pmatrix} MV & MU \\ NV & MV \end{pmatrix} \in \mathbf{RH}_+^\infty$.

Conversely, assume $H(G,C) \in \mathbf{RH}_+^\infty$. Since

$$(I-G(z)C(z))^{-1}G(z)C(z) = (I-G(z)C(z))^{-1}G(z)C(z)(I-(I-G(z)C(z))) \in \mathbf{RH}_+^\infty,$$

it follows that $\Delta(z)^{-1}N(z)U(z) \in \mathbf{RH}_+^\infty$. This implies

$$\begin{pmatrix} M(z) \\ N(z) \end{pmatrix} \Delta(z)^{-1} \left(V(z) \ U(z) \right) \in \mathbf{RH}_+^\infty.$$

Since coprimeness in $\mathbf{RH}_+^\infty$ is equivalent to the Bezout identities, it follows that $\Delta(z)^{-1} \in \mathbf{RH}_+^\infty$. ∎

Corollary 10.21. *Let $G(z) = N(z)/M(z)$ be a coprime factorization over $\mathbf{RH}_+^\infty$. Then there exists a stabilizing compensator $C(z)$ if and only if $C(z) = U(z)/V(z)$ and $M(z)V(z) - N(z)U(z) = I$.*

Proof. If there exist $U(z), V(z)$ such that $M(z)V(z) - N(z)U(z) = I$, then $C(z) = U(z)/V(z)$ is easily checked to be a stabilizing controller.

Conversely, if $C(z) = U_1(z)/V_1(z)$ is a stabilizing controller, then by Proposition 10.20, $\Delta(z) = M(z)V_1(z) - N(z)U_1(z)$ is invertible in $\mathbf{RH}_+^\infty$. Defining $U(z) = U_1(z)\Delta(z)^{-1}$ and $V(z) = V_1(z)\Delta(z)^{-1}$, we obtain $C(z) = U(z)/V(z)$ and $M(z)V(z) - N(z)U(z) = I$. ∎

The following result, going by the name of the **Youla–Kucera parametrization**, is the cornerstone of modern stabilization theory.

Theorem 10.22. *Let $G(z) = N(z)/M(z)$ be a coprime factorization. Let $U(z), V(z)$ be any solution of the Bezout equation $M(z)V(z) - N(z)U(z) = I$. Then the set of all stabilizing controllers is given by*

$$\left\{ C(z) = \frac{U(z) + M(z)Q(z)}{V(z) + N(z)Q(z)} \mid Q(z) \in \mathbf{RH}_+^\infty, V(z) + N(z)Q(z) \neq 0 \right\}.$$

Proof. Suppose $C(z) = \frac{U(z) + M(z)Q(z)}{V(z) + N(z)Q(z)}$ with $Q(z) \in \mathbf{RH}_+^\infty$. We check that $M(z)(V(z) + N(z)Q(z)) - N(z)(U(z) + M(z)Q(z)) = M(z)V(z) - N(z)U(z) = I$. Thus, by Corollary 10.21, $C(z)$ is a stabilizing compensator.

Conversely, assume $C(z)$ stabilizes $G(z)$. We know, by Corollary 10.21, that $C(z) = U_1(z)/V_1(z)$ with $M(z)V_1(z) - N(z)U_1(z) = I$. By subtraction we get $M(z)(V_1(z) - V(z)) = N(z)(U_1(z) - U(z))$. Since $M(z), N(z)$ are coprime, $M(z)$ divides $U_1(z) - U(z)$, so there exists $Q(z) \in \mathbf{RH}_+^\infty$ such that $U_1(z) = U(z) + M(z)Q(z)$. This immediately implies $V_1(z) = V(z) + N(z)Q(z)$. ∎

10.6 Exercises

1. (The **Hautus test**). Consider the pair (A, b) with A a real $n \times n$ matrix and b an n vector.

   a. Show that (A, b) is reachable if and only if rank $\left(zI - A \ b \right) = n$ for all $z \in \mathbb{C}$.
   b. Show that (A, b) is stabilizable by state feedback if and only if for all z in the closed right half-plane, we have rank $\left(zI - A \ b \right) = n$.

2. Let $\phi_n(z), \psi_n(z)$ be the monic orthogonal polynomials of the first and second kind respectively, associated with the positive sequence $\{c_n\}$, which were defined in exercise 8.6b. Show that

$$\frac{1}{2}\frac{\psi_n}{\phi_n} = \left(\begin{array}{c|c} A_n & B_n \\ \hline C_n & D_n \end{array}\right),$$

where the matrices A_n, B_n, C_n, D_n are given by

$$A_n = \begin{pmatrix} \gamma_1 & \gamma_2(1-\gamma_1^2) & \gamma_3\Pi_{i=1}^2(1-\gamma_i^2) & \cdot & \cdot & \cdot & \gamma_n\Pi_{i=1}^{n-1}(1-\gamma_i^2) \\ 1 & -\gamma_2\gamma_1 & \gamma_2(1-\gamma_1^2) & \cdot & \cdot & \cdot & -\gamma_n\gamma_1\Pi_{i=2}^{n-1}(1-\gamma_i^2) \\ & 1 & -\gamma_3\gamma_2 & & & & \\ & & 1 & \cdot & & & \\ & & & & \cdot & & \\ & & & & & 1 & -\gamma_n\gamma_{n-2}(1-\gamma_{n-1}^2) \\ & & & & & 1 & -\gamma_n\gamma_{n-1} \end{pmatrix},$$

$$B_n = \begin{pmatrix} 1 \\ 0 \\ \cdot \\ \cdot \\ \cdot \\ 0 \end{pmatrix}, \qquad C_n = \left(\gamma_1 \quad \gamma_2(1-\gamma_1^2) \cdots \gamma_n\Pi_{i=1}^{n-1}(1-\gamma_i^2)\right),$$

$$D_n = \tfrac{1}{2}.$$

3. Show that $\det A_n = (-1)^{n-1}\gamma_n$.
4. Show that the Toeplitz matrix can be diagonalized via the following congruence transformation:

$$\begin{pmatrix} 1 & & & \\ \phi_{1,0} & 1 & & \\ & & \cdot & \\ & & & \cdot \\ \phi_{n,0} & & \phi_{n,n-1} & 1 \end{pmatrix} \begin{pmatrix} c_0 & \cdots & c_n \\ \cdot & & \cdot \\ \cdot & & \cdot \\ \cdot & & \cdot \\ \cdot & & \cdot \\ c_n & \cdots & c_0 \end{pmatrix} \begin{pmatrix} 1 & \phi_{1,0} & & \phi_{n,0} \\ & 1 & & \phi_{n,1} \\ & & \cdot & \\ & & & \cdot\ \phi_{n,n-1} \\ & & & 1 \end{pmatrix},$$

$$= c_0 \begin{pmatrix} h_0 & & & \\ & h_1 & & \\ & & \cdot & \\ & & & \cdot \\ & & & & h_n \end{pmatrix} = c_0 \begin{pmatrix} 1 & & & \\ & (1-\gamma_1^2) & & \\ & & \cdot & \\ & & & \cdot \\ & & & (1-\gamma_{n-1}^2)\cdots(1-\gamma_1^2) \end{pmatrix}.$$

5. Show that

$$\det T_n = c_0(1 - \gamma_1^2)^{n-1}(1 - \gamma_2^2)^{n-2}\cdots(1 - \gamma_{n-1}^2). \qquad (10.38)$$

6. Define a sequence of kernel functions by

$$K_n(z,w) = \sum_{j=0}^{n} \frac{\phi_j(z)\overline{\phi_j(w)}}{h_j}.$$

Show that the kernels $K_n(z,w)$ have the following properties:

a. We have

$$K_n(z,w) = \overline{K_n(w,z)}.$$

b. The matrix $K_n = (k_{ij})$, defined through $K_n(z,w) = \sum_{i=0}^{n}\sum_{j=0}^{n} K_{ij} z^i \overline{w}^j$, is Hermitian, i.e., $k_{ij} = \overline{k_{ji}}$.

c. The kernels K_n are reproducing kernels, i.e., for each $f \in X_{\phi_{n+1}}$ we have $\langle f, K(\cdot, w)\rangle_C = f(w)$.

d. The reproducing kernels have the following determinantal representation:

$$K_n(z,w) = -\frac{1}{\det T_n}\begin{vmatrix} c_0 & \cdots & c_n & 1 \\ & \cdots & & \overline{w} \\ & \cdots & & \cdot \\ & \cdots & & \cdot \\ c_n & \cdots & c_0 & \overline{w}^n \\ 1 & z & \cdots z^n & 0 \end{vmatrix}.$$

e. The following representation for the reproducing kernels holds:

$$K_n(z,w) = \sum_{j=0}^{n} \frac{\phi_j(z)\overline{\phi_j(w)}}{h_j} = \frac{\phi_n^\sharp(z)\overline{\phi_n^\sharp(w)} - z\overline{w}\phi_n(z)\overline{\phi_n(w)}}{h_n(1 - z\overline{w})}.$$

This representation of the reproducing kernels $K_n(z,w)$ is called the **Christoffel–Darboux formula**.

7. Show that all the zeros of the orthogonal polynomials ϕ_n are located in the open unit disk.

8. Prove the following Gohberg–Semencul formula

$$K_n = \frac{1}{h_n}[\phi_n^{\sharp}(S)\phi_n^{\sharp}(\tilde{S}) - S\phi_n(\tilde{S})\phi_n(S)\tilde{S}]$$

$$= \left[\begin{pmatrix} 1 & & \\ \phi_{n,n-1} & \cdot & \\ \cdot & \cdot & \\ \cdot & & \cdot \\ \phi_{n,1} & \cdot\cdot & \phi_{n,n-1} & 1 \end{pmatrix} \begin{pmatrix} 1 & \phi_{n,n-1} & \cdot\cdot & \phi_{n,1} \\ & \cdot & & \cdot\cdot\cdot \\ & & \cdot\cdot\cdot \\ & & & \cdot & \phi_{n,n-1} \\ & & & & 1 \end{pmatrix} \right.$$

$$\left. - \begin{pmatrix} 0 & & \\ \phi_{n,0} & & \\ \cdot & & \cdot \\ \cdot & & \cdot \\ \phi_{n,n-1} & \cdot\cdot & \phi_{n,0} & 0 \end{pmatrix} \begin{pmatrix} 0 & \phi_{n,0} & \cdot\cdot & \phi_{n,n-1} \\ \cdot & & \cdot\cdot\cdot \\ & & \cdot\cdot\cdot \\ & & & \cdot & \phi_{n,0} \\ & & & & 0 \end{pmatrix} \right].$$

9. With

$$K_n(z,w) = \Sigma_{j=0}^n \frac{\phi_j(z)\overline{\phi_j(w)}}{h_j} = \sum_{i=0}^n \sum_{j=0}^n K_{ij} z^i \overline{w}^j,$$

show that $K_n = T_n^{-1}$.

10.7   Notes and Remarks

The study of systems, linear and nonlinear, is the work of many people. However, the contributions of Kalman to this topic were, and remain, central. His insight connecting an applied engineering subject to abstract module theory is of fundamental importance. This insight led eventually to the wide use of polynomial techniques in linear system theory. An early exposition of this is given in Chapter 11 of Kalman, Falb, and Arbib (1969). Another approach to systems problems, using polynomial techniques, was developed by Rosenbrock (1970), who also proved the ultimate result on pole placement by state feedback. The connection between the abstract module approach employed by Kalman and Rosenbrock's method of polynomial system matrices was clarified in Fuhrmann (1977). A more recent approach was initiated in Willems (1986, 1989, 1991). This approach moves away from the input/output point of view and concentrates on system equations.

Chapter 11
Rational Hardy Spaces

11.1 Introduction

We devote this chapter to a description of the setting in which we will study, in Chapter 12, model reduction problems. These will include approximation in Hankel norm as well as approximation by balanced truncation. Since optimization is involved, we choose to work in rational Hardy spaces. This fits with our focus on finite-dimensional linear systems.

The natural setting for the development of AAK theory is that of Hardy spaces, whose study is part of complex and functional analysis. However, to keep our exposition algebraic, we shall assume the rationality of functions and the finite dimensionality of model spaces. Moreover, to stay in line with the exposition in previous chapters, we shall treat only the case of scalar-valued transfer functions. There is another choice we are making, and that is to study Hankel operators on Hardy spaces corresponding to the open left and right half-planes, rather than the spaces corresponding to the unit disk and its exterior. Thus in effect we are studying continuous-time systems, though we shall not delve into this in great detail. The advantage of this choice is greater symmetry. The price we pay for this is that to a certain extent shift operators have to be replaced by translation semigroups. Even this can be corrected. Indeed, once we get the identification of the Hankel operator ranges with rational models, shift operators reenter the picture.

Of course, with the choices we made, the results are not the most general. However, we gain the advantage of simplicity. As a result, this material can be explained to an undergraduate without any reference to more advanced areas of mathematics such as functional analysis and operator theory. In fact, the material presented in this chapter covers, albeit in simplified form, some of the most important results in modern operator theory and system theory. Thus it can also be taken as an introduction to these topics.

The chapter is structured as follows. In Section 11.2 we collect basic information on Hankel operators, invariant subspaces, and their representation via Beurling's theorem. Next we introduce model intertwining operators. We do this using the

P.A. Fuhrmann, *A Polynomial Approach to Linear Algebra*, Universitext,
DOI 10.1007/978-1-4614-0338-8_11, © Springer Science+Business Media, LLC 2012

frequency-domain representation of the right-translation semigroup. We study the
basic properties of intertwining maps and in particular their invertibility properties.
The important point here is the connection of invertibility to the solvability of an
H_+^∞ Bezout equation. We follow this by defining Hankel operators. For the case of a
rational, antistable function we give specific, Beurling-type, representations for the
cokernel and the image of the corresponding Hankel operator. Of importance is the
connection between Hankel operators and intertwining maps. This connection, coupled with invertibility properties of intertwining maps, is the key to duality theory.

11.2   Hardy Spaces and Their Maps

11.2.1   Rational Hardy Spaces

Our principal interest in the following will be centered around the study of Hankel
operators with rational and bounded symbols. In the algebraic approach, taken in
Chapter 8, the Hankel operators were defined using the direct sum decomposition
$\mathbb{F}((z^{-1})) = \mathbb{F}[z] \oplus z^{-1}\mathbb{F}[[z^{-1}]]$, or alternatively $\mathbb{F}(z) = \mathbb{F}[z] \oplus \mathbb{F}_{spr}(z)$, where $\mathbb{F}_{spr}(z)$
is the space of all strictly proper rational functions. The way to think of this direct
sum decomposition is that it is a decomposition based on the location of singularities
of rational functions. Functions in $\mathbb{F}_{spr}(z)$ have only finite singularities and behave
nicely at ∞, whereas polynomials behave nicely at all points but have a singularity
at ∞.

   We shall now consider analogous spaces. Since we are passing to an area
bordering on analysis, we restrict the field to be the complex field $\mathbb{C}$. The two
domains where we shall consider singularities are the open left and right half-planes.
This choice is dictated by considerations of symmetry, which somewhat simplify the
development of the theory. In particular, many results concerning duality are made
more easily accessible

   Our setting will be that of **Hardy spaces**. Thus, H_+^2 is the Hilbert space of all
analytic functions in the open right half-plane $\mathbb{C}_+$, with

$$||f||^2 = \sup_{x>0} \frac{1}{\pi} \int_{-\infty}^{\infty} |f(x+iy)|^2 dy.$$

The space H_-^2 is similarly defined in the open left half plane. It is a theorem of Fatou
that guarantees the existence of boundary values of $H_\pm^2$-functions on the imaginary
axis. Thus the spaces $H_\pm^2$ can be considered closed subspaces of $L^2(i\mathbf{R})$, the space
of Lebesgue square integrable functions on the imaginary axis. It follows from the
Fourier-Plancherel and Paley-Wiener theorems that

$$L^2(i\mathbb{R}) = H_+^2 \oplus H_-^2,$$

with H^2_+ and H^2_- the Fourier-Plancherel transforms of $L^2(0,\infty)$ and $L^2(-\infty,0)$ respectively. Also, H^∞_+ and H^∞_- will denote the spaces of bounded analytic functions in the open right and left half-planes respectively. These spaces can be considered subspaces of $L^\infty(i\mathbb{R})$, the space of Lebesgue measurable and essentially bounded functions on the imaginary axis. We will define

$$f^*(z) = \overline{f(-\bar{z})}. \tag{11.1}$$

Note that on the imaginary axis we have $f^*(it) = \overline{f(it)}$. Also, since we shall deal with inner product spaces, we will reserve the use of $M \oplus N$ for the orthogonal direct sum of the spaces M and N, whereas we shall use $M \dotplus N$ for the algebraic direct sum.

This describes the general analytic setting. Since we want to keep the exposition elementary, that is, as algebraic as possible, we shall not work with these spaces but rather with the subsets of these spaces consisting of their rational elements. We use the letter **R** in front of the various spaces to denote **rational**. The following definition introduces the spaces of interest.

Definition 11.1. We denote by $\mathbf{RL}^\infty, \mathbf{RH}^\infty_+, \mathbf{RH}^\infty_-, \mathbf{RL}^2, \mathbf{RH}^2_+, \mathbf{RH}^2_-$ the sets of all rational functions that are contained in $L^\infty, H^\infty_+, H^\infty_-, L^2, H^2_+, H^2_-$ respectively.

The following proposition, the proof of which is obvious, gives a characterization of the elements of these spaces.

Proposition 11.2. *In terms of polynomials, the spaces of Definition 11.1 have the following characterizations:*

$$\mathbf{RL}^\infty = \left\{ \frac{p(z)}{q(z)} \Big| p(z), q(z) \in \mathbb{C}[z], \deg p \le \deg q, q(\zeta) \ne 0, \ \zeta \in i\mathbf{R} \right\},$$

$$\mathbf{RH}^\infty_+ = \left\{ \frac{p(z)}{q(z)} \in \mathbf{RL}^\infty \Big| q(z) \ stable \right\},$$

$$\mathbf{RH}^\infty_- = \left\{ \frac{p(z)}{q(z)} \in \mathbf{RL}^\infty \Big| q(z) \ antistable \right\},$$

$$\mathbf{RL}^2 = \left\{ \frac{p(z)}{q(z)} \in \mathbf{RL}^\infty \Big| \deg p < \deg q \right\},$$

$$\mathbf{RH}^2_+ = \left\{ \frac{p(z)}{q(z)} \in \mathbf{RL}^2 \Big| q(z) \ stable \right\},$$

$$\mathbf{RH}^2_- = \left\{ \frac{p(z)}{q(z)} \in \mathbf{RL}^2 \Big| q(z) \ antistable \right\}. \tag{11.2}$$

We note that all the spaces introduced here are obviously infinite-dimensional, and $\mathbf{RL}^2$ is an inner product space, with the inner product given, for $f(z), g(z) \in \mathbf{RL}^2$, by

$$(f,g) = \frac{1}{2\pi} \int_{-\infty}^{\infty} f(it)\overline{g(it)}dt. \tag{11.3}$$

We observe that, assuming the functions $f(z), g(z)$ to be rational, the integral can be evaluated by computing residues, using partial fraction decompositions and replacing the integral on the imaginary axis by integration over large enough semi circles.

Proposition 11.3. *We have the following additive decompositions:*

$$\mathbf{RL}^\infty = \mathbf{RH}_+^\infty + \mathbf{RH}_-^\infty, \tag{11.4}$$

$$\mathbf{RL}^2 = \mathbf{RH}_+^2 \oplus \mathbf{RH}_-^2. \tag{11.5}$$

Proof. Follows by applying partial fraction decomposition. The sum in (11.4) is not a direct sum, since the constants appear in both summands. The fact that the direct sum in (11.5) is orthogonal follows from the most elementary properties of complex contour integration and the partial fraction decomposition, the details of which are omitted. ∎

We shall denote by P_+ and P_- the orthogonal projections of $\mathbf{RL}^2$ on $\mathbf{RH}_+^2$ and $\mathbf{RH}_-^2$ respectively that correspond to the direct sum (11.5). We note that if $f(z) \in \mathbf{RL}^2$ and $f(z) = \frac{p(z)}{q(z)}$ with $p(z), q(z)$ coprime polynomials, then $q(z)$ has no zeros on the imaginary axis. Thus, we have the factorization $q(z) = q_-(z)q_+(z)$ with $q_-(z)$ stable and $q_+(z)$ antistable. Thus there exists a unique partial fraction decomposition $f(z) = \frac{p(z)}{q(z)} = \frac{p_1(z)}{q_-(z)} + \frac{p_2(z)}{q_+(z)}$, with $\deg p_1 < \deg q_-$ and $\deg p_2 < \deg q_+$. Clearly, $\frac{p_1(z)}{q_-(z)} \in \mathbf{RH}_+^2$ and $\frac{p_2(z)}{q_+(z)} \in \mathbf{RH}_-^2$, and we have

$$P_+ \frac{p}{q} = \frac{p_1}{q_-},$$

$$P_- \frac{p}{q} = \frac{p_2}{q_+}. \tag{11.6}$$

The space $\mathbf{RL}^\infty$ is clearly a commutative ring with identity. The spaces $\mathbf{RH}_+^\infty$ and $\mathbf{RH}_-^\infty$ are subrings. So

$$\mathbf{RH}_+^\infty = \left\{ \frac{p(z)}{q(z)} \mid q(z) \in S, \deg p \le \deg q \right\}.$$

Thus $\mathbf{RH}_+^\infty$ is the set of all rational functions that are uniformly bounded in the closed right half plane. This is clearly a commutative ring with identity.

Given $f(z) = p(z)/q(z) \in \mathbf{RH}_+^\infty$ and a factorization $p(z) = p_+(z)p_-(z)$ into the stable factor $p_-(z)$ and antistable factor $p_+(z)$, we use the notation $\nabla f = p_+$ and set $\pi_+ = \deg \nabla f$. As in Chapter 1, we define the **relative degree** ρ by $\rho(\frac{p}{q}) = \deg q - \deg p$, and $\rho(0) = -\infty$. We define a **degree function** $\delta : \mathbf{RH}_+^\infty \longrightarrow \mathbb{Z}_+$ by

$$\delta(f) = \begin{cases} \deg \nabla f + \rho(f), & f \neq 0, \\ -\infty, & f = 0. \end{cases} \tag{11.7}$$

Obviously $f(z) \in \mathbf{RH}_+^\infty$ is invertible in $\mathbf{RH}_+^\infty$ if and only if it can be represented as the quotient of two stable polynomials of equal degree. This happens if and only if $\delta(f) = 0$.

Let us fix now a monic, stable polynomial $\sigma(z)$. To be specific we can choose $\sigma(z) = z + 1$. Given $f(z) = p(z)/q(z) \in \mathbf{RH}_+^\infty$, then $f(z)$ can be brought, by multiplication by an invertible element, to the form $\frac{p_+}{\sigma^{v+\pi}}$, where $v = \deg q - \deg p$.

Proposition 11.4. *Given* $f_i(z) = \frac{p_i(z)}{q_i(z)} \in \mathbf{RH}_+^\infty$, $i = 1, 2$, *with* $\nabla(\frac{p_i}{q_i}) = p_i^+$ *and* $\pi_{i,+} = \deg p_i^+$, *we set* $v_i = \deg q_i = \deg p_i$. *Let* $p_+(z)$ *be the greatest common divisor of* $p_1^+(z), p_2^+(z)$ *and set* $\pi_+ = \deg p_+$ *and* $v = \min\{v_1, v_2\}$. *Then* $\frac{p_+(z)}{\sigma(z)^{v+\pi_+}}$ *is a greatest common divisor of* $\frac{p_i(z)}{q_i(z)}$, $i = 1, 2$.

Proof. Without loss of generality we can assume

$$f_1(z) = \frac{p_1^+}{\sigma^{v_1+\pi_{1,+}}}, \quad f_2(z) = \frac{p_2^+}{\sigma^{v_2+\pi_{2,+}}}.$$

Writing $p_i^+(z) = p_+(z)\hat{p}_i^+(z)$ we have $f_i = \frac{p_+}{\sigma^{v+\pi_+}} \cdot \frac{\hat{p}_i^+}{\sigma^{v_i-v+\pi_{i,+}-\pi_+}}$. Now

$$\rho\left(\frac{\hat{p}_i^+}{\sigma^{v_i-v+\pi_{i,+}-\pi_+}}\right) = v_i - v + \pi_{i,+} - \pi_+ - \pi_{i,+} + \deg p_+ = v_i - v \geq 0.$$

This shows that $\frac{\hat{p}_i^+}{\sigma^{v_i-v+\pi_{i,+}-\pi_+}}$ is proper and hence $\frac{p_+}{\sigma^{v+\pi_+}}$ is a common factor.

Let now $\frac{r(z)}{s(z)}$ be any other common factor of $\frac{p_i(z)}{q_i(z)}$, $i = 1, 2$. Without loss of generality we can assume that $r(z)$ is antistable. Let

$$\frac{p_i(z)}{q_i(z)} = \frac{r(z)}{s(z)} \cdot \frac{u_i(z)}{t_i(z)}.$$

From the equality $st_i p_i^+ = ru_i \sigma^{v_i+\pi_{i,+}}$, it follows that $r_+(z)$, the antistable factor of $r(z)$, divides $p_i^+(z)$ and hence divides also $p_+(z)$, the greatest common divisor of the $p_i^+(z)$. Let us write $p_+(z) = r_+(z)\hat{r}_+(z)$. Since $u(z)/t(z)$ is proper, we have $\deg s - \deg r \leq \deg q_i - \deg p_i = v_i$ and hence $\deg s - \deg r \leq v$. Now we write

$$\frac{p_+}{\sigma^{v+\pi_+}} = \frac{r_+\hat{r}_+}{\sigma^{v+\pi_+}} = \frac{r_+\hat{r}_+r_-}{s} \cdot \frac{s}{\sigma^{v+\pi_+}r_-} = \frac{r}{s} \cdot \frac{s}{\sigma^{v+\pi_+}r_-}.$$

We compute now the relative degree of the right factor:

$$v + \pi_+ + \deg r_- - \deg s = v + \deg r_- + \deg p_+ - \deg s + (\pi_+ - \deg p_+)$$
$$= v + \deg r - \deg s \geq 0.$$

This shows that $\frac{s}{\sigma^{v+\pi_+} r_-}$ is proper, and hence $\frac{p_+}{\sigma^{v+\pi_+}}$ is indeed a greatest common divisor of $\frac{p_1(z)}{q_1(z)}$ and $\frac{p_2(z)}{q_2(z)}$.   ∎

The next theorem studiies the solvability of the Bezout equation in $\mathbf{RH}_+^\infty$.

Theorem 11.5. *Let* $f_1(z), f_2(z) \in \mathbf{RH}_+^\infty$ *be coprime. Then there exist* $g_i(z) \in \mathbf{RH}_+^\infty$ *such that*

$$g_1(z)f_1(z) + g_2(z)f_2(z) = 1. \tag{11.8}$$

Proof. We may assume without loss of generality that

$$f_1(z) = \frac{p_1(z)}{\sigma(z)^{\pi_1}}, \quad f_2(z) = \frac{p_2(z)}{\sigma(z)^{v+\pi_2}}$$

with $p_i(z)$ coprime, antistable polynomials and $\pi_i = \deg p_i$.

Clearly, there exist polynomials $\alpha_1(z), \alpha_2(z)$ for which $\alpha_1 p_1 + \alpha_2 p_2 = \sigma^{\pi_1 + \pi_2 + v}$, and we may assume without loss of generality that $\deg \alpha_2 < \deg p_1$. Dividing by the right-hand side, we obtain

$$\frac{\alpha_1}{\sigma^{\pi_2 + v}} \cdot \frac{p_1}{\sigma^{\pi_1}} + \frac{\alpha_2}{\sigma^{\pi_1}} \cdot \frac{p_2}{\sigma^{\pi_2 + v}} = 1.$$

Now $\frac{p_2(z)}{\sigma(z)^{\pi_2 + v}}$ is proper, whereas $\frac{\alpha_2(z)}{\sigma(z)^{\pi_1}}$ is strictly proper by the condition $\deg \alpha_2 < \deg p_1$. So the product of these functions is strictly proper. Next, we note that the relative degree of $\frac{p_1(z)}{\sigma(z)^{\pi_1}}$ is zero, which forces $\frac{\alpha_1(z)}{\sigma(z)^{\pi_2 + v}}$ to be proper. Defining

$$g_1(z) = \frac{\alpha_1(z)}{\sigma(z)^{\pi_2 + v}}, \quad g_2(z) = \frac{\alpha_2(z)}{\sigma(z)^{\pi_1}},$$

we have, by the stability of $\sigma(z)$, that $g_i(z) \in \mathbf{RH}_+^\infty$ and that the Bezout identity (11.8) is satisfied.   ∎

Corollary 11.6. *Let* $f_1(z), f_2(z) \in \mathbf{RH}_+^\infty$ *and let* $f(z)$ *be a greatest common divisor of* $f_1(z), f_2(z)$. *Then there exist* $g_i(z) \in \mathbf{RH}_+^\infty$ *such that*

$$g_1(z)f_1(z) + g_2(z)f_2(z) = f(z). \tag{11.9}$$

Theorem 11.7. *The ring* $\mathbf{RH}_+^\infty$ *is a principal ideal domain.*

Proof. It suffices to show that any ideal $J \subset \mathbf{RH}_+^\infty$ is principal. This is clearly the case for the zero ideal. So we may as well assume that J is a nonzero ideal. Now any nonzero element $f(z) \in J$ can be written as $f(z) = \frac{p(z)}{q(z)} = \frac{p_+(z)p_-(z)}{q}$ with $p(z), q(z)$ coprime and $q(z)$ stable and $p(z)$ factored into its stable factor $p_-(z)$ and antistable factor $p_+(z)$.

Of all nonzero elements in J we choose $f(z)$ for which $\pi_+ = \deg \nabla f$ is minimal. Let $\mathscr{C} = \{f \in J \mid \deg f = \pi_+\}$. In $\mathscr{C}$ we choose an arbitrary element $g(z)$ of minimal relative degree. We will show that $g(z)$ is a generator of J.

To this end let $f(z)$ be an arbitrary element in J. Let $h(z)$ be a greatest common divisor of $f(z)$ and $g(z)$. By Corollary 11.6, there exist $k(z), l(z) \in \mathbf{RH}^\infty_+$ for which $k(z)f(z) + l(z)g(z) = h(z)$.

Since $\deg \nabla g$ is minimal, it follows that $\deg \nabla g \le \deg \nabla h$. On the other hand, since $h(z)$ divides $g(z)$, we have $\nabla h | \nabla g$ and hence $\deg \nabla g \ge \deg \nabla h$. So the equality $\deg \nabla g = \deg \nabla h$ follows. This shows that $h(z) \in \mathscr{C}$ and hence $\rho(g) \le \rho(h)$, since $g(z)$ is the element of $\mathscr{C}$ of minimal relative degree. Again the division relation $h(z) \mid g(z)$ implies $\rho(g) \ge \rho(h)$. Hence we have the equality $\rho(g) = \rho(h)$. It follows that $g(z)$ and $h(z)$ differ at most by a factor that is invertible in $\mathbf{RH}^\infty_+$, so $g(z)$ divides $f(z)$. ∎

Actually, it can be shown that with the degree function defined in (11.7), $\mathbf{RH}^\infty_+$ is a Euclidean domain if we define $\delta(f) = \rho(f) + \deg \nabla f$. Thus $\delta(f)$ counts the number of zeros of $f(z)$ in the closed right half-plane together with the zeros at infinity. We omit the details.

The following theorem introduces important module structures. The proofs of the statements are elementary and we omit them.

Theorem 11.8. *1.* $\mathbf{RL}^\infty$ *is a ring with identity under the usual operation of addition and multiplication of rational functions, with* $\mathbf{RH}^\infty_+$ *and* $\mathbf{RH}^\infty_-$ *subrings.*

2. $\mathbf{RL}^2$ *is is a linear space but also a module over the ring* $\mathbf{RL}^\infty$, *with the module structure defined, for* $\psi(z) \in \mathbf{RL}^\infty$ *and* $f(z) \in \mathbf{RL}^2$, *by*

$$(\psi \cdot f)(z) = \psi(z)f(z), \tag{11.10}$$

as well as over the rings $\mathbf{RH}^\infty_+$ *and* $\mathbf{RH}^\infty_-$.

3. With the $\mathbf{RH}^\infty_+$ *induced module structure on* $\mathbf{RL}^2$, $\mathbf{RH}^2_+$ *is a submodule.*

4. $\mathbf{RH}^2_-$ *has an* $\mathbf{RH}^\infty_+$ *module structure given by*

$$\psi \cdot h = P_-(\psi h), \qquad h(z) \in \mathbf{RH}^2_-.$$

We wish to point out that since $z \notin \mathbf{RL}^\infty$, the multiplication by z operator is not defined in these spaces. This forces us to some minor departures from the algebraic versions of some of the statements.

With all these spaces at hand, we can proceed to introduce several classes of operators.

Definition 11.9. Let $\psi(z) \in \mathbf{RL}^\infty$.

1. The operator $L_\psi : \mathbf{RL}^2 \longrightarrow \mathbf{RL}^2$ defined by

$$L_\psi f = \psi f \tag{11.11}$$

is called the **Laurent operator** with symbol $\psi(z)$.
2. The operator $T_\psi : \mathbf{RH}_+^2 \longrightarrow \mathbf{RH}_+^2$ defined by

$$T_\psi f = P_+ \psi f \tag{11.12}$$

is called the **Toeplitz operator** with symbol $\psi(z)$. If $\psi(z) \in \mathbf{RH}_+^\infty$, T_ψ will be called an **analytic Toeplitz operator**.
3. The operator $H_\psi : \mathbf{RH}_+^2 \longrightarrow \mathbf{RH}_-^2$ defined by

$$H_\psi f = P_- \psi f \tag{11.13}$$

is called the **Hankel operator** with symbol $\psi(z)$. Similarly, the operator $\hat{H}_\psi :$ $\mathbf{RH}_-^2 \longrightarrow \mathbf{RH}_+^2$,

$$\hat{H}_\psi f = P_+ \psi f, \tag{11.14}$$

is called the **reverse Hankel operator** with **symbol** $\psi(z)$.

The next proposition studies the duality properties of these operators.

Proposition 11.10. *Relative to the inner product in $\mathbf{RL}^2$ and the induced inner products in $\mathbf{RH}_\pm^2$, we have for $\psi(z) \in \mathbf{RL}^\infty$ and with $\psi^*(z)$ is defined by (11.1),*

$$L_\psi^* = L_{\psi^*}, \tag{11.15}$$

$$T_\psi^* = T_{\psi^*}, \tag{11.16}$$

and

$$H_\psi^* = \hat{H}_{\psi^*}. \tag{11.17}$$

Proof. By elementary computations.   ∎

The next proposition introduces an orthogonal projection operator in $\mathbf{RH}_+^2$ that links to the projection operator π^q, defined in (5.35). This in preparation for Proposition 11.13, which links analytic and algebraic invariance concepts.

Proposition 11.11. *Given a stable polynomial $q(z)$, then*

1. The map $P_q : \mathbf{RH}_+^2 \longrightarrow \mathbf{RH}_+^2$ defined by

$$P_q f = \frac{q^*}{q} P_- \frac{q}{q^*} f, \qquad f(z) \in \mathbf{RH}_+^2, \tag{11.18}$$

is an orthogonal projection.

2. We have

$$\operatorname{Ker} P_q = \frac{q^*}{q} \mathbf{RH}_+^2 \tag{11.19}$$

and

$$\operatorname{Im} P_q = X^q. \tag{11.20}$$

3. We have the orthogonal direct sum decomposition

$$\mathbf{RH}_+^2 = X^q \oplus \frac{q^*}{q} \mathbf{RH}_+^2, \tag{11.21}$$

as well as

$$(X^q)^\perp = \frac{q^*}{q} \mathbf{RH}_+^2. \tag{11.22}$$

4. We have $\dim X^q = \dim \left\{ \frac{q^*}{q} \mathbf{RH}_+^2 \right\}^\perp = \deg q$.

Proof. 1. We compute, for $f(z) \in \mathbf{RH}_+^2$,

$$P_q^2 f = \frac{q^*}{q} P_- \frac{q}{q^*} \frac{q^*}{q} P_- \frac{q}{q^*} f = \frac{q^*}{q} P_-^2 \frac{q}{q^*} f = \frac{q^*}{q} P_-^2 \frac{q}{q^*} f = \frac{q^*}{q} P_- \frac{q}{q^*} f = P_q f.$$

So P_q is indeed a projection. For $f(z), g(z) \in \mathbf{RH}_+^2$ we have

$$(P_q f, g) = \left(\frac{q^*}{q} P_- \frac{q}{q^*} f, g \right) = \left(P_- \frac{q}{q^*} f, \frac{q}{q^*} g \right)$$

$$= \left(\frac{q}{q^*} f, P_- \frac{q}{q^*} g \right) = \left(f, \frac{q^*}{q} P_- \frac{q}{q^*} g \right) = (f, P_q g).$$

This shows that $P_q^* = P_q$, i.e., P_q is an orthogonal projection.

2. Assume $f(z) \in \frac{q^*(z)}{q(z)} \mathbf{RH}_+^2$. Then $f(z) = \frac{q^*(z)}{q(z)} g(z)$ and

$$P_q f = \frac{q^*}{q} P_- \frac{q}{q^*} \frac{q^*}{q} g = \frac{q^*}{q} P_- g = 0.$$

Thus $\frac{q^*}{q} \mathbf{RH}_+^2 \subset \operatorname{Ker} P_q$.

Conversely, let $f(z) \in \operatorname{Ker} P_q$. Then $\frac{q^*}{q} P_- \frac{q}{q^*} f = 0$. This shows that $\frac{q(z)}{q^*(z)} f(z) \in \mathbf{RH}_+^2$ and hence $f(z) \in \frac{q^*}{q} \mathbf{RH}_+^2$ or $\operatorname{Ker} P_q \subset \frac{q^*}{q} \mathbf{RH}_+^2$. Thus equality (11.19) follows.

Assume $f(z) = p(z)/q(z) \in X^q$. Then

$$P_q f = \frac{q^*}{q} P_- \frac{q}{q^*} \frac{p}{q} = \frac{q^*}{q} P_- \frac{p}{q^*} = \frac{p}{q} = f.$$

So $X^q \subset \operatorname{Im} P_q$.

Conversely, suppose $f(z) \in \mathrm{Im} P_q$, i.e., $f(z) = \frac{q^*(z)}{q(z)} h(z)$ with $h = P_- \frac{q}{q^*} f_1 \in$ $\mathbf{RH}_+^2$. This representation shows that $f(z) \in X^q$. Thus $\mathrm{Im} P_q \subset X^q$ and equality follows.

3. Follows from $I = P_q + (I - P_q)$.
4. Follows from Proposition 5.3.   ∎

11.2.2   Invariant Subspaces

Before the introduction of Hankel operators, we digress a bit on invariant subspaces of $\mathbf{RH}_+^2$. Since we are using the half-planes for our definition of the $\mathbf{RH}_+^2$ spaces, we do not have the shift operators conveniently at our disposal. This forces us to a slight departure from the usual convention.

Definition 11.12. 1. A subspace $\mathcal{M} \subset \mathbf{RH}_+^2$ is called an **invariant subspace** if for each $\psi(z) \in \mathbf{RH}_+^\infty$, we have

$$T_\psi \mathcal{M} \subset \mathcal{M}.$$

2. A subspace $\mathcal{M} \subset \mathbf{RH}_+^2$ is called a **backward invariant subspace** if for each $\psi(z) \in \mathbf{RH}_+^\infty$, we have

$$T_\psi^* \mathcal{M} \subset \mathcal{M}.$$

Since $\mathbf{RH}_+^2$ is a subspace of $\mathbf{R}_-(z)$, the space of strictly proper rational functions, we have in it two notions of backward invariance. One is analytic, given by Definition 11.12. The other is algebraic, namely invariance with respect to the shift S_-, defined in equation (5.32), restricted to $\mathbf{R}_-(z)$. The following proposition shows that in this case, analysis and algebra meet and the two notions of invariance coincide.

Proposition 11.13. *Let $\mathcal{M} \subset \mathbf{RH}_+^2$ be a subspace. Then $\mathcal{M}$ is a backward invariant subspace if and only if $\mathcal{M}$ is S_--invariant.*

Proof. Assume first that $\mathcal{M}$ to be S_--invariant. Let $f(z) = p(z)/q(z) \in \mathcal{M}$ with $p(z), q(z)$ coprime and $q(z)$ stable. We show first that $X^q \subset \mathcal{M}$. By the coprimeness of $p(z)$ and $q(z)$, there exist polynomials $a(z), b(z) \in \mathbb{C}[z]$ that solve the Bezout equation $a(z)p(z) + b(z)q(z) = 1$. Then

$$a(S_-)f = \pi_- a \cdot \frac{p}{q} = \pi_- \frac{ap + bq}{q} = \pi_- \cdot \frac{1}{q} = \frac{1}{q}.$$

We conclude that $1/q(z) \in \mathcal{M}$ and hence $z^i/q(z) \in \mathcal{M}$ for $i = 1, \ldots, \deg q - 1$. Thus we have $X^q \subset \mathcal{M}$. For $f(z)$ as before, and $\psi(z) = r(z)/s(z) \in \mathbf{RH}_+^\infty$, we compute now, taking partial fractions and using the fact that $q(z)$ and $s^*(z)$ are coprime,

$$\frac{r^*(z)}{s^*(z)} \frac{p(z)}{q(z)} = \frac{v(z)}{s^*(z)} + \frac{t(z)}{q(z)}.$$

and hence

$$P_+ \psi^* f = P_+ \frac{r^*}{s^*} \frac{p}{q} = \frac{t}{q} \in X^q \subset \mathcal{M}.$$

This shows that $\mathcal{M}$ is backward invariant.

Conversely, let $\mathcal{M} \subset \mathbf{RH}_+^2$ be backward invariant, and let $f(z) = p(z)/q(z) \in \mathcal{M}$. Clearly, there exists a scalar α such that $zp(z)/q(z) = \alpha + t(z)/q(z)$, and hence $S_- f = t/q$ with $\deg t < \deg q$. To show algebraic invariance, we have to show the existence of a proper rational stable function $\psi(z) = r(z)/s(z)$ for which $P_+ \frac{r^*}{s^*} \frac{p}{q} = \frac{t}{q}$. This, of course, is equivalent to the partial fraction decomposition

$$\frac{r^*(z)}{s^*(z)} \frac{p(z)}{q(z)} = \frac{t(z)}{q(z)} + \frac{v(z)}{s^*(z)}.$$

Let us choose now $s(z)$ to be an arbitrary stable polynomial satisfying $\deg s = \deg q - 1$. Let $e^* = \pi_q z s^*$, i.e., there exists a constant γ such that $e^*(z) = zs^*(z) - \gamma q(z)$ with $\deg e < \deg q$. We compute now

$$\frac{e^*(z)}{s^*(z)} \frac{p(z)}{q(z)} = \frac{zs^*(z) - \gamma q(z)}{s^*(z)} \frac{p(z)}{q(z)} = \frac{zp(z)}{q(z)} - \frac{\gamma p(z)}{s^*(z)}$$

$$= \alpha + \frac{t(z)}{q(z)} - \frac{\gamma p(z)}{s^*(z)} = \frac{t(z)}{q(z)} + \frac{\alpha s^*(z) - \gamma p(z)}{s^*(z)}$$

$$= \frac{t(z)}{q(z)} + \frac{v(z)}{s^*(z)}$$

and hence

$$P_+ \frac{e^*}{s^*} \frac{p}{q} = \frac{t(z)}{q(z)},$$

and the proof is complete.   ∎

Clearly, orthogonal complements of invariant subspaces are backward-invariant subspaces. However, in inner product spaces, we may have proper subspaces whose orthogonal complement is trivial. The next result is of this type.

Proposition 11.14. *Let $\mathcal{M} \subset \mathbf{RH}_+^2$ be an infinite-dimensional backward-invariant subspace. Then $\mathcal{M}^\perp = \{0\}$.*

Proof. Let $f(z) \in \mathcal{M}^\perp$. Since $f(z)$ is rational, it has an irreducible representation $f(z) = e(z)/d(z)$, with $\deg e < \deg d = \delta$. Since, by assumption, $\mathcal{M}$ is infinite-dimensional, we can find δ linearly independent functions, $g_1(z), \ldots, g_\delta(z)$ in $\mathcal{M}$. Let $\mathcal{M}_1 \subset \mathcal{M}$ be the smallest backward-invariant subspace of $\mathbf{RH}_+^2$ containing all the $g_i(z)$. We claim that $\mathcal{M}_1$ is finite-dimensional. For this, it suffices to show that if $g(z) = r(z)/s(z)$ and $\phi(z) \in \mathbf{RH}_+^\infty$, then $T_\phi^* = t(z)/s(z) \in X^s$. Assume therefore that $\phi(z) = a(z)/b(z)$. Then by partial fraction decomposition,

$$\phi^*(z) g(z) = \frac{a^*(z)}{b^*(z)} \cdot \frac{r(z)}{s(z)} = \frac{\alpha(z)}{s(z)} + \frac{\beta(z)}{b^*(z)}$$

and hence $T_\phi^* g = P_+ \phi^* g = \frac{\alpha(z)}{s(z)} \in X^s$. Now if $g_i(z) = r_i(z)/s_i(z)$, then clearly $\mathscr{M}_1 \subset X^{s_1 \cdots s_\delta}$, and so $\mathscr{M}_1$ is finite-dimensional. By Proposition 5.24, there exists a polynomial $q(z)$ for which $\mathscr{M}_1 = X^q$. Since $g_1(z), \ldots, g_\delta(z) \in \mathscr{M}_1$, it follows that $\delta \leq \dim \mathscr{M}_1 = \deg q$. Now, by Proposition 11.11, $\mathscr{M}_1^\perp = \frac{q^*}{q} \mathbf{RH}_+^2$. So, for $f(z) = e(z)/d(z)$, we have the representation $f(z) = \frac{q^*(z)}{q(z)} f_1(z)$ with $f_1(z) \in \mathbf{RH}_+^2$. Since $q^*(z)$ is antistable, there can be no pole-zero cancellation between the zeros of $q^*(z)$ and those of the denominator of $f_1(z)$. This clearly implies that $q^*(z)$ divides $e(z)$. But $\deg e < \delta \leq \deg q$. Necessarily $e(z) = 0$ and, of course, also $f(z) = 0$. ∎

Definition 11.15. 1. A function $m(z) \in \mathbf{RL}^\infty$ is called an **all-pass** function if it satisfies $m^* m = 1$ on the imaginary axis, i.e., $|m(it)| = 1$.
2. A function $m(z) \in \mathbf{RH}_+^\infty$ is called **inner** if it is also an all-pass function.

Rational inner functions have a simple characterization.

Proposition 11.16. $m(z) \in \mathbf{RH}_+^\infty$ *is an inner function if and only if it has a representation*

$$m(z) = \alpha \frac{p^*(z)}{p(z)} \tag{11.23}$$

for some stable polynomial $p(z)$ and $\alpha \in \mathbb{C}$, with $|\alpha| = 1$.

Proof. Assume $p(z)$ is a stable polynomial. Then $m(z)$, defined by (11.23), is clearly in $\mathbf{RH}_+^\infty$. Moreover, on the imaginary axis, we have

$$m^*(z) m(z) = \overline{\left(\frac{p^*(z)}{p(z)} \right)} \left(\frac{p^*(z)}{p(z)} \right) = 1.$$

Conversely, assume $m(z) \in \mathbf{RH}_+^\infty$ is inner. Let $m(z) = r(z)/p(z)$ with $p(z)$ a stable polynomial. We may assume without loss of generality that $r(z)$ and $p(z)$ are coprime. Since $m(z)$ is inner, we get $\frac{r(z)}{p(z)} \frac{r^*(z)}{p^*(z)} = 1$, or $r(z) r^*(z) = p(z) p^*(z)$. Using our coprimeness assumption, we have $r(z) \mid p^*(z) \mid r(z)$. This implies that $r(z) = \alpha p^*(z)$, with $|\alpha| = 1$. ∎

Note that $m(z) \in \mathbf{RH}_+^\infty$ is inner if and only if

$$m^*(z) = m(z)^{-1}. \tag{11.24}$$

We proceed to prove the basic tool for all that follows, namely the representation of invariant subspaces. We give an algebraic version of his theorem that is better suited to our needs.

Theorem 11.17 (Beurling). $\mathscr{M} \subset \mathbf{RH}_+^2$ *is a nonzero invariant subspace if and only if*

$$\mathscr{M} = m \mathbf{RH}_+^2,$$

where $m(z)$ is a rational inner function.

Proof. If $m(z)$ is inner, then clearly $\mathscr{M} = m\mathbf{RH}^2_+$ is an invariant subspace.

To prove the converse, assume $\mathscr{M} \subset \mathbf{RH}^2_+$ is a nonzero invariant subspace. By Proposition 11.14, it follows that $\mathscr{M}^\perp$ is necessarily finite-dimensional. Since $\mathscr{M}^\perp$ is backward invariant, by Proposition 11.13, it is also S_--invariant. We invoke now Proposition 5.24 to conclude that $\mathscr{M}^\perp = X^q$ for some polynomial $q(z)$. Using Proposition 11.11, it follows that $\mathscr{M} = \frac{q^*}{q}\mathbf{RH}^2_+$. ∎

Two functions $f_1(z), f_2(z) \in \mathbf{RH}^\infty_+$ are called **coprime** if their greatest common inner factor is the 1. Clearly, this is equivalent to the existence of $\delta > 0$ for which $|f_1(z)| + |f_2(z)| \geq \delta$.

Proposition 11.18. *Let $f_1(z), f_2(z) \in \mathbf{RH}^\infty_+$. Then $f_1(z), f_2(z)$ are coprime if and only if there exist functions $a_1(z), a_2(z) \in \mathbf{RH}^\infty_+$ such that the Bezout identity holds,*

$$a_1(z)f_1(z) + a_2(z)f_2(z) = 1. \tag{11.25}$$

Proof. Clearly, if (11.25) holds then $f_1(z), f_2(z)$ cannot have a nontrivial common inner factor.

Conversely, assume that $f_1(z), f_2(z)$ are coprime. Then $f_1\mathbf{RH}^2_+, f_2\mathbf{RH}^2_+$ are both invariant subspaces of $\mathbf{RH}^2_+$ and so is their sum. Therefore there exists a rational inner function $m(z)$ such that

$$f_1\mathbf{RH}^2_+ + f_2\mathbf{RH}^2_+ = m\mathbf{RH}^2_+. \tag{11.26}$$

This implies that $m(z)$ is a common inner factor of the $f_i(z)$. Hence necessarily we have $m(z) = 1$. Let us choose $\alpha > 0$. Then $\frac{1}{z+\alpha} \in \mathbf{RH}^2_+$. Equation (11.26), with $m = 1$, shows that there exist strictly proper functions $b_1(z), b_2(z) \in \mathbf{RH}^2_+$ such that $b_1(z)f_1(z) + b_2(z)f_2(z) = \frac{1}{z+\alpha}$. Defining $a_i(z) = (z+\alpha)b_i(z)$, we have $a_i(z) \in \mathbf{RH}^\infty_+$, and they satisfy (11.25). ∎

11.2.3 Model Operators and Intertwining Maps

The $\mathbf{RH}^\infty_+$ module structure on $\mathbf{RH}^2_+$, defined by (11.10), induces a similar module structure on backward-invariant subspaces.

Given an inner function $m(z) \in \mathbf{RH}^\infty_+$, we consider the backward-invariant subspace

$$H(m) = \{m\mathbf{RH}^2_+\}^\perp = \mathbf{RH}^2_+ \ominus m\mathbf{RH}^2_+. \tag{11.27}$$

Similarly, we define

$$\overline{H}(m^*) = \{m^*\mathbf{RH}^2_-\}^\perp = \mathbf{RH}^2_- \ominus m^*\mathbf{RH}^2_-. \tag{11.28}$$

The following proposition not only gives a useful characterization for elements of $H(m)$, but actually shows that in the rational case, the set of invariant subspaces of the form $H(m)$ actually coincides with the set of rational models X^d, where we have $m(z) = \frac{d^*(z)}{d(z)}$. This makes this set of spaces a meeting point of algebra and analysis.

Proposition 11.19. *Let $m(z) \in \mathbf{RH}_+^\infty$ be inner. Then*

1. *We have the orthogonal direct sum*

$$\mathbf{RH}_+^2 = H(m) \oplus m\mathbf{RH}_+^2. \tag{11.29}$$

2. *$f(z) \in H(m)$ if and only if $m^*(z)f(z) \in \mathbf{RH}_-^2$.*
3. *The orthogonal projection $P_{H(m)}$ of $\mathbf{RH}_+^2$ onto $H(m)$ has the representation*

$$P_{H(m)}f = mP_-m^*f, \qquad f(z) \in \mathbf{RH}_+^2. \tag{11.30}$$

4. *Let $m(z)$ have the representation*

$$m(z) = \alpha \frac{d^*(z)}{d(z)}, \tag{11.31}$$

for some stable polynomial $d(z)$. Then as sets, we have

$$H(m) = X^d. \tag{11.32}$$

Proof. 1. Follows from (11.27).
2. Since $m(z)$ is inner, multiplication by $m^*(z)$ is unitary in $\mathbf{RL}_+^2$ and therefore preserves orthogonality. Since $m^*m\mathbf{RH}_+^2 = \mathbf{RH}_+^2$, the direct sum (11.29) implies that $m^*H(m)$ is orthogonal to $\mathbf{RH}_+^2$.
3. Given $f(z) \in \mathbf{RH}_+^2$, let $f(z) = g(z) + m(z)h(z)$ be its decomposition corresponding to the direct sum representation (11.29). This implies $m^*(z)f(z) = m^*(z)g(z) + h(z)$, with $m^*(z)g(z) \in \mathbf{RH}_-^2$. Applying the orthogonal projection P_- to the previous equality, we obtain $P_-m^*f = m^*g$ and hence (11.30) follows.
4. Assume $f(z) = \frac{n(z)}{d(z)}$, with $d(z)$ stable. Then, with $m(z)$ defined by (11.31), we compute

$$m^*f = \frac{d(z)}{d^*(z)}\frac{n(z)}{d(z)} = \frac{n(z)}{d^*(z)} \in \mathbf{RH}_-^2.$$

This shows that $X^d \subset H(m)$.

Conversely, assume $f(z) \in H(m)$. Since $f(z)$ is rational, it has a polynomial coprime factorization $f(z) = \frac{p(z)}{q(z)}$, with $q(z)$ stable. Now, by Part 2, we have $\frac{d(z)}{d^*(z)}\frac{p(z)}{q(z)} \in \mathbf{RH}_-^2$. Taking a partial fraction decomposition

$$\frac{d(z)}{d^*(z)}\frac{p(z)}{q(z)} = \frac{n(z)}{d^*(z)} + \frac{r(z)}{q(z)},$$

we conclude that, necessarily, $r(z) = 0$. From this it follows that $f(z) = \frac{n(z)}{d(z)}$, i.e., $f(z) \in X^d$. Thus $H(m) \subset X^d$, and (11.32) is proved. $\blacksquare$

The $\mathbf{RH}^\infty_+$-module structure on $\mathbf{RH}^2_+$ induces an $\mathbf{RH}^\infty_+$-module structure on invariant subspaces. This is done by defining the module structure to be given by the following.

Definition 11.20. 1. Given an inner function $m(z) \in \mathbf{RH}^\infty_+$, we define the $\mathbf{RH}^\infty_+$-module structure on $H(m)$ by

$$\psi \cdot f = P_{H(m)}(\psi f), \qquad f(z) \in H(m), \tag{11.33}$$

for all $\psi(z) \in \mathbf{RH}^\infty_+$.
2. For each $\psi(z) \in \mathbf{RH}^\infty_+$, the map $T_\psi : H(m) \longrightarrow H(m)$ is defined by

$$T_\psi f = \psi \cdot f = P_{H(m)} \psi f, \qquad f(z) \in H(m). \tag{11.34}$$

We will refer to operators of the form T_ψ as **model operators**.

Thus, the $\mathbf{RH}^\infty_+$-module structure on the invariant subspaces $H(m)$ is identical to the module structure induced by the algebra of analytic Toeplitz operators. The next proposition shows that we have indeed a module structure on $H(m)$ and use it to explore the lattice of its submodules.

Proposition 11.21. *Given an inner function $m(z) \in \mathbf{RH}^\infty_+$, then*

1. For all $\phi(z), \psi(z) \in \mathbf{RH}^\infty_+$, we have

$$T_\phi T_\psi = T_\psi T_\phi. \tag{11.35}$$

2. A subspace $\mathcal{M} \subset H(m)$ is a submodule with respect to the $\mathbf{RH}^\infty_+$-module structure on $H(m)$ if and only if there exists a factorization

$$m(z) = m_1(z) m_2(z) \tag{11.36}$$

into inner factors for which we have the representation

$$\mathcal{M} = m_1 H(m_2). \tag{11.37}$$

Proof. 1. It suffices to show that for $\psi(z) \in \mathbf{RH}^\infty_+$, the inclusion $T_\psi \operatorname{Ker} P_{H(m)} \subset \operatorname{Ker} P_{H(m)}$ holds. By Proposition 11.19, the general element of $\operatorname{Ker} P_{H(m)}$ is of the form $m(z) f(z)$ with $f(z) \in \mathbf{RH}^2_+$. So, the required inclusion follows immediately from $\psi(z)(m(z) f(z)) = m(z)(\psi(z) f(z))$.

2. Assume a factorization (11.36) exists and $\mathcal{M}$ is given by (11.37). First, we show that $m_1 H(m_2) \subset H(m)$. Indeed, if $f(z) \in m_1 H(m_2)$, then $f(z) = m_1(z)g(z)$, with $g(z) \in H(m_2)$. Using the characterization given in Proposition 11.19, we compute

$$m^* f = m_2^* m_1^* (m_1 g) = m_2^* g \in \mathbf{RH}_-^2,$$

i.e., $f(z) \in H(m)$.

Next, $f(z) \in \mathcal{M}$ implies $f(z) = m_1(z)g(z)$, with $g(z) \in H(m_2)$. For $\psi(z) \in \mathbf{RH}_+^\infty$, we compute

$$\psi \cdot f = T_\psi f = P_{H(m)}(\psi f) = m P_- m^* \psi m_1 g$$

$$= m_1 m_2 P_- m_2^* m_1^* m_1 \psi g = m_1 P_{H(m_2)}(\psi g) \in m_1 H(m_2),$$

i.e., $m_1 H(m_2)$ is a submodule of $H(m)$.

Conversely, assume $\mathcal{M}$ is a submodule of $H(m)$. This implies that $\mathcal{M} + m\mathbf{RH}_+^2$ is a submodule of $\mathbf{RH}_+^2$ with respect to the $\mathbf{RH}_+^\infty$-module structure. By Theorem 11.17, we have

$$\mathcal{M} + m\mathbf{RH}_+^2 = m_1 \mathbf{RH}_+^2, \tag{11.38}$$

for some inner function $m_1(z)$. Since (11.38) implies the inclusion $m\mathbf{RH}_+^2 \subset m_1 \mathbf{RH}_+^2$, it follows that a factorization (11.37) exists. Now, for $f(z) \in \mathcal{M}$, (11.38) implies a representation $f(z) = m_1(z)g(z)$. Since $f(z) \in H(m)$, it follows that

$$m^* f = m_2^* m_1^* m_1 g = m_2^* g \in \mathbf{RH}_-^2,$$

i.e., $g(z) \in H(m_2)$. This establishes the representation (11.37). ∎

There is a natural division relation for inner functions. If $m(z) = m_1(z)m_2(z)$ with $m(z), m_1(z), m_2(z)$ inner functions in $\mathbf{RH}_+^\infty$, then we say that $m_1(z)$ divides $m(z)$. The greatest common inner divisor of $m_1(z), m_2(z)$ is a common inner divisor that is divided by every other common inner divisor. The least common inner multiple is similarly defined.

$\mathbf{RH}_+^\infty$-submodules of $\mathbf{RH}_+^2$ are closed under sums and intersections. These lattice operations can be easily interpreted in terms of arithmetic operations on inner functions.

Proposition 11.22. *Let $m(z), m_i(z) \in \mathbf{RH}_+^\infty$, $i = 1, \dots, s$, be rational inner functions. Then*

1. We have $m\mathbf{RH}_+^2 \subset m_1 \mathbf{RH}_+^2$ if and only if $m_1(z)$ divides $m(z)$.
2. We have

$$\sum_{i=1}^s m_i \mathbf{RH}_+^2 = m\mathbf{RH}_+^2,$$

where $m(z)$ is the greatest common inner divisor of all $m_i(z)$, $i = 1, \dots, s$.

3. We have

$$\cap_{i=1}^{s} m_i \mathbf{RH}_+^2 = m\mathbf{RH}_+^2,$$

where $m(z)$ is the least common inner multiple of all $m_i(z)$.

Proof. The proof follows along the same lines as that of Proposition 1.46. We omit the details. ∎

The previous discussion has its counterpart in any backward-invariant subspace $H(m)$.

Proposition 11.23. *Let $m(z) \in \mathbf{RH}_+^\infty$ be an inner function. We consider $H(m)$ with the $\mathbf{RH}_+^\infty$-module structure defined in (11.33). Then*

1. Let $m(z) = m_i(z)n_i(z)$ be factorizations into inner factors. Then

$$m_1 H(n_1) \subset m_2 H(n_2) \tag{11.39}$$

if and only if $m_2(z)$ is an inner factor of $m_1(z)$, or equivalently, $n_1(z)$ is an inner factor of $n_2(z)$.

2. Given factorizations $m(z) = m_i(z)n_i(z)$, $i = 1, \ldots, s$, into inner factors, then we have

$$m_0 H(n_0) = \cap_{i=1}^{s} m_i H(n_i), \tag{11.40}$$

where $m_0(z)$ is the least common inner multiple of all $m_i(z)$, and $n_0(z)$ is the greatest common inner divisor of all $n_i(z)$.

3. Given factorizations $m(z) = m_i(z)n_i(z)$, $i = 1, \ldots, s$, into inner factors, then we have

$$m_0 H(n_0) = \sum_{i=1}^{s} m_i H(n_i) \tag{11.41}$$

where m_0 is the greatest common inner divisor of all m_i and n_0 is the least common inner multiple of all n_i.

4. We have

$$H(m) = \sum_{i=1}^{s} m_i H(n_i) \tag{11.42}$$

if and only if the $m_i(z)$ are coprime.

5. $\sum_{i=1}^{s} m_i H(n_i)$ is an algebraic direct sum if and only if the $n_i(z)$ are mutually coprime.

6. We have the algebraic direct sum decomposition

$$H(m) = m_1 H(n_1) \dot{+} \cdots \dot{+} m_s H(n_s) \tag{11.43}$$

if and only if the n_i are mutually coprime and $m(z) = \Pi_{i=1}^{s} n_i(z)$.

Proof. The proof follows along the same lines as that of Proposition 5.7. We omit the details. ∎

The equality (11.32), namely $H(m) = X^d$, with $m(z) = \frac{d^*(z)}{d(z)}$, shows that $H(m)$ carries two different module structures, one over $\mathbf{RH}_+^\infty$, defined by (11.33), and the other over $\mathbb{C}[z]$, defined by (5.34). In both cases, the lattices are determined by factorizations. This is the content of the following proposition.

Proposition 11.24. *Let $m(z)$ have the representation*

$$m(z) = \frac{d^*(z)}{d(z)}, \tag{11.44}$$

for some stable polynomial $d(z)$. Then the lattices of invariant subspaces, with respect to the two module structures described above, are isomorphic.

Proof. We note that factorizations of $m(z)$ are related to factorizations of $d(z)$. If $d(z) = d_1(z)d_2(z)$, the $d_i(z)$ are stable. Defining $m_i(z) = \frac{d_i^*(z)}{d_i(z)}$, we have the factorization $m(z) = m_1(z)m_2(z)$. The argument is reversible. The isomorphism of lattices follows from the correspondence of X^{d_2} and $m_1 H(m_2)$.   ∎

Note that in view of Proposition 11.24, the results of Propositions 11.23 and 11.22 could have been derived from the analogous results on polynomial and rational models.

The next theorem sums up duality properties of intertwining operators of the form T_ψ.

Theorem 11.25. *Let $\psi(z), m(z) \in \mathbf{RH}_+^\infty$ with $m(z)$ an inner function, and let T_ψ be defined by (11.34). Then*

1. Its adjoint, T_ψ^*, is given by

$$T_\psi^* f = P_+ \psi^* f, \quad for \ f(z) \in H(m). \tag{11.45}$$

2. The operator $\tau_m : \mathbf{RL}^2 \longrightarrow \mathbf{RL}^2$ defined by

$$\tau_m f := m f^* \tag{11.46}$$

is unitary.

3. The operators T_{ψ^*} and T_ψ^* are unitarily equivalent. Specifically, we have

$$T_\psi \tau_m = \tau_m T_\psi^*. \tag{11.47}$$

Proof. 1. For $f(z), g(z) \in H(m)$, we compute

$$
\begin{aligned}
(T_\psi f, g) &= (P_{H(m)} \psi f, g) = (m P_- m^* \psi f, g) \\
&= (P_- m^* \psi f, m^* g) = (m^* \psi f, P_- m^* g) \\
&= (m^* \psi f, m^* g) = (\psi f, g) = (f, \psi^* g) \\
&= (P_+ f, \psi^* g) = (f, P_+ \psi^* g) = (f, T_\psi^* g).
\end{aligned}
$$

Here we used the fact that $g(z) \in H(m)$ if and only if $m^*(z)g(z) \in \mathbf{RH}_-^2$.

2. Clearly the map τ_m, as a map in $\mathbf{RL}^2$, is unitary. From the orthogonal direct sum decomposition

$$\mathbf{RL}^2 = \mathbf{RH}_-^2 \oplus H(m) \oplus m\mathbf{RH}_+^2$$

it follows, by conjugation, that

$$\mathbf{RL}^2 = m^*\mathbf{RH}_-^2 \oplus \{\mathbf{RH}_-^2 \ominus m^*\mathbf{RH}_-^2\} \oplus \mathbf{RH}_+^2.$$

Hence $m\{\mathbf{RH}_-^2 \ominus m^*\mathbf{RH}_-^2\} = H(m)$.

3. We compute, with $f(z) \in H(m)$,

$$T_\psi \tau_m f = T_\psi m f^* = P_{H(m)} \psi m f^* = m P_- m^* \psi m f^* = m P_- \psi f^*.$$

Now

$$\tau_m T_\psi^* = \tau_m (P_+ \psi^* f) = m(P_+ \psi^* f)^* = m P_- \psi f^*.$$

Comparing the two expressions, (11.47) follows.   ∎

We proceed to study the invertibility properties of the maps T_ψ that are the counterpart of Theorem 5.18. This will be instrumental in the analysis of Hankel operators restricted to their cokernels.

Theorem 11.26. *Let* $\psi(z), m(z) \in \mathbf{RH}_+^\infty$ *with* $m(z)$ *an inner function. The following statements are equivalent:*

1. *The operator* $T_\psi : H(m) \longrightarrow H(m)$, *defined in (11.34), is invertible.*
2. *There exists* $\delta > 0$ *such that*

$$|\psi(z)| + |m(z)| \geq \delta, \quad \text{for all } z \text{ with } \operatorname{Re} z > 0. \tag{11.48}$$

3. *There exist* $\xi(z), \eta(z) \in \mathbf{RH}_+^\infty$ *that solve the Bezout equation*

$$\xi(z)\psi(z) + \eta(z)m(z) = 1. \tag{11.49}$$

In this case we have

$$T_\psi^{-1} = T_\xi.$$

Proof. $1 \Rightarrow 2$ We prove this by contradiction. Being inner, $m(z)$ has no zero at ∞; hence if no such δ exists, then necessarily there exists a point α in the open right half-plane $\mathbb{C}_+$ such that $\psi(\alpha) = m(\alpha) = 0$. This implies the existence of factorizations

$$m(z) = \pi_\alpha(z)m_\alpha(z),$$
$$\psi(z) = \psi_\alpha(z)m_\alpha(z),$$

where $m_\alpha(z) = \frac{z-\alpha}{z+\overline{\alpha}}$. To the first factorization there corresponds the one-dimensional invariant subspace $\pi_\alpha H(m_\alpha) \subset H(m)$. The second factorization implies the inclusion $\operatorname{Ker} T_{m_\alpha} \subset \pi_\alpha H(m_\alpha)$. With $l_\alpha(z) = \frac{1}{z+\overline{\alpha}} \in H(m_\alpha)$, we have

$$T_{m_\alpha} \pi_\alpha l_\alpha = P_{H(m)} m_\alpha \pi_\alpha l_\alpha = 0.$$

Next, we compute

$$T_\psi(\pi_\alpha l_\alpha) = P_{H(m)} \psi \pi_\alpha l_\alpha = P_{H(m)} \psi_\alpha m_\alpha \pi_\alpha l_\alpha = P_{H(m)} \psi_\alpha P_{H(m)} m l_\alpha = 0.$$

So $\pi_\alpha l_\alpha \in \operatorname{Ker} T_\psi$ and T_ψ cannot be invertible.

$2 \Rightarrow 3$ That coprimeness implies the solvability of the Bezout equation was proved in Proposition 11.18.

$3 \Rightarrow 1$ Assume there exist $\xi(z), \eta(z) \in \mathbf{RH}_+^\infty$ that solve the Bezout equation (11.49). We will show that $T_\psi^{-1} = T_\xi$. To this end, let $f(z) \in H(m)$. Then

$$T_\xi T_\psi f = P_{H(m)} \xi P_{H(m)} \psi f = P_{H(m)} \xi \psi f = P_{H(m)} (1 - m\eta) f = f.$$

Here we used the fact that $\xi \operatorname{Ker} P_{H(m)} = \xi m \mathbf{RH}_+^2 \subset m \mathbf{RH}_+^2 = \operatorname{Ker} P_{H(m)}$. ∎

In Theorem 11.8, we introduced natural $\mathbf{RH}_+^\infty$-module structures in both $\mathbf{RH}_+^2$ and $\mathbf{RH}_-^2$. Given an inner function $m(z) \in \mathbf{RH}_+^\infty$, $m \mathbf{RH}_+^2$ is an $\mathbf{RH}_+^\infty$ submodule, hence it follows that there is a naturally induced $\mathbf{RH}_+^\infty$-module structure on the quotient $\mathbf{RH}_+^2 / m \mathbf{RH}_+^2$. Using the orthogonal decomposition (11.29), we have the $\mathbf{RH}_+^\infty$-module isomorphism

$$\mathbf{RH}_+^2 / m \mathbf{RH}_+^2 \simeq H(m). \tag{11.50}$$

Thus $H(m)$ is a $\mathbf{RH}_+^\infty$ module, with the module structure induced by the $\mathbf{RH}_+^\infty$ module structure of $\mathbf{RH}_+^2$, i.e.,

$$\psi \cdot f = P_{H(m)}(\psi f), \qquad f(z) \in H(m). \tag{11.51}$$

11.2.4   Intertwining Maps and Interpolation

Definition 11.27. Let $m(z)$ be an inner function. We say that a map $X : H(m) \longrightarrow H(m)$ is an **intertwining map** if for all $\psi(z) \in \mathbf{RH}_+^\infty$, we have

$$X T_\psi = T_\psi X. \tag{11.52}$$

Because of the commutativity property (11.35), all operators T_ψ, defined in (11.34), are intertwining maps. Putting it differently, (11.35) expresses the fact that the maps T_ψ are $\mathbf{RH}_+^\infty$ homomorphisms in $H(m)$ with the module structure defined by (11.51). Next, we will show that being of the form T_ψ is not only sufficient for being an intertwining map, but also necessary. We will do so by showing that there is a close connection between intertwining maps and a rational interpolation problem. We begin this by exploring this connection for a special case, namely that of inner functions with distinct zeros, which leads to a representation theorem for intertwining maps. This will be extended, in Theorem 11.33, to the general case.

Next, we introduce the $\mathbf{RH}_+^\infty$-interpolation problem.

Definition 11.28. 1. Given $\xi_i \in \mathbb{C}$, and $\lambda_i \in \mathbb{C}_+$, $i = 1, \ldots, s$, find $\psi(z) \in \mathbf{RH}_+^\infty$ for which

$$\text{IP}: \quad \psi(\lambda_i) = \xi_i, \qquad i = 1, \ldots, s, \tag{11.53}$$

we say that $\psi(z)$ solves the **interpolation problem** IP.

2. Given $\xi_j \in \mathbb{C}$, and $\lambda_j \in \Pi_+$, $j = 1, \ldots, s$, find $\psi_i(z) \in \mathbf{RH}_+^\infty$ for which

$$\text{IP}(i): \quad \psi_i(\lambda_j) = \begin{cases} 0, & j \neq i, \\ \xi_i, & j = i, \end{cases} \tag{11.54}$$

we say that $\psi_i(z)$ solves the **interpolation problem** IP(i).

Clearly, the second part of the definition is geared to a Lagrange-like solution of the interpolation problem (11.53). The next proposition addresses these interpolation problems.

Proposition 11.29. *Given $\xi_i \in \mathbb{C}$ and $\lambda_i \in \mathbb{C}_+$, $i = 1, \ldots, s$, define the functions* $m(z), \pi_{\lambda_i}(z), m_{\lambda_i}(z), l_{\lambda_i}(z)$ *by*

$$m(z) = \Pi_{i=1}^s \frac{z - \lambda_i}{z + \overline{\lambda}_i},$$

$$\pi_{\lambda_i}(z) = \Pi_{j \neq i} \frac{z - \lambda_j}{z + \overline{\lambda}_j},$$

$$m_{\lambda_i}(z) = \frac{z - \lambda_i}{z + \overline{\lambda}_i},$$

$$l_{\lambda_i}(z) = \frac{1}{z + \overline{\lambda}_i}. \tag{11.55}$$

Then

1. There exist solutions $\psi_i(z)$ of the interpolation problems IP(i). A solution is given by

$$\psi_i(z) = \xi_i \pi_{\lambda_i}(\lambda_i)^{-1} \pi_{\lambda_i}(z). \tag{11.56}$$

2. A solution of the interpolation problems IP is given by

$$\psi(z) = \sum_{i=1}^s \xi_i \pi_{\lambda_i}(\lambda_i)^{-1} \pi_{\lambda_i}(z). \tag{11.57}$$

Proof. 1. For any $\psi_i(z) \in \mathbf{RH}_+^\infty$, (11.61) holds. Therefore, $\psi_i(\lambda_j) = 0$ for all $j \neq i$ if and only if $\psi(z)$ is a multiple of $\pi_i(z)$. We assume therefore that, for some constant α_i, we have $\psi_i(z) = \alpha_i \pi_i(z)$. The second interpolation condition in IP(i) reduces to $\xi_i = \psi_i(\lambda_i) = \alpha_i \pi_{\lambda_i}(\lambda_i)$, i.e., $\alpha_i = \xi_i \pi_{\lambda_i}(\lambda_i)^{-1}$. Thus (11.56) follows.
2. This follows easily from the first part. ∎

Note that $\psi_i(z)$, defined in (11.56), is proper and has McMillan degree $s - 1$. On the other hand, since the least common inner multiple of all $\pi_i(z)$ is $m(z)$, it follows that $\psi(z)$, given in (11.57), has McMillan degree $\leq s$.

Proposition 11.30. *Given an inner function $m(z) \in \mathbf{RH}_+^\infty$ having all its zeros, λ_i, $i = 1,\ldots,s$, distinct, we define functions $m(z), \pi_{\lambda_i}(z), m_{\lambda_i}(z), l_{\lambda_i}(z)$ by (11.55). Then*

1. We have the factorizations

$$m(z) = \pi_{\lambda_i}(z) m_{\lambda_i}(z). \tag{11.58}$$

2. The space $H(m_{\lambda_i})$ is spanned by the function $l_{\lambda_i}(z) = \frac{1}{z + \overline{\lambda_i}}$. We have

$$\mathbf{RH}_+^2 = H(m_{\lambda_i}) \oplus m_{\lambda_i} \mathbf{RH}_+^2. \tag{11.59}$$

3. We have the algebraic direct sum decomposition in terms of invariant subspaces

$$H(m) = \pi_{\lambda_1} H(m_{\lambda_1}) \dotplus \cdots \dotplus \pi_{\lambda_s} H(m_{\lambda_s}), \tag{11.60}$$

and $\{\pi_{\lambda_i}(z) l_{\lambda_i}(z)\}_{i=1}^s$ forms a basis for $H(m)$.
4. For any $\psi(z) \in \mathbf{RH}_+^\infty$, we have

$$T_\psi(\pi_{\lambda_i}(z) l_{\lambda_i}(z)) = \psi(\lambda_i) \pi_{\lambda_i}(z) l_{\lambda_i}(z). \tag{11.61}$$

5. A map $X : H(m) \longrightarrow H(m)$ is an intertwining map if and only if there exist $\xi_i \in \mathbb{C}$ such that

$$X(\pi_{\lambda_i} l_{\lambda_i}) = \xi_i \pi_{\lambda_i}(z) l_{\lambda_i}(z). \tag{11.62}$$

6. A map $X : H(m) \longrightarrow H(m)$ is an intertwining map if and only if there exists $\psi(z) \in \mathbf{RH}_+^\infty$ such that

$$X = T_\psi. \tag{11.63}$$

7. If $\psi(z) \in \mathbf{RH}_+^\infty$, we have, for $T_\psi : H(m) \longrightarrow H(m)$ defined by (11.34), $\|T_\psi\| \leq \|\psi\|_\infty$.

Proof. 1. This is obvious.
2. We compute, applying Cauchy's theorem, for $f(z) \in \mathbf{RH}_+^2$,

$$\begin{aligned}
(f, l_i) &= \frac{1}{2\pi} \int_{-\infty}^{\infty} f(it) \overline{\left(\frac{1}{it + \overline{\lambda_i}}\right)} dt = \frac{1}{2\pi} \int_{-\infty}^{\infty} f(it) \left(\frac{1}{-it + \lambda_i}\right) dt \\
&= \frac{1}{2\pi i} \int_{\gamma} \left(\frac{f(\zeta)}{\zeta - \lambda_i}\right) d\zeta = f(\lambda_i).
\end{aligned}$$

Here the contour γ is a positively oriented semicircle of sufficiently large radius.

3. Follows from Proposition 11.23.

4. We compute, using the representation (11.30) of $P_{H(m)}$,

$$
\begin{aligned}
T_\psi(\pi_{\lambda_i} l_{\lambda_i}) &= m P_- m^* \psi \pi_{\lambda_i} l_{\lambda_i} = \pi_{\lambda_i} m_{\lambda_i} P_- m_{\lambda_i}^* \pi_{\lambda_i}^* \psi \pi_{\lambda_i} l_{\lambda_i} \\
&= \pi_i P_{H(m_i)} \psi l_{\lambda_i} = \pi_{\lambda_i} P_{H(m_{\lambda_i})} \left(\frac{\psi(z) - \psi(\lambda_i)}{z + \overline{\lambda}_i} + \frac{\psi(\lambda_i)}{z + \overline{\lambda}_i} \right) \\
&= \psi(\lambda_i)(\pi_{\lambda_i}(z) l_{\lambda_i}(z)).
\end{aligned}
$$

Here we used the fact that $\frac{\psi(z) - \psi(\lambda_i)}{z + \overline{\lambda}_i} \in m_{\lambda_i} \mathbf{RH}_+^2$.

5. Assume first that there exist $\xi_i \in \mathbb{C}$ such that (11.62) holds. Then for every $\psi(z) \in \mathbf{RH}_+^\infty$, we compute

$$
\begin{aligned}
X T_\psi(\pi_{\lambda_i}(z) l_{\lambda_i}(z)) &= X(\psi(\lambda_i) \pi_{\lambda_i}(z) l_{\lambda_i}(z)) = \psi(\lambda_i) X(\pi_{\lambda_i}(z) l_{\lambda_i}(z)) \\
&= \psi(\lambda_i) \xi_i (\pi_{\lambda_i}(z) l_{\lambda_i}(z)) = T_\psi \xi_i (\pi_{\lambda_i}(z) l_{\lambda_i}(z)) \\
&= T_\psi X(\pi_{\lambda_i}(z) l_{\lambda_i}(z)).
\end{aligned}
$$

Since $\{\pi_{\lambda_i}(z) l_{\lambda_i}(z)\}$ form a basis of $H(m)$, it follows that $X T_\psi = T_\psi X$, i.e., X is an intertwining map.

Conversely, assume X is an intertwining map. We note that we have $\operatorname{Ker} T_{m_{\lambda_i}} = \pi_{\lambda_i} H(m_{\lambda_i})$ for

$$
\begin{aligned}
T_{m_{\lambda_i}}(\pi_{\lambda_i}(z) l_{\lambda_i}(z)) &= P_{H(m)}(\pi_{\lambda_i}(z) l_{\lambda_i}(z)) = m P_- m^*(\pi_{\lambda_i}(z) l_{\lambda_i}(z)) \\
&= \pi_{\lambda_i} m_{\lambda_i} P_- m_{\lambda_i}^* \pi_{\lambda_i}^* \pi_{\lambda_i}(z) l_{\lambda_i}(z) = \pi_{\lambda_i} P_{H(m_{\lambda_i})} m_{\lambda_i}(z) l_{\lambda_i}(z) = 0.
\end{aligned}
$$

Since $X T_{m_{\lambda_i}} = T_{m_{\lambda_i}} X$, it follows that

$$
T_{m_{\lambda_i}} X(\pi_{\lambda_i}(z) l_{\lambda_i}(z)) = X T_{m_{\lambda_i}}(\pi_{\lambda_i}(z) l_{\lambda_i}(z)) = 0.
$$

So $X(\pi_{\lambda_i}(z) l_{\lambda_i}(z)) \in \pi_{\lambda_i} H(m_{\lambda_i})$, i.e., there exist $\xi_i \in \mathbb{C}$ such that (11.62) holds.

6. If X has the representation $X = T_\psi$, then by (11.35) it is intertwining.

To prove the converse, assume to begin with that all the zeros λ_i of $m(z)$ are distinct. Thus, we have the factorization $m(z) = \Pi_{j=1}^s m_{\lambda_j}(z)$, where $m_{\lambda_i}(z) = \frac{z - \lambda_i}{z + \overline{\lambda}_i}$. By Proposition 11.21, we have $l_{\lambda_i}(z) = \frac{1}{z + \overline{\lambda}_i} \in H(m)$, and since they are linearly independent, these functions form a basis of $H(m)$. Using the duality results of Theorem 11.25, also $\{ (\Pi_{j \neq i} m_{\lambda_j}(z)) l_{\lambda_i}(z) \}$ is a basis for $H(m)$. This implies the algebraic direct sum representation

$$
H(m) = \left(\Pi_{j \neq 1} m_{\lambda_j} \right) H(m_{\lambda_1}) \dotplus \cdots \dotplus \left(\Pi_{j \neq s} m_{\lambda_j} \right) H(m_{\lambda_s}). \tag{11.64}
$$

Let now $X : H(m) \longrightarrow H(m)$ be an intertwining map. Since $\operatorname{Ker} T_{m_i} = \pi_{\lambda_i} H(m_{\lambda_i})$, we have for $f(z) = \pi_{\lambda_i}(z) l_{\lambda_i}(z)$,

$$T_{m_i} X (\pi_{\lambda_i} l_{\lambda_i}) = X T_{m_i} (\pi_{\lambda_i} l_{\lambda_i}) = 0.$$

This means that $X \operatorname{Ker} T_{m_i} \subset \operatorname{Ker} T_{m_i}$. In turn, this implies the existence of constants ξ_i for which $X(\pi_{\lambda_i} l_{\lambda_i}) = \xi_i \pi_{\lambda_i} l_{\lambda_i}$. Thus $X = T_\psi$ for any $\psi(z) \in \mathbf{RH}_+^\infty$ that solves the interpolation problem (11.53). In particular, $\psi(z)$ given by (11.57) is such a solution.

7. Follows from the facts that $P_{H(m)}$ is an orthogonal projection and that for $\psi(z) \in \mathbf{RH}_+^\infty$ and $f(z) \in \mathbf{RH}_+^2$, we have

$$\|T_\psi f\|_2 = \|P_{H(m)} \psi f\|_2 \leq \|\psi f\|_2 \leq \|\psi\|_\infty \|f\|_2.$$

∎

We note that, due to (11.61), the direct sum representation (11.60) is a spectral decomposition for all maps T_ψ.

We proceed to extend Proposition 11.30 to the case that the inner function $m(z)$ may have multiple zeros. However, before doing that, we introduce and study higher-order interpolation problems.

Definition 11.31. 1. Given $\lambda_i \in \mathbb{C}_+$, $\nu_i \in \mathbb{N}$, and $\{\xi_{i,t} \in \mathbb{C}\}_{t=0}^{\nu_i-1}$, $i = 1,\dots,k$. If $\psi(z) \in \mathbf{RH}_+^\infty$ has the local expansions

$$\psi(z) = \sum_{t=0}^{\nu_i-1} \psi_{i,t} m_{\lambda_i}(z)^t + m_{\lambda_i}(z)^{\nu_i} \rho_{\nu_i}(z), \qquad i = 1,\dots,k, \tag{11.65}$$

and satisfies the interpolation conditions

$$\text{HOIP:} \qquad \psi_{i,t} = \xi_{i,t}, \qquad i = 1,\dots,k,\ t = 0,\dots,\nu_i-1, \tag{11.66}$$

we say that $\psi(z)$ solves the **high-order interpolation problem** HOIP.

2. Given distinct $\lambda_i \in \mathbb{C}_+$, $\nu_i \in \mathbb{N}$, and $\{\xi_{i,t} \in \mathbb{C}\}_{t=0}^{\nu_i-1}$, $i = 1,\dots,k$, find $\psi_i(z) \in \mathbf{RH}_+^\infty$, $i = 1,\dots,k$, having the local expansions

$$\psi_i(z) = \begin{cases} m_{\lambda_j}(z)^{\nu_j} \rho_{i,\nu_j}(z), & j \neq i, \\ \sum_{t=0}^{\nu_i-1} \psi_{i,t} m_{\lambda_i}(z)^t + m_{\lambda_i}(z)^{\nu_i} \rho_{i,\nu_i}(z), & j = i, \end{cases} \tag{11.67}$$

and satisfies the interpolation conditions

$$\text{HOIP}(i): \qquad \psi_{i,t} = \begin{cases} \xi_{i,t}, & i = j,\ t = 0,\dots,\nu_i-1, \\ 0, & i \neq j, \end{cases} \tag{11.68}$$

we say that $\psi_i(z)$ solves the **high-order interpolation problem** HOIP(i).

Proposition 11.32. *Given distinct $\lambda_i \in \mathbb{C}_+$, $v_i \in \mathbb{N}$, and $\{\xi_{i_j} \in \mathbb{C}\}_{j=0}^{v_i-1}$, $i = 1, \ldots, k$, define the functions $m(z), \pi_{\lambda_i}(z), m_{\lambda_i}(z), l_{\lambda_i}(z)$ by*

$$m(z) = \Pi_{i=1}^k m_{\lambda_i}(z)^{v_i},$$

$$m_{\lambda_i}(z) = \frac{z - \lambda_i}{z + \overline{\lambda}_i},$$

$$\pi_{\lambda_i}(z) = \Pi_{j \neq i} m_{\lambda_j}(z)^{v_j},$$

$$l_{\lambda_i}(z) = \frac{1}{z + \overline{\lambda}_i}. \tag{11.69}$$

Then

1. For $i = 1, \ldots, k$, we have the factorizations

$$\pi_{\lambda_i}(z) m_{\lambda_i}(z)^{v_i} = m_{\lambda_i}(z)^{v_i} \pi_{\lambda_i}(z), \tag{11.70}$$

with $m_{\lambda_i}(z)^{v_i}, \pi_{\lambda_i}(z)$ coprime.
2. There exist $a_{\lambda_i}(z), b_{\lambda_i}(z) \in \mathbf{RH}_+^\infty$ solving the Bezout equation

$$a_{\lambda_i}(z) \pi_{\lambda_i}(z) + b_{\lambda_i}(z) m_{\lambda_i}(z)^{v_i} = 1. \tag{11.71}$$

3. Define the intertwining map $X_i : H(m_{\lambda_i}^{v_i}) \longrightarrow H(m_{\lambda_i}^{v_i})$ by

$$f_i(z) = X_i g_i = P_{H(m_{\lambda_i}^{v_i})} \pi_{\lambda_i} g_i, \qquad g_i(z) \in H(m_{\lambda_i}^{v_i}). \tag{11.72}$$

Then X_i is invertible with its inverse, $X_i^{-1} : H(m_{\lambda_i}^{v_i}) \longrightarrow H(m_{\lambda_i}^{v_i})$, given by

$$g_i(z) = X_i^{-1} f_i = P_{H(m_{\lambda_i}^{v_i})} a_{\lambda_i} f_i, \qquad f_i(z) \in H(m_{\lambda_i}^{v_i}), \tag{11.73}$$

where $a_{\lambda_i}(z)$ is determined by the solution to the Bezout equation (11.71).
4. A solution to HOIP(i) is given by

$$\psi_i(z) = \pi_{\lambda_i}(z) g_i(z), \tag{11.74}$$

with $v_{\lambda_i} l_{\lambda_i} = X_i^{-1}(w_{\lambda_i} l_{\lambda_i})$ determined by (11.73).
5. A solution of the interpolation problems HOIP is given by

$$\psi(z) = \sum_{i=1}^k \psi_i(z), \tag{11.75}$$

with $\psi_i(z)$ given by (11.74).

Proof. 1. This is immediate.

2. Follows, using Proposition 11.18, from the coprimeness of $\pi_{\lambda_i}(z)$ and $m_{\lambda_i}(z)^{\nu_i}$.

3. Follows from Theorem 11.26.

4. For $\psi_i(z)$ to satisfy the homogeneous interpolation conditions of (11.68), it is necessary and sufficient that we have the factorizations $\psi_i(z) = m_{\lambda_j}^{\nu_j}(z)u_{\lambda_j}(z)$ for all $j \neq i$. Since the λ_j are distinct, this implies the factorization $\psi_i(z) = \pi_{\lambda_i}(z)v_{\lambda_i}(z)$ for some $v_{\lambda_i}(z) \in \mathbf{RH}_+^\infty$. So, all that remains is to choose $v_{\lambda_i}(z)$ so as to satisfy the last interpolation constraint of (11.68).

Note that $m_{\lambda_i}(z)^{-\nu_i+1}(\pi_{\lambda_i}(z)v_{\lambda_i}(z) - w_{\lambda_i}(z)) \in \mathbf{RH}_-^\infty$ if and only if

$$P_{H(m_{\lambda_i}^{\nu_i})}(\pi_{\lambda_i}(z)v_{\lambda_i}(z) - w_{\lambda_i}(z))l_{\lambda_i}(z) \in \mathbf{RH}_-^2.$$

However, the last condition can be interpreted as $X_i(v_{\lambda_i}l_{\lambda_i}) = w_{\lambda_i}l_{\lambda_i}$; hence $v_{\lambda_i}l_{\lambda_i} = X_i^{-1}(w_{\lambda_i}l_{\lambda_i})$ and from this the result follows.

5. Follows from the previous part.   ∎

The following theorem, characterizing intertwining maps, is the algebraic analogue of the commutant lifting theorem.

Theorem 11.33. *Given an inner function $m(z) \in \mathbf{RH}_+^\infty$ having the primary inner decomposition*

$$m(z) = \Pi_{i=1}^k m_{\lambda_i}(z)^{\nu_i}, \tag{11.76}$$

where $m_{\lambda_j}(z) = \frac{z-\lambda_j}{z+\overline{\lambda_j}}$ and the inner functions $m_{\lambda_j}(z)$ are mutually coprime, we define

$$\pi_{\lambda_i}(z) = \Pi_{j\neq i}^s m_{\lambda_j}(z)^{\nu_j},$$
$$l_{\lambda_i}(z) = \frac{1}{z+\overline{\lambda_i}}. \tag{11.77}$$

Then

1. For $i = 1, \ldots, k$, we have the factorizations

$$m(z) = \pi_{\lambda_i}(z)m_{\lambda_i}(z)^{\nu_i}. \tag{11.78}$$

2. We have the algebraic direct sum decomposition of invariant subspaces

$$H(m) = \pi_{\lambda_1}H(m_{\lambda_1}^{\nu_1}) \dotplus \cdots \dotplus \pi_{\lambda_k}H(m_{\lambda_k}^{\nu_k}). \tag{11.79}$$

3. We have

$$\operatorname{Ker} T_{m_{\lambda_i}^j} = \pi_{\lambda_i}m_{\lambda_i}^{\nu_i-j}H(m_{\lambda_i}^j), \tag{11.80}$$

and a basis $\mathscr{B}_{\lambda_i,j}$ for this invariant subspace is given by the following set of vectors

$$\left\{ \pi_{\lambda_i}(z)m_{\lambda_i}(z)^{\nu_i-j}l_{\lambda_i}(z), \pi_{\lambda_i}(z)m_{\lambda_i}(z)^{\nu_i-j+1}l_{\lambda_i}(z), \ldots, \pi_{\lambda_i}(z)m_{\lambda_i}(z)^{\nu_i-1}l_{\lambda_i}(z) \right\}. \tag{11.81}$$

Moreover, we have

$$\dim \pi_{\lambda_i}m_{\lambda_i}^{\nu_i-j}H(m_{\lambda_i}^{j}) = j. \tag{11.82}$$

4. For $j=1,\ldots,\nu_i$, a basis for $\operatorname{Ker} T_{m_{\lambda_i}^{j}}$ is given by

$$\mathscr{B}_{\lambda_i,j} = \left\{ \pi_{\lambda_i}(z)m_{\lambda_i}(z)^{\nu_i-j}l_{\lambda_i}(z), \ldots, \pi_{\lambda_i}(z)m_{\lambda_i}(z)^{\nu_i-1}l_{\lambda_i}(z) \right\}. \tag{11.83}$$

5. Let $m_\lambda(z) = \frac{z-\lambda}{z+\overline{\lambda}}$ be an inner function and let $\psi(z) \in \mathbf{RH}_+^\infty$. Then for each $\nu \geq 0$, there exist uniquely determined $\psi_{\lambda,t} \in \mathbb{C}$, $t=0,\ldots,\nu-1$, and $\rho_\nu(z) \in \mathbf{RH}_+^\infty$ such that we have the representation

$$\psi(z) = \sum_{t=0}^{\nu-1} \psi_{\lambda,t}m_\lambda(z)^t + m_\lambda(z)^\nu \rho_\nu(z). \tag{11.84}$$

6. Let $\psi(z) \in \mathbf{RH}_+^\infty$ have the local representations (11.84). Then

$$T_\psi \pi_{\lambda_i}(z)m_{\lambda_i}(z)^{\nu_i-j}l_{\lambda_i}(z) = \sum_{t=0}^{j-1} \psi_{t,\lambda_i}\pi_{\lambda_i}(z)m_{\lambda_i}(z)^{\nu_i-j+t}l_{\lambda_i}(z). \tag{11.85}$$

Therefore, the matrix representation of $T_\psi \mid \operatorname{Ker} T_{m_{\lambda_i}^{j}}$, with respect to the basis $\mathscr{B}_{\lambda_i,j}$ given in (11.83), is

$$\begin{pmatrix} \psi_{\lambda_i,0} & 0 & . & . & 0 \\ . & & & & . \\ . & & . & & . \\ . & & & . & 0 \\ \psi_{\lambda_i,j-1} & . & . & . & \psi_{\lambda_i,0} \end{pmatrix}. \tag{11.86}$$

7. A map $X : H(m) \longrightarrow H(m)$ is an intertwining map if and only if there exists $\psi(z) \in \mathbf{RH}_+^\infty$ for which

$$X = T_\psi. \tag{11.87}$$

Proof. 1. Follows from (11.76) and the definition of $\pi_{\lambda_i}(z)$ given in (11.77).
2. Follows from the mutual coprimeness of the $m_i(z)$.
3. Equality (11.80) follows from the factorization $m(z) = (\pi_{\lambda_i}(z)m_{\lambda_i}(z)^{\nu_i-j})m_{\lambda_i}(z)^{j}$.

Using the characterization of Proposition 11.19, all elements $\pi_{\lambda_i}(z)m_{\lambda_i}(z)^{\nu_i-t}$
$l_{\lambda_i}(z)$, $i = 1,\ldots,k$, $t = 1,\ldots,j$, are in $\pi_{\lambda_i}m_{\lambda_i}(z)^{\nu_i-j}H(m_{\lambda_i}^j)$. Their linear independence is easily verfied, and since $\dim H(m_{\lambda_i}^j) = j$, it follows that $\mathscr{B}_{\lambda_i,j}$ is indeed
a basis for $\pi_{\lambda_i}(z)m_{\lambda_i}(z)^{\nu_i-j}H(m_{\lambda_i}^j)$.

4. The proof is similar to the proof of the previous part.

5. Let $\psi(z) \in \mathbf{RH}_+^\infty$ and let $l_\lambda(z)$ be defined by (11.77). We note that $\mathscr{B}_\lambda = \{m_\lambda(z)^t l_\lambda(z)\}_{t=0}^{\nu-1}$ is a basis for $H(m_\lambda^\nu)$. Obviously, $\psi(z)l_\lambda(z) \in \mathbf{RH}_+^2$, and therefore it has a decomposition with respect to the direct sum $\mathbf{RH}_+^2 = H(m_\lambda^\nu) \oplus m_\lambda^\nu \mathbf{RH}_+^2$. So there exist $\psi_t \in \mathbb{C}$ and $\theta(z) \in \mathbf{RH}_+^2$ for which we have $\psi(z)l_\lambda(z) = \sum_{t=0}^{\nu-1} \psi_t m_\lambda(z)^t + m_\lambda(z)^\nu \theta_\nu(z)$. Since $m_\lambda(z)^\nu \theta_\nu(z) \in \mathbf{RH}_+^2$, it follows that $\rho_\nu(z) = \theta_\nu(z)l_\lambda(z)^{-1} \in \mathbf{RH}_+^\infty$, and (11.84) follows.

6. Follows from the local expansion (11.84) and

$$T_{m_{\lambda_i}^t}\, \pi_{\lambda_i}(z)m_{\lambda_i}(z)^s l_{\lambda_i}(z) = \begin{cases} 0, & t+s \geq \nu_i, \\ \\ \pi_{\lambda_i}(z)m_{\lambda_i}(z)^{t+s}l_{\lambda_i}(z), & t+s < \nu_i. \end{cases} \tag{11.88}$$

7. As in the proof of Proposition 11.30, if X has the representation $X = T_\psi$, then by (11.35), it is intertwining.

To prove the converse, assume $X : H(m) \longrightarrow H(m)$ is intertwining. We compute, for $f(z) \in \operatorname{Ker} T_{m_{\lambda_i}^j}$,

$$0 = X T_{m_{\lambda_i}^j} f = T_{m_{\lambda_i}^j} X f,$$

which shows that

$$X \pi_{\lambda_i}(z)m_{\lambda_i}(z)^{\nu_i-j}H(m_{\lambda_i}^j) \subset \pi_{\lambda_i}(z)m_{\lambda_i}(z)^{\nu_i-j}H(m_{\lambda_i}^j). \tag{11.89}$$

In turn, this implies the existence of constants $\xi_{\lambda_i,t}$, $t = 0,\ldots,j-1$, for which

$$X \pi_{\lambda_i}(z)m_{\lambda_i}(z)^{\nu_i-j}l_{\lambda_i}(z) = \sum_{t=0}^{j-1} \xi_{\lambda_i,t}\pi_{\lambda_i}(z)m_{\lambda_i}(z)^{\nu_i-j}l_{\lambda_i}(z). \tag{11.90}$$

Choosing $\psi(z)$ to be the solution of the high-order interpolation problem (11.65)-(11.66), with $\psi_{\lambda_i,t} = \xi_{\lambda_i,t}$, we get the representation (11.87).   ∎

We can interpret (11.84) as a division rule in $\mathbf{RH}_+^\infty$, and we will refer to $\sum_{t=0}^{\nu-1} \psi_t m_\lambda(z)^t$ as the **remainder**. Note that $m_\lambda(z)^{-\nu}\sum_{t=0}^{\nu-1}\psi_t m_\lambda(z)^t \in \mathbf{RH}_-^\infty$.

11.2.5   RH$_+^\infty$-*Chinese Remainder Theorem*

We have already seen many analogies between polynomial and rational models on the one hand and invariant subspaces of $\mathbf{RH}_+^2$ on the other. Results applying coprimeness of inner functions follow the lines of results applying polynomial coprimeness. This applies to direct sum representations, as well as to the spectral analysis of intertwining maps. This analogy leads us to an analytic version of the Chinese remainder theorem, a result that can be immediately applied to rational interpolation problems.

Theorem 11.34. *1. Let $m_i(z)$ be mutually coprime inner functions, $m(z) = \Pi_{i=1}^s m_i(z)$, and let $g_i(z) \in H(m_i)$. Then there exists a unique $f(z) \in H(m)$ for which*

$$f(z) - g_i(z) \in m_i \mathbf{RH}_+^2. \tag{11.91}$$

2. Let $m_i(z)$ be mutually coprime inner functions, $m(z) = \Pi_{i=1}^s m_i(z)$. Let $w_i(z) \in \mathbf{RH}_+^\infty$ be such that $m_i(z)^{-1} w_i(z) \in \mathbf{RH}_-^\infty$. Then there exists a unique $v(z) \in \mathbf{RH}_+^\infty$ such that

a. $m(z)^{-1} v(z) \in \mathbf{RH}_-^\infty$,
b. $v(z) - w_i(z) \in m_i(z) \mathbf{RH}_+^\infty$.

Proof. 1. We define $\pi_i(z) = \Pi_{j \neq i} m_j(z)$. This implies the factorizations

$$m(z) = \pi_i(z) m_i(z). \tag{11.92}$$

In turn, we have the direct sum representation

$$H(m) = \pi_1 H(m_1) \dotplus \cdots \dotplus \pi_k H(m_k), \tag{11.93}$$

and $f(z) \in H(m)$ has a unique representation of the form $f(z) = \sum_{j=1}^s \pi_j(z) f_j(z)$, with $f_j(z) \in H(m_j)$. If $f(z)$ satisfies (11.91), then $P_- m_i^{-1} (\sum_{j=1}^s \pi_j f_j - g_j) = 0$, which implies $P_{H(m_i)} \pi_i f_i = X_i f_i = g_i$, where $X_i : H(m_i) \longrightarrow H(m_i)$ is defined by $X_i f_i = P_{H(m_i)} \pi_i f_i$. This implies $f_i = X_i^{-1} g_i$. In order to invert X_i, we apply Theorem 11.26, to infer that $X_i^{-1} : H(m_i) \longrightarrow H(m_i)$ is given by $X_i^{-1} g_i = P_{H(m_i)} a_i g_i$, where $a_i(z)$ arises from the solution of the $\mathbf{RH}_+^\infty$ Bezout equation $a_i(z) \pi_i(z) + b_i(z) m_i(z) = 1$.

2. Let now $e(z) \in H(m)$ be a cyclic vector. For example, we can take, for any α in the open left half-plane, $e(z) = P_{H(m)} \frac{1}{z-\alpha} = \frac{1 - m(z) m^*(\alpha)}{z - \alpha} \in H(m)$. Clearly, $m(z)^{-1} v(z) \in \mathbf{RH}_-^\infty$ if and only if $m(z)^{-1} v(z) e(z) \in \mathbf{RH}_-^2$, which is equivalent to $v(z) e(z) \in H(m)$. In that case, we have a decomposition with respect to the direct sum representation (11.93), namely

$$v(z) e(z) = \sum_{j=1}^k \pi_j(z) f_j(z),$$

with $f_j(z) \in H(m_j)$. Using $v(z) - w_i(z) \in m_i(z)\mathbf{RH}^\infty_+$, we have

$$m_i(z)^{-1}\left(\sum_{j=1}^{k} \pi_j(z)f_j(z) - w_i(z)e_i(z)\right) \in \mathbf{RH}^2_+,$$

which implies

$$0 = m_i P_- m_i^{-1}(\textstyle\sum_{j=1}^{k} \pi_j(z)f_j(z) - w_i(z)e_i(z)) = P_{H(m_i)} \sum_{j=1}^{k} \pi_j(z)f_j(z) - w_i(z)e_i(z)$$
$$= P_{H(m_i)}\pi_i(z)f_i(z) - w_i(z)e_i(z) = (X_i f_i)(z) - w_i(z)e_i(z).$$

In order to evaluate $f_i(z)$, we have to invert the maps $X_i : H(m_i) \longrightarrow H(m_i)$, defined by $X_i f = P_{H(m_i)}\pi_i f$, and by Theorem 11.26, this is accomplished by solving the Bezout equation $a_i(z)\pi_i(z) + b_i(z)m_i(z) = 1$.   ∎

11.2.6   Analytic Hankel Operators and Intertwining Maps

Our next topic is the detailed study of Hankel operators, introduced in Defininition 11.9. We shall study their kernel and image and their relation to Beurling's theorem. We shall also describe the connection between Hankel operators and intertwining maps.

Definition 11.35. Given a function $\phi(z) \in \mathbf{RL}^\infty(i\mathbb{R})$, the **Hankel operator** $H_\phi :$ $\mathbf{RH}^2_+ \longrightarrow \mathbf{RH}^2_-$ is defined by

$$H_\phi f = P_-(\phi f), \quad f \in \mathbf{RH}^2_+. \tag{11.94}$$

The adjoint operator $(H_\phi)^* : \mathbf{RH}^2_- \longrightarrow \mathbf{RH}^2_+$ is given by

$$(H_\phi)^* f = P_+(\phi^* f), \quad f \in \mathbf{RH}^2_-. \tag{11.95}$$

Thus, assume $\phi(z) = \frac{n(z)}{d(z)} \in \mathbf{RH}^\infty_-$ and $n(z) \wedge d(z) = 1$. Our assumption implies that $d(z)$ is antistable. In spite of the slight ambiguity, we will write $n = \deg d$. It will always be clear from the context what n means. This leads to

$$\phi(z) = \frac{n(z)}{d(z)} = \frac{n(z)}{d^*(z)}\frac{d^*(z)}{d(z)} = \left(\frac{d(z)}{d^*(z)}\right)^{-1}\frac{n(z)}{d^*(z)}.$$

Thus

$$\phi(z) = m^*(z)\eta(z) = m(z)^{-1}\eta(z), \tag{11.96}$$

with $\eta(z) = \frac{n(z)}{d^*(z)}$ and $m(z) = \frac{d(z)}{d^*(z)}$, is a **coprime factorization** over $\mathbf{RH}^\infty_+$. This particular coprime factorization, where the denominator is an inner function, is called the **Douglas, Shapiro, and Shields (DSS) factorization.**

The next theorem discusses the functional equation of Hankel operators.

Theorem 11.36. *Let* $\phi(z) \in \mathbf{RL}^\infty$.

1. *For every* $\psi(z) \in \mathbf{RH}_+^\infty$ *the Hankel operator* $H_\phi : \mathbf{RH}_+^2 \longrightarrow \mathbf{RH}_-^2$ *satisfies the* **Hankel functional equation**

$$P_- \psi H_\phi f = H_\phi \psi f, \quad f \in \mathbf{RH}_+^2. \tag{11.97}$$

2. $\operatorname{Ker} H_\phi$ *is an invariant subspace, i.e., for* $f(z) \in \operatorname{Ker} H_\phi$ *and* $\psi(z) \in \mathbf{RH}_+^\infty$ *we have* $\psi(z) f(z) \in \operatorname{Ker} H_\phi$.

Proof. 1. We compute

$$P_- \psi H_\phi f = P_- \psi P_- \phi f = P_- \psi \phi f = P_- \phi \psi f = H_\phi \psi f.$$

2. Follows from the Hankel functional equation. ∎

We have an $\mathbf{RH}_+^\infty$ module structure on any finite-dimensional backward-invariant subspace $H(m)$. The $\mathbf{RH}_+^\infty$ module homomorphisms are the maps $X : H(m) \longrightarrow H(m)$ that satisfy $X T_\psi = T_\psi X$ for every $\psi(z) \in \mathbf{RH}_+^\infty$. At the same time the abstract Hankel operators, i.e., the operators $H : \mathbf{RH}_+^2 \longrightarrow \mathbf{RH}_+^2$ that satisfy the Hankel functional equation (11.97), are also $\mathbf{RH}_+^\infty$ module homomorphisms. The next result relates these two classes of module homomorphisms.

Theorem 11.37. *A map* $H : \mathbf{RH}_+^2 \longrightarrow \mathbf{RH}_-^2$ *is a Hankel operator with a nontrivial kernel* $m\mathbf{RH}_+^2$, *with* $m(z)$ *rational inner, if and only if we have*

$$H = m^* X P_{H(m)}, \tag{11.98}$$

or equivalently, the following diagram is commutative:

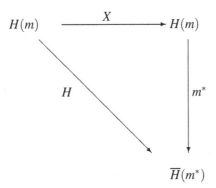

Here $X : H(m) \longrightarrow H(m)$ *is an intertwining map, i.e., satisfies*

$$X T_\psi = T_\psi X$$

for every $\psi(z) \in \mathbf{RH}_+^\infty$.

Proof. Assume $X : H(m) \longrightarrow H(m)$ is an intertwining map. We define $H : \mathbf{RH}_+^2 \longrightarrow \mathbf{RH}_-^2$ by (11.98). Let now $\psi(z) \in \mathbf{RH}_+^\infty$ and $f(z) \in \mathbf{RH}_+^2$. Then

$$
\begin{aligned}
H\psi f &= m^* X P_{H(m)} \psi f = m^* X P_{H(m)} \psi P_{H(m)} f \\
&= m^* X T_\psi P_{H(m)} f = m^* T_\psi X P_{H(m)} f = m^* P_{H(m)} \psi X P_{H(m)} f \\
&= m^* P_{H(m)} \psi X P_{H(m)} f = m^* m P_- m^* \psi X P_{H(m)} f \\
&= P_- \psi m^* X P_{H(m)} f = P_- \psi H f.
\end{aligned}
$$

So H satisfies the Hankel functional equation.

Conversely, if H is a Hankel operator with kernel $m\mathbf{RH}_+^2$, then we define $X : H(m) \longrightarrow H(m)$ by $X = mH|_{H(m)}$. For $f(z) \in H(m)$ and $\psi(z) \in \mathbf{RH}_+^\infty$, we compute

$$
\begin{aligned}
X T_\psi f &= m H P_{H(m)} \psi f = m H \psi f = m P_- \psi H f \\
&= m P_- m^* m \psi H f = P_{H(m)} \psi m H f = T_\psi X f.
\end{aligned}
$$

So X is intertwining. Obviously, (11.98) holds.   ∎

The following theorem shows that the Hankel functional equation characterizes the class of Hankel operators.

Theorem 11.38. *Let $H : \mathbf{RH}_+^2 \longrightarrow \mathbf{RH}_-^2$ satisfy the Hankel functional equation (11.97). Then there exists a function $\psi(z) \in \mathbf{RH}_-^\infty$ for which*

$$
H = H_\psi.
$$

Proof. **First proof:**
Choosing $\alpha \in \mathbb{C}_+$, then $\frac{1}{z+\alpha} \in \mathbf{RH}_+^2$, we clearly have

$$
\mathbf{RH}_+^2 = \left\{ \frac{\theta(z)}{z+\alpha} \,\middle|\, \theta(z) \in \mathbf{RH}_+^\infty \right\}. \tag{11.99}
$$

Note that with α, β in the open right half-plane, $\frac{z+\alpha}{z+\beta}$ is a unit in $\mathbf{RH}_+^\infty$. If we had $H = H_\psi$, with $\psi(z) \in \mathbf{RH}_-^\infty$, then we would have as well

$$
H \frac{1}{z+\alpha} = P_- \frac{\psi(z)}{z+\alpha} = P_- \frac{\psi(z) - \psi(-\alpha) + \psi(-\alpha)}{z+\alpha}
$$

$$
= \frac{\psi(z) - \psi(-\alpha)}{z+\alpha} = \phi(z).
$$

Therefore, we define

$$
\psi(z) = \phi(z)(z+\alpha) + \psi(-\alpha).
$$

Obviously $\psi(z) \in \mathbf{RH}^{\infty}_{-}$. We clearly have

$$P_{-}\psi \frac{1}{z+\alpha} = P_{-} \frac{\phi(z)(z+\alpha) + \psi(-\alpha)}{z+\alpha} = \phi(z).$$

Therefore, for any $\theta(z) \in \mathbf{RH}^{\infty}_{+}$,

$$H \frac{\theta}{z+\alpha} = P_{-}\theta H \frac{1}{z+\alpha} = P_{-}\theta P_{-}\psi \frac{1}{z+\alpha}$$

$$= P_{-}\theta \psi \frac{1}{z+\alpha} = P_{-}\psi \frac{\theta}{z+\alpha}.$$

So $Hf = P_{-}\psi f = H_{\psi}f$ for every $f(z) \in \mathbf{RH}^2_{+}$. This shows that $H = H_{\psi}$.

Second proof:
By our assumption, $\operatorname{Ker} H$ is a nontrivial invariant subspace of $\mathbf{RH}^2_{+}$. Thus, by Theorem 11.37, there exists a rational inner function $m(z)$ for which $\operatorname{Ker} H = m\mathbf{RH}^2_{+}$. Since for any $f(z) \in \mathbf{RH}^2_{+}$, we have $m(z)f(z) \in \operatorname{Ker} H$, it follows from the functional equation of Hankel operators that $P_{-}mHf = H(mf) = 0$. This implies $\operatorname{Im} H = \overline{H}(m^{*})$. Taking the restriction of H to a map from $H(m)$ to $\overline{H}(m^{*})$, it follows that the map $X : H(m) \longrightarrow H(m)$, defined by $X = mH$, is an intertwining map. Applying Theorem 11.33, it follows that $X = T_{\Theta}$ for some $\Theta(z) \in \mathbf{RH}^{\infty}_{+}$, which, without loss of generality, can be taken to satisfy $m(z)^{*}\Theta(z) \in \mathbf{RH}^{\infty}_{-}$. From (11.98), it follows that for $f(z) \in H(m)$,

$$Hf = m(z)^{*}Xf = m(z)^{*}T_{\Theta}f = m(z)^{*}P_{H(m)}\Theta f = H_{m(z)^{*}\Theta}f = H_{\psi}f,$$

where $\psi(z) = m(z)^{*}\Theta(z) \in \mathbf{RH}^{\infty}_{-}$. ∎

It follows from Beurling's theorem that $\operatorname{Ker} H_{\phi} = m\mathbf{RH}^2_{+}$ for some inner function $m \in \mathbf{RH}^{\infty}_{+}$. Since we are dealing with the rational case, the next theorem can make this more specific. It gives a characterization of the kernel and image of a Hankel operator and clarifies the connection between them and polynomial and rational models.

Theorem 11.39. *Let $d(z)$ be antistable and $\phi(z) = \frac{n(z)}{d(z)} \in \mathbf{RH}^{\infty}_{-}$. Then*

1. *We have $\operatorname{Ker} H_{\phi} \supset \frac{d}{d^{*}}\mathbf{RH}^2_{+}$, with equality holding if and only if $n(z)$ and $d(z)$ are coprime.*
2. *If $n(z)$ and $d(z)$ are coprime, then*

$$\{\operatorname{Ker} H_{\phi}\}^{\perp} = \left\{ \frac{d}{d^{*}}\mathbf{RH}^2_{+} \right\}^{\perp} = X^{d^{*}}. \tag{11.100}$$

3. *We have* $\operatorname{Im} H_\phi \subset \mathbf{RH}^2_- \ominus \dfrac{d^*}{d} \mathbf{RH}^2_- = X^d$ *with equality holding if and only if* $n(z)$
   and $d(z)$ *are coprime.*
4. *If* $n(z)$ *and* $d(z)$ *are coprime, then*

$$\dim \operatorname{Im} H_\phi = \deg d. \tag{11.101}$$

Proof. 1. $\{\operatorname{Ker} H_\phi\}^\perp$ contains only rational functions. Let $f(z) = \frac{p(z)}{q(z)} \in \{\frac{d}{d^*} \mathbf{RH}^2_+\}^\perp$.
Then $\frac{d^*(z)p(z)}{d(z)q(z)} \in \mathbf{RH}^2_-$. So $q(z) \mid d^*(z)p(z)$. But, since $p(z) \wedge q(z) = 1$, it follows
that $q(z) \mid d^*(z)$, i.e., $d^*(z) = q(z)r(z)$. Hence $f(z) = \frac{r(z)p(z)}{d^*(z)} \in X^{d^*}$.
    Conversely, let $\frac{p(z)}{d^*(z)} \in X^{d^*}$. Then $\frac{p(z)}{d^*(z)} = \frac{p(z)}{d(z)} \frac{d(z)}{d^*(z)}$ or $\frac{d^*(z)}{d(z)} \frac{p(z)}{d^*(z)} \in \mathbf{RH}^2_-$.
Therefore we conclude that $\frac{p(z)}{d^*(z)} \in \{\frac{d}{d^*} \mathbf{RH}^2_+\}^\perp$.
2. -4. These statements follow from Proposition 11.11.   ∎

As a corollary, we can state the following theorem.

Theorem 11.40 (Kronecker). *Let* $\phi(z) \in \mathbf{RL}^\infty$. *Then* $\operatorname{rank} H_\phi = k$ *if and only if*
$\phi(z)$ *has* k *poles, counting multiplicities, in the open right half-plane.*

Proof. Since $\phi(z) \in \mathbf{RL}^\infty$, it cannot have any poles on the imaginary axis. Taking
a partial fraction decomposition into stable and antistable parts, we may assume
without loss of generality that $\phi(z)$ is antistable.   ∎

The previous theorem, though of an elementary nature, is central to all further
development, since it provides the direct link between the infinite-dimensional
object, namely the Hankel operator, and the well-developed theory of polynomial
and rational models. This link will be continuously exploited.

Hankel operators in general and those with rational symbol in particular are
never invertible. Still, we may want to invert the Hankel operator as a map from
its cokernel, i.e., the orthogonal complement of its kernel, to its image. We saw
that such a restriction of a Hankel operator is of considerable interest because of
its connection to intertwining maps of model operators. Theorem 11.26 gave a
full characterization of invertibility properties of intertwining maps. These will be
applied, in Chapter 12, to the inversion of the restricted Hankel operators, which will
turn out to be of great importance in the development of a duality theory, linking
model reduction with robust stabilization.

11.3 Exercises

1. **Spectral factorization**: Show that for $g(z) \in \mathbf{RH}^\infty_+$ we have $g(i\omega) \geq 0$ for all
   $\omega \in \mathbb{R}$ if and only if for some $h(z) \in \mathbf{RH}^\infty_+$, we have $g(z) = h^*(z)h(z)$. Here
   $h^*(z) = \overline{h(-\bar{z})}$. Show that if $g(z)$ is nonzero on the extended imaginary axis, then
   $h(z)$ can be chosen such that also $h(z)^{-1} \in \mathbf{RH}^\infty_+$.

2. **Normalized coprime factorization**: Let $g(z) \in \mathbf{RH}_-^\infty$. Show that there exist $n(z), m(z) \in \mathbf{RH}_+^\infty$ for which $g(z) = n(z)m(z)^{-1}$ and $n^*(z)n(z) + m^*(z)m(z) = I$.

11.4 Notes and Remarks

The topics covered in this chapter have their roots early in the Twentieth century in the work of Carathéodory, Fejér, and Schur. In two classic papers, Schur (1917, 1918), gives a complete study of contractive analytic functions in the unit disk. The problem is attacked by a variety of methods, including what has become known as the Schur algorithm as well as the use of quadratic forms. One of the early problems considered was the minimum H^∞-norm extension of polynomials. This in turn led to the Nevanlinna-Pick interpolation problem. Nehari's theorem was proved in Nehari (1957). Many of the interpolation problems can be recast as best-approximation problems, which are naturally motivated by computational considerations.

A new impetus for the development of operator theory, based on complex analysis and the use of functional models, can be traced to the work of Livsic, Potapov, Debranges, Sz.-Nagy and Foias (1970), Lax and Phillips (1967), and many others. The characterization of invariant subspaces of H^2 by Beurling (1949) proved to be of central importance. Theorem 11.17 is a mild algebraic version. For a nice geometric proof of Beurling's theorem, see Helson (1964). Of great impact was the work of Sarason (1967) on the connection between modern operator theory and generalized interpolation. This led in turn to the commutant lifting theorem, proved in full generality by Sz.-Nagy and Foias.

Proposition 11.18 has a very simple proof. However, if we remove the assumption of rationality and work in the H^∞ setting, then the equivalence of the coprimeness condition $\sum_{i=1}^s |a_i(z)| \geq \delta > 0$ for all z in the open right half-plane and the existence of an H^∞ solution to the Bezout identity is a deep result in analysis, referred to as the corona theorem, due to Carleson (1962). The corona theorem was applied to the characterization of the invertibility of intertwining maps in Fuhrmann (1968a,b). The study of the DSS coprime factorization (11.96) is due to Douglas, Shapiro, and Shields (1971). For a matrix-valued version of the DSS factorization, see Fuhrmann (1975).

Chapter 12
Model Reduction

12.1 Introduction

Analytic Hankel operators are generally defined in the time domain, and via the
Fourier transform their frequency domain representation is obtained. We will skip
this part and introduce Hankel operators directly as frequency-domain objects. Our
choice is to develop the theory of continuous-time systems. This means that the
relevant frequency domain spaces are the Hardy spaces of the left and right half-
planes. Thus we will study Hankel operators defined on half-plane Hardy spaces
rather than on those of the unit disk as was done by Adamjan, Arov, and Krein
(1968a,b, 1971, 1978). In this, we follow the choice of Glover (1984). This choice
seems to be a very convenient one, since all results on duality simplify significantly,
due to the greater symmetry between the two half-planes in comparison to the unit
disk and its exterior.

In this, the last chapter of this book, we have chosen as our theme the theory of
Hankel norm approximation problems. This theory has come to be generally known
as the AAK theory, in honor of the pioneering and very influential contribution made
by Adamjan, Arov, and Krein.

The reason for the choice of this topic is twofold. First and foremost, it is
an extremely interesting and rich circle of ideas and allows us the consideration
of several distinct problems within a unified framework. The second reason is
just as important. We shall use the problems under discussion as a vehicle for
the development of many results that are counterparts in the setting of Hardy
spaces of results obtained previously in polynomial terms. This development can
be considered the construction of a bridge that connects algebra and analysis.

In Section 12.2.1, we do a detailed analysis of Schmidt pairs of a Hankel operator
with scalar, rational symbol. Some important lemmas are derived in our setting from
an algebraic point of view. These lemmas lead to a polynomial formulation of the
singular-value singular-vector equation of the Hankel operator. This equation, to
which we refer as the fundamental polynomial equation (FPE), is easily reduced,
using the theory of polynomial models, to a standard eigenvalue problem. Using

P.A. Fuhrmann, *A Polynomial Approach to Linear Algebra*, Universitext,
DOI 10.1007/978-1-4614-0338-8_12, © Springer Science+Business Media, LLC 2012

nothing more than the polynomial division algorithm, the subspace of all singular vectors corresponding to a given singular value is parametrized via the minimal-degree solution of the FPE. We obtain a connection between the minimal-degree solution and the multiplicity of the singular value.

The FPE can be transformed, using a simple algebraic manipulation, to a form that leads immediately to lower bound estimates on the number of antistable zeros of $p_k(z)$, the minimal-degree solution corresponding to the kth Hankel singular value. This lower bound is shown to actually coincide with the degree of the minimal-degree solution for the special case of the smallest singular value. Thus this polynomial turns out to be antistable. Another algebraic manipulation of the FPE leads to a Bezout equation over $\mathbf{RH}_+^\infty$. This provides the key to duality considerations.

Section 12.2.3 has duality theory as its main theme. Using the previously obtained Bezout equation, we invert the intertwining map corresponding to the initial Hankel operator. The inverse intertwining map is related to a new Hankel operator that has inverse singular values to those of the original one. Moreover we can compute the Schmidt pairs corresponding to this Hankel operator in terms of the original Schmidt pairs. The estimates on the number of antistable zeros of the minimum-degree solutions of the FPE that were obtained for the original Hankel operator Schmidt vectors are applied now to the new Hankel operator Schmidt vectors. Thus we obtain a second set of inequalities. The two sets of inequalities, taken together, lead to precise information on the number of antistable zeros of the minimal-degree solutions corresponding to all singular values. Utilizing this information leads to the solution of the Nehari problem as well as that of the general Hankel norm approximation problem.

Section 12.2.6 is a brief introduction to Nevanlinna-Pick interpolation, i.e., interpolation by functions that minimize the $\mathbf{RH}_+^\infty$-norm.

In Section 12.2.7, we study the singular values and singular vectors of both the Nehari complement and the best Hankel norm approximant. In both cases the singular values are a subset of the set of singular values of the original Hankel operator, and the corresponding singular vectors are obtained by orthogonal projections on the respective model spaces.

Section 12.2.9, using the analysis carried out in Section 12.2.3, is devoted to a nontrivial duality theory that connects robust stabilization with model reduction problems.

12.2   Hankel Norm Approximation

Real-life problems tend to be complex, and as a result, the modeling of systems may result in high complexity (usually measured in dimension), models. Complexity can be measured in various ways, e.g., by McMillan degree or number of stable or antistable poles. In order to facilitate computational algorithms, we want to approximate complex linear systems by systems of lower complexity, while retaining the essential features of the original system. These problems fall under the general term of **model reduction**.

Since we saw, in Chapter 7, that approximating a linear tranformation is related to singular values and singular vectors, and as the external behavior of a system is described by the Hankel operator associated with its transfer function, it is only to be expected that the analysis of the singular values and singular vectors of this Hankel operator will play a central role.

12.2.1 Schmidt Pairs of Hankel Operators

We saw, in Section 7.4, that singular values of linear operators are closely related to the problem of best approximation by operators of lower rank. That this basic method could be applied to the approximation of Hankel operators, by Hankel operators of lower ranks, through the detailed analysis of singular values and the corresponding Schmidt pairs is a fundamental contribution of Adamjan, Arov, and Krein.

We recall that given a linear operator A on an inner product space, μ is a singular value of A if there exists a nonzero vector f such that $A^*Af = \mu^2 f$. Rather than solve the previous equation, we let $g = \frac{1}{\mu}Af$ and go over to the equivalent system

$$
\begin{aligned}
Af &= \mu g, \\
A^*g &= \mu f,
\end{aligned}
\tag{12.1}
$$

i.e., μ is a singular value of both A and A^*.

For the rational case, we present an algebraic derivation of the basic results on Hankel singular values and vectors.

We recall Definition 11.1, i.e.,

$$
f^*(z) = \overline{f(-\bar{z})}.
\tag{12.2}
$$

Proposition 12.1. *Assume, without loss of generality, that $\phi(z) = \frac{n(z)}{d(z)} \in \mathbf{RH}_-^\infty$ is strictly proper, with $n(z)$ and $d(z)$ coprime polynomials with real coefficients. Let $H_\phi : \mathbf{RH}_+^2 \longrightarrow \mathbf{RH}_-^2$ be the Hankel operator, defined by (11.94).*

1. The pair $\{\frac{p(z)}{d^*(z)}, \frac{\hat{p}(z)}{d(z)}\}$ is a Schmidt pair for H_ϕ if and only if there exist polynomials $\pi(z)$ and $\xi(z)$ for which

$$
n(z)p(z) = \mu d^*(z)\hat{p}(z) + d(z)\pi(z),
\tag{12.3}
$$

$$
n^*(z)\hat{p}(z) = \mu d(z)p(z) + d^*(z)\xi(z).
\tag{12.4}
$$

    We will say that a pair of polynomials $(p(z), \hat{p}(z))$, with $\deg p, \deg \hat{p} < \deg d$, is a **solution pair** if there exist polynomials $\pi(z)$ and $\xi(z)$ such that equations (12.3) and (12.4) are satisfied.

2. Let $\{\frac{p(z)}{d^*(z)}, \frac{\hat{p}(z)}{d(z)}\}$ and $\{\frac{q(z)}{d^*(z)}, \frac{\hat{q}(z)}{d(z)}\}$ be two Schmidt pairs of the Hankel operator $H_{\frac{n}{d}}$, corresponding to the same singular value μ. Then

$$\frac{p(z)}{\hat{p}(z)} = \frac{q(z)}{\hat{q}(z)},$$

i.e., this ratio is independent of the Schmidt pair.

3. Let $\{\frac{p(z)}{d^*(z)}, \frac{\hat{p}(z)}{d(z)}\}$ be a Schmidt pair associated with the singular value μ. Then $\frac{p(z)}{\hat{p}(z)}$ is unimodular, that is we have $\|\frac{p(z)}{\hat{p}(z)}\| = 1$ on the imaginary axis. Such a function is also called an **all-pass function**.

4. Let μ be a singular value of the Hankel operator $H_{\frac{n}{d}}$. Then there exists a unique, up to a constant factor, solution pair $(p(z), \hat{p}(z))$ of minimal degree. The set of all solution pairs is given by

$$\{(q(z), \hat{q}(z)) \mid q(z) = p(z)a(z), \ \hat{q}(z) = \hat{p}(z)a(z) \ , \deg a < \deg q - \deg p\}.$$

5. Let $p(z), q(z) \in \mathbb{C}[z]$ be coprime polynomials such that $\frac{p(z)}{q(z)}$ is all-pass. Then $q(z) = \alpha p^*(z)$, with $|\alpha| = 1$.

6. Let $p(z), q(z)$ be polynomials such that $p(z) \wedge q(z) = 1$ and $\frac{p(z)}{q(z)}$ is all-pass. Then, with $r(z) = p(z) \wedge \hat{p}(z)$, we have $p(z) = r(z)s(z)$ and $\hat{p}(z) = \alpha r(z)s^*(z)$.

Proof. 1. In view of equation (7.19), in order to compute the singular vectors of the Hankel operator H_ϕ, we have to solve

$$H_\phi f = \mu g,$$

$$H_\phi^* g = \mu f. \tag{12.5}$$

Since by Theorem 11.39, we have $\text{Ker} \, H_\phi = \frac{d}{d^*}\mathbf{RH}_+^2$ and $\text{Im} \, H_\phi = \{\frac{d^*}{d}\mathbf{RH}_+^2\}^\perp$, we may as well consider H_ϕ as a map from $H(m)$ to $\overline{H}(m^*)$. Therefore a Schmidt pair for a singular value μ is necessarily of the form $\{\frac{p(z)}{d^*(z)}, \frac{\hat{p}(z)}{d(z)}\}$. Thus (12.5) translates into

$$P_- \frac{n}{d} \frac{p}{d^*} = \mu \frac{\hat{p}}{d},$$

$$P_+ \frac{n^*}{d^*} \frac{\hat{p}}{d} = \mu \frac{p}{d^*}.$$

This means that there exist polynomials $\pi(z)$ and $\xi(z)$ such that we have the following partial fraction decomposition:

$$\frac{n(z)}{d(z)}\frac{p(z)}{d^*(z)} = \mu\frac{\hat{p}(z)}{d(z)} + \frac{\pi(z)}{d^*(z)},$$

$$\frac{n^*(z)}{d^*(z)}\frac{\hat{p}(z)}{d(z)} = \mu\frac{p(z)}{d^*(z)} + \frac{\xi(z)}{d(z)}.$$

2. The polynomials $p(z)$, $\hat{p}(z)$ correspond to one Schmidt pair, and let the polynomials $q(z)$, $\hat{q}(z)$ correspond to another Schmidt pair, i.e., we have

$$n(z)q(z) = \mu d^*(z)\hat{q}(z) + d(z)\rho(z) \tag{12.6}$$

and

$$n^*(z)\hat{q}(z) = \mu d(z)q(z) + d^*(z)\eta(z). \tag{12.7}$$

Now, from equations (12.3) and (12.7) we get

$$0 = \mu d(z)(p(z)\hat{q}(z) - q(z)\hat{p}(z)) + d^*(z)(\xi(z)\hat{q}(z) - \eta(z)\hat{p}(z)).$$

Since $d(z)$ and $d^*(z)$ are coprime, it follows that $d^*(z) \mid (p(z)\hat{q}(z) - q(z)\hat{p}(z))$. On the other hand, from equations (12.3) and (12.6), we get

$$0 = \mu d^*(z)(\hat{p}(z)q(z) - \hat{q}(z)p(z)) + d(z)(\pi(z)q(z) - \rho(z)p(z)),$$

and hence that $d(z) \mid (\hat{p}(z)q(z) - \hat{q}(z)p(z))$. Now both $d(z)$ and $d^*(z)$ divide $\hat{p}(z)q(z) - \hat{q}(z)p(z)$, and by the coprimeness of $d(z)$ and $d^*(z)$, so does $d(z)d^*(z)$. Since $\deg(\hat{p}q - \hat{q}p) < \deg d + \deg d^*$, it follows that

$$\hat{p}(z)q(z) - \hat{q}(z)p(z) = 0,$$

or equivalently,

$$\frac{p(z)}{\hat{p}(z)} = \frac{q(z)}{\hat{q}(z)},$$

i.e., $\frac{p(z)}{\hat{p}(z)}$ is independent of the particular Schmidt pair associated to the singular value μ.

3. Going back to equation (12.4) and the dual of (12.3), we have

$$n^*(z)\hat{p}(z) = \mu d(z)p(z) + d^*(z)\xi(z),$$

$$n^*(z)p^*(z) = \mu d(z)(\hat{p}(z))^* + d^*(z)\pi^*(z).$$

It follows that

$$0 = \mu d(z)(p(z)p^*(z) - \hat{p}(z)(\hat{p}(z))^*) + d^*(z)(\xi(z)p^*(z) - \pi^*(z)\hat{p}(z)),$$

and hence $d^*(z) \mid (p(z)p^*(z) - \hat{p}(z)(\hat{p}(z))^*)$. By symmetry also $d(z) \mid (p(z)$ $p^*(z) - \hat{p}(z)\hat{p}^*(z))$, and so, arguing as before, necessarily, $p(z)p^*(z) - \hat{p}(z)$ $(\hat{p})^*(z) = 0$. This can be rewritten as $\frac{p(z)}{\hat{p}(z)} \frac{p^*(z)}{\hat{p}^*(z)} = 1$, i.e., $\frac{p(z)}{\hat{p}(z)}$ is all-pass.

4. Clearly, if μ is a singular value of the Hankel operator, then a nonzero solution pair $(p(z), \hat{p}(z))$ of minimal degree exists. Let $(q(z), \hat{q}(z))$ be any other solution pair with $\deg q$, $\deg \hat{q} < \deg d$. By the division rule for polynomials, we have $q(z) = a(z)p(z) + r(z)$ with $\deg r < \deg p$. Similarly, $\hat{q}(z) = \hat{a}(z)\hat{p}(z) + \hat{r}(z)$ with $\deg \hat{r} < \deg \hat{p}$. From equation (12.3) we get

$$n(z)(a(z)p(z)) = \mu d^*(z)(a(z)\hat{p}(z)) + d(z)(a(z)\pi(z)), \qquad (12.8)$$

whereas equation (12.6) yields

$$n(z)(a(z)p(z) + r(z)) = \mu d^*(z)(\hat{a}(z)\hat{p}(z) + \hat{r}(z)) + d(z)(\tau(z)). \qquad (12.9)$$

By subtraction we obtain

$$n(z)r(z) = \mu d^*(z)((\hat{a}(z) - a(z))\hat{p}(z) + \hat{r}(z)) + d(z)(\tau(z) - a(z)\pi(z)). \quad (12.10)$$

Similarly, from equation (12.7) we get

$$n^*(z)(\hat{a}(z)\hat{p}(z) + \hat{r}(z)) = \mu d(z)(a(z)p(z) + r(z)) + d^*(z)\xi(z), \qquad (12.11)$$

whereas equation (12.4) yields

$$n^*(z)(a(z)\hat{p}(z)) = \mu d(z)(a(z)p(z)) + d^*(z)(a(z)\xi(z)). \qquad (12.12)$$

Subtracting one from the other gives

$$n^*(z)((\hat{a}(z) - a(z))\hat{p}(z) + \hat{r}(z)) = \mu d(z)r(z) + d^*(z)(\eta(z) - a(z)\xi(z)). \quad (12.13)$$

Equations (12.10) and (12.13) imply that $\{\frac{r(z)}{d^*(z)}, \frac{(\hat{a}(z) - a(z))\hat{p}(z) + \hat{r}(z))}{d(z)}\}$ is a μ-Schmidt pair. Since necessarily $\deg r = \deg(\hat{a} - a)\hat{p} + \hat{r})$, we get $\hat{a}(z) = a(z)$. Finally, since we assumed $(p(z), \hat{p}(z))$ to be of minimal degree, we must have $r(z) = \hat{r}(z) = 0$.

Conversely, if $a(z)$ is any polynomial satisfying $\deg a < \deg d - \deg p$, then from equations (12.3) and (12.4), it follows by multiplication that $(p(z)a(z), \hat{p}(z)a(z))$ is also a solution pair.

5. Since $\frac{p(z)}{q(z)}$ is all-pass, it follows that $\frac{p(z)}{q(z)} \frac{p^*(z)}{q^*(z)} = 1$, or $p(z)p^*(z) = q(z)q^*(z)$. Since the polynomials $p(z)$ and $q(z)$ are coprime, it follows that $p(z) \mid q^*(z)$. By the same token, we have $q^*(z) \mid p(z)$ and hence $q^*(z) = \pm p(z)$.

6. Write $p(z) = r(z)s(z)$, $\hat{p}(z) = r(z)\hat{s}(z)$. Then $s(z) \wedge \hat{s}(z) = 1$ and $\frac{s(z)}{\hat{s}(z)}$ is all-pass. The result follows by applying Part 5.   ∎

Equation (12.3), considered as an equation modulo the polynomial $d(z)$, is not an eigenvalue equation, since there are too many unknowns. Specifically, we have to find the coefficients of both $p(z)$ and $\hat{p}(z)$. To overcome this difficulty, we study in more detail the structure of Schmidt pairs of Hankel operators.

The importance of the next theorem is due to the fact that it reduces the analysis of Schmidt pairs to one polynomial. This leads to an equation that is easily reduced to an eigenvalue problem.

Theorem 12.2. *Let* $\phi(z) = \frac{n(z)}{d(z)} \in \mathbf{RH}^\infty_-$ *be strictly proper, with* $n(z)$ *and* $d(z)$ *coprime polynomials with real coefficients. Let* μ *be a singular value of* H_ϕ *and let* $(p(z), \hat{p}(z))$ *be a nonzero, minimal degree solution pair of equations (12.3) and (12.4). Then* $p(z)$ *is a solution of*

$$n(z)p(z) = \lambda d^*(z)p^*(z) + d(z)\pi(z), \qquad (12.14)$$

with λ *real and* $|\lambda| = \mu$.

Proof. Let $(p(z), \hat{p}(z))$ be a nonzero minimal-degree solution pair of equations (12.3) and (12.4). By taking their adjoints, we can easily see that $(\hat{p}^*(z), p^*(z))$ is also a nonzero minimal-degree solution pair. By uniqueness of such a solution, i.e., by Proposition 12.1, we have

$$\hat{p}^*(z) = \varepsilon p(z). \qquad (12.15)$$

Since $\frac{\hat{p}(z)}{p(z)}$ is all-pass and both polynomials are real, we have $\varepsilon = \pm 1$. Let us put $\lambda = \varepsilon\mu$. Then (12.15) can be rewritten as $\hat{p}(z) = \varepsilon p^*(z)$, and so (12.14) follows from (12.3). ∎

We will refer to equation (12.14) as the **fundamental polynomial equation** (FPE), and it will be the basis for all further derivations.

Corollary 12.3. *Let* μ_i *be a singular value of* H_ϕ *and let* $p_i(z)$ *be the minimal-degree solution of the fundamental polynomial equation, i.e.,*

$$n(z)p_i(z) = \lambda_i d^*(z)p_i^*(z) + d(z)\pi_i(z).$$

Then

1.
$$\deg p_i = \deg p_i^* = \deg \pi_i.$$

2. Putting $p_i(z) = \sum_{j=0}^{n-1} p_{i,j} z^j$ *and* $\pi_i(z) = \sum_{j=0}^{n-1} \pi_{i,j} z^j$, *we have the equality*

$$\pi_{i,n-1}(z) = \lambda_i p_{i,n-1}(z). \qquad (12.16)$$

Corollary 12.4. *Let $p(z)$ be a minimal-degree solution of equation (12.14). Then*

1. *The set of all singular vectors of the Hankel operator $H_{\frac{n}{d}}$ corresponding to the singular value μ, is given by*

$$\mathrm{Ker}\left(H_{\frac{n}{d}}^* H_{\frac{n}{d}} - \mu^2 I\right) = \left\{ \frac{p(z)a(z)}{d^*(z)} \mid a(z) \in \mathbb{C}[z], \ \deg a < \deg d - \deg p \right\}.$$

(12.17)

2. *The multiplicity of $\mu = \|H_\phi\|$ as a singular value of H_ϕ is equal to $m = \deg d - \deg p$, where $p(z)$ is the minimum-degree solution of (12.14).*
3. *There exists a constant c such that $c + \frac{n(z)}{d(z)}$ is a constant multiple of an antistable all-pass function if and only if $\mu_1 = \cdots = \mu_n$.*

Proof. We will prove (3) only. Assume all singular values are equal to μ. Thus the multiplicity of μ is $\deg d$. Hence the minimal-degree solution $p(z)$ of (12.14) is a constant and so is $\pi(z)$. Putting $c = -\frac{\pi}{p}$, then (12.14) can be rewritten as

$$\frac{n(z)}{d(z)} + c = \lambda \frac{d^*(z)p^*(z)}{d(z)p(z)},$$

and this is a multiple of an antistable all-pass function.

Conversely assume, without loss of generality, that $\frac{n(z)}{d(z)} + c$ is antistable all-pass. Then the induced Hankel operator is isometric, and all its singular values are equal to 1. ∎

The following simple proposition is important in the study of zeros of singular vectors.

Proposition 12.5. *Let μ_k be a singular value of H_ϕ and let $p_k(z)$ be the minimal degree solution of*

$$n(z)p_k(z) = \lambda_k d^*(z)p_k^*(z) + d(z)\pi_k(z).$$

Then

1. *The polynomials $p_k(z)$ and $p_k^*(z)$ are coprime.*
2. *The polynomial $p_k(z)$ has no imaginary-axis zeros.*

Proof. 1. Let $e(z) = p_k(z) \wedge p_k^*(z)$. Without loss of generality, we may assume that $e(z) = e^*(z)$. The polynomial $e(z)$ has no imaginary-axis zeros, for that would imply that $e(z)$ and $\pi_k(z)$ have a nontrivial common divisor. Thus the fundamental polynomial equation could be divided by a suitable polynomial factor, this in contradiction to the assumption that $p_k(z)$ is a minimal-degree solution.
2. This clearly follows from the first part. ∎

12.2.2 Reduction to Eigenvalue Equation

The fundamental polynomial equation is easily reduced, using polynomial models, to either a generalized eigenvalue equation or a regular eigenvalue equation. Starting from (12.14), we apply the standard functional calculus and the fact that $d(S_d) = 0$, i.e., the Cayley–Hamilton theorem, to obtain

$$n(S_d)p_i = \lambda_i d^*(S_d)p_i^*. \qquad (12.18)$$

Now $d(z)$, $d^*(z)$ are coprime, since $d(z)$ is antistable and $d^*(z)$ is stable. Thus, by Theorem 5.18, $d^*(S_d)$ is invertible. In fact, the inverse of $d^*(S_d)$ is easily computed through the unique solution of the Bezout equation $a(z)d(z) + b(z)d^*(z) = 1$, satisfying the constraints $\deg a$, $\deg b < \deg d$. In this case, the polynomials $a(z)$ and $b(z)$ are uniquely determined, which, by virtue of symmetry, forces the equality $a(z) = b^*(z)$. Hence

$$b^*(z)d(z) + b(z)d^*(z) = 1. \qquad (12.19)$$

From this we get $b(S_d)d^*(S_d) = I$ or $d^*(S_d)^{-1} = b(S_d)$.

Because of the symmetry in the Bezout equation (12.19), we expect that some reduction in the computational complexity should be possible. This indeed turns out to be the case.

Given an arbitrary polynomial $f(z)$, we let

$$f_+(z^2) = \frac{f(z) + f^*(z)}{2},$$

$$f_-(z^2) = \frac{f(z) - f^*(z)}{2z}.$$

The Bezout equation can be rewritten as

$$(b_+(z^2) - zb_-(z^2))(d_+(z^2) + zd_-(z^2)) + (b_+(z^2) + zb_-(z^2))(d_+(z^2) - zd_-(z^2)) = 1,$$

or

$$2(b_+(z^2)d_+(z^2) - z^2 b_-(z^2)d_-(z^2)) = 1.$$

We can, of course, solve the lower-degree Bezout equation

$$2(b_+(z)d_+(z) - zb_-(z)d_-(z)) = 1.$$

This is possible because, by the assumption that $d(z)$ is antistable, $d_+(z)$ and $zd_-(z)$ are coprime. Putting $b(z) = b_+(z^2) + zb_-(z^2)$, we get a solution to the Bezout equation (12.19).

Going back to equation (12.18), we have

$$n(S_d)b(S_d)p_i = \lambda_i p_i^*. \tag{12.20}$$

To simplify, we let $r = \pi_d(bn) = (bn) \mod (d)$. Then (12.20) is equivalent to

$$r(S_d)p_i = \lambda_i p_i^*. \tag{12.21}$$

If $K : X_d \longrightarrow X_d$ is given by $(Kp)(z) = p^*(z)$, then (12.21) is equivalent to the generalized eigenvalue equation

$$r(S_d)p_i = \lambda_i K p_i.$$

Since K is obviously invertible and $K^{-1} = K$, the last equation transforms into the regular eigenvalue equation

$$K r(S_d)p_i = \lambda_i p_i.$$

To get a matrix equation, one can take the matrix representation with respect to any choice of basis in X_d.

12.2.3   Zeros of Singular Vectors and a Bezout equation

We begin now the study of the zero location of the numerator polynomials of singular vectors. This, of course, is the same as the study of the zeros of minimal-degree solutions of equation (12.14). The following proposition provides a lower bound on the number of zeros the minimal-degree solutions of (12.14) can have in the open left half plane. However, we also show that the lower bound is sharp in one special case, and that leads to the solvability of a special Bezout equation. Using the connection between Hankel operators and model intertwining maps given in Theorem 11.37 and the invertibility of intertwining maps given in Theorem 11.26, opens the road to a full duality theory.

Proposition 12.6. *Let* $\phi(z) = \frac{n(z)}{d(z)} \in \mathbf{RH}_-^\infty$, *be strictly proper, with* $n(z)$ *and* $d(z)$ *coprime polynomials with real coefficients.*

1. Let μ_k *be a singular value of* H_ϕ *satisfying*

$$\mu_1 \geq \cdots \geq \mu_{k-1} > \mu_k = \cdots = \mu_{k+v-1} > \mu_{k+v} \geq \cdots \geq \mu_n,$$

i.e., μ_k *is a singular value of multiplicity* v. *Let* $p_k(z)$ *be the minimum-degree solution of (12.14) corresponding to* μ_k. *Then the number of antistable zeros of* $p_k(z)$ *is* $\geq k - 1$.

2. If μ_n *is the smallest singular value of* H_ϕ *and is of multiplicity* v, *i.e.,*

$$\mu_1 \geq \cdots \geq \mu_{n-\nu} > \mu_{n-\nu+1} = \cdots = \mu_n,$$

and $p_{n-\nu+1}(z)$ is the corresponding minimum-degree solution of (12.14), then all the zeros of $p_{n-\nu+1}(z)$ are antistable.

3. The following Bezout identity in $\mathbf{RH}_+^\infty$ holds:

$$\frac{n(z)}{d^*(z)} \left(\frac{1}{\lambda_n} \frac{p_{n-\nu+1}(z)}{p_{n-\nu+1}^*(z)} \right) - \frac{d(z)}{d^*(z)} \left(\frac{1}{\lambda_n} \frac{\pi_{n-\nu+1}(z)}{p_{n-\nu+1}^*(z)} \right) = 1. \qquad (12.22)$$

Proof. 1. From equation (12.14), i.e., $n(z)p_k(z) = \lambda_k d^*(z)p_k^*(z) + d(z)\pi_k(z)$, we get, dividing by $d(z)p_k(z)$,

$$\frac{n(z)}{d(z)} - \frac{\pi_k(z)}{p_k(z)} = \lambda_k \frac{d^*(z)p_k^*(z)}{d(z)p_k(z)},$$

which implies, of course, that

$$\left\| H_{\frac{n}{d}} - H_{\frac{\pi_k}{p_k}} \right\| \leq \left\| \frac{n}{d} - \frac{\pi_k}{p_k} \right\|_\infty = \mu_k \left\| \frac{d^* p_k^*}{d p_k} \right\|_\infty = \mu_k.$$

This means, by the definition of singular values, that $\operatorname{rank} H_{\frac{\pi_k}{p_k}} \geq k - 1$. But this implies, by Kronecker's theorem, that the number of antistables poles of $\frac{\pi_k(z)}{p_k(z)}$ which is the same as the number of antistable zeros of $p_k(z)$, is $\geq k - 1$.

2. If μ_n, the smallest singular value, has multiplicity ν, and $p_{n-\nu+1}(z)$ is the minimal-degree solution of equation (12.14), then it has degree $n - \nu$. But by the previous part it must have at least $n - \nu$ antistable zeros. This implies that all the zeros of $p_{n-\nu+1}(z)$ are antistable.

3. From equation (12.14) we obtain, dividing by $\lambda_n d^*(z)p_{n-\nu+1}^*(z)$, the Bezout identity (12.22). Since the polynomials $p_{n-\nu+1}(z)$ and $d(z)$ are antistable, all four functions appearing in the Bezout equation (12.22) are in $\mathbf{RH}_+^\infty$. ∎

The previous result is extremely important from our point of view. It shifts the focus from the largest singular value, the starting point in all derivations so far, to the smallest singular value. Certainly, the derivation is elementary, inasmuch as we use only the definition of singular values and Kronecker's theorem. The great advantage is that at this stage we can solve an important Bezout equation that is the key to duality theory.

We have now at hand all that is needed to obtain the optimal Hankel norm approximant corresponding to the smallest singular value. We shall delay this analysis to a later stage and develop duality theory first.

We will say that two inner product space operators $T : H_1 \longrightarrow H_2$ and $T' : H_3 \longrightarrow H_4$ are **equivalent** if there exist unitary operators $U : H_1 \longrightarrow H_3$ and $V : H_2 \longrightarrow H_4$ such that $VT = T'U$. Clearly, this is an equivalence relation.

Lemma 12.7. *Let* $T : H_1 \longrightarrow H_2$ *and* $T' : H_3 \longrightarrow H_4$ *be equivalent. Then* T *and* T' *have the same singular values.*

Proof. Let $T^*Tx = \mu^2 x$. Since $VT = T'U$, it follows that

$$U^*T'^*T'Ux = T^*V^*VTx = T^*Tx = \mu^2 x,$$

or $T'^*T'(Ux) = \mu^2(Ux)$.   ∎

The following proposition is bordering on the trivial, and no proof need be given. However, when applied to Hankel operators, it has far-reaching implications. In fact, it provides a key to duality theory and leads eventually to the proof of the central results.

Proposition 12.8. *Let* T *be an invertible linear transformation. Then if* x *is a singular vector of the operator* T *corresponding to the singular value* μ, *i.e.,* $T^*Tx = \mu^2 x$, *then*

$$T^{-1}(T^{-1})^* x = \mu^{-2} x,$$

i.e., x *is also a singular vector for* $(T^{-1})^*$ *corresponding to the singular value* μ^{-1}.

In view of this proposition, it is of interest to compute $[(H_\phi | H(m))^{-1}]^*$. Before proceeding with this, we compute the inverse of a related, intertwining, operator T_θ in $H(m)$. This is a special case of Theorem 11.26. Note that, since $\|T_\theta^{-1}\| = \mu_n^{-1}$, there exists, by Theorem 11.26, $\xi(z) \in H_+^\infty$ such that $T_\theta^{-1} = T_\xi$ and $\|\xi\|_\infty = \mu_n^{-1}$. The next theorem provides this $\xi(z)$.

Theorem 12.9. *Let* $\phi(z) = \frac{n(z)}{d(z)} \in \mathbf{RH}_-^\infty$, *with* $n(z)$ *and* $d(z)$ *coprime polynomials with real coefficients and let* $m(z) = \frac{d(z)}{d^*(z)}$. *Then* $\theta(z) = \frac{n(z)}{d^*(z)} \in \mathbf{RH}_+^\infty$ *and the operator* $T_\theta : H(m) \longrightarrow H(m)$, *defined by equation (11.34), is invertible and its inverse given by* $T_{\frac{1}{\lambda_n} \frac{p_n}{p_n^*}}$, *where* λ_n *is the last signed singular value of* H_ϕ *and* $p_n(z)$ *is the minimal-degree solution of the FPE*

$$n(z)p_n(z) = \lambda_n d^*(z) p_n^*(z) + d(z)\pi_n(z). \tag{12.23}$$

Proof. From the previous equation we obtain the Bezout equation

$$\frac{n(z)}{d^*(z)} \left(\frac{1}{\lambda_n} \frac{p_n(z)}{p_n^*(z)} \right) - \frac{d(z)}{d^*(z)} \left(\frac{\pi_n(z)}{\lambda_n p_n^*(z)} \right) = 1. \tag{12.24}$$

which shows that $\frac{p_n(z)}{p_n^*(z)} \in \mathbf{RH}_+^\infty$. This, by Theorem 11.26, implies the result.   ∎

We know, see Corollary 10.21, that stabilizing controllers are related to solutions of Bezout equations over $\mathbf{RH}_+^\infty$. Thus we expect equation (12.22) to lead to a stabilizing controller. The next corollary is a result of this type.

Corollary 12.10. *Let* $\phi(z) = \frac{n(z)}{d(z)} \in \mathbf{RH}_-^\infty$, *with* $n(z)$ *and* $d(z)$ *coprime polynomials with real coefficients. The controller* $k(z) = \frac{p_n(z)}{\pi_n(z)}$ *stabilizes* $\phi(z)$. *If the multiplicity of* μ_n *is* ν, *there exists a stabilizing controller of degree* $n - \nu$.

Proof. Since $p_n(z)$ is antistable, we get from (12.14) that $n(z)p_n(z) - d(z)\pi_n(z) = \lambda_n d^*(z)p_n^*(z)$ is stable. We compute

$$\frac{\phi(z)}{1 - k(z)\phi(z)} = \frac{\dfrac{n(z)}{d(z)}}{1 - \dfrac{p_n(z)}{\pi_n(z)}\dfrac{n(z)}{d(z)}} = \frac{-n(z)\pi_n(z)}{n(z)p_n(z) - d(z)\pi_n(z)} = \frac{-n(z)\pi_n(z)}{\lambda_n d^*(z)p_n^*(z)} \in \mathbf{RH}_+^\infty.$$

∎

Theorem 12.11. *Let* $\phi(z) = \frac{n(z)}{d(z)} \in \mathbf{RH}_-^\infty$, *with* $n(z)$ *and* $d(z)$ *coprime polynomials with real coefficients. Let* $H : X^{d^*} \longrightarrow X^d$ *be defined by* $H = H_\phi \mid X^{d^*}$. *Then*

1. $H^{-1} : X^d \longrightarrow X^{d^*}$ *is given by*

$$H^{-1}h = \frac{1}{\lambda_n}\frac{d}{d^*}P_-\frac{p_n}{p_n^*}h. \tag{12.25}$$

2. $(H^{-1})^* : X^{d^*} \longrightarrow X^d$ *is given by*

$$(H^{-1})^*f = \frac{1}{\lambda_n}\frac{d^*}{d}P_+\frac{p_n^*}{p_n}f. \tag{12.26}$$

Proof. 1. Let $m(z) = \frac{d(z)}{d^*(z)}$ and let $T : X^{d^*} \longrightarrow X^{d^*}$ be the map given by $T = mH_{\frac{n}{d}}$. Thus we have the following commutative diagram:

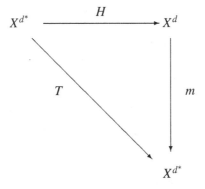

We compute

$$Tf = \frac{d}{d^*}P_-\frac{n}{d}f = \frac{d}{d^*}P_-\frac{d^*}{d}\frac{n}{d^*}f = P_{H(\frac{d}{d^*})}\frac{n}{d^*}f = P_{X^{d^*}}\frac{n}{d^*}f,$$

i.e., $T = T_\theta$, where $\theta(z) = \frac{n(z)}{d^*(z)}$. Now from $T_\theta = mH$ we have, by Theorem 12.9, $T_\theta^{-1} = T_{\frac{1}{\lambda_n}\frac{p_n}{p_n^*}}$. So for $h(z) \in X^d$,

$$H^{-1}h = \frac{1}{\lambda_n}P_{H(\frac{d}{d^*})}\frac{p_n}{p_n^*}\frac{d}{d^*}h = \frac{1}{\lambda_n}\frac{d}{d^*}P_-\frac{d^*}{d}\frac{p_n}{p_n^*}\frac{d}{d^*}h = \frac{1}{\lambda_n}\frac{d}{d^*}P_-\frac{p_n}{p_n^*}h.$$

2. The previous equation can also be rewritten as

$$H^{-1}h = T_{\frac{1}{\lambda_n}\frac{p_n}{p_n^*}}mh.$$

Therefore, using Theorem 11.25, we have, for $f(z) \in X^{d^*}$,

$$(H^{-1})^*f = m^*(T_{\frac{1}{\lambda_n}\frac{p_n}{p_n^*}})^*f = \frac{d^*}{d}P_+\frac{1}{\lambda_n}\frac{p_n^*}{p_n}f = \frac{1}{\lambda_n}\frac{d^*}{d}P_+\frac{p_n}{p_n^*}f.$$

∎

Corollary 12.12. *There exist polynomials $\alpha_i(z)$, of degree $\leq n-2$, such that*

$$\lambda_i p_n^*(z)p_i(z) - \lambda_n p_n(z)p_i^*(z) = \lambda_i d^*(z)\alpha_i(z), \quad i = 1,\ldots,n-1. \tag{12.27}$$

This holds also formally for $i = n$ with $\alpha_n(z) = 0$.

Proof. Since

$$H_{\frac{n}{d}}\frac{p_i}{d^*} = \lambda_i\frac{p_i^*}{d},$$

it follows that

$$(H_{\frac{n}{d}}^{-1})^*\frac{p_i}{d^*} = \lambda_i^{-1}\frac{p_i^*}{d}.$$

So, using equation (12.26), we have

$$\frac{1}{\lambda_n}\frac{d^*}{d}P_+\frac{p_n^*}{p_n}\frac{p_i}{d^*} = \frac{1}{\lambda_i}\frac{p_i^*}{d},$$

i.e.,

$$\frac{\lambda_n}{\lambda_i}\frac{p_i^*}{d^*} = P_+\frac{p_n^*}{p_n}\frac{p_i}{d^*}.$$

This implies, using a partial fraction decomposition, the existence of polynomials $\alpha_i(z)$, $i=1,\ldots,n$, such that $\deg \alpha_i < \deg p_n = n-1$, and $\frac{p_n^*(z)}{p_n(z)}\frac{p_i(z)}{d^*(z)} = \frac{\lambda_n}{\lambda_i}\frac{p_i^*(z)}{d^*(z)} + \frac{\alpha_i(z)}{p_n(z)}$, i.e., (12.27) follows. $\blacksquare$

We observed in Theorem 12.11 that for the Hankel operator H_ϕ, the map $(H_\phi^{-1})^*$ is not a Hankel map. However, there is an equivalent Hankel map. We sum this up in the following.

Theorem 12.13. *Let $\phi(z) = \frac{n(z)}{d(z)} \in \mathbf{RH}_-^\infty$ with $n(z)$ and $d(z)$ coprime polynomials with real coefficients. Let $H : X^{d^*} \longrightarrow X^d$ be defined by $H = H_\phi|X^{d^*}$. Then*

1. *The operator $(H_\phi^{-1})^*$ is equivalent to the Hankel operator $H_{\frac{1}{\lambda_n}\frac{d^* p_n}{dp_n^*}}$.*

2. *The Hankel operator $H_{\frac{1}{\lambda_n}\frac{d^* p_n}{dp_n^*}}$ has singular values $\mu_1^{-1} < \cdots < \mu_n^{-1}$, and its Schmidt pairs are $\{\frac{p_i^*(z)}{d^*(z)}, \frac{p_i(z)}{d(z)}\}$.*

Proof. 1. We saw, in (12.26), that

$$(H_{\frac{n}{d}}^{-1})^* = \frac{d^*}{d} T_{\frac{1}{\lambda_n}\frac{p_n}{p_n^*}}^*.$$

Since multiplication by $\frac{d^*(z)}{d(z)}$ is a unitary map of X^{d^*} onto X^d, the operator $(H_{\frac{n}{d}}^{-1})^*$ has, by Lemma 12.7, the same singular values as $T_{\frac{1}{\lambda_n}\frac{p_n}{p_n^*}}^*$. These are the same as those of the adjoint operator $T_{\frac{1}{\lambda_n}\frac{p_n}{p_n^*}}$. However, the last operator is equivalent to the Hankel operator $H_{\frac{1}{\lambda_n}\frac{d^* p_n}{dp_n^*}}$. Indeed, noting that multiplication by the all-pass function $\frac{d^*}{d}$ is a unitary map from X^{d^*} to X^d, we compute

$$\frac{d^*}{d}T_{\frac{1}{\lambda_n}\frac{p_n}{p_n^*}}f = \frac{d^*}{d}P_{H(\frac{d}{d^*})}\frac{1}{\lambda_n}\frac{p_n}{p_n^*}f = \frac{d^*}{d}P_-\frac{d}{d^*}\frac{1}{\lambda_n}\frac{p_n}{p_n^*}f = H_{\frac{1}{\lambda_n}\frac{d^* p_n}{dp_n^*}}f.$$

2. Next, we show that this Hankel operator has singular values $\mu_1^{-1} < \cdots < \mu_n^{-1}$ and its Schmidt pairs are $\{\frac{p_i^*(z)}{d^*(z)}, \frac{p_i(z)}{d(z)}\}$. Indeed

$$H_{\frac{d^* p_n}{dp_n^*}}\frac{p_i^*}{d^*} = P_-\frac{d^* p_n}{dp_n^*}\frac{p_i^*}{d^*} = P_-\frac{p_n p_i^*}{dp_n^*}.$$

Now from equation (12.27), we get $p_n(z)p_i^*(z) = \frac{\lambda_i}{\lambda_n}p_n^*(z)p_i(z) - \frac{\lambda_i}{\lambda_n}d^*(z)\alpha_i(z)$, or, taking the dual of that equation, $p_n(z)p_i^*(z) = \frac{\lambda_n}{\lambda_i}p_n^*(z)p_i(z) + d(z)\alpha_i^*(z)$. So

$$\frac{p_n(z)p_i^*(z)}{d(z)p_n^*(z)} = \frac{\lambda_n}{\lambda_i}\frac{p_n^*(z)p_i(z)}{d(z)p_n^*(z)} + \frac{d(z)\alpha_i^*(z)}{d(z)p_n^*(z)} = \frac{\lambda_n}{\lambda_i}\frac{p_i(z)}{d(z)} + \frac{\alpha_i^*(z)}{p_n^*(z)},$$

which implies $P_-\frac{p_n p_i^*}{dp_n^*} = \frac{\lambda_n}{\lambda_i}\frac{p_i}{d}$ and therefore $\frac{1}{\lambda_n}H_{\frac{d^* p_n}{dp_n^*}}\frac{p_i^*}{d^*} = \frac{1}{\lambda_i}\frac{p_i}{d}$.

$\blacksquare$

12.2.4   More on Zeros of Singular Vectors

The duality results obtained previously allow us to complete our study of the
zero structure of minimal-degree solutions of the fundamental polynomial equation
(12.14). This in turn leads to an elementary proof of the central theorem in the AAK
theory.

Theorem 12.14 (Adamjan, Arov, and Krein). *Let* $\phi(z) = \frac{n(z)}{d(z)} \in \mathbf{RH}_-^\infty$, *with* $n(z)$
and $d(z)$ *coprime polynomials with real coefficients.*

1. *Let* μ_k *be a singular value of* H_ϕ *satisfying*

$$\mu_1 \geq \cdots \geq \mu_{k-1} > \mu_k = \cdots = \mu_{k+\nu-1} > \mu_{k+\nu} \geq \cdots \geq \mu_n,$$

   i.e., μ_k *is a singular value of multiplicity* ν. *Let* $p_k(z)$ *be the minimum-degree
   solution of (12.14) corresponding to* μ_k. *Then the number of antistable zeros of*
   $p_k(z)$ *is exactly* $k-1$.
2. *If* μ_1 *is the largest singular value of* H_ϕ *and is of multiplicity* ν, *i.e.,*

$$\mu_1 = \cdots = \mu_\nu > \mu_{\nu+1} \geq \cdots \geq \mu_n,$$

   and $p_1(z)$ *is the corresponding minimum-degree solution of (12.14), then all the
   zeros of* $p_1(z)$ *are stable. This is equivalent to saying that* $\frac{p_1(z)}{d^*(z)}$ *is outer.*

Proof. 1. We saw, in the proof of Proposition 12.6, that the number of antistable
   zeros of $p_k(z)$ is $\geq k-1$. Now by Theorem 12.13, $p_k^*(z)$ is the minimum-degree
   solution of the fundamental equation corresponding to the transfer function
   $\frac{1}{\lambda_n} \frac{d^*(z) p_n(z)}{d(z) p_n^*(z)}$ and the singular value $\mu_{k+\nu-1}^{-1} = \cdots = \mu_k^{-1}$. Clearly we have

$$\mu_n^{-1} \geq \cdots \geq \mu_{k+\nu}^{-1} > \mu_{k+\nu-1}^{-1} = \cdots = \mu_k^{-1} > \mu_{k-1}^{-1} \geq \cdots \geq \mu_1^{-1}.$$

   In particular, applying Proposition 12.6, the number of antistable zeros of $p_k^*(z)$
   is $\geq n-k-\nu+1$. Since the degree of $p_k^*(z)$ is $n-\nu$, it follows that the number
   of stable zeros of $p_k^*(z)$ is $\leq k-1$. However, this is the same as saying that the
   number of antistable zeros of $p_k(z)$ is $\leq k-1$. Combining the two inequalities,
   it follows that the number of antistable zeros of $p_k(z)$ is exactly $k-1$.
2. The first part implies that the minimum-degree solution of (12.14) has only stable
   zeros.   ∎

We now come to apply some results of the previous sections to the case of
Hankel norm approximation. We use here the characterization, obtained in Section
7.4, of singular values $\mu_1 \geq \mu_2 \geq \cdots$ of a linear transformation $A : V_1 \longrightarrow V_2$ as
approximation numbers, namely

$$\mu_k = \inf\{\|A - A_k\| \mid \operatorname{rank} A_k \leq k-1\}.$$

We shall denote by $\mathbf{RH}^{\infty}_{[\mathbf{k}-1]}$ the set of all rational functions in $\mathbf{RL}^{\infty}$ that have at most $k-1$ antistable poles.

Theorem 12.15 (Adamjan, Arov, and Krein). *Let $\phi(z) = \frac{n(z)}{d(z)} \in \mathbf{RH}^{\infty}_{-}$ be a scalar strictly proper transfer function, with $n(z)$ and $d(z)$ coprime polynomials and $d(z)$ monic of degree n. Assume that*

$$\mu_1 \geq \cdots \geq \mu_{k-1} > \mu_k = \cdots = \mu_{k+v-1} > \mu_{k+v} \geq \cdots \geq \mu_n > 0$$

are the singular values of H_{ϕ}. Then

$$\mu_k = \inf \left\{ \|H_{\phi} - A\| \,|\, \mathrm{rank}\, A \leq k-1 \right\}$$

$$= \inf \left\{ \|H_{\phi} - H_{\psi}\| \,|\, \mathrm{rank}\, H_{\psi} \leq k-1 \right\}$$

$$= \inf \left\{ \|\phi - \psi\|_{\infty} \,|\, \psi \in \mathbf{RH}^{\infty}_{[\mathbf{k}-1]} \right\}.$$

Moreover, the infimum is attained on a unique function $\psi_k(z) = \phi(z) - \frac{(H_{\phi} f_k)(z)}{f_k(z)} = \phi(z) - \mu_k \frac{g_k(z)}{f_k(z)}$, where $\{f_k(z), g_k(z)\}$ is an arbitrary Schmidt pair of H_{ϕ} that corresponds to μ_k.

Proof. Given $\psi(z) \in \mathbf{RH}^{\infty}_{[\mathbf{k}-1]}$, we have, by Kronecker's theorem, that $\mathrm{rank}\, H_{\psi} = k-1$. Therefore we clearly have

$$\mu_k = \inf \left\{ \|H_{\phi} - A\| \,|\, \mathrm{rank}\, A \leq k-1 \right\}$$

$$\leq \inf \left\{ \|H_{\phi} - H_{\psi}\| \,|\, \mathrm{rank}\, H_{\psi} \leq k-1 \right\}$$

$$\leq \inf \left\{ \|\phi - \psi\|_{\infty} \,|\, \psi \in \mathbf{RH}^{\infty}_{[\mathbf{k}-1]} \right\}.$$

Therefore, the proof will be complete if we can exhibit a function $\psi_k(z) \in \mathbf{RH}^{\infty}_{[\mathbf{k}-1]}$ for which the equality $\mu_k = \|\phi - \psi\|_{\infty}$ holds. To this end let $p_k(z)$ be the minimal-degree solution of (12.14), and define $\psi_k(z) = \frac{\pi_k(z)}{p_k(z)}$. From the fundamental polynomial equation

$$n(z)p_k(z) = \lambda_k d^*(z)p_k^*(z) + d(z)\pi_k(z)$$

we get, dividing by $d(z)p_k(z)$, that

$$\frac{n(z)}{d(z)} - \frac{\pi_k(z)}{p_k(z)} = \lambda_k \frac{d^*(z)p_k^*(z)}{d(z)p_k(z)}. \tag{12.28}$$

This is, of course, equivalent to

$$\psi_k(z) = \frac{\pi_k(z)}{p_k(z)} = \frac{n(z)}{d(z)} - \lambda_k \frac{d^*(z)p_k^*(z)}{d(z)p_k(z)} = \phi(z) - \frac{H_\phi f_k}{f_k},$$

since for $f_k(z) = \frac{p_k^*(z)}{d(z)}$ we have $H_\phi f_k = \lambda_k \frac{p_k(z)}{d^*(z)}$, and by Proposition 12.1, the ratio $\frac{H_\phi f_k}{f_k}$ is independent of the particular Schmidt pair. So from (12.28), we obtain the following norm estimate:

$$\|\phi - \psi\|_\infty = \|\frac{n}{d} - \frac{\pi_k}{p_k}\|_\infty = \|\lambda_k \frac{d^* p_k^*}{d p_k}\|_\infty = \mu_k. \qquad (12.29)$$

Moreover, $\frac{\pi_k(z)}{p_k(z)} \in \mathbf{RH}^\infty_{[k-1]}$, as $p_k(z)$ has exactly $k-1$ antistable zeros.

■

Corollary 12.16. *The polynomials $\pi_k(z)$ and $p_k(z)$, defined in Theorem 12.15, have no common antistable zeros.*

Proof. Follows from the fact that $\operatorname{rank} H_{\frac{\pi_k}{p_k}} \geq k-1$.   ■

The error estimates (12.28) and (12.29) are important from the point of view of model reduction. We note that in the generic case of distinct singular values, the polynomials $\pi_k(z), p_k(z)$ have degree $n-1$, so the quotient $\frac{\pi_k(z)}{p_k(z)}$, which is proper, has $2n-1$ free parameters. From (12.28) it follows that $\frac{\pi_k(z)}{p_k(z)}$ interpolates the values of $\phi(z) = \frac{n(z)}{d(z)}$ at the $2n-1$ zeros of $d^*(z)p_k^*(z)$. In the special case of $k=n$, the polynomials $d^*(z), p_k^*(z)$ are both stable, so all interpolation points are in the open left half-plane. This observation is important, inasmuch as it places interpolation as a potential tool in model reduction. The choice of interpolation points is crucial for controlling the error. Our method not only establishes the optimal Hankel norm approximant but also gives a tight bound on the L^∞ error.

12.2.5   Nehari's Theorem

Nehari's theorem is one of the simplest model-reduction problems, i.e., the problem of approximating a given system by a system of smaller complexity. It states that if one wants to approximate an antistable function $g(z)$ by a stable function, then the smallest error norm that can be achieved is precisely the Hankel norm of $g(z)$. We are ready to give now a simple proof of Nehari's theorem in our rational context.

Theorem 12.17 (Nehari). *Given a rational function $\phi(z) = \frac{n(z)}{d(z)} \in \mathbf{RH}^\infty_-$ and $n(z) \wedge d(z) = 1$, then*

$$\sigma_1 = \|H_\phi\| = \inf\{\|\phi - q\|_\infty \mid q(z) \in \mathbf{RH}_+^\infty\},$$

and this infimum is attained on a unique function $q(z) = \phi(z) - \sigma_1 \frac{g(z)}{f(z)}$, *where* $\{f(z), g(z)\}$ *is an arbitrary* σ_1-*Schmidt pair of* H_ϕ.

Proof. Let $\sigma_1 = \|H_\phi\|$. It follows from equation (11.94) and the fact that for $q(z) \in \mathbf{RH}_+^\infty$, we have $H_q = 0$, that

$$\sigma_1 = \|H_\phi\| = \|H_\phi - H_q\| = \|H_{\phi - q}\| \leq \|\phi - q\|_\infty,$$

and so $\sigma_1 \leq \inf_{q \in \mathbf{RH}_+^\infty} \|\phi - q\|_\infty$.

To complete the proof we will show that there exists $q(z) \in \mathbf{RH}_+^\infty$ for which equality holds. We saw, in Theorem 12.14, that for $\sigma_1 = \|H_\phi\|$ there exists a stable solution $p_1(z)$ of

$$n(z) p_1(z) = \lambda_1 d^*(z) p_1^*(z) + d(z) \pi_1.$$

Dividing this equation by $d(z) p_1(z)$, we get

$$\frac{n(z)}{d(z)} - \frac{\pi_1(z)}{p_1(z)} = \lambda_1 \frac{d^*(z) p_1^*(z)}{d(z) p_1(z)}.$$

So, with $q(z) = \frac{\pi_1(z)}{p_1(z)} = \frac{n(z)}{d(z)} - \lambda_1 \frac{d^*(z) p_1^*(z)}{d(z) p_1(z)} \in \mathbf{RH}_+^\infty$, we get

$$\|\phi - q\|_\infty = \sigma_1 = \|H_\phi\|.$$

■

12.2.6 Nevanlinna–Pick Interpolation

We discuss now briefly the connection between Nehari's theorem in the rational case and the finite Nevanlinna–Pick interpolation problem, which is described next.

Definition 12.18. Given points $\lambda_1, \ldots, \lambda_s$ in the open right half-plane and complex numbers $c_1, \ldots, c_s$, then $\psi(z) \in \mathbf{RH}_+^\infty$ is a **Nevanlinna–Pick interpolant** if it is a function of minimum $\mathbf{RH}_+^\infty$ norm that satisfies

$$\psi(\lambda_i) = c_i, \quad i = 1, \ldots, s.$$

We can state the following.

Theorem 12.19. *Given the Nevanlinna–Pick interpolation problem of Definition 12.18, we define*

$$d(z) = \prod_{i=1}^s (z - \lambda_i), \qquad (12.30)$$

and let $n(z)$ be the unique minimal-degree polynomial satisfying the interpolation constraints

$$n(\lambda_i) = d^*(\lambda_i)c_i.$$

Let $p_1(z)$ be the minimal-degree solution of

$$n(z)p_1(z) = \lambda_1 d^*(z)p_1^*(z) + d(z)\pi_1(z)$$

corresponding to the largest singular value σ_1. Then the Nevanlinna–Pick interpolant is given by

$$\psi(z) = \lambda_1 \frac{p_1^*(z)}{p_1(z)}. \tag{12.31}$$

Proof. Note that since the polynomial $d(z)$, defined by (12.30), is clearly antistable, therefore $d^*(z)$ is stable. We proceed to construct a special $\mathbf{RH}_+^\infty$ interpolant. Let $n(z)$ be the unique polynomial, with $\deg n < \deg d = s$, that satisfies the following interpolation constraints:

$$n(\lambda_i) = d^*(\lambda_i)c_i, \qquad i = 1,\dots,s. \tag{12.32}$$

This polynomial interpolant can be easily constructed by Lagrange interpolation or any other equivalent method. We note that since $d^*(z)$ is stable, $d^*(\lambda_i) \neq 0$ for $i = 1,\dots,s$ and $\frac{n(z)}{d^*(z)} \in \mathbf{RH}_+^\infty$. Moreover, equation (12.32) implies

$$\frac{n(\lambda_i)}{d^*(\lambda_i)} = c_i, \qquad i = 1,\dots,s, \tag{12.33}$$

i.e., $\frac{n(z)}{d^*(z)}$ is an $\mathbf{RH}_+^\infty$ interpolant. Any other interpolant is necessarily of the form $\frac{n(z)}{d^*(z)} - \frac{d(z)}{d^*(z)}\theta(z)$ for some $\theta(z) \in \mathbf{RH}_+^\infty$.

To find $\inf_{\theta \in \mathbf{RH}_+^\infty} \|\frac{n}{d^*} - \frac{d}{d^*}\theta\|_\infty$, is equivalent, noting that $\frac{d(z)}{d^*(z)}$ is inner, to finding $\inf_{\theta(z) \in \mathbf{RH}_+^\infty} \|\frac{n}{d} - \theta\|_\infty$. However, this is just the content of Nehari's theorem, and we have

$$\inf_{\theta \in \mathbf{RH}_+^\infty} \left\|\frac{n}{d} - \theta\right\|_\infty = \sigma_1.$$

Moreover, the minimizing function is given by $\theta(z) = \frac{\pi_1(z)}{p_1(z)} = \frac{n(z)}{d(z)} - \lambda_1 \frac{d^*(z)p_1^*(z)}{d(z)p_1(z)}$.

Going back to the interpolation problem, we get for the Nevanlinna–Pick interpolant

$$\psi(z) = \frac{n(z)}{d^*(z)} - \frac{d(z)}{d^*(z)}\theta(z) = \frac{d(z)}{d^*(z)} - \lambda_1 \frac{d^*(z)p_1^*(z)}{d(z)p_1(z)} = \lambda_1 \frac{p_1^*(z)}{p_1(z)}.$$

■

12.2.7 Hankel Approximant Singular Values and Vectors

Our aim in this section is, given a Hankel operator with rational symbol $\phi(z)$, to study the geometry of singular values and singular vectors corresponding to the Hankel operators with symbols equal to the best Hankel norm approximant and the Nehari complement. For the simplicity of exposition, in the rest of this chapter we will make the genericity assumption that for $\phi(z) = \frac{n(z)}{d(z)} \in \mathbf{RH}_-^\infty$, all the singular values of $H_{\frac{n}{d}}$ are simple, i.e., have multiplicity 1. In our approach, the geometric aspects of the Hankel norm approximation method become clear, for we shall see that the singular vectors of the approximant are the orthogonal projections of the initial singular vectors on the appropriate model space. They correspond to the original singular values except the smallest one.

Theorem 12.20. *Let* $\phi(z) = \frac{n(z)}{d(z)} \in \mathbf{RH}_-^\infty$, *and let* $p_i(z)$ *be the minimal-degree solutions of*

$$n(z)p_i(z) = \lambda_i d^*(z)p_i^*(z) + d(z)\pi_i(z).$$

Consider the best Hankel norm approximant $\frac{\pi_n(z)}{p_n(z)} = \frac{n(z)}{d(z)} - \lambda_n \frac{d^*(z)p_n^*(z)}{d(z)p_n(z)}$ *that corresponds to the smallest nonzero singular value. Then*

1. $\frac{\pi_n}{p_n} \in \mathbf{RH}_-^\infty$ *and* $H_{\frac{\pi_n}{p_n}}$ *has the singuar values* $\sigma_i = |\lambda_i|$, $i = 1,\ldots,n-1$, *and the*
   σ_i-*Schmidt pairs of* $H_{\frac{\pi_n}{p_n}}$ *are given by* $\{\frac{\alpha_i(z)}{p_n^*(z)}, \frac{\alpha_i^*(z)}{p_n(z)}\}$, *where the polynomials* $\alpha_i(z)$ *are given by*

$$\lambda_i p_n^*(z)p_i(z) - \lambda_n p_n(z)p_i^*(z) = \lambda_i d^*(z)\alpha_i(z). \tag{12.34}$$

2. *Moreover, we have*

$$\frac{\alpha_i}{p_n^*} = P_{X^{p_n^*}}\frac{p_i}{d^*}, \tag{12.35}$$

   i.e., the singular vectors of $H_{\frac{\pi_n}{p_n}}$ *are projections of the singular vectors of* $H_{\frac{n}{d}}$ *onto* $X^{p_n^*}$, *the orthogonal complement of* $\operatorname{Ker} H_{\frac{\pi_n}{p_n}} = \frac{p_n}{p_n^*}\mathbf{RH}_+^2$.

3. *We have*

$$\left\| \frac{\alpha_i}{p_n^*} \right\|^2 = \left(1 - \left|\frac{\lambda_n}{\lambda_i}\right|^2\right) \left\| \frac{p_i}{d^*} \right\|^2. \tag{12.36}$$

Proof. 1. Rewrite equation (12.27) as

$$\lambda_i \frac{p_i(z)}{d^*(z)} - \lambda_n \frac{p_n(z)p_i^*(z)}{d^*(z)p_n^*(z)} = \lambda_i \frac{\alpha_i(z)}{p_n^*(z)}. \tag{12.37}$$

So

$$\lambda_i \frac{\pi_n(z)}{p_n(z)} \frac{p_i(z)}{d^*(z)} - \lambda_n \frac{\pi_n(z)}{p_n(z)} \frac{p_n(z)p_i^*(z)}{d^*(z)p_n^*(z)} = \lambda_i \frac{\pi_n(z)}{p_n(z)} \frac{\alpha_i(z)}{p_n^*(z)}.$$

Projecting on $\mathbf{RH}^2_-$, and recalling that $p_n(z)$ is antistable, we get

$$P_- \frac{\pi_n}{p_n} \frac{p_i}{d^*} = P_- \frac{\pi_n}{p_n} \frac{\alpha_i}{p_n^*},$$

which implies $\frac{p_i(z)}{d^*(z)} - \frac{\alpha_i(z)}{p_n^*(z)} \in \operatorname{Ker} H_{\frac{\pi_n}{p_n}}$. This is also clear from

$$\frac{p_i(z)}{d^*(z)} - \frac{\alpha_i(z)}{p_n^*(z)} = \frac{\lambda_n}{\lambda_i} \frac{p_n(z)}{p_n^*(z)} \frac{p_i(z)}{d^*(z)} \in \frac{p_n}{p_n^*} \mathbf{RH}^2_+ = \operatorname{Ker} H_{\frac{\pi_n}{p_n}}.$$

From

$$\frac{\pi_n(z)}{p_n(z)} = \frac{n(z)}{d(z)} - \lambda_n \frac{d^*(z)p_n(z)^*(z)}{d(z)p_n(z)},$$

we compute

$$H_{\frac{\pi_n}{p_n}} \frac{p_i}{d^*} = H_{(\frac{n}{d} - \lambda_n \frac{d^* p_n^*}{d p_n})} \frac{p_i}{d^*} = H_{\frac{n}{d}} \frac{p_i}{d^*} - \lambda_n H_{\frac{d^* p_n^*}{d p_n}} \frac{p_i}{d^*}$$

$$= \lambda_i \frac{p_i^*}{d} - \lambda_n P_- \frac{d^* p_n^*}{d p_n} \frac{p_i}{d^*} = \lambda_i \frac{p_i^*}{d} - P_- \frac{\lambda_n p_n^* p_i}{d p_n}$$

$$= \lambda_i \frac{p_i^*}{d} - \frac{\lambda_n p_n^* p_i}{d p_n} = \lambda_i \frac{p_i^*}{d} - \frac{\{\lambda_i p_n p_i^* - \lambda_i d^* \alpha_i\}}{d p_n}$$

$$= \lambda_i \frac{d \alpha_i^*}{d p_n} = \lambda_i \frac{\alpha_i^*}{p_n}.$$

Finally, we get

$$H_{\frac{\pi_n}{p_n}} \frac{\alpha_i}{p_n^*} = \lambda_i \frac{\alpha_i^*}{p_n}.$$

2. Note that equation (12.37) can be rewritten as

$$\frac{p_i(z)}{d^*(z)} = \frac{\alpha_i(z)}{p_n^*(z)} + \frac{\lambda_n}{\lambda_i} \frac{p_n(z)}{p_n^*(z)} \frac{p_i^*(z)}{d^*(z)}. \tag{12.38}$$

Since $\frac{p_i^*(z)}{d^*(z)} \in \mathbf{RH}^2_+$, this yields, by projecting on $X^{p_n^*} = \{\frac{p_n}{p_n^*} \mathbf{RH}^2_+\}^{\perp}$, equation (12.35).

3. Follows from equation (12.38), using orthogonality and computing norms.   ∎

12.2.8 Orthogonality Relations

We present next the derivation of some polynomial identities arising out of the singular value/singular vector equations. We proceed to interpret these relations as orthogonality relations between singular vectors associated with different singular values. Furthermore, the same equations provide useful orthogonal decompositions of singular vectors.

Equation (12.34), in fact more general relations, could be derived directly by simple computations, and this we proceed to explain. Starting from the singular value equations, i.e.,

$$n(z)p_i(z) = \lambda_i d^*(z)p_i^*(z) + d(z)\pi_i(z),$$

$$n(z)p_j(z) = \lambda_j d^*(z)p_j^*(z) + d(z)\pi_j(z),$$

we get

$$0 = d^*(z)\{\lambda_i p_i^*(z)p_j(z) - \lambda_j p_i(z)p_j^*(z)\} + d(z)\{\pi_i(z)p_j(z) - \pi_j(z)p_i(z)\}. \quad (12.39)$$

Since $d(z)$ and $d^*(z)$ are coprime, there exist polynomials $\alpha_{ij}(z)$, of degree $\leq n-2$, for which

$$\lambda_i p_i^*(z)p_j(z) - \lambda_j p_i(z)p_j^*(z) = d(z)\alpha_{ij}(z) \quad (12.40)$$

and $\pi_i(z)p_j(z) - \pi_j(z)p_i(z) = -d^*(z)\alpha_{ij}(z)$.

These equations have a very nice interpretation as orthogonality relations. In fact, we know that for any self-adjoint operator, eigenvectors corresponding to different eigenvalues are orthogonal. In particular, this applies to singular vectors. Thus, under our genericity assumption, we have for $i \neq j$,

$$\left(\frac{p_i}{d^*}, \frac{p_j}{d^*}\right)_{\mathbf{RH}_+^2} = 0. \quad (12.41)$$

This orthogonality relation could be derived directly from the polynomial equations by contour integration in the complex plane. Indeed, equation (12.40) could be rewritten as

$$\frac{p_i^*(z)}{d(z)}\frac{p_j(z)}{d^*(z)} - \frac{\lambda_j}{\lambda_i}\frac{p_i(z)}{d^*(z)}\frac{p_j^*(z)}{d(z)} = \frac{\alpha_{ij}}{d^*(z)}.$$

This equation can be integrated over the boundary of a half-disk of sufficiently large radius R, centered at origin and that lies in the right half-plane. Since $d^*(z)$ is stable, the integral on the right-hand side is zero. A standard estimate, using the fact that $\deg p_i^* p_j \leq 2n-2$, leads in the limit, as $R \to \infty$, to

$$\left(\frac{p_j}{d^*}, \frac{p_i}{d^*}\right)_{\mathbf{RH}_+^2} = \frac{\lambda_j}{\lambda_i}\left(\frac{p_i}{d^*}, \frac{p_j}{d^*}\right)_{\mathbf{RH}_+^2}.$$

This indeed implies the orthogonality relation (12.41).

Equation (12.40) can be rewritten as

$$p_i^*(z)p_j(z) = \frac{\lambda_j}{\lambda_i}p_i(z)p_j^*(z) + \frac{1}{\lambda_i}d(z)\alpha_{ij}(z).$$

However, if we rewrite (12.40) as $\lambda_j p_i(z)p_j^*(z) = \lambda_i p_i^*(z)p_j(z) - d(z)\alpha_{ij}(z)$, then, after conjugation, we get $p_i^*(z)p_j(z) = \frac{\lambda_i}{\lambda_j}p_i(z)p_j^*(z) - \frac{1}{\lambda_j}d^*(z)\alpha_{ij}^*(z)$. Equating the two expressions leads to

$$\left(\frac{\lambda_i}{\lambda_j} - \frac{\lambda_j}{\lambda_i}\right)p_i(z)p_j^*(z) = \frac{1}{\lambda_i}d(z)\alpha_{ij}(z) + \frac{1}{\lambda_j}d^*(z)\alpha_{ij}^*(z).$$

Putting $j = i$, we get $\alpha_{ii}(z) = 0$. Otherwise, we have

$$p_i(z)p_j^*(z) = \frac{1}{\lambda_i^2 - \lambda_j^2}\{\lambda_j d(z)\alpha_{ij}(z) - \lambda_i d^*(z)\alpha_{ij}^*(z)\}. \tag{12.42}$$

Conjugating this last equation and interchanging indices leads to

$$p_i(z)p_j^*(z) = \frac{1}{\lambda_j^2 - \lambda_i^2}\{\lambda_j d(z)\alpha_{ji}(z) - \lambda_i d^*(z)\alpha_{ji}^*(z)\}. \tag{12.43}$$

Comparing the two expressions leads to

$$\alpha_{ji}(z) = -\alpha_{ij}(z). \tag{12.44}$$

We continue by studying two special cases. For the case $j = n$ we put $\alpha_i = \frac{\alpha_{in}^*}{\lambda_i}$ to obtain

$$\lambda_i p_n^*(z)p_i(z) - \lambda_n p_n(z)p_i^*(z) = \lambda_i d^*(z)\alpha_i(z), \quad i = 1,\ldots,n-1, \tag{12.45}$$

or

$$\lambda_i p_i^*(z)p_n(z) - \lambda_n p_i(z)p_n^*(z) = \lambda_i d(z)\alpha_i^*(z), \quad i = 1,\ldots,n-1. \tag{12.46}$$

From equation (12.39) it follows that $\pi_n(z)p_i(z) - \pi_i(z)p_n(z) = \lambda_i d^*(z)\alpha_i^*(z)$, or equivalently, $\pi_n^*(z)p_i^*(z) - \pi_i^*(z)p_n^*(z) = \lambda_i d(z)\alpha_i(z)$. If we specialize now to the case $i = 1$, we obtain. $\pi_n(z)p_1(z) - \pi_1(z)p_n(z) = \lambda_1 d^*(z)\alpha_1^*(z)$, which, after dividing through by $p_1(z)p_n(z)$ and conjugating, yields

$$\frac{\pi_n^*(z)}{p_n^*(z)} - \frac{\pi_1^*(z)}{p_1^*(z)} = \lambda_1 \frac{d(z)\alpha_1(z)}{p_1^*(z)p_n^*(z)}. \tag{12.47}$$

Similarly, starting from equation (12.40) and putting $i = 1$, we get $\lambda_1 p_1^*(z) p_i(z) - \lambda_i p_1(z) p_i^*(z) = d(z) \alpha_{1i}(z)$. Putting also $\beta_i(z) = \lambda_1^{-1} \alpha_{1i}(z)$, we get

$$\lambda_1 p_1^*(z) p_i(z) - \lambda_i p_1(z) p_i^*(z) = \lambda_1 d(z) \beta_i(z), \qquad (12.48)$$

and of course $\beta_1(z) = 0$. This can be rewritten as

$$p_1^*(z) p_i(z) - \frac{\lambda_i}{\lambda_1} p_1(z) p_i^*(z) = d(z) \beta_i(z), \qquad (12.49)$$

or

$$p_1(z) p_i^*(z) - \frac{\lambda_i}{\lambda_1} p_1^*(z) p_i(z) = d^*(z) \beta_i^*(z). \qquad (12.50)$$

This is equivalent to

$$\frac{p_i^*(z)}{d^*(z)} = \frac{\beta_i^*(z)}{p_1(z)} + \frac{\lambda_i}{\lambda_1} \frac{p_1^*(z)}{p_1(z)} \frac{p_i}{d^*}. \qquad (12.51)$$

We note that equation (12.51) is nothing but the orthogonal decomposition of $\frac{p_i^*(z)}{d^*(z)}$ relative to the orthogonal direct sum $\mathbf{RH}_+^2 = X^{p_1} \oplus \frac{p_1^*}{p_1} \mathbf{RH}_+^2$. Therefore we have

$$\left\| \frac{\beta_i^*}{p_1} \right\|^2 = \left\| \frac{p_i^*}{d^*} \right\|^2 - \frac{|\lambda_i|^2}{|\lambda_1|^2} \left\| \frac{p_i}{d^*} \right\|^2 = \left\| \frac{p_i}{d^*} \right\|^2 \left(1 - \frac{|\lambda_i|^2}{|\lambda_1|^2} \right). \qquad (12.52)$$

Notice that if we specialize equation (12.46) to the case $i = 1$ and equation (12.48) to the case $i = n$, we obtain the relation $\beta_n(z) = \alpha_1^*(z)$.

12.2.9 Duality in Hankel Norm Approximation

In the present subsection we will shed some light on intrinsic duality properties of the problems of Hankel norm approximation and Nehari extensions. Results strongly suggesting such an underlying duality have appeared previously. This, in fact, turns out to be the case, though the duality analysis is far from being obvious. There are three operations that we shall apply to a given, antistable, transfer function, namely, inversion of the restricted Hankel operator, taking the adjoint map and finally one-sided multiplication by unitary operators. The last two operations do not change the singular values, whereas the first operation inverts them. In the process, we will prove a result dual to Theorem 12.20. While this analysis, after leading to the form of the Schmidt pairs in Theorem 12.22, is not necessary for the proof, it is of independent interest, and its omission would leave all intuition out of the exposition.

The analysis of duality can be summed up in the following scheme, which exhibits the relevant Hankel operators, their singular values, and the corresponding Schmidt pairs:

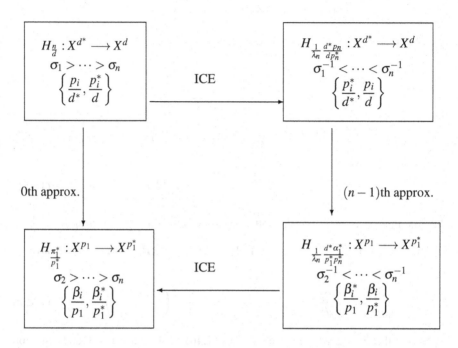

We would like to analyze the truncation that corresponds to the largest singular value. To this end we invert (I) the Hankel operator $H_{\frac{n}{d}}$ and conjugate (C) it, i.e., take its adjoint, as in Theorem 12.11. This operation preserves Schmidt pairs and inverts singular values. However, the operator so obtained is not a Hankel operator. This we correct by replacing it with an equivalent Hankel operator (E). This preserves singular values but changes the Schmidt pairs. Thus ICE in the previous diagram stands for a sequence of these three operations. To the Hankel operator so obtained, i.e., to $H_{\frac{1}{\lambda_n}\frac{d^*p_n}{dp_n^*}}$, we apply Theorem 12.20, which leads to the Hankel operator $H_{\frac{1}{\lambda_n}\frac{d^*\alpha_1^*}{p_1^*p_n^*}}$. This is done in Theorem 12.21. To this Hankel operator we apply again the sequence of three operations ICE, and this leads to Theorem 12.22. We proceed to study this Hankel map.

Theorem 12.21. *For the Hankel operator $H_{\frac{1}{\lambda_n}\frac{d^*p_n}{dp_n^*}}$, the Hankel norm approximant corresponding to the least singular value, i.e., to σ_1^{-1}, is $\frac{1}{\lambda_n}\frac{d^*\alpha_1^*}{p_1^*p_n^*}$.*

For the Hankel operator $H_{\frac{1}{\lambda_n}\frac{d^*\alpha_1^*}{p_1^*p_n^*}}$ we have

1. $\text{Ker}\, H_{\frac{1}{\lambda_n}\frac{d^*\alpha_1^*}{p_1^*p_n^*}} = \frac{p_1^*}{p_1}\mathbf{RH}_+^2$.

2. $X^{p_1} = \{\frac{p_1^*}{p_1}\mathbf{RH}_+^2\}^\perp$.

3. *The singular values of* $H_{\frac{1}{\lambda_n}\frac{d^*\alpha_1^*}{p_1^*p_n^*}}$ *are* $\sigma_2^{-1} < \cdots < \sigma_n^{-1}$.

4. *The Schmidt pairs of* $H_{\frac{1}{\lambda_n}\frac{d^*\alpha_1^*}{p_1^*p_n^*}}$ *are* $\{\frac{\beta_i^*(z)}{p_1(z)}, \frac{\beta_i(z)}{p_1^*(z)}\}$, *where*

$$\frac{\beta_i^*}{p_1} = P_{X^{p_1}}\frac{p_i^*}{d^*}, \tag{12.53}$$

or

$$\frac{\beta_i^*(z)}{p_1(z)} = \frac{p_i^*(z)}{d^*(z)} - \frac{\lambda_i}{\lambda_1}\frac{p_1^*(z)}{p_1(z)}\frac{p_i(z)}{d^*(z)}.$$

Proof. By Theorem 12.13, the Schmidt pairs for $H_{\frac{1}{\lambda_n}\frac{d^*p_n}{dp_n^*}}$ are $\{\frac{p_i^*(z)}{d^*(z)}, \frac{p_i(z)}{d(z)}\}$. Therefore the best Hankel norm approximant associated to σ_1^{-1} is, using also equation (12.45),

$$\frac{1}{\lambda_n}\frac{d^*(z)p_n(z)}{d(z)p_n^*(z)} - \frac{1}{\lambda_1}\frac{p_1(z)}{d(z)}\Big/\frac{p_1^*(z)}{d^*(z)} = \frac{1}{\lambda_n}\frac{d^*(z)p_n(z)}{d(z)p_n^*(z)} - \frac{1}{\lambda_1}\frac{d^*(z)p_1(z)}{d(z)p_1^*(z)}$$

$$= \frac{1}{\lambda_1\lambda_n}\frac{d^{*(z)}\{\lambda_1 p_n(z)p_1^*(z) - \lambda_n p_1(z)p_n^*(z)\}}{d(z)p_1^*(z)p_n^*(z)} = \frac{1}{\lambda_1\lambda_n}\frac{d^*(z)\{\lambda_1 d(z)\alpha_1^*(z)\}}{d(z)p_1^*(z)p_n^*(z)}$$

$$= \frac{1}{\lambda_n}\frac{d^*(z)\alpha_1^*(z)}{p_1^*(z)p_n^*(z)}.$$

1. Let $f(z) \in \frac{p_1^*}{p_1}\mathbf{RH}_+^2$, i.e., $f(z) = \frac{p_1^*(z)}{p_1(z)}g(z)$, for some $g(z) \in \mathbf{RH}_+^2$. Then

$$P_-\frac{1}{\lambda_n}\frac{d^*\alpha_1^*}{p_1^*p_n^*}\frac{p_1^*}{p_1}g = \frac{1}{\lambda_n}P_-\frac{d^*\alpha_1^*}{p_1 p_n^*}g = 0,$$

since both $\frac{d^*(z)\alpha_1^*(z)}{p_1(z)p_n^*(z)}$ and $g(z)$ are in $\mathbf{RH}_+^2$.

Conversely, let $f(z) \in \text{Ker}\, H_{\frac{1}{\lambda_n}\frac{d^*\alpha_1^*}{p_1^*p_n^*}}$ i.e., $P_-\frac{d^*\alpha_1^*}{p_1^*p_n^*}f = 0$. This implies, $p_1^*(z) \mid d^*(z)\alpha_1(z)f(z)$. Now $p_1(z)$ and $d(z)$ are coprime, since the first polynomial is stable, whereas the second is antistable. This implies naturally the coprimeness of $p_1^*(z)$ and $d^*(z)$. Also we have

$$\lambda_1 p_1^*(z)p_n(z) - \lambda_n p_1(z)p_n^*(z) = \lambda_1 d(z)\alpha_1^*(z).$$

If $p_1^*(z)$ and $\alpha_1^*(z)$ are not coprime, then by the previous equation, $p_1^*(z)$ has a common factor with $p_1(z)p_n^*(z)$. However, $p_1(z)$ and $p_n(z)$ are coprime, since the first is stable and the second antistable. So are $p_1(z)$ and $p_1^*(z)$, and for the same reason. Therefore, we must have that $\frac{f(z)}{p_1^*(z)}$ is analytic in the right half-plane. So $\frac{p_1(z)}{p_1^*(z)}f(z) \in \mathbf{RH}_+^2$, i.e., $f(z) \in \frac{p_1^*}{p_1}\mathbf{RH}_+^2$,

2. Follows from the previous part.
3. This is a consequence of Theorem 12.20.
4. Follows also from Theorem 12.20, since the singular vectors of $H_{\frac{1}{\lambda_n}\frac{d^*\alpha_1^*}{p_1^*p_n^*}}$ are given by $P_{XP_1}\frac{p_i^*}{d^*}$. This can be computed. Indeed, starting from equation (12.48), we have

$$\beta_i(z) = \frac{\lambda_1 p_1^*(z)p_i(z) - \lambda_i p_1(z)p_i^*(z)}{\lambda_1 d(z)}.$$

We compute

$$\lambda_i \beta_n(z)p_i^*(z) - \lambda_n \beta_i(z)p_n^*(z)$$

$$= \lambda_i \left\{ \frac{\lambda_1 p_1^*(z)p_n(z) - \lambda_n p_1(z)p_n^*(z)}{\lambda_1 d(z)} \right\} p_i^*(z)$$

$$- \lambda_n \left\{ \frac{\lambda_1 p_1^*(z)p_i(z) - \lambda_i p_1(z)p_i^*(z)}{\lambda_1 d(z)} \right\} p_n^*(z)$$

$$= \frac{\lambda_1 p_1^*(z)}{\lambda_1 d(z)} \{ \lambda_i p_i^*(z)p_n(z) - \lambda_n p_i(z)p_n^*(z) \}$$

$$= \frac{p_1^*(z)}{d(z)} \{ \lambda_i d(z)\alpha_i^*(z) \} \{ \lambda_i p_i^*(z)p_n(z) - \lambda_n p_i(z)p_n^*(z) \}$$

$$= \lambda_i p_1^*(z)\alpha_i^*(z).$$

Recalling that

$$\frac{\beta_i^*(z)}{p_1(z)} = \frac{p_i^*(z)}{d^*(z)} - \frac{\lambda_i}{\lambda_1}\frac{p_1^*(z)}{p_1(z)}\frac{p_i(z)}{d^*(z)} \tag{12.54}$$

and $\beta_n(z) = \alpha_1^*(z)$, we have

$$H_{\frac{1}{\lambda_n}\frac{d^*\alpha_1^*}{p_1^*p_n^*}}\frac{\beta_i^*}{p_1} - \frac{1}{\lambda_i}\frac{\beta_i}{p_1^*} = P_- \frac{1}{p_1^*p_n^*p_1} \left\{ \frac{1}{\lambda_n}d^*\beta_n\beta_i^* - \frac{1}{\lambda_i}p_1p_n^*\beta_i \right\}.$$

Thus, it suffices to show that the last term is zero. From equation (12.50), we have

$$d^*(z)\beta_i^*(z) = p_1(z)p_i^*(z) - \frac{\lambda_i}{\lambda_1}p_1^*(z)p_i(z),$$

hence

$$P_- \frac{1}{p_1^*(z)p_n^*(z)p_1(z)} \left\{ \frac{1}{\lambda_n} d^*(z)\beta_n(z)\beta_i^*(z) - \frac{1}{\lambda_i} p_1(z)p_n^*(z)\beta_i(z) \right\}$$

$$= P_- \frac{1}{p_1^*(z)p_n^*(z)p_1(z)} \left\{ \frac{1}{\lambda_n} p_1(z)p_i^*(z)\beta_n(z) - \frac{\lambda_i}{\lambda_1} p_i(z)p_1^*(z)\beta_n(z) \right.$$

$$\left. - \frac{1}{\lambda_i} p_1(z)p_n^*(z)\beta_i(z) \right\}$$

$$= P_- \frac{1}{p_1^*p_n^*p_1} \left\{ \frac{p_1}{\lambda_i\lambda_n}(\lambda_i p_i^*\beta_n - \lambda_n p_n^*\beta_i) - \frac{\lambda_i}{\lambda_1} p_i p_1^*\beta_n \right\}$$

$$= P_- \frac{1}{p_1^*p_n^*p_1} \left\{ \frac{p_1}{\lambda_i\lambda_n}\lambda_1\lambda_i p_1^*\alpha_i^* - \frac{\lambda_i}{\lambda_1} p_i p_1^*\beta_n \right\}$$

$$= P_- \frac{1}{p_n^*p_1} \left\{ \frac{\lambda_1}{\lambda_n} p_1\alpha_i^* - \frac{\lambda_i}{\lambda_1} p_i\beta_n \right\} = 0,$$

since $p_1(z)p_n^*(z)$ is stable.

∎

Theorem 12.22. *Let* $\phi(z) = \frac{n(z)}{d(z)} \in \mathbf{RH}_-^\infty$ *and* $n(z) \wedge d(z) = 1$, *i.e.,* $d(z)$ *is antistable.*
Let $\frac{\pi_1(z)}{p_1(z)}$ *be the optimal causal, i.e.,* $\mathbf{RH}_+^\infty$, *approximant, to* $\frac{n(z)}{d(z)}$. *Then*

1. $\frac{n(z)}{d(z)} - \frac{\pi_1(z)}{p_1(z)}$ *is all-pass.*
2. *The singular values of the Hankel operator* $H_{\frac{\pi_1^*}{p_1^*}}$ *are* $\sigma_2 > \cdots > \sigma_n$, *and the*
   corresponding Schmidt pairs of $H_{\frac{\pi_1^*}{p_1^*}}$ *are* $\{\frac{\beta_i(z)}{p_1(z)}, \frac{\beta_i^*(z)}{p_1^*(z)}\}$, *where the* $\beta_i(z)$ *are*
   defined by

$$\beta_i(z) = \frac{\lambda_1 p_1^*(z)p_i(z) - \lambda_i p_1(z)p_i^*(z)}{\lambda_1 d(z)}. \tag{12.55}$$

Proof. 1. Follows from the equality

$$\frac{n(z)}{d(z)} - \frac{\pi_1(z)}{p_1(z)} = \lambda_1 \frac{d^*(z)p_1^*(z)}{d(z)p_1(z)}.$$

2. We saw, in equation (12.47), that $\frac{\pi_1^*(z)}{p_1^*(z)} = \frac{\pi_n^*(z)}{p_n^*(z)} - \lambda_1 \frac{d(z)\alpha_1(z)}{p_1^*(z)p_n^*(z)}$. Since $\frac{\pi_n^*(z)}{p_n^*(z)} \in \mathbf{RH}_-^\infty$,
the associated Hankel operator is zero. Hence $H_{\frac{\pi_1^*}{p_1^*}} = H_{-\lambda_1 \frac{d\alpha_1}{p_1^*p_n^*}}$. Thus we have to

show that

$$H_{-\lambda_1 \frac{d\alpha_1}{p_1^* p_n^*}} \frac{\beta_i}{p_1} = \lambda_i \frac{\beta_i^*}{p_1^*}.$$

To this end, we start from equation (12.49), which, multiplied by α_1, yields

$$d(z)\alpha_1(z)\beta_i(z) = p_1^*(z)p_i(z)\alpha_1(z) - \frac{\lambda_i}{\lambda_1}p_1(z)p_i^*(z)\alpha_1(z). \qquad (12.56)$$

This in turn implies

$$\frac{d(z)\alpha_1(z)}{p_1^*(z)p_n^*(z)} \frac{\beta_i(z)}{p_1(z)} = \frac{p_i(z)\alpha_1(z)}{p_1(z)p_n^*(z)} - \frac{\lambda_i}{\lambda_1} \frac{\alpha_1(z)p_i^*(z)}{p_1^*(z)p_n^*(z)}. \qquad (12.57)$$

Since $\frac{p_i(z)\alpha_1(z)}{p_1(z)p_n^*(z)} \in H_+^2$, we have

$$-\lambda_1 P_- \frac{d\alpha_1}{p_1^* p_n^*} \frac{\beta_i}{p_1} = \lambda_i P_- \frac{\alpha_1 p_i^*}{p_1^* p_n^*}. \qquad (12.58)$$

All we have to do is to obtain a partial fraction decomposition of the last term. To this end, we go back to equation (12.49), from which we get

$$d(z)\beta_i(z)p_n(z) = p_1^*(z)p_i(z)p_n(z) - \frac{\lambda_i}{\lambda_1}p_1(z)p_i^*(z)p_n(z),$$

$$\qquad\qquad (12.59)$$

$$d(z)\beta_n(z)p_i(z) = p_1^*(z)p_n(z)p_i(z) - \frac{\lambda_i}{\lambda_1}p_1(z)p_n^*(z)p_i(z).$$

Hence

$$d(z)(\beta_n(z)p_i(z) - d(z)\beta_i(z)p_n(z)) = \frac{p_1(z)}{\lambda_1}\{\lambda_i p_i^*(z)p_n(z) - \lambda_n p_n^*(z)p_i(z)\}$$

$$= \frac{p_1}{\lambda_1}\{\lambda_i d(z)\alpha_i^*(z)\}$$

$$\qquad\qquad (12.60)$$

and

$$\beta_n p_i(z) - d(z)\beta_i p_n(z) = \frac{\lambda_i}{\lambda_1}p_1(z)\alpha_i^*(z). \qquad (12.61)$$

Now

$$\beta_n^*(z)p_i^*(z) - \beta_i^*(z)p_n^*(z) = \alpha_1(z)p_i^*(z) - \beta_i^*(z)p_n^*(z) = \frac{\lambda_i}{\lambda_1}p_1^*(z)\alpha_i(z).$$

Dividing through by $p_1^*(z)p_n^*(z)$, we get

$$\frac{\alpha_1(z)p_i^*(z)}{p_1^*(z)p_n^*(z)} = \frac{\beta_i^*(z)}{p_1^*(z)} + \frac{\lambda_i}{\lambda_1}\frac{\alpha_i(z)}{p_n^*(z)},$$

and from this it follows that $P_-\dfrac{\alpha_1 p_i^*}{p_1^* p_n^*} = \dfrac{\beta_i^*}{p_1^*}$. Using equation (12.58), we have

$$-\lambda_1 P_-\frac{d\alpha_1}{p_1^* p_n^*}\frac{\beta_i}{p_1} = \lambda_i \frac{\beta_i^*}{p_1^*}, \qquad (12.62)$$

and this completes the proof. ∎

There is an alternative way of looking at duality, and this is summed up in the following diagram:

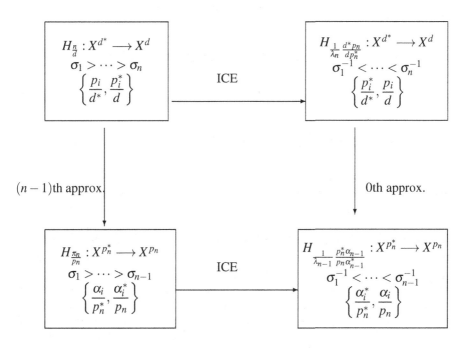

We will not go into the details except for the following.

Theorem 12.23. *The Hankel operator* $\dfrac{1}{\lambda_{n-1}}H_{\frac{p_n^*\alpha_{n-1}}{p_n\alpha_{n-1}^*}} : X^{p_n^*} \longrightarrow X^{p_n}$ *has singular values* $\sigma_1^{-1} < \cdots < \sigma_{n-1}^{-1}$ *and Schmidt pairs* $\{\frac{\alpha_i^*(z)}{p_n^*(z)}, \frac{\alpha_i(z)}{p_n(z)}\}$.

Proof. Starting from

$$\pi_n(z)\alpha_i(z) = \lambda_i p_n^*(z)\alpha_i^*(z) + p_n(z)\zeta_i(z),$$

$$\pi_n(z)\alpha_j(z) = \lambda_j p_n^*(z)\alpha_j^*(z) + p_n(z)\zeta_j(z),$$

we get $0 = p_n^*(z)\{\lambda_i \alpha_j(z)\alpha_i^*(z) - \lambda_j \alpha_i(z)\alpha_j^*(z)\} + p_n(z)\{\alpha_j(z)\zeta_i(z) - \alpha_i(z)\zeta_j(z)\}$.
For $j = n - 1$ we can write $\lambda_i \alpha_{n-1}(z)\alpha_i^*(z) - \lambda_{n-1}\alpha_i(z)\alpha_{n-1}^*(z) = \lambda_i p_n(z)\kappa_i(z)$, or
$\alpha_{n-1}(z)\alpha_i^*(z) = p_n(z)\kappa_i(z) + \frac{\lambda_{n-1}}{\lambda_i}\alpha_i(z)\alpha_{n-1}^*(z)$, i.e.,

$$\frac{\alpha_i^*(z)}{p_n(z)} = \frac{\kappa_i(z)}{\alpha_{n-1}(z)} + \frac{\lambda_{n-1}}{\lambda_i}\frac{\alpha_{n-1}^*(z)}{\alpha_{n-1}(z)}\frac{\alpha_i(z)}{p_n(z)}.$$

Thus we have

$$\frac{1}{\lambda_{n-1}}H_{\frac{p_n^*\alpha_{n-1}}{p_n\alpha_{n-1}^*}}\frac{\alpha_i^*}{p_n^*} = \frac{1}{\lambda_{n-1}}P_- \frac{1}{p_n\alpha_{n-1}^*}\left\{p_n\kappa_i + \frac{\lambda_{n-1}}{\lambda_i}\alpha_i\alpha_{n-1}^*\right\}$$

$$= \frac{1}{\lambda_{n-1}}P_- \frac{\kappa_i}{\alpha_{n-1}^*} + \frac{1}{\lambda_i}P_- \frac{\alpha_i}{p_n} = \frac{1}{\lambda_i}\frac{\alpha_i}{p_n}.$$

■

12.3   Model Reduction: A Circle of Ideas

In this section we explore a circle of ideas related to model reduction, i.e., to the approximation of a large-scale linear system by a lower-order system, easier to compute with, but such that the approximation error is kept under control. In the spirit of this book, we will treat only the scalar case, i.e., the case of a SISO system. Recently, several different methods have been explored for the purpose of system approximation. These include, apart from Hankel norm approximation, approximation by interpolation, the application of the Sylvester equation, and truncation or projection methods of which balanced truncation is a special case. Not much effort has been invested in the study of the connections between the different methods. Our aim will be to show how the polynomial Sylvester equation can be used to illuminate the interrelation between the different methods. We will show how the polynomial Sylvester equation can be used directly for model-reduction purposes using either the interpolation method or the projection method. We will show also how the important methods of Hankel norm approximation and balanced truncation relate to the polynomial Sylvester equation.

12.3.1   The Sylvester Equation and Interpolation

The simplest case of model reduction by interpolation is that of approximation at ∞ up to a certain order. Given a strictly proper rational function $g(z)$ having the expansion $g(z) = \sum_{j=1}^{\infty}\frac{g_j}{z^j}$ at ∞, we look for a lower-degree strictly proper rational function $\bar{g}(z)$ that matches the Markov parameters of $g(z)$, i.e., the coefficients $\{g_j\}_{j=1}^{\infty}$, up to a certain order. A realization of $\bar{g}(z)$ is called a **partial realization** of $g(z)$.

Theorem 12.24. *Let $g(z) = p(z)/q(z)$ be stricly proper with $q(z), p(z) \in \mathbb{F}[z]$ coprime polynomials. Let $\overline{q}(z), \overline{p}(z) \in \mathbb{R}[z]$ be the unique solution of the polynomial Bezout–Sylvester equation*

$$q(z)\overline{p}(z) - p(z)\overline{q}(z) + 1 = 0 \qquad (12.63)$$

that satisfies the degree constraints

$$\deg \overline{q} < \deg q, \qquad \deg \overline{p} < \deg p. \qquad (12.64)$$

Then a realization of $\overline{g}(z) = \overline{p}(z)/\overline{q}(z)$ is a partial realization of $g(z)$ that approximates $g(z)$ at ∞ up to order $n + \overline{n}$.

Proof. The Bezout–Sylvester equation (12.63) implies, for the error transfer function,

$$e(z) = g(z) - \overline{g}(z) = \frac{p(z)}{q(z)} - \frac{\overline{p}(z)}{\overline{q}(z)} = \frac{1}{q(z)\overline{q}(z)}.$$

Since $\deg q(z) = n$ and $\deg \overline{q}(z) = \overline{n}$, this implies that $e(z)$ has a zero of order $n + \overline{n}$ at ∞. Hence $\overline{g}(z)$ interpolates the value of $g(z)$ and its first $n + \overline{n} - 1$ derivatives at ∞. ∎

It has been known since antiquity that the Bezout–Sylvester equation (12.63) can be solved using the Euclidean algorithm, i.e., by recursively using the division rule of polynomials. The approximant can be obtained by use of the Lanczos polynomials, introduced in Chapter 8, which are the orthogonal polynomials corresponding to a Hankel matrix associated with $g(z)$.

So far, we have used only the Bezout–Sylvester equation. If we replace it by the polynomial Sylvester equation, namely by

$$q(z)\overline{p}(z) - p(z)\overline{q}(z) + r(z) = 0, \qquad (12.65)$$

where we assume $\rho = \deg r < \deg q$, then again there exists a unique solution that satisfies the degree constraints (12.64). Thus for the error transfer function we obtain

$$e(z) = g(z) - \overline{g}(z) = \frac{p(z)}{q(z)} - \frac{\overline{p}(z)}{\overline{q}(z)} = \frac{r(z)}{q(z)\overline{q}(z)}. \qquad (12.66)$$

This shows that $\overline{g}(z)$ can be obtained by interpolating the values of $g(z)$, and its derivatives up to appropriate orders, depending on the zeros of $r(z)$, as well as at ∞ to the order of $n + \overline{n} - \rho$.

12.3.2   The Sylvester Equation and the Projection Method

Another approach to model reduction is to project the system on a subspace of the state space. The projection, in general, is an oblique projection.

Proposition 12.25. *Given a transfer function* $g(z)$, *of McMillan degree* n, *having the minimal realization* $g(z) = D + C(zI - A)^{-1}B$, *in the state space* $\mathscr{X}$, *let* $\overline{\mathscr{X}}$ *be a linear space of lower dimension. Let* $\mathscr{W} : \mathscr{X} \longrightarrow \overline{\mathscr{X}}$ *be surjective and* $\mathscr{Y} : \overline{\mathscr{X}} \longrightarrow \mathscr{X}$ *injective linear maps for which we have*

$$\mathscr{W}\mathscr{Y} = I. \tag{12.67}$$

Define linear maps $\overline{A}, \overline{B}, \overline{C}, \overline{D}$ *by*

$$\overline{A} = \mathscr{W}A\mathscr{Y}, \quad \overline{B} = \mathscr{W}B,$$
$$\overline{C} = C\mathscr{Y}, \quad \overline{D} = D. \tag{12.68}$$

Then we have

1. *The map* $P : \mathscr{X} \longrightarrow \mathscr{X}$, *defined by*

$$P = \mathscr{Y}\mathscr{W}, \tag{12.69}$$

   is a projection.
2. *We have*

$$\operatorname{Ker} P = \operatorname{Ker} \mathscr{W},$$
$$\operatorname{Im} P = \operatorname{Im} \mathscr{Y}. \tag{12.70}$$

3. *We have the direct sum decomposition*

$$\mathscr{X} = \operatorname{Ker} \mathscr{W} = \operatorname{Ker} \mathscr{W} \oplus \operatorname{Im} \mathscr{Y}. \tag{12.71}$$

4. *Let* $\{f_1, \ldots, f_{n-k}\}$ *be a basis for* $\operatorname{Ker} \mathscr{W}$ *and* $\{e_1, \ldots, e_k\}$ *a basis for* $\overline{\mathscr{X}}$. *Then* $\{f_1, \ldots, f_{n-k}, \mathscr{Y}e_1, \ldots, \mathscr{Y}e_k\}$ *is a basis for* $\mathscr{X}$, *and with respect to these bases, we have the matrix representations*

$$\mathscr{W} = \begin{pmatrix} 0 & I \end{pmatrix}, \quad \mathscr{Y} = \begin{pmatrix} 0 \\ I \end{pmatrix}. \tag{12.72}$$

5. *If*

$$A = \begin{pmatrix} A_{11} & A_{12} \\ A_{21} & A_{22} \end{pmatrix}, \quad B = \begin{pmatrix} B_1 \\ B_2 \end{pmatrix}, \quad C = \begin{pmatrix} C_1 & C_2 \end{pmatrix} \tag{12.73}$$

   are the matrix representations with respect to the above bases, then we have

$$\overline{A} = A_{11}, \quad \overline{B} = B_1, \quad \overline{C} = C_1. \tag{12.74}$$

Proof. 1. We compute, using (12.67),

$$P^2 = (\mathscr{Y}\mathscr{W})(\mathscr{Y}\mathscr{W}) = \mathscr{Y}(\mathscr{W}\mathscr{Y})\mathscr{W} = \mathscr{Y}\mathscr{W} = P,$$

i.e., $P = \mathscr{Y}\mathscr{W}$ is a projection.

2. The factorization (12.69) implies the inclusion $\operatorname{Ker}\mathscr{W} \subset \operatorname{Ker}P$. The injectivity of $\mathscr{Y}$ proves the opposite inclusion.

Similarly, from (12.69), we conclude that $\operatorname{Im}P \subset \operatorname{Im}\mathscr{Y}$. The surjectivity of $\mathscr{W}$ shows the opposite inclusion.

3. Follows from (12.70).

4. That $\{\mathscr{Y}e_1,\dots,\mathscr{Y}e_k\}$ is a linearly independent set is a consequence of the injectivity of $\mathscr{Y}$. That it spans $\operatorname{Im}\mathscr{Y}$ is immediate from the fact that $\{e_1,\dots,e_k\}$ spans $\overline{\mathscr{X}}$. We note that $\mathscr{W}f_j = 0$ for $j = 1,\dots,n-k$, while, using (12.67), $\mathscr{W}(\mathscr{Y}e_i) = e_i$ for $i = 1,\dots,k$. The representation of $\mathscr{Y}$ is immediate.

5. Follows from the matrix representations (12.72) and (12.73). ∎

The previous proposition shows that the approximating system $\left(\begin{array}{c|c}\overline{A} & \overline{B} \\ \hline \overline{C} & \overline{D}\end{array}\right)$ is a projection of the original system $\left(\begin{array}{c|c}A & B \\ \hline C & D\end{array}\right)$. It goes without saying that the properties of the approximating system depend on the choice of the maps $\mathscr{W},\mathscr{Y}$. Unless we get good bounds for the error, the projection method remains but a formal procedure.

Our next result in this section is the clarification of the connection between the solution of the polynomial Sylvester equation (12.63) and the projection method. This we achieve by interpreting the polynomial data from a geometric point of view. As usual, the interpretation uses polynomial models and the shift realization. Furthermore, the maps $\mathscr{W},\mathscr{Y}$, used in Proposition 12.25, are defined using the polynomials $\overline{p}(z),\overline{q}(z)$, obtained in solving (12.63).

With $g(z) = p(z)/q(z)$ we associate the shift realization, in the state space X_q,

$$\Sigma_{q^{-1}p} := \begin{cases} A &= S_q, \\ B\alpha = p\alpha, \quad \alpha \in \mathbb{F}, \\ Cf &= (q^{-1}f)_{-1}, \end{cases} \tag{12.75}$$

whereas with $\overline{g}(z) = \overline{p}(z)/\overline{q}(z)$, we associate the shift realization

$$\Sigma_{\overline{p}\overline{q}^{-1}} := \begin{cases} \overline{A} &= S_{\overline{q}}, \\ \overline{B}\alpha = \alpha, \quad \alpha \in \mathbb{F}, \\ \overline{C}g &= (\overline{p}\overline{q}^{-1}g)_{-1}, \end{cases} \tag{12.76}$$

with both realizations constructed in Theorem 10.13.

Theorem 12.26. *Let* $q(z),p(z) \in \mathbb{F}[z]$ *be coprime polynomials. Let* $\overline{q}(z),\overline{p}(z) \in \mathbb{F}[z]$ *be the unique solution of the polynomial Sylvester equation (12.63) that satisfies*

*the degree constraints (12.64). For the realizations defined by (12.75) and (12.76),
define linear maps $\mathscr{Y} : X_{\bar{q}} \longrightarrow X_q$ and $\mathscr{W} : X_q \longrightarrow X_{\bar{q}}$ as follows:*

$$\mathscr{Y} = p(S_q)\pi_q,$$
$$\mathscr{W} = \pi_{\bar{q}}\bar{q}(S_q). \tag{12.77}$$

*With such choices, $\mathscr{W}$ is surjective, $\mathscr{Y}$ injective, and equations (12.63) and (12.68)
are satisfied.*

Proof. Since as sets, we have $X_{\bar{q}} \subset X_q$, the projection $\pi_q : X_{\bar{q}} \longrightarrow X_q$ is clearly injective.

The polynomial Sylvester equation (12.63) implies the coprimeness of $q(z)$ and $\bar{q}(z)$. So by Theorem 5.18, $\bar{q}(S_q)$ is invertible. Since $\pi_{\bar{q}} : X_q \longrightarrow X_{\bar{q}}$ is surjective, the surjectivity of $\mathscr{W}$ follows.

We compute, for $g(z) \in X_{\bar{q}}$ and using (12.63),

$$\mathscr{W}\mathscr{Y}g = \pi_{\bar{q}}\bar{q}(S_q)p(S_q)\pi_q g = \pi_{\bar{q}}\pi_q\bar{q}p\pi_q g$$

$$= \pi_{\bar{q}}\pi_q(1 + q\bar{p})g = \pi_{\bar{q}}g = g.$$

This proves (12.67).

To prove equations (12.68), we begin by computing, with $g \in X_{\bar{q}}$,

$$\mathscr{W}A\mathscr{Y}g = \pi_{\bar{q}}\bar{q}(S_q)S_q p(S_q)\pi_q g = \pi_{\bar{q}}\pi_q z\bar{q}pg$$

$$= \pi_{\bar{q}}\pi_q z(1 + \bar{p}q)g = \pi_{\bar{q}}\pi_q zg = S_{\bar{q}}g = \bar{A}g.$$

Here we used the fact that since $\deg\bar{q} < \deg q$, we have for $g(z) \in X_{\bar{q}}$ that $zg(z) \in X_q$, and hence $\pi_q zg = zg$.

Similarly,

$$\mathscr{W}B\alpha = \pi_{\bar{q}}\pi_q\bar{q}\pi_q p\alpha = \pi_{\bar{q}}\pi_q\bar{q}p\alpha$$

$$= \pi_{\bar{q}}\pi_q(1 + q\bar{p})\alpha = \pi_{\bar{q}}\alpha = \alpha = \bar{B}\alpha.$$

Finally, using $q(z)^{-1}p(z) = \bar{p}(z)\bar{q}(z)^{-1} + q(z)^{-1}\bar{q}(z)^{-1}$, which is a consequence of the polynomial Sylvester equation (12.63), we compute

$$C\mathscr{Y}g = (q^{-1}\pi_q pg)_{-1} = (q^{-1}q\pi_- q^{-1}pg)_{-1}$$

$$= (q^{-1}pg)_{-1} = ((\bar{p}\bar{q}^{-1} + q^{-1}\bar{q}^{-1})g)_{-1}$$

$$= (\bar{p}\bar{q}^{-1}g)_{-1} = \bar{C}g.$$

∎

The polynomial Sylvester equation (12.63) is, under the isomorphism

$$X_{q(z)\bar{q}(w)} \simeq \mathrm{Hom}_{\mathbb{F}}(X_{\bar{q}}, X_q), \tag{12.78}$$

equivalent to a standard Sylvester equation. This is summed up by the following proposition.

Proposition 12.27. *Let $q(z), p(z) \in \mathbb{F}[z]$ be coprime polynomials. Let $\overline{q}(z), \overline{p}(z) \in \mathbb{F}[z]$ be the unique solutions of the polynomial Sylvester equation (12.63) that satisfy the degree constraints (12.64). Define $y(z,w) \in X_{q(z)\overline{q}(w)}$ by*

$$y(z,w) = \frac{q(z)\overline{p}(w) - p(z)\overline{q}(w) + 1}{z - w}, \tag{12.79}$$

and define $\mathcal{Y} : X_{\overline{q}} \longrightarrow X_q$ by

$$\mathcal{Y}g = \langle g(\cdot), y(z, \cdot) \rangle, \qquad g \in X_{\overline{q}}. \tag{12.80}$$

Let $\mathcal{E} : X_{\overline{q}} \longrightarrow X_q$ be defined, for $g(z) \in X_{\overline{q}}$, by

$$\mathcal{E}g = \langle g, 1 \rangle = (\overline{q}^{-1}g)_{-1} = \overline{\xi}_g. \tag{12.81}$$

Then $\mathcal{Y}$ is a solution of the Sylvester equation

$$S_q \mathcal{Y} - \mathcal{Y} S_{\overline{q}} = \mathcal{E}. \tag{12.82}$$

Proof. In view of equation (12.63), the polynomials $q(z), \overline{q}(z)$ are coprime; hence equation (12.82) is solvable and the solution is unique. To see that $\mathcal{Y}$, given by (12.80), is that solution, we compute

$$(S_q \mathcal{Y} - \mathcal{Y} S_{\overline{q}})g = S_q p(S_q) I_{\overline{q},q} g - p(S_q) I_{\overline{q},q} S_{\overline{q}} g = p(S_q)(S_q g - I_{\overline{q},q} S_{\overline{q}} g)$$

$$= p(S_q)\left((zg - q\xi_g) - (zg - \overline{q}\overline{\xi}_g)\right) = p(S_q)\overline{q}\overline{\xi}_g = \pi_q p\overline{q}\overline{\xi}_g$$

$$= q\pi_- q^{-1} p\overline{q}\overline{\xi}_g = q\pi_-(\overline{p} + q^{-1})\overline{\xi}_g = \pi_q \overline{\xi}_g = \overline{\xi}_g$$

$$= \mathcal{E}g.$$

Here we used the polynomial Sylvester equation (12.63), equation (12.81), and the representation $S_q g = zg - q(z)\xi_g$ of the shift action. ∎

12.4 Exercises

1. Define the map $J : \mathbf{RL}^2 \longrightarrow \mathbf{RL}^2$ by $Jf(z) = f^*(z) = \overline{f(-\overline{z})}$. Clearly this is a unitary map in $\mathbf{RL}^2$, and it satisfies $J\mathbf{RH}^2_{\pm} = \mathbf{RH}^2_{\mp}$. Let H_ϕ be a Hankel operator.

   a. Show that $JP_- = P_+ J$, and $JH_\phi = H_\phi^* J \mid \mathbf{RH}^2_+$.
   b. If $\{f, g\}$ is a Schmidt pair of H_ϕ, then $\{Jf, Jg\}$ is a Schmidt pair of H_ϕ^*.
   c. Let $\sigma > 0$. Show that

i. The map $\hat{U} : \mathbf{RH}_+^2 \longrightarrow \mathbf{RH}_+^2$ defined by

$$\hat{U}f = \frac{1}{\sigma}H^*Jf$$

is a bounded linear operator in $\mathbf{RH}_+^2$.

ii. $\mathrm{Ker}\,(H^*H - \sigma^2 I)$ is an invariant subspace for $\hat{U}$.

iii. The map $U : \mathrm{Ker}\,(H^*H - \sigma^2 I) \longrightarrow \mathrm{Ker}\,(H^*H - \sigma^2 I)$ defined by $U = \hat{U} \mid \mathrm{Ker}\,(H^*H - \sigma^2 I)$ satisfies $U = U^* = U^{-1}$.

iv. Defining

$$K_+ = \frac{I+U}{2}, \quad K_- = \frac{I-U}{2},$$

show that $K_\pm$ are orthogonal projections and

$$\mathrm{Ker}\,(I-U) = \mathrm{Im}\,K_+,$$

$$\mathrm{Ker}\,(I+U) = \mathrm{Im}\,K_-,$$

and

$$\mathrm{Ker}\,(H^*H - \sigma^2 I) = \mathrm{Im}\,K_+ \oplus \mathrm{Im}\,K_-.$$

Also $U = K_+ - K_-$, which is the spectral decomposition of U, i.e., U is a signature operator.

2. Let σ be a singular value of the Hankel operator H_ϕ, let J be defined as before, and let $p(z)$ be the minimal-degree solution of (12.14). Assume $\deg d = n$ and $\deg p = m$. If $\varepsilon = \frac{\lambda}{\sigma}$, prove the following:

$$\dim \mathrm{Ker}\,(H_\phi - \lambda J) = \left[\frac{n-m+1}{2}\right],$$

$$\dim \mathrm{Ker}\,(H_\phi + \lambda J) = \left[\frac{n-m}{2}\right],$$

$$\dim \mathrm{Ker}\,(H_\phi^* H_\phi - \sigma^2 I) = n - m,$$

$$\dim \mathrm{Ker}\,(H_\phi - \lambda J) - \dim \mathrm{Ker}\,(H_\phi + \lambda J) = \begin{cases} 0, & n-m \text{ even}, \\ 1, & n-m \text{ odd}. \end{cases}$$

3. A minimal realization $\left(\begin{array}{c|c} A & B \\ \hline C & D \end{array}\right)$ of an asymptotically stable (antistable) transfer function $G(z)$ is called **balanced** if there exists a diagonal matrix diag $(\sigma_1, \ldots, \sigma_n)$ such that

$$A\Sigma + \Sigma\tilde{A} = -B\tilde{B},$$
$$\tilde{A}\Sigma + \Sigma A = -\tilde{C}C$$

(with the minus signs removed in the antistable case). The matrix Σ is called the **gramian** of the system and its diagonal entries are called the **system singular values**. Let $\phi(z) = \frac{n(z)}{d(z)} \in \mathbf{RH}_-^\infty$. Assume all singular values of $H_{\frac{n}{d}}$ are distinct. Let $p_i(z)$ be the minimal-degree solutions of the FPE, normalized so that $\| \frac{p_i}{d^*} \|^2 = \sigma_i$. Show that

a. The system singular values are equal to the singular values of $H_{\frac{n}{d}}$.

b. The function $\phi(z)$ has a balanced realization of the form

$$
\begin{aligned}
A &= \left(\frac{\varepsilon_j b_i b_j}{\lambda_i + \lambda_j} \right), \\
B &= (b_1, \ldots, b_n)^{\tilde{}}, \\
C &= (\varepsilon_1 b_1, \ldots, \varepsilon_n b_n), \\
D &= \phi(\infty),
\end{aligned}
$$

with

$$
\begin{aligned}
b_i &= (-1)^n \varepsilon_i p_{i,n-1}, \\
c_i &= (-1)^{n-1} p_{i,n-1} = -\varepsilon_i b_i.
\end{aligned}
$$

c. The balanced realization is sign-symmetric. Specifically, with $\varepsilon_i = \frac{\lambda_i}{\sigma_i}$ and $J = \mathrm{diag}(\varepsilon_1, \ldots, \varepsilon_n)$, we have

$$
JA = \tilde{A}J, \quad JB = \tilde{C}.
$$

d. Relative to a conformal block decomposition

$$
\Sigma = \begin{pmatrix} \Sigma_1 & 0 \\ 0 & \Sigma_2 \end{pmatrix}, \quad A = \begin{pmatrix} A_{11} & A_{12} \\ A_{21} & A_{22} \end{pmatrix},
$$

we have

$$
B\tilde{B} = \begin{pmatrix} A_{11}\Sigma_1 + \Sigma_1\tilde{A}_{11} & A_{12}\Sigma_2 + \Sigma_1\tilde{A}_{21} \\ A_{21}\Sigma_1 + \Sigma_2\tilde{A}_{12} & A_{22}\Sigma_2 + \Sigma_2\tilde{A}_{22} \end{pmatrix}.
$$

e. With respect to the constructed balanced realization, we have the following representation:

$$
\frac{p_i^*(z)}{d(z)} = C(zI - A)^{-1} e_i.
$$

4. Let $\phi(z) = \frac{n(z)}{d(z)} \in \mathbf{RH}_-^\infty$ and let $\frac{\pi_1(z)}{p_1(z)}$ of Theorem 12.17 be the Nehari extension of $\frac{n(z)}{d(z)}$. With respect to the balanced realization of $\phi(z)$, given in Exercise 3, show that $\frac{\pi_1(z)}{p_1(z)}$ admits a balanced realization $\left(\begin{array}{c|c} A_N & B_N \\ \hline C_N & D_N \end{array} \right)$, with

$$\begin{cases} A_N = -\left(\dfrac{\varepsilon_j \mu_i \mu_j b_i b_j}{\lambda_i + \lambda_j}\right), \\[2ex] B_N = (\mu_2 b_2, \ldots, \mu_n b_n), \\[2ex] C_N = (\mu_2 \varepsilon_2 b_2, \ldots, \mu_n \varepsilon_n b_n), \\[2ex] D_N = \lambda_1, \end{cases}$$

where

$$\mu_i = \sqrt{\left(\frac{\lambda_1 - \lambda_i}{\lambda_1 + \lambda_i}\right)} \quad \text{for } i = 2, \ldots, n.$$

12.5 Notes and Remarks

Independently of Sarason's work on H^∞ interpolation, Krein and his students undertook a detailed study of Hankel operators, motivated by classical extension and approximation problems. This was consolidated in a series of articles that became known as **AAK theory**; see Adamjan, Arov, and Krein (1968a,b, 1971, 1978).

The relevance of AAK theory to control problems was recognized by J.W. Helton and P. Dewilde. It was immediately and widely taken up, not least due to the influence of the work of G. Zames, which brought a resurgence of frequency-domain methods. A most influential contribution to the state-space aspects of the Hankel norm approximation problems was given in the classic paper Glover (1984). The content of this chapter is based mostly on Fuhrmann (1991, 1994). Young (1988) is a very readable account of elementary Hilbert space operator theory and contains an infinite-dimensional version of AAK theory. Nikolskii (1985) is a comprehensive study of the shift operator.

The result we refer to in this chapter as Kronecker's theorem was not proved in this form. Actually, Kronecker proved that an infinite Hankel matrix has finite rank if and only if its generating function is rational. Hardy spaces were introduced later.

For more on the partial realization problem, we refer the reader to Gragg and Lindquist (1983). An extensive monograph on model reduction is Antoulas (2005).

We note that Hankel norm approximation is related to a polynomial Sylvester equation and the optimal approximant determined by interpolation. However, we have to depart from the convention adopted before of taking the solution of the Sylvester equation that satisfies the degree constraints (12.64). We proceed to explain the reason for this. It has been stated in the literature, see for example Sorensen and Antoulas (2002) or Gallivan, Vandendorpe, and Van Dooren (2003), that for model reduction we may assume that if the transfer function of the system is strictly proper, then so is the approximant. Our analysis of Hankel norm approximation shows that this is obviously false when norm bounds for the error function are involved. In fact, in the Hankel norm approximation case, the error function turns out to be a constant multiple of an all-pass function, so the optimal approximant cannot be strictly proper. A simple example is $g(z) = 2(z-1)^{-1} \in$ **RH**$_-^{\infty}$ with $\|g\|_{\infty} = 2$. A 0th-order approximant then is -1 with $\|e\|_{\infty} = 1$:

$$\|e\|_{\infty} = \|2(z-1)^{-1} - (-1)\|_{\infty} = \|(z+1)(z-1)^{-1}\|_{\infty} = 1.$$

An important, SVD-related, method of model reduction is balanced truncation, introduced in Moore (1981), see also Glover (1984). Exercise 3 gives a short description of the method. We indicate, again in the scalar rational case, how balanced truncation reduces to an interpolation result. Assume $g(z) = n(z)/d(z)$ is a strictly proper antistable function, with $n(z), d(z)$ coprime, and having a minimal realization $g(z) = \left(\begin{array}{c|c} A & b \\ \hline c & 0 \end{array} \right)$. It has been shown that there exists a state-space isomorphism that brings the realization into balanced form, namely one for which the Lyapunov equations

$$\begin{aligned} AX + XA^* &= bb^*, \\ A^*X + XA &= c^*c, \end{aligned} \tag{12.83}$$

have a common diagonal and positive solution $\Sigma = \text{diag}(\sigma_1, \ldots, \sigma_n)$. The σ_i, decreasingly ordered, are the Hankel singular values of $g(z)$. If the set of singular values σ_i satisfies $\sigma_1 \geq \cdots \geq \sigma_k >> \sigma_{k+1} \geq \cdots \geq \sigma_n$, then we write accordingly $\Sigma = \text{diag}(\sigma_1, \ldots, \sigma_n) = \text{diag}(\sigma_1, \ldots, \sigma_k) \oplus \text{diag}(\sigma_{k+1}, \ldots, \sigma_n)$ and partition the realization conformally to get

$$g(z) = \left(\begin{array}{cc|c} A_{11} & A_{12} & b_1 \\ A_{21} & A_{22} & b_1 \\ \hline c_1 & c_2 & 0 \end{array} \right).$$

By truncation, or equivalently by projection, of this realization, the transfer function $g_b(z) = \left(\begin{array}{c|c} A_{11} & b_1 \\ \hline c_1 & 0 \end{array} \right)$ is obtained. It is easy to check that $g_b(z)$ is antistable too and $g_b(z) = n_b(z)/d_b(z)$ serves as an approximation of $g(z)$ for which an error estimate is available. In fact, if we partition $\Sigma = \text{diag}(\sigma_1, \ldots, \sigma_n)$ as $\Sigma = \text{diag}(\sigma_1, \ldots, \sigma_{n-1}) \oplus \sigma_n$, and take $p_n^*(z)/d^*(z)$ to be the Hankel singular vector corresponding to the singular value σ_n, then $d_b(z)$ is antistable too. It can be shown, see Fuhrmann (1991, 1994) for the details, that the Glover approximation error satisfies

$$e(z) = g(z) - g_b(z) = \lambda_n \frac{p_n^*(z)}{p_n(z)} \left(\frac{d^*(z)}{d(z)} + \frac{d_b^*(z)}{d_b(z)} \right), \tag{12.84}$$

which implies $\|e\|_{\infty} \leq 2\sigma_n$. Furthermore, we have

$$d_b(z)n(z) - d(z)n_b(z) = -\varepsilon_n(p_n^*(z))^2, \tag{12.85}$$

which implies the error estimate

$$e(z) = \varepsilon_n \frac{(p_n^*(z))^2}{d(z)d_b(z)}. \tag{12.86}$$

Equation (12.86) shows that the balanced truncation that corresponds to the smallest singular value can be obtained by second-order interpolation at the zeros of $p_n^*(z)$.

References

Adamjan, V.M., Arov, D.Z., and Krein, M.G. (1968a) "Infinite Hankel matrices and generalized problems of Carathéodory-Fejér and F. Riesz," *Funct. Anal. Appl.* 2, 1–18.

Adamjan, V.M., Arov, D.Z., and Krein, M.G. (1968b) "Infinite Hankel matrices and generalized problems of Carathéodory-Fejér and I. Schur," *Funct. Anal. Appl.* 2, 269–281.

Adamjan, V.M., Arov, D.Z., and Krein, M.G. (1971) "Analytic properties of Schmidt pairs for a Hankel operator and the generalized Schur-Takagi problem," *Math. USSR Sbornik* 15, 31–73.

Adamjan, V.M., Arov, D.Z., and Krein, M.G. (1978) "Infinite Hankel block matrices and related extension problems," Amer. Math. Soc. Transl., series 2, Vol. 111, 133–156.

Antoulas, A.C. (2005) *Approximation of Large Scale Dynamical Systems*, SIAM, Philadelphia.

Axler, S. (1995) "Down with determinants," *Amer. Math. Monthly*, 102, 139–154.

Beurling, A. (1949) "On two problems concerning linear transformations in Hilbert space," *Acta Math.* 81, 239–255.

Brechenmacher, F. (2007) "Algebraic generality vs arithmetic generality in the controversy between C. Jordan and L. Kronecker (1874)." http://arxiv.org/ftp/arxiv/papers/0712/0712.2566.pdf.

Carleson, L. (1962) "Interpolation by bounded analytic functions and the corona problem," *Ann. of Math.* 76, 547–559.

Davis, P.J. (1979) *Circulant Matrices*, J. Wiley, New York.

Douglas, R.G., Shapiro, H.S. & Shields, A.L. (1971) "Cyclic vectors and invariant subspaces for the backward shift," *Ann. Inst. Fourier, Grenoble* 20, 1, 37–76.

Dunford, N. and Schwartz, J.T. (1958) *Linear Operators, Part I*, Interscience, New York.

Dunford, N. and Schwartz, J.T. (1963) *Linear Operators, Part II*, Interscience, New York.

Duren, P. (1970) *Theory of H^p Spaces*, Academic Press, New York.

Fuhrmann, P.A. (1968a) "On the corona problem and its application to spectral problems in Hilbert space," *Trans. Amer. Math. Soc.* 132, 55–67.

Fuhrmann, P.A. (1968b) "A functional calculus in Hilbert space based on operator valued analytic functions," *Israel J. Math.* 6, 267–278.

Fuhrmann, P.A. (1975) "On Hankel operator ranges, meromorphic pseudo-continuation and factorization of operator valued analytic functions," *J. Lon. Math. Soc.* (2) 13, 323–327.

Fuhrmann, P.A. (1976) "Algebraic system theory: An analyst's point of view," *J. Franklin Inst.* 301, 521–540.

Fuhrmann, P.A. (1977) "On strict system equivalence and similarity," *Int. J. Contr.* 25, 5–10.

Fuhrmann, P.A. (1981) *Linear Systems and Operators in Hilbert Space*, McGraw-Hill, New York.

Fuhrmann, P.A. (1981b) "Polynomial models and algebraic stability criteria," *Proceedings of Joint Workshop on Synthesis of Linear and Nonlinear Systems*, Bielefeld June 1981, 78–90.

Fuhrmann, P.A. (1991) "A polynomial approach to Hankel norm and balanced approximations," *Lin. Alg. Appl.* 146, 133–220.

Fuhrmann, P.A. (1994) "An algebraic approach to Hankel norm approximation problems," in *Differential Equations, Dynamical Systems, and Control Science*, the *L. Markus Festschrift*, Edited by K.D. Elworthy, W.N. Everitt, and E.B. Lee, M. Dekker, New York, 523–549.

Fuhrmann, P.A. (1994b) "A duality theory for robust control and model reduction," *Lin. Alg. Appl.* 203–204, 471–578.

Fuhrmann, P.A. (2002) "A study of behaviors," *Lin. Alg. Appl.* (2002), 351–352, 303–380.

Fuhrmann, P.A. and Helmke, U. (2010) "Tensored polynomial models," *Lin. Alg. Appl.* (2010), 432, 678–721.

Gallivan, K., Vandendorpe, A. and Van Dooren, P. (2003) "Model reduction via truncation: an interpolation point of view," *Lin. Alg. Appl.* 375, 115–134.

Gantmacher, F.R. (1959) *The Theory of Matrices*, Chelsea, New York.

Garnett, J.B. (1981) *Bounded Analytic Functions*, Academic Press, New York.

Glover, K. (1984) "All optimal Hankel-norm approximations and their L^∞-error bounds," *Int. J. Contr.* 39, 1115–1193.

Glover, K. (1986) "Robust stabilization of linear multivariable systems, relations to approximation," *Int. J. Contr.* 43, 741–766.

Gohberg, I.C. and Krein, M.G. (1969) *Introduction to the Theory of Nonselfadjoint Operators*, Amer. Math. Soc., Providence.

Gohberg, I.C. and Semencul, A.A. (1972) "On the inversion of finite Toephtz matrices and their continuous analogs" (in Russian), *Mat. Issled.* 7, 201–233.

Gragg, W.B. and Lindquist, A. (1983) "On the partial realization problem," *Lin. Alg. Appl.* 50, 277–319.

Grassmann, H. (1844) *Die Lineale Ausdehnungslehre*, Wigand, Leipzig.

Halmos, P.R. (1950) "Normal dilations and extensions of operators," *Summa Brasil.* 2, 125–134.

Halmos, P.R. (1958) *Finite-Dimensional Vector Spaces*, Van Nostrand, Princeton.

Helmke, U. and Fuhrmann, P. A. (1989) "Bezoutians," *Lin. Alg. Appl.* 122–124, 1039–1097.

Helson, H. (1964) *Lectures on Invariant Subspaces*, Academic Press, New York.

Hermite, C. (1856) "Sur le nombre des racines d'une equation algebrique comprise entre des limites donnés," *J. Reine Angew. Math.* 52, 39–51.

Higham, N.J. (2008) "Cayley, Sylvester, and early matrix theory," *Lin. Alg. Appl.* 428, 39–43.

Hoffman, K. (1962) *Banach Spaces of Analytic Functions*, Prentice Hall, Englewood Cliffs.

Hoffman, K. and Kunze, R. (1961) *Linear Algebra*, Prentice-Hall, Englewood Cliffs.

Householder, A.S. (1970) "Bezoutians, elimination and localization," *SIAM Review* 12, 73–78.

Hungerford, T.W. (1974) *Algebra*, Springer-Verlag, New York.

Hurwitz, A. (1895) "Über die bedingungen, unter welchen eine Gleichung nur Wurzeln mit negativen reelen Teilen besitzt," *Math. Annal.* 46, 273–284.

Joseph, G.G. (2000) *The Crest of the Peacock: Non-European Roots of Mathematics*, Princeton University Press.

Kailath, T. (1980) *Linear Systems*, Prentice-Hall, Englewood Cliffs.

Kalman, R.E. (1969) "Algebraic characterization of polynomials whose zeros lie in algebraic domains," *Proc. Nat. Acad. Sci.* 64, 818–823.

Kalman, R.E. (1970) "New algebraic methods in stability theory," *Proceeding V. International Conference on Nonlinear Oscillations, Kiev.*

Kalman R.E., Falb P., and Arbib M. (1969) *Topics in Mathematical System Theory*, McGraw Hill.

Kravitsky N. (1980) "On the discriminant function of two noncommuting nonselfadjoint operators," *Integr. Eq. and Oper. Th.* 3, 97–124.

Krein, M.G. and Naimark, M.A. (1936) "The method of symmetric and Hermitian forms in the theory of the separation of the roots of algebraic equations," English translation in *Linear and Multilinear Algebra* 10 (1981), 265–308.

Lander, F. I. (1974) "On the discriminant function of two noncommuting nonselfadjoint operators," (in Russian), *Mat. Issled.* IX(32), 69–87.

Lang, S. (1965) *Algebra*, Addison-Wesley, Reading.

Lax, P.D. and Phillips, R.S. (1967) *Scattering Theory*, Academic Press, New York.

Livsic, M.S. (1983) "Cayley–Hamilton theorem, vector bundles and divisors of commuting operators," *Integr. Eq. and Oper. Th.* 6, 250–273.

Liapunov, A.M. (1893) "Problemè général de la stabilité de mouvement", *Ann. Fac. Sci. Toulouse* 9 (1907), 203–474. (French translation of the Russian paper published in *Comm. Soc. Math. Kharkow*).

Mac Lane, S. and Birkhoff, G. (1967) *Algebra*, Macmillan, New York.

Magnus, A. (1962) "Expansions of power series into P-fractions," *Math. Z.* 80, 209–216.

Malcev, A.I. (1963) *Foundations of Linear Algebra*, W.H. Freeman & Co., San Francisco.

Maxwell, J.C. (1868) "On governors," *Proc. Roy. Soc.* Ser. A, 16, 270–283.

Moore, B.C. (1981) "Principal component analysis in linear systems: Controllability, observability and model reduction," *IEEE Trans. Automat. Contr.* 26, 17–32.

Nehari, Z. (1957) "On bounded bilinear forms," *Ann. of Math.* 65, 153–162.

Nikolskii, N.K. (1985) *Treatise on the Shift Operator*, Springer-Verlag, Berlin.

Peano, G. (1888) *Calcolo geometrico secondo l'Ausdehnungslehre di H. Grassmann, preceduto dalle operazioni della logica deduttiva*, Bocca, Turin.

Prasolov, V.V. (1994) *Problems and Theorems in Linear Algebra*, Trans. of Math. Monog. v. 134, Amer. Math. Soc., Providence.

Rosenbrock, H.H. (1970) *State-Space and Multivariable Theory*, John Wiley, New York.

Rota, G.C. (1960) "On models for linear operators," *Comm. Pure and Appl. Math.* 13, 469–472.

Routh, E.J. (1877) *A Treatise on the Stability of a Given State of Motion*, Macmillan, London.

Sarason, D. (1967) "Generalized interpolation in H^∞," *Trans. Amer. Math. Soc.* 127, 179–203.

Schoenberg, I.J. (1987) "The Chinese remainder problem and polynomial interpolation," *The College Mathematics Journal*, 18, 320–322.

Schur, I. (1917,1918) "Über Potenzreihen die im Innern des Einheitskreises beschränkt sind," *J. Reine Angew. Math.* 147, 205–232; 148, 122–145.

Sorensen, D.C. and Antoulas, A.C. (2002) "The Sylvester equation and approximate balanced reduction," *Lin. Alg. Appl.* 351–352, 671–700.

Sz.-Nagy, B. and Foias, C. (1970) *Harmonic Analysis of Operators on Hilbert Space*, North Holland, Amsterdam.

Vidyasagar, M. (1985) *Control System Synthesis: A Coprime Factorization Approach*, M.I.T. Press, Cambridge MA.

van der Waerden, B.L. (1931) *Moderne Algebra*, Springer-Verlag, Berlin.

Willems, J.C. (1986) "From time series to linear systems. Part I: Finite-dimensional linear time invariant systems," *Automatica*, 22, 561–580.

Willems, J.C. (1989) "Models for dynamics," *Dynamics Reported*, 2, 171–269, U. Kirchgraber and H.O. Walther (eds.), Wiley-Teubner.

Willems, J.C. (1991) "Paradigms and puzzles in the theory of dynamical systems," *IEEE Trans. Autom. Contr.* AC-36, 259–294.

Willems, J.C. and Fuhrmann, P.A. (1992) "Stability theory for high order systems," *Lin. Alg. Appl.* 167, 131–149.

Wimmer, H. (1990) "On the history of the Bezoutian and the resultant matrix," *Lin. Alg. Appl.* 128, 27–34.

Young, N. (1988) *An Introduction to Hilbert Space*, Cambridge University Press, Cambridge.

Index

Symbols

$\mathbb{F}[z]$-Kronecker product, 260
$\mathbb{F}[z]$-Kronecker product model, 260

A

AAK theory, 400
abelian group, 3
addition, 8
adjoint, 167
adjoint transformation, 86
all-pass function, 364
analytic Toeplitz operator, 332
annihilator, 81
atoms, 237

B

backward invariant subspace, 334
backward shift operator, 30, 113
balanced realization, 398
Barnett factorization, 213
barycentric representation, 132
basis, 28, 38
basis transformation matrix, 47
Bessel inequality, 163
Beurling's theorem, 336
Bezout equation, 15, 17
Bezout form, 210
Bezout identity, 10
Bezout map, 272
Bezoutian, 210
bijective map, 2
bilateral shift, 29
bilinear form, 79, 196, 254
bilinear pairing, 79

binary operation, 3
binary relation, 1

C

canonical embedding, 84
canonical factorization, 2
canonical map, 256
canonical projection, 7
Cauchy determinant, 66
Cauchy index, 248
causality, 299
Cayley transform, 177
Cayley–Hamilton theorem, 129
characteristic polynomial, 93
characteristic vector, 93
Chinese remainder theorem, 118, 353
Christoffel–Darboux formula, 323
circulant matrix, 109
classical adjoint, 56, 60
codimension, 43
coefficients of a linear combination, 36
cofactor, 56
commutant, 136
commutative ring, 8
companion matrices, 100
composition, 2, 71
congruent, 197
continued fraction representation, 237
control basis, 99
controllability realization, 310
controller realization, 310
coordinate vector, 45
coordinates, 45
coprime factorization, 22, 354
cosets, 4

P.A. Fuhrmann, *A Polynomial Approach to Linear Algebra*, Universitext,
DOI 10.1007/978-1-4614-0338-8, © Springer Science+Business Media, LLC 2012